Engineering Acoustics

Second Edition

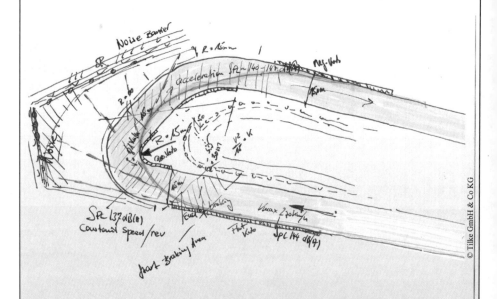

Michael Möser

Engineering Acoustics

An Introduction to Noise Control

Second Edition

Translated by S. Zimmermann and R. Ellis

Prof. Dr. Michael Möser
TU Berlin
Fak. III Prozesswissenschaften
Inst. Anlagen- und Prozesstechnik
Einsteinufer 25
10587 Berlin
Germany
moes0338@mailbox.tu-berlin.de

ISBN 978-3-540-92722-8 e-ISBN 978-3-540-92723-5
DOI 10.1007/978-3-540-92723-5
Springer Dordrecht Heidelberg London New York

Library of Congress Control Number: 2009926691

Cover design: eStudio Calamar S.L., Heidelberg

Printed on acid-free paper

Springer is part of Springer Science+Business Media (www.springer.com)

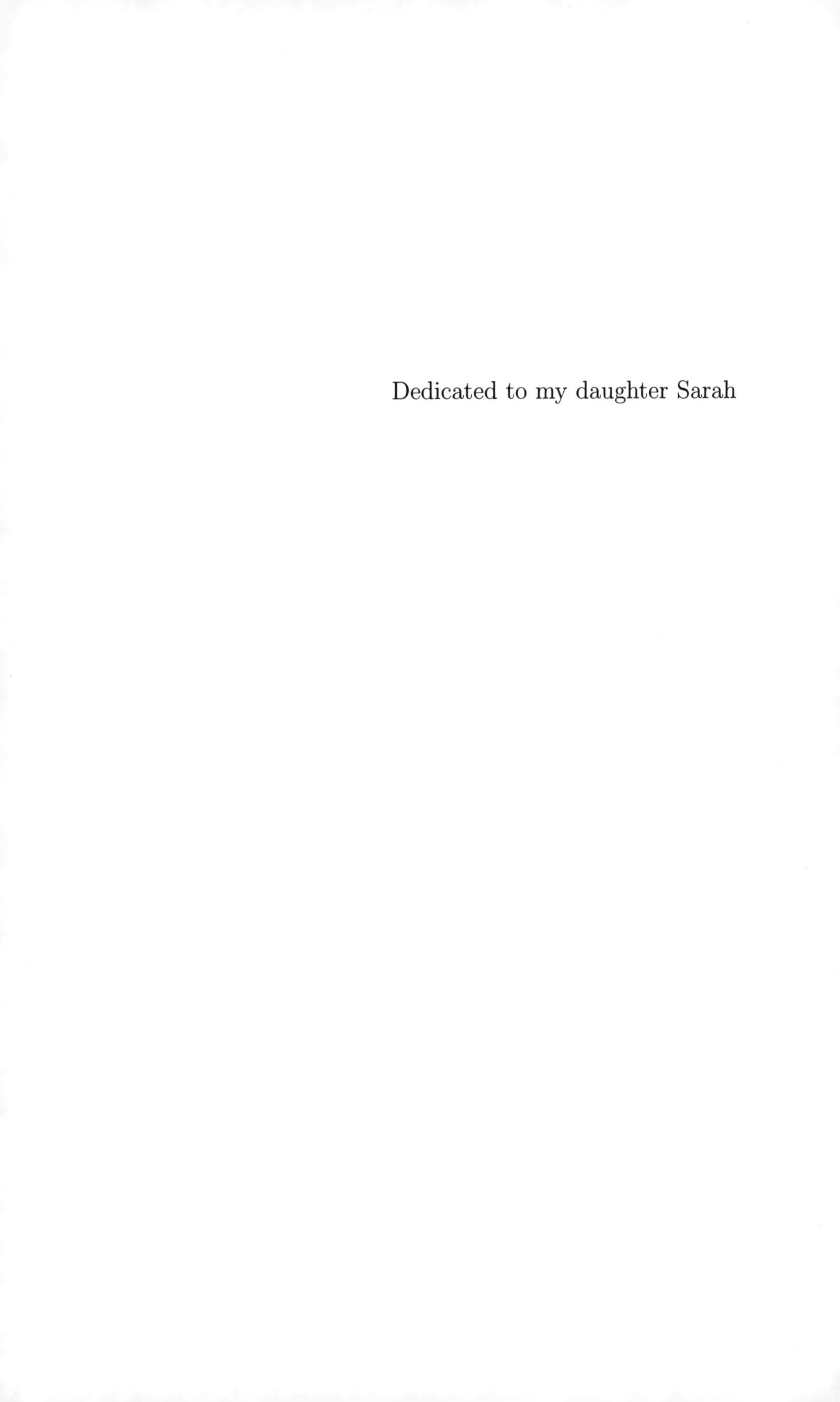

Dedicated to my daughter Sarah

Preface

This book presents the English translation of the author's German edition 'Technische Akustik' (the 8th edition published by Springer Verlag in 2009). In whatever language, 'Engineering Acoustics' sees itself as a teaching textbook that could serve as a tool for autodidactic studies and as a compendium of lectures and courses as well. Readers are addressed who already possess a certain training in physical and mathematical thinking and in expressing ideas and explanations using mathematical formulas. On the other hand no highly specified knowledge is vital: readers with no more than the usual skills – like taking derivatives and solving simple integrals – are assumed. The appendix gives a short introduction on the use of complex amplitudes in acoustics and the reasons for their use. It is in general one of the author's most important aims not only to describe *how* the topic and its description develops but also *why* a specific way is chosen. Often difficulties in understanding do not consist in comprehending the single steps but in the question why they are done in that - and in no other way.

Moreover the explanations do not restrict themselves to the mathematical formulas. No doubt that formulas give the most unambiguous description of matters, and they show problems and their solutions in quantity also, but more remains to be done. Only the illustrative explanation relying on the reader's imagination produces understanding and comprehension. Textbooks should make learning – often difficult enough – as easy as possible, and this certainly does not imply to reduce the level.

In many respects this book is obliged to Lothar Cremer. For example, parts of the author's own knowledge originate from Cremer's very first 'Vorlesungen über Technische Akustik'. Important discoveries of Cremer are included in this new edition and it's translations. Examples are, the optimum impedance for mufflers and the coincidence effect which leads to a satisfying explanation for sound transmission through walls; perhaps Cremer's most important discovery.

This book tries to present the foundations of that what nowadays seems necessary to make our environment quieter. All chapters between 'elastic isolation' – the 5th – and 'diffraction' – the 10th – directly or indirectly address the question, how to reduce the sound level in the most important environs of everyday life indoors and outdoors – in buildings and in the open air. This requires the understanding of some principal features first. To fully comprehend the physics of sound transmission through walls for example, implies the understanding of bending wave propagation on plates. Because of that reason chapters on 'the media' precede the chapters on the noise reduction methods. The (short) chapter on sound perception serves as an introduction. The last chapter deals with the most important receiving and source instruments: microphones and loudspeakers. Specific measurement procedures are already discussed in many other chapters. The chapter 'absorption' for example begins with a discussion of how to measure the absorption coefficient.

The translation of this book was done by Stefan Zimmermann and Rebecca Ellis. The cooperation with them was interesting, satisfying, and excellent. Many thanks to them for all their efforts and patience with me.

Berlin, February 2009 *Michael Möser*

Contents

1 Perception of sound ... 1
1.1 Octave and third-octave band filters ... 7
1.2 Hearing levels ... 9
1.3 A-Weighting ... 10
1.4 Noise fluctuating over time ... 12
1.5 Summary ... 13
1.6 Further reading ... 13
1.7 Practice exercises ... 14

2 Fundamentals of wave propagation ... 17
2.1 Thermodynamics of sound fields in gases ... 18
2.2 One-dimensional sound fields ... 24
2.2.1 Basic equations ... 24
2.2.2 Progressive waves ... 29
2.2.3 Complex notation ... 34
2.2.4 Standing waves and resonance phenomena ... 34
2.3 Three-dimensional sound fields ... 38
2.4 Energy and power transport ... 40
2.5 Intensity measurements ... 44
2.5.1 Time domain ... 45
2.5.2 Frequency domain ... 45
2.5.3 Measurement error and limitations ... 49
2.5.4 Standards ... 53
2.6 Wave propagation in a moving medium ... 53
2.7 Raised waves ... 59
2.8 Summary ... 62
2.9 Further reading ... 63
2.10 Practice exercises ... 63

3 Propagation and radiation of sound ... 67
3.1 Omnidirectional sound radiation of point sources ... 67
3.2 Omnidirectional sound radiation of line sources ... 68
3.3 Volume velocity sources ... 70
3.4 Sound field of two sources ... 72
3.5 Loudspeaker arrays ... 85
3.5.1 One-dimensional piston ... 87
3.5.2 Formation of main and side lobes ... 91
3.5.3 Electronic beam steering ... 96
3.5.4 Far-field conditions ... 100
3.6 Sound radiation from plane surfaces ... 103
3.6.1 Sound field on the axis of a circular piston ... 106
3.7 Summary ... 110
3.8 Further reading ... 111
3.9 Practice exercises ... 111

4 Structure-borne sound ... 117
4.1 Introduction ... 117
4.2 Bending waves in beams ... 120
4.3 Propagation of bending waves ... 124
4.4 Beam resonances ... 126
4.4.1 Supported beams ... 127
4.4.2 Bilaterally mounted beams ... 129
4.4.3 Bilaterally suspended beams ... 132
4.5 Bending waves in plates ... 133
4.5.1 Plate vibrations ... 136
4.6 Summary ... 140
4.7 Further reading ... 141
4.8 Practice exercises ... 141

5 Elastic isolation ... 143
5.1 Elastic bearings on rigid foundations ... 145
5.2 Designing elastic bearings ... 150
5.3 Influence of foundations with a compliance ... 153
5.3.1 Foundation impedance ... 153
5.3.2 The effect of foundation impedance ... 154
5.4 Determining the transfer path ... 161
5.5 Determining the loss factor ... 162
5.6 Dynamic mass ... 164
5.7 Conclusion ... 166
5.8 Summary ... 167
5.9 Further reading ... 167
5.10 Practice exercises ... 167

6 **Sound absorbers** ... 171
6.1 Sound propagation in the impedance tube ... 171
6.1.1 Tubes with rectangular cross sections ... 176
6.1.2 Tubes with circular cross sections ... 176
6.2 Measurements in the impedance tube ... 178
6.2.1 Mini-max procedure ... 180
6.3 Wall impedance ... 183
6.4 Theory of locally reacting absorbers ... 186
6.5 Specific absorbent structures ... 191
6.5.1 The 'infinitely thick' porous sheet ... 191
6.5.2 The porous sheet of finite thickness ... 194
6.5.3 The porous curtain ... 198
6.5.4 Resonance absorbers ... 201
6.6 Oblique sound incidence ... 208
6.7 Summary ... 211
6.8 Further reading ... 212
6.9 Practice exercises ... 213

7 **Fundamentals of room acoustics** ... 217
7.1 Diffuse sound field ... 221
7.1.1 Reverberation ... 224
7.1.2 Steady-state conditions ... 226
7.1.3 Measurement of the absorption coefficient in the reverberation room ... 231
7.2 Summary ... 232
7.3 Further reading ... 233
7.4 Practice exercises ... 233

8 **Building acoustics** ... 237
8.1 Measurement of airborne transmission loss ... 239
8.2 Airborne transmission loss of single-leaf partitions ... 241
8.3 Double-leaf partitions (flexible additional linings) ... 252
8.4 Impact sound reduction ... 259
8.4.1 Measuring impact sound levels ... 259
8.4.2 Improvements ... 260
8.5 Summary ... 263
8.6 Further reading ... 264
8.7 Practice exercises ... 265

9 **Silencers** ... 267
9.1 Changes in the cross-section of rigid ducts ... 268
9.1.1 Abrupt change in cross-section ... 268
9.1.2 Duct junctions ... 270
9.1.3 Expansion chambers ... 275
9.1.4 Chamber combinations ... 279

9.2 Lined ducts ... 284
9.2.1 Ducts with rigid walls ... 286
9.2.2 Ducts with soft boundaries ... 288
9.2.3 Silencers with arbitrary impedance boundaries ... 290
9.3 Summary ... 308
9.4 Further reading ... 308
9.5 Practice exercises ... 308

10 Noise barriers ... 311
10.1 Diffraction by a rigid screen ... 312
10.2 Approximation of insertion loss ... 329
10.3 The importance of height in noise barriers ... 332
10.4 Sound barriers ... 333
10.5 Absorbent noise barriers ... 335
10.6 Transmission through the barrier ... 338
10.7 Conclusion ... 339
10.8 Summary ... 339
10.9 Further reading ... 339
10.10 Practice exercises ... 340
10.11 Appendix: MATLAB program for Fresnel integrals ... 342

11 Electro-acoustic converters for airborne sound ... 345
11.1 Condenser microphones ... 347
11.2 Microphone directivity ... 355
11.3 Electrodynamic microphones ... 358
11.4 Electrodynamic loudspeakers ... 362
11.5 Acoustic antennae ... 365
11.5.1 Microphone arrays ... 367
11.5.2 Two-dimensional sensor arrangements ... 374
11.6 Summary ... 380
11.7 Further reading ... 380
11.8 Practice exercises ... 381

12 Fundamentals of Active Noise Control ... 383
12.1 The Influence of Replication Errors ... 387
12.1.1 Perpendicularly Interfering Waves ... 389
12.2 Reflection and Absorption ... 390
12.3 Active Stabilization of Self-Induced Vibrations ... 395
12.4 Summary ... 403
12.5 Further Reading ... 404
12.6 Practice Exercises ... 404

13 Aspects and Properties of Transmitters 407
13.1 Properties of Transmitters 408
13.1.1 Linearity 408
13.1.2 Time Invariance 409
13.2 Description using the Impulse Response 410
13.3 The Invariance Principle 414
13.4 Fourier Decomposition 415
13.4.1 Fourier Series 416
13.4.2 Fourier Transform 424
13.4.3 The Transmission Function and the Convolution Law .. 426
13.4.4 Symmetries 428
13.4.5 Impulse Responses and Hilbert Transformation 430
13.5 Fourier Acoustics: Wavelength Decomposition 432
13.5.1 Radiation from Planes 434
13.5.2 Emission of Bending Waves 436
13.5.3 Acoustic Holography 438
13.5.4 Three-dimensional Sound Fields 439
13.6 Summary 442
13.7 Further Reading 443
13.8 Practice Exercises 444

A Level Arithmetics 449
A.1 Decadic Logarithm 449
A.2 Level Inversion 450
A.3 Level Summation 451

B Complex Pointers 453
B.1 Introduction to Complex Pointer Arithmetics 453
B.2 Using Complex Pointers in Acoustics 455

C Solutions to the Practice Exercises 459
C.1 Practice Exercises Chapter 1 459
C.2 Practice Exercises Chapter 2 463
C.3 Practice Exercises Chapter 3 468
C.4 Practice Exercises Chapter 4 480
C.5 Practice Exercises Chapter 5 486
C.6 Practice Exercises Chapter 6 489
C.7 Practice Exercises Chapter 7 497
C.8 Practice Exercises Chapter 8 501
C.9 Practice Exercises Chapter 9 503
C.10 Practice Exercises Chapter 10 505
C.11 Practice Exercises Chapter 11 508
C.12 Practice Exercises Chapter 12 510
C.13 Practice Exercises Chapter 13 514

Index 527

1 Perception of sound

The perception of sound incidents requires the presence of some simple physical effects. A sound source oscillates and brings the surrounding air into motion. The compressability and mass of the air cause these oscillations to be transmitted to the listener's ear.

Small pressure fluctuations, referred to as sound pressure p, occur in air (or gas or fluid) and which are superimposed to the atmospheric pressure p_0. A spatially distributed sound field radiates from the source with different instantaneous sound pressures at each moment. The sound pressure is the most important quantity to describe sound fields and os always space- and time-dependent.

The observed sound incident at a point has two main distinguishing attributes: 'timbre' and 'loudness'. The physical quantity for loudness is sound pressure and the quantity for timbre is frequency f, measured in cycles per second, or Hertz (Hz). The frequency range of technical interest covers more than the range that is audible by the human ear, which is referred to as hearing level. The hearing range starts at about 16 Hz and ranges up to 16000 Hz (or 16 kHz). The infrasound, which is located below that frequency range, is less important for air-borne sound problems, but becomes relevant when dealing with vibrations of structures (e.g. in vibration control of machinery). Ultrasound begins above the audible frequency range. It is used in applications ranging from acoustic modelling techniques to medical diagnosis and non-destructive material testing.

The boundaries of the audible frequency range dealt with in this book cannot be defined precisely. The upper limit varies individually, depending on factors like age, and also in cases of extensive workplace noise exposure or the misuse of musical devices. The value of 16 kHz refers to a healthy, human being who is about 20 years old. With increasing age, the upper limit decreases by about 1 kHz per decade.

The lower limit is likewise not easy to define and corresponds to flickering. At very low frequencies a series of single sound incidents (e.g. a series of impulses) can be distinguished as well. If the frequency increases above the

M. Möser, *Engineering Acoustics*, DOI 10.1007/978-3-540-92723-5_1,

flickering frequency of (about) 16 Hz, single incidents are no longer perceived individually, but seem to merge into a single noise. This transition can be found, for example, when it slowly starts to rain: the knocking of single rain drops at the windows can be heard until the noise at a certain density of rain merges into a continuous crackling. Note that the audible limit for the perception of flickering occurs at the same frequency at which a series of single images in a film start to appear as continuous motion.

The term 'frequency' in acoustics is bound to pure tones, meaning a sinusoidal wave form in the time-domain. Such a mathematically well-defined incident can only rarely be observed in natural sound incidents. Even the sound of a musical instrument contains several colourations: the superposition of several harmonic (pure) tones produces the typical sound of the instrument (see Fig. 1.1 for examples). An arbitrary wave form can generally be

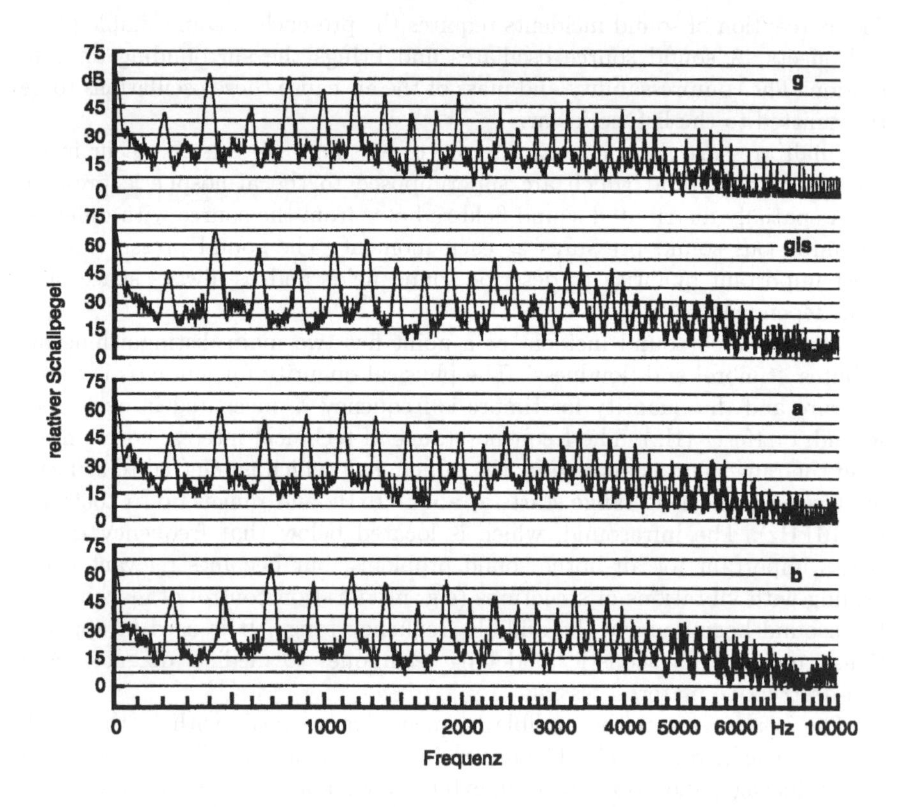

Fig. 1.1. Sound spectra of a violin played at different notes (from: Meyer, J.: "Akustik und musikalische Aufführungspraxis". Verlag Erwin Bochinsky, Frankfurt 1995). Relative sound pressure level versus frequency.

represented by its frequency components extracted through spectrum analysis, similar to the analysis of light. Arbitrary signals can be represented by a sum of harmonics (with different amplitudes and frequencies). The association of decomposed time signals directly leads to the representation of the acoustic properties of transducers by their frequency response functions (as, for example, those of walls and ceilings in building acoustics, see Chap. 8). If, for instance, the frequency-dependent transmission loss of a wall is known, it is easy to imagine how it reacts to the transmission of certain sound incidents like, for example, speech. The transmission loss is nearly always bad at low frequencies and good at high frequencies: speech is therefore not only transmitted 'quieter' but also 'dull' through the wall. The more intuitive association that arbitrary signals can be represented by their harmonics will be sufficient throughout this book in the aforementioned hypothetical case. The expansion of a given signal into a series of harmonics, the so-called Fourier series and Fourier integrals, is based on a solid mathematical foundation of proof (see the last chapter of this book).

The subjective human impression of the sound pitch is perceived in such a way that a tonal difference of two pairs of tones is perceived equally if the ratio (and not the difference) of the two frequency pairs is equal. The tonal difference between the pair made of f_{a1} and f_{a2} and the pair made of f_{b1} and f_{b2} is perceived equally if the ratio

$$\frac{f_{\mathrm{a1}}}{f_{\mathrm{a2}}} = \frac{f_{\mathrm{b1}}}{f_{\mathrm{b2}}}$$

is valid. The transition from 100 Hz to 125 Hz and from 1000 Hz to 1250 Hz is, for example, perceived as an equal change in pitch. This law of 'relative tonal impression' is reflected in the subdivision of the scale into octaves (a doubling in frequency) and other intervals like second, third, fourth and fifth, etc. used for a long time in music. All of these stand for the ratio in frequency and not for the 'absolute increase in Hz'.

This law of 'tonal impression,' which more generally means that a stimulus R has to be increased by a certain percentage to be perceived as an equal change in perception, is not restricted to the tonal impression of the human being. It is true for other human senses as well. In 1834, Weber conducted experiments using weights in 1834 and found that the difference between two masses laid on the hand of a test subject was only perceived equally, when a mass of 14 g was increased by 1 g and a mass of 28 g was increased by 2 g. This experiment and the aforementioned tonal perception leads to the assumption that the increment of a perception ΔE for these and other physical stimuli is proportional to the ratio of the absolute increase of the stimulus ΔR and the stimulus R

$$\Delta E = k\frac{\Delta R}{R}, \tag{1.1}$$

where k is a proportionality constant. For the perception of pitch the stimulus $R = f$ represents the frequency, for the perception of weight $R = m$ represents the mass on the hand.

This law of relative variation (1.1) is also true for the perception of loudness. If a test subject is repeatedly presented with sound incidents consisting of pairs of sound pressure incidents p and $2p$ and $5p$ and $10p$ respectively, the perceived difference in loudness should be equal. The perception of both pitch and loudness should at least roughly follow the law of relative variation (1.1).

As mentioned earlier (1.1), there is a relativity law which governs variations in stimulus ΔR and in perception ΔE. It is, of course, also interesting to examine the relation between R and E. Given that it is, at best, problematic, if not presumably impossible, to quantify perceptions, the principal characteristics of the $E(R)$ function should be clarified. These 'perception characteristics' are easily constructed from the variation law, if two points of stimulus R and perception E are chosen as shown in Fig. 1.2. A threshold stimulus R_0 is defined, at which the perception starts: stimuli $R < R_0$ below the threshold are not perceivable. A minimal stimulus is needed to achieve perception at all. The second point is chosen arbitrarily to be twice the threshold $R = 2R_0$ and the (arbitrary) perception E_0 is assigned. The further characteristics re-

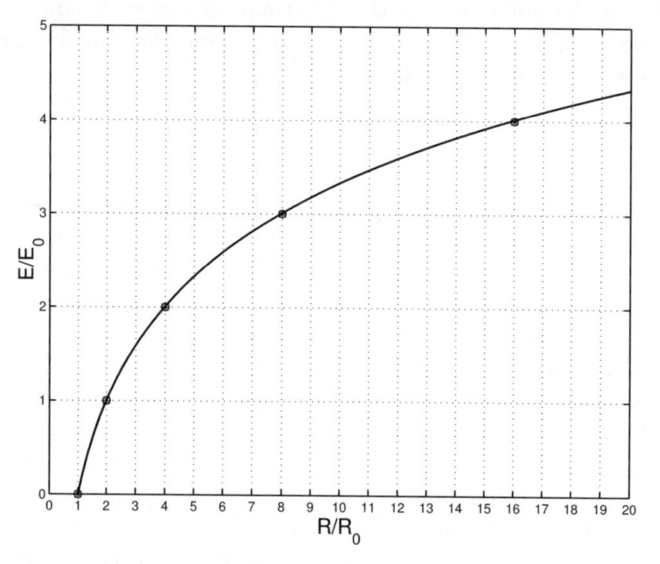

Fig. 1.2. Qualitative relation between stimulus R and perception E

sult from examining the perceptions $2E_0$, $3E_0$, $4E_0$, etc. The perception $2E_0$ is assigned twice the stimulus of E_0, therefore related to $R = 4R_0$. Just as $E = 3E_0$ is related to the stimulus $R = 8R_0$ the perception $4E_0$ is related to $R = 16R_0$, etc. As can be seen from Fig. 1.2 the gradient of the curve $E = E(R)$ decreases with increasing stimulus R. The greater the perception, the greater the increase of the stimulus has to be to achieve another increment of perception (for example E_0).

The functional relation $E = E(R)$ can certainly be determined from the variation law (1.1) by moving towards infinitesimal small variations $\mathrm{d}E$ and $\mathrm{d}R$:

$$\mathrm{d}E = k\frac{\mathrm{d}R}{R} .$$

Integration yields

$$E = 2.3k\,\lg(R/R_0). \tag{1.2}$$

Bare in mind that the logarithms of different bases are proportional, e.g. $\ln x = 2.3\lg x$. The perception of loudness is therefore proportional to the logarithm of the physical stimulus –in this case, sound pressure. This relation, validated at least roughly by numerous investigations, is also known as the Weber-Fechner-Law.

The sensual perception according to a logarithmic law (for the characteristics see Fig. 1.2 again) is a very sensible development of the 'human species'. Stimuli close to the threshold $R = R_0$ are emphasized and therefore 'well perceivable', whereas very large stimuli are highly attenuated in their perception; the logarithmic characteristics act as a sort of 'overload protection'. A wide range of physical values can thus be experienced (without pain) and several decades of physical orders of magnitude are covered. The history of the species shows that those perceptions necessary to survive in the given environment, which also cover a wide range of physical values, follow the Weber-Fechner-Law. This is not true for the comparatively smaller range of temperature perception. Variations of a tenth or a hundredth of a degree are by no means of interest to the individual. In contrast, the perception of light needs to cover several decades of order of magnitude. Surviving in the darkest night is as important as the ability to see in the sunlight of a very bright day. And the perception of weight covers a range starting from smallest masses of about 1 g up to loads of several 10000 g. The perception of loudness follows the logarithmic Weber-Fechner-Law, because the human ear is facing the problem of perceiving very quiet sounds, like the falling of the leaves in quiet surroundings, as well as very loud sounds, like the roaring sound of a waterfall in close vicinity. As a matter of fact, humans are able to perceive sound pressures in the range of $20 \cdot 10^{-6} N/m^2$ to approximately $200\,\mathrm{N/m^2}$, where the upper limit roughly depicts the pain threshold. About ten decades of loudness are covered, which represents an exceedingly large physical interval. To illustrate, this range in equivalent distances would cover an interval between 1 mm and 10 km. The amazing ear is able to perceive this range. Imagine the impossibility of an optical instrument (like a magnifying glass), to be able to operate in the millimeter range as well as in the kilometer range!

When technically quantifying sound pressure, it is more handy to use a logarithmic measure instead of the physical sound pressure itself to represent this wide range. The sound pressure level L is internationally defined as

$$L = 20\lg\left(\frac{p}{p_0}\right) = 10\lg\left(\frac{p}{p_0}\right)^2 , \tag{1.3}$$

with $p_0 = 20\,10^{-6}\,\mathrm{N/m^2}$, as an expressive and easy to use measure. The reference value p_0 roughly corresponds to the hearing threshold (at a frequency of 1 kHz, because the hearing threshold is frequency-dependent, as will be shown in the next section), so that 0 dB denotes the 'just perceivable' or 'just not perceivable' sound event. If not otherwise stated, the sound pressure p stands for the root mean square (rms-value) of the time domain signal. The specification in decibels (dB) is not related to a specific unit. It indicates the use of the logarithmic law. The factor 20 (or 10) in (1.3) is chosen in such a way that 1 dB corresponds to the difference threshold between two sound pressures: if two sound incidents differ by 1 dB they can just be perceived differently.

The physical sound pressure covering 7 decades is mapped to a 140 dB scale by assigning sound pressure levels, as can be seen in Table 1.1. Some examples for noise levels occurring in situations of every day life are also shown.

Table 1.1. Relationship between absolute sound pressure and sound pressure level

Sound pressure p ($\mathrm{N/m^2}$, rms)	Sound pressure level L (dB)	Situation/description
$2\,10^{-5}$	0	hearing threshold
$2\,10^{-4}$	20	forest, slow winds
$2\,10^{-3}$	40	library
$2\,10^{-2}$	60	office
$2\,10^{-1}$	80	busy street
$2\,10^{0}$	100	pneumatic hammer, siren
$2\,10^{1}$	120	jet plane during take-off
$2\,10^{2}$	140	threshold of pain, hearing loss

It should be noted that sound pressures related to the highest sound pressure levels are still remarkably smaller than the static atmospheric pressure of about $10^5\,\mathrm{N/m^2}$. The rms value of the sound pressure at 140 dB is only $200\,\mathrm{N/m^2}$ and therefore 1/500 of atmospheric pressure.

The big advantage when using sound pressure levels is that they roughly represent a measure of the perceived loudness. However, think twice when calculating with sound pressure levels and be careful in your calculations. For instance: How high is the total sound pressure level of several single sources with known sound pressure levels? The derivation of the summation of sound pressure levels (where the levels are in fact *not* summed) gives an answer to the question for incoherent sources (and can be found more detailed in Appendix A)

$$L_{\mathrm{tot}} = 10 \lg \left(\sum_{i=1}^{N} 10^{L_i/10} \right) , \tag{1.4}$$

where N is the total number of incoherent sources with level L_i. Three vehicles, for example, with equal sound pressure levels produce a total sound pressure level

$$L_{\text{tot}} = 10 \lg \left(3\, 10^{L_i/10}\right) = 10 \lg 10^{L_i/10} + 10 \lg 3 = L_i + 4.8\,\text{dB}$$

which is 4.8 dB higher than the individual sound pressure level (and not three times higher than the individual sound pressure level).

1.1 Octave and third-octave band filters

In some cases a high spectral resolution is needed to decompose time domain signals. This may be the case when determining, for example, the possibly narrow-banded resonance peaks of a resonator, where one is interested in the actual bandwidth of the peak (see Chap. 5.5). Such a high spectral resolution can, for example, be achieved by the commonly used FFT-Analysis (FFT: Fast Fourier Transform). The FFT is not dealt with here, the interested reader can find more details for example in the work of Oppenheim and Schafer "Digital Signal Processing" (Prentice Hall, Englewood Cliffs New Jersey 1975).

In most cases, a high spectral resolution is neither desired nor necessary. If, for example, an estimate of the spectral composition of vehicle or railway noise is needed, it is wise to subdivide the frequency range into a small number of coarse intervals. Larger intervals do not express the finer details. They contain a higher random error rate and cannot be reproduced very accurately. Using broader frequency bands ensures a good reproducibility (provided that, for example, the traffic conditions do not change). Broadband signals are also often used for measurement purposes. This is the case in measurements of room acoustics and building acoustics, which use (mainly white) noise as excitation signal. Spectral details are not only of no interest, they furthermore would divert the attention from the validity of the results.

Measurements of the spectral components of time domain signals are realized using filters. These filters are electronic circuits which let a supplied voltage pass only in a certain frequency band. The filter is characterized by its bandwidth Δf, the lower and upper limiting frequency f_l and f_u, respectively and the center frequency f_c (Fig. 1.3). The bandwidth is determined by the difference of f_u and f_l, $\Delta f = f_u - f_l$. Only filters with a constant relative bandwidth are used for acoustic purposes. The bandwidth is proportional to the center frequency of the filter. With increasing center frequency the bandwidth is also increasing. The most important representatives of filters with constant relative bandwidth are the octave and third-octave band filters. Their center frequency is determined by

$$f_c = \sqrt{f_l f_u}$$

The characteristic filter frequencies are known, if the ratio of the limiting frequencies f_l and f_u is given.

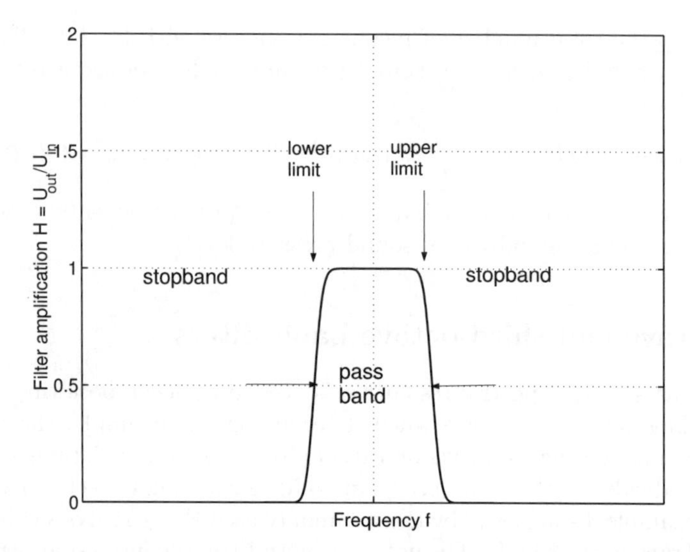

Fig. 1.3. Typical frequency response function of a filter (bandpass)

Octave bandwidth

$$f_u = 2f_l \, ,$$

which results in $f_c = \sqrt{2}f_l$ and $\Delta f = f_u - f_l = f_l = f_c/\sqrt{2}$.

Third-octave bandwidth

$$f_u = \sqrt[3]{2}f_l = 1.26f_l \, ,$$

which results in $f_c = \sqrt[6]{2}f_l = 1.12f_l$ and $\Delta f = 0.26f_l$.

The third-octave band filters are named that way, because three adjacent filters form an octave band filter ($\sqrt[3]{2}\sqrt[3]{2}\sqrt[3]{2} = 2$). The limiting frequencies are standardized in the international regulations EN 60651 and 60652.

When measuring sound levels one must state which filters were used during the measurement. The (coarser) octave band filters have a broader pass band than the (narrower) third-octave band filters which let contributions of a higher frequency range pass. Therefore octave band levels are always greater than third-octave band levels. The advantage of third-octave band measurements is the finer resolution (more data points in the same frequency range) of the spectrum.

By using the level summation (1.4), the octave band levels can be calculated using third-octave band measurements. In the same way, the levels of broader frequency bands may be calculated with the aid of level summation (1.4). The (non-weighted) linear level is often given. It contains all attributes of the frequency range between 16 Hz and 20000 Hz and can either be measured directly using an appropriate filter or determined by the level addition of the third-octave or octave band levels in the frequency band (when converting from octave bands, $N = 11$ and the center frequencies of the filters are 16 Hz, 31.5 Hz, 63 Hz, 125 Hz, 250 Hz, 500 Hz, 1 kHz, 2 kHz, 4 kHz, 8 kHz and 16 kHz). The linear level is always higher than the individual levels, by which it is calculated.

1.2 Hearing levels

Results of acoustic measurements are also often specified using another single value called the 'A-weighted sound pressure level'. Some basic principles of the frequency dependence of the sensitivity of human hearing are now explained, as the measurement procedure for the A-weighted level is roughly based on this.

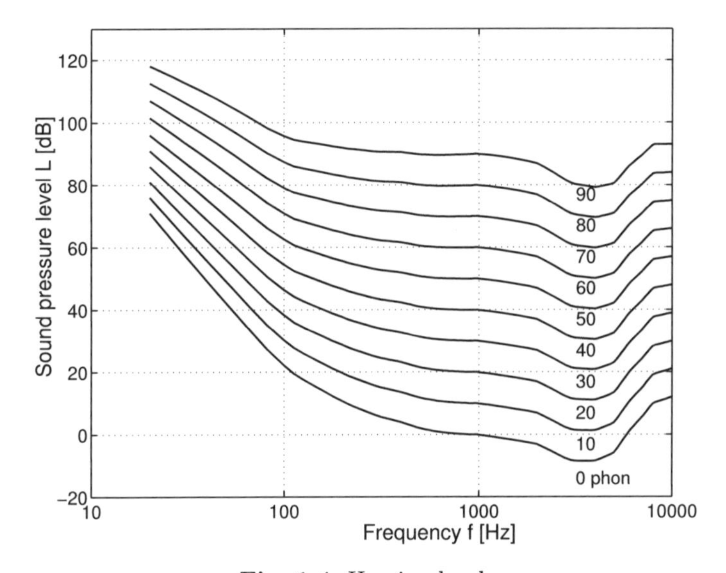

Fig. 1.4. Hearing levels

The sensitivity of the human ear is strongly dependent on the tonal pitch. The frequency dependence is depicted in Fig. 1.4. The figure is based on the findings from audiometric testing. The curves of perceived equal loudness (which have the unit 'phon') are drawn in a sound pressure level versus frequency plot. One can imagine the development of these curves as follows: a test subject compares a 1 kHz tone of a certain level to a second tone of another frequency and has to adjust the level of the second tone in such a way that it is perceived with equal loudness. The curve of one hearing level is obtained by varying the frequency of the second tone and is simply defined by the level of the 1 kHz tone. The array of curves obtained by varying the level of the 1 kHz tone is called hearing levels. It reveals, for example, that a 50 Hz tone with an actual sound pressure level of 80 dB is perceived with the same loudness as a 1 kHz tone with 60 dB. The ear is more sensitive in the middle frequency range than at very high or very low frequencies.

1.3 A-Weighting

The relationship between the objective quantity sound pressure or sound pressure level, respectively, and the subjective quantity loudness is in fact quite complicated, as can be seen in the hearing levels shown in Fig. 1.4. The frequency dependence of the human ear's sensitivity, for example, is also level-dependent. The curves with a higher level are significantly flatter than the curves with smaller levels. The subjective perception 'loudness' is not only depending on frequency, but also on the bandwidth of the sound incident. The development of measurement equipment accounting for all properties of the human ear could only be realized with a very large effort.

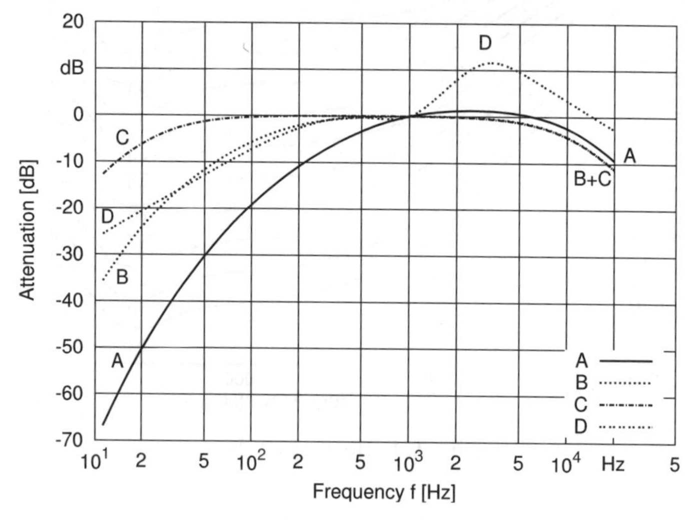

Fig. 1.5. Frequency response functions of A-, B-, C- and D-weighting filters

A frequency-weighted sound pressure level is used both nationally and internationally, which accounts for the basic aspects of the human ear's sensitivity and can be realized with reasonable effort. This so-called 'A-weighted sound pressure level' includes contributions of the whole audible frequency range. In practical applications the dB(A)-value is measured using the A-filter. The frequency response function of the A-filter is drawn in Fig. 1.5. The A-filter characteristics roughly represent the inverse of the hearing level curve with 30 dB at 1 kHz. The lower frequencies and the very high frequencies are devaluated compared to the middle frequency range when determining the dB(A)-value. As a matter of fact, the A-weighted level can also be determined from measured third-octave band levels. The levels given in Fig. 1.5 are added to the third-octave band levels and the total sound pressure level, now A-weighted, is calculated according to the law of level summation (1.4). The A-weighting function is standardized in EN 60651.

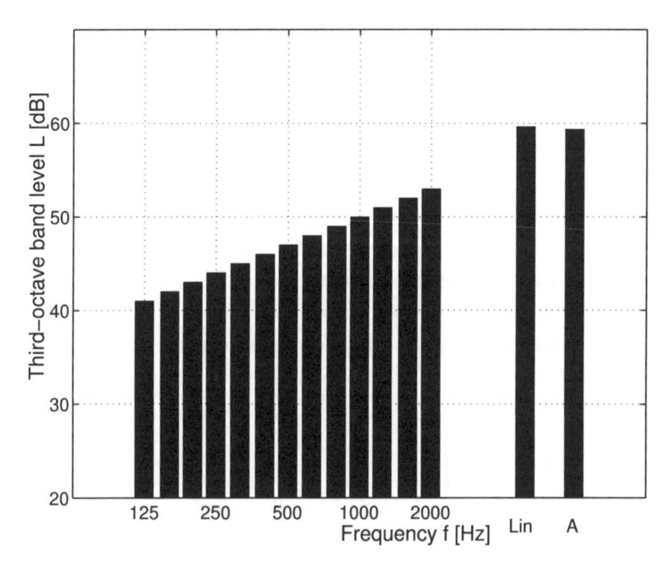

Fig. 1.6. Third-octave band, non-weighted and A-weighted levels of band-limited white noise

A practical example for the aforementioned level summation is given in Fig. 1.6 by means of a white noise signal. The third-octave band levels, the non-weighted (Lin) and the A-weighted (A) total sound pressure level are determined. The third-octave band levels increase by 1 dB for each band with increasing frequency. The linear (non-weighted) total sound pressure level is higher than each individual third-octave band level, the A-weighted level is only slightly smaller than the non-weighted level.

It should be noted that exceptions for certain noise problems (especially for vehicle and aircraft noise) exist, where other weighting functions (B, C and D) are used (see also Fig. 1.5). Regulations by law still commonly insist on the dB(A)-value.

Linearly determined single-number values, regardless of the filter used to produce them, are somewhat problematic, because considerable differences in individual perceptions do not become apparent. Fig. 1.4 clearly shows, for example, that 90 dB of level difference are needed at 1 kHz to increase the perception from 0 to 90 phon; at the lower frequency limit at 20 Hz only 50 dB are needed. A simple frequency weighting is not enough to prevent certain possible inequities. On the other hand, simple and easy-to-use evaluation procedures are indispensable.

1.4 Noise fluctuating over time

It is easy to determine the noise level of constant, steady noise, such as from an engine with constant rpm, a vacuum cleaner, etc. Due to their uniform formations, such noise can be sufficiently described by the A-level (or third-octave level, if so desired.)

How, on the other hand, can one measure intermittent signals, such as speech, music or traffic noise? Of course, one can use the level-over-time notation, but this description falls short, because a notation of various noise events along a time continuum makes an otherwise simple quantitative comparison of a variety of noise scenarios, such as traffic on different highways, quite difficult. In order to obtain simple comparative values, one must take the mean value over a realistic average time period.

The most conventional and simplest method is the so-called 'energy-equivalent continuous sound level' L_{eq}. It reflects the sound pressure square over a long mean time

$$L_{eq} = 10\lg\left(\frac{1}{T}\int_0^T \frac{p_{eff}^2(t)}{p_0^2}\right) dt = 10\lg\left(\frac{1}{T}\int_0^T 10^{L(t)/10}\right) dt \tag{1.5}$$

($p_0 = 20\ 10^{-6}\ N/m^2$). Hereby $p_{eff}(t)$ indicates the time domain function of the rms-value and $L(t) = 10lg(p_{eff}(t)/p_0)^2$, the level gradient over time. The square of a time-dependent signal function is also referred to as 'signal energy', the energy-equivalent continuous sound level denotes the average signal energy; this explains the somewhat verbose terminology. For sound pressure signals obtained using an A-filter, third-octave filters, or the like, we use an A-weighted energy-equivalent level.

Depending on the need or application, any amount of integration time T can be used, ranging from a few seconds to several hours. There are comprehensive bodies of legislation which outline norms set by a maximum level L_{eq}, which is permissible over a certain time reference period ranging up to several hours. For instance, the reference time period 'night' is normally between the hours of 10 p.m. and 6 a.m., an eight-hour time period. Normally, for measuring purposes, a much smaller mean time frame is used in order to reduce the effect of background noise. The level L_{eq} is then reconstructed based on the number of total noise events and applied over a longer period of time. For example, let L_{eq} be an energy-equivalent sound level to be verified for a city railway next to a street. The mean time period for this measurement would then be approximately how long it takes for the train to pass by the street. That is, we are looking for the energy-equivalent continuous sound level for an average period that corresponds to the time it takes for the train to pass by the street, such as of 30 seconds $L_{eq}(30s)$. Suppose the train passes by every 5 minutes without a break. In this case, we can easily calculate the long-term L_{eq} (measured over several hours, as may correspond to the reference periods 'day' or 'night', for instance) by $L_{eq}(long) = L_{eq}(30s) - 10lg(5min/30s) = L_{eq}(30s) - 10\,dB$.

Applying mean values is often the most sensible and essential method for determining or verifying maximum permissible noise levels. However, mean values, per definition, omit single events over a time-based continuum and can thus blur the distinctions between possibly very different situations. A light rail train passing by once every hour could well emit similar sound levels L_{eq} as characterized by the permanent noise level of a busy street over a very long reference period, for example. The long-term effects of both sources combined may, in fact, result in one of the sources being completely obliterated from the L_{eq} altogether (refer to Exercise 5).

The energy-equivalent continuous sound level serves as the simplest way to characterize sound levels fluctuating over time. Cumulative frequency levels can be determined by the peak time-related characteristic measurements, a method which is instrumental in the statistical analysis of sound levels.

1.5 Summary

Sound perception is governed by relativity. Changes are perceived to be the same when the stimulus increases by a certain percentage. This has led to the conclusion of the Weber-Fechner law, according to which perception is proportional to the logarithm of the stimulus. The physical sound pressures are therefore expressed through their logarithmic counterparts using sound levels of a pseudo-unit, the decibel (dB). The entire span of sound pressure relevant for human hearing, encompassing about 7 powers of ten, is reflected in a clearly defined scale from about $0\,dB$ (hearing threshold) up to approximately $140\,dB$ (threshold of pain). A-weighting is scaled to the human ear in order to roughly capture the frequency response function of hearing. A-weighted sound levels are expressed in the pseudo-unit $dB(A)$.

Sounds at intermittent time intervals are quantified using mean time values. One such quantification method approximates 'energy-equivalent permanent sound levels.'

1.6 Further reading

Stanley A. Gelfand's "Hearing – an Introduction to Psychological and Physiological Acoustics" (Marcel Dekker, New York 1998) is a physiologically oriented work and contains a detailed description of the anatomy of the human ear and the conduction of stimuli.

1.7 Practice exercises

Problem 1

An A-weighted sound level of 50 $dB(A)$ originating from a neighboring factory was registered at an emission control center. A pump is planned for installation 50 m away from the emission control center. How high can the A-weighted decibel level, resulting from the pump alone, be registered at the emission control center so that the overall sound level does not exceed 55 $dB(A)$?

Problem 2

A noise contains only the frequency components listed in the table below:

f/Hz	$L_{Thirdoctave}/dB$	Δ_i/dB
400	78	-4,8
500	76	-3,2
630	74	-1,9
800	75	-0,8
1000	74	0
1250	73	0,6

Calculate

- both non-weighted octave levels
- the non-weighted overall decibel level and
- the A-weighted overall decibel level.

The corresponding A-weights are given in the last column of the table.

Problem 3

A noise consisting of what is known as white noise is defined by a 1 dB-increase from third-octave to third-octave (see Figure 1.6). How much does the octave level increase from octave to octave? How much higher is the overall decibel level in proportion to the smallest third-octave when N third-octaves are contained in the noise? State the numerical value for $N = 10$.

Problem 4

A noise consisting of what is known as pink noise is defined by equal decibel levels for all the third-octaves it contains. How much does the octave level increase from octave to octave and what are their values? How much higher is the overall decibel level when there are N thirds contained in the noise? State the numerical value for $N = 10$.

Problem 5

The energy-equivalent permanent sound level is registered at $55\,dB(A)$ at an emission control center near a street during the reference time period 'day,' lasting 16 hours. A new high-speed train track is scheduled to be constructed near the sight. The 2-minute measurement of a sound level L_{eq} of a passing train is $75\,dB(A)$. The train passes by every 2 hours.

How high is the energy-equivalent permanent sound level measured over a longer period of time (in this case, reference time interval 'day')

- a) of the train alone and
- b) of both sound sources combined?

Problem 6

A city train travels every 5 minutes from 6 a.m. to 10 p.m. At night, between 10 p.m. and 2 a.m., it travels every 20 minutes, with a break from 2-6 a.m. in between. A single train passes within 30 seconds and for this time duration, the sound level registers at $L_{eq}(30s) = 78\,dB(A)$. How high is the energy-equivalent permanent sound level for the reference time intervals 'day' and 'night'?

Problem 7

The sound pressure level L of a particular event, such as the emission of a city train as in the previous example, can under certain circumstances only be measured against a given background noise, such as traffic. Assume that the background noise differs from the sound event to be measured at a decibel level ΔL. How high is the actual combined noise level? State the general equation for measuring errors and the numerical value for $\Delta L = 6\,dB$, $\Delta L = 10\,dB$ and $\Delta L = 20\,dB$.

Problem 8

As in Problem 7, sound emission is to be measured in the presence of a noise disturbance. How far away does the the noise source have to be in order to obtain a measuring error of $0.1\,dB$?

Problem 9

Sometimes filters of a relatively constant bandwidth are employed to take finer measurements in sixth-octaves increments as opposed to the customary third-octave increments. State the equations for

- the consecutive center frequencies
- the bandwidth and
- the limiting frequencies.

Problem 10

In a calculated measurement where an octave and all of its constituent thirds are given, it appears that one of the thirds may have been a measurement error. How can the result be checked against the other three values that are assumed to be correct?

2

Fundamentals of wave propagation

The most important qualitative statements about wave propagation can be deduced by everyday life experience. When observing temporary, frequently repetitive sound incidences, such as a child bouncing a ball, hammering at a construction site, etc., a time delay can easily be observed between the optical perception and the arrival of the acoustic signal. This time delay increases with increasing distance between the observer and the source. Apart from the facts that

- the sound pressure level decreases with increasing distance and
- sound sources have a radiation pattern and
- echoes accumulate, for instance, at large reflecting surfaces (like house walls) or, more generally, if the 'acoustic environment' (ground, trees, bushes, etc.) is left out of our considerations,

the only thing that distinguishes different observation points is the time delay. Indeed, sound incidences sound the same from any vantage point, as the frequency components are the same. The wave form of a sound field (in a gas) is not altered during propagation. The propagation is called 'non-dispersive', because the form of the signal is not altered during wave transmission. In contrast, the propagation of bending waves in beams and plates, for instance, is dispersive in gases (see Chap. 4). The fact that sound fields do not alter their wave form during transportation is not trivial. Non-dispersive wave propagation in air is not only an essential physical property of sound travel; imagine if sound incidences were composed differently at various distances. Such a case would render communication impossible!

This chapter attempts to describe and to explain the physical properties of wave propagation in gases. First, it seems reasonable to clarify the physical quantities and their basic relations, which are needed to describe sound fields. This chapter should likewise serve as a means to refresh basic knowledge in thermodynamics. The following deliberations are based on the assumption of perfect gases. This assumption in respect to air-borne sound in the audible

M. Möser, *Engineering Acoustics*, DOI 10.1007/978-3-540-92723-5_2,

frequency range is justified by extensive experimental evidence with highly significant correlation.

2.1 Thermodynamics of sound fields in gases

The physical condition of a perfect gas, starting with a given, constant mass M, can be described by

- its volume $V_{t}ot$ it fills
- its density $\rho_{t}ot$
- its inner pressure $p_{t}ot$ and
- its temperature $T_{t}ot$.

When conducting theoretical experiments with a small and constant mass of a gas, bound, for example, by a small enclosure with uniform constant pressure and uniform constant density, the state descriptions of volume, temperature, and pressure are the most illuminating. The density $\rho_{t}ot = M/V_{t}ot$ then appears as a redundant quantity which can be determined by the volume. It is enough to describe the gaseous state of large (sometimes infinite) masses and volumes–relevant when dealing with sound fields–in terms of pressure, density and temperature, but for the purposes of review in the basics of thermodynamics, the following theoretical investigations outlined below will focus on constant gas masses. Sometimes, however, the following derivations are based on , since the origins of thermodynamics should be refreshed as mentioned earlier. The principles from the following discussions will then be appropriately applied to the important sound field quantities.

As a matter of fact, the question arises as to how the quantities used for describing a gaseous state are related. The criteria of a constant mass of a gas (when put, for example, into a vessel with a variable volume) should be met and can be approximately described as such:

- heating of the gas with constant volume results in an increased pressure $p_{t}ot \sim T_{t}ot$
- the pressure of the gas is inversely proportional to the volume $p_{t}ot \sim 1/V_{t}ot$.

These and other such statements can be summarized in the Boyle-Mariotte equation, if one accounts for the fact that an increased mass (with constant pressure and constant temperature) needs an increased volume. It is given by

$$p_{t}otV_{t}ot = \frac{M}{M_{\mathrm{mol}}} RT_{t}ot, \tag{2.1}$$

where M_{mol} is a material constant, the so-called 'molar mass'. The molar mass M_{mol} defines the 'molecular mass in grams' of the corresponding element (see the periodic system of elements), e.g. $M_{\mathrm{mol}}(N_2) = 28\,\mathrm{g}$ and $M_{\mathrm{mol}}(O_2) = 32\,\mathrm{g}$ which results in $M_{\mathrm{mol}}(\mathrm{air}) = 28.8\,\mathrm{g}$ (air consists of approximately 20% oxygen and 80% nitrogen). $R = 8.314\,\mathrm{N\,m/K}$ is the universal gas constant (K=Kelvin is the unit of the absolute temperature, $0°\mathrm{C} = 273\,\mathrm{K}$).

As mentioned earlier, sound fields are better described by using densities. Equation (2.1) is therefore transformed for 'acoustic purposes' to

$$p_t ot = \frac{R}{M_{\mathrm{mol}}} \rho_t ot T_t ot \ . \tag{2.2}$$

A graphical representation of (2.2) can easily be illustrated by isotherms, where curves with $T_t ot$ = const. are straight lines in the $p_t ot$-$\rho_t ot$-plane. It represents an array of characteristic lines (see Fig. 2.1). To describe the actual path that the three variables of state take in that array of characteristic lines, a second piece of information is needed. The Boyle-Mariotte equation does not completely express how a variation of one variable of state (e.g. carried out in the experiment) influences the other. If, for example, the volume of a gas is compressed (by pressing a piston into a vessel), it is also possible that the temperature or the pressure can change. The Boyle-Mariotte equation does not include that detailed information. It only states that the ratio of the two quantities is altered. Additional observations are needed to clarify this. Experience shows that the speed of the compression and the environ-

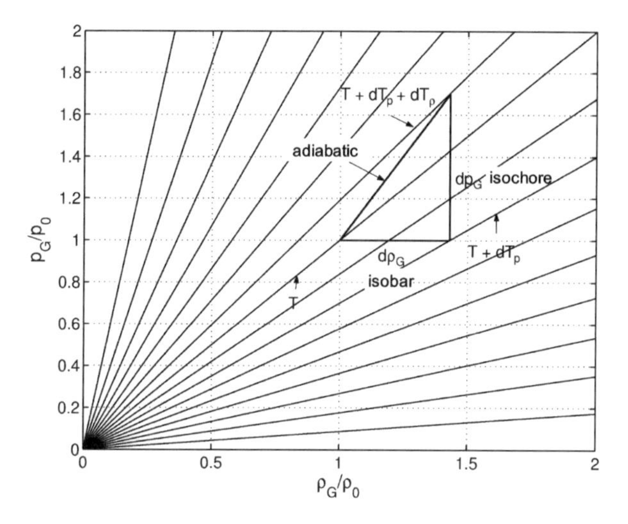

Fig. 2.1. Isotherms with the composition of adiabatic compression by one isobar and one isochor step

ment in which the compression takes place are of major importance. If the compression is done very quickly in a piston (or in an insulated environment without thermal conduction) then a temperature increase in the gas can be observed. Usually, thermal conduction is a very slow process that takes a long time (and is even impossible in a thermally non-conducting, insulated environment). Therefore, the observed increase in temperature is not achieved by heat consumed by the exterior, but rather, only results from the interior compression. If the volume compression is done very slowly and takes place in an environment with good thermal conduction, so that a temperature difference between the interior and the exterior can be compensated, the inner temperature can stay constant. In other words, thermal conduction is a crucial prerequisite for isothermal compression.

As already mentioned, thermal conduction is a slow process. Thus, isothermal compensations take a long time. In contrast, sound fields are subject to fast changes (apart from the lowest frequencies). It can, therefore, be assumed that sound-related processes happen without the participation of thermal conduction in the gas. In other words, when dealing with sound fields the gas can (nearly) always be assumed to lack thermal conduction. Thermal transportation processes play a minor role. This change in the gaseous state without thermal conduction is called 'adiabatic'. The fact that sound related processes are adiabatic also means that they cannot be isothermal, which would imply that thermal conduction takes place. The temperature of the gas and likewise the pressure and the density must therefore be subject to changes in time and space. Apart from scaling factors, these three variables of state even have the same time- and space-dependence, as will be shown shortly.

To derive adiabatic equations of state, one could refer to the literature. Nevertheless, it is given here because the derivation is neither difficult nor very extensive. As a starting point of the observation, imagine the adiabatic process taking place with a net thermal consumption of zero. It is composed of two steps: one with constant density and one with constant pressure (see also Fig. 2.1). All changes are assumed to be infinitesimal. The steps are then inevitably related to the temperature changes $\mathrm{d}T_p$ (p_tot=const.) and $\mathrm{d}T_\varrho$ (ϱ_tot=const.). In both steps heat can be transferred; only the total adiabatic process has to manage without any thermal conduction. When the total adiabatic process is composed, the sum of all thermal transfers has to be zero

$$\mathrm{d}E_p = -\mathrm{d}E_\varrho. \tag{2.3}$$

The isobar step consumes the heat

$$\mathrm{d}E_p = Mc_p\mathrm{d}T_p, \tag{2.4}$$

where c_p is the specific heat constant for constant pressure. The isochor step consumes the heat (note that ϱ=const. and V=const. is equivalent when the mass is constant)

$$\mathrm{d}E_\varrho = Mc_V\mathrm{d}T_\varrho, \tag{2.5}$$

where c_V is the specific heat constant for constant volume. For the adiabatic process defined in (2.3),

$$\frac{\mathrm{d}T_\varrho}{\mathrm{d}T_p} = -\kappa, \tag{2.6}$$

with

$$\kappa = \frac{c_p}{c_V} . \qquad (2.7)$$

The infinitely small temperature changes for constant pressure and constant density can now be expressed by the corresponding changes of the pressure (for the isochor step) and the density (for the isobar step). Equation (2.2) is solved for the gas temperature

$$T_tot = \frac{M_{\mathrm{mol}}}{R} \frac{p_tot}{\rho_tot} ,$$

which results in

$$\frac{\mathrm{d}T_p}{\mathrm{d}\rho_tot} = -\frac{M_{\mathrm{mol}}}{R} \frac{p_tot}{\rho_tot^2}$$

and

$$\frac{\mathrm{d}T_\varrho}{\mathrm{d}p_tot} = \frac{M_{\mathrm{mol}}}{R} \frac{1}{\rho_tot} .$$

Equation (2.6) is thus synonymous with

$$\frac{\mathrm{d}T_\varrho}{\mathrm{d}T_p} = \frac{\frac{\mathrm{d}p_tot}{\rho_tot}}{\frac{p_tot \mathrm{d}\rho_tot}{\rho_tot^2}} = -\frac{\rho_tot}{p_tot} \frac{\mathrm{d}p_tot}{\mathrm{d}\rho_tot} = -\kappa$$

or with

$$\frac{\mathrm{d}p_tot}{p_tot} = \kappa \frac{\mathrm{d}\rho_tot}{\rho_tot} . \qquad (2.8)$$

Integration of both sides yields

$$\ln \frac{p_tot}{p_0} = \kappa \ln \frac{\rho_tot}{\varrho_0} = \ln \left(\frac{\rho_tot}{\varrho_0} \right)^\kappa ,$$

which finally gives the adiabatic equation of state

$$\frac{p_tot}{p_0} = \left(\frac{\rho_tot}{\varrho_0} \right)^\kappa . \qquad (2.9)$$

The integration constants are chosen in such a way that (2.9) is fulfilled for the static quantities p_0 and ϱ_0. Equation (2.9) describes the relationship between pressure and density in a perfect gas 'without thermal conduction', as already mentioned. For perfect, diatomic gases, which are of interest in acoustics only, $\kappa = 1.4$.

It only remains to adapt the relations found in the Boyle-Mariotte equation (2.2) and in the adiabatic equation of state (2.9) for the lucid description of sound fields. The acoustic quantities represent small time- (and space-) dependent changes which are superimposed on the static quantities. It is thus reasonable to split the total quantities (thus the index tot) into a static part and an alternating part

$$p_tot = p_0 + p \tag{2.10a}$$

$$\rho_tot = \varrho_0 + \varrho \tag{2.10b}$$

$$T_tot = T_0 + T, \tag{2.10c}$$

where p_0, ϱ_0 and T_0 are the static quantities 'without any sound' and p, ϱ and T are the alterations due to the sound field. The superimposed sound field related quantities are designated as sound density, sound temperature and sound pressure. These quantities are actually tiny compared to the static quantities. As mentioned in Chap. 1, the rms-value of the sound pressure due to (dangerously high) levels of 100 dB is only 2 N/m^2. The atmospheric pressure is about 100000 N/m^2! The static quantities as well as the total quantities, but not the sound field related quantities alone, because they represent only a part of the total quantity, must certainly fulfil the Boyle-Mariotte equation (2.2). The opposite is true, which is shown by inserting (2.10–c) into (2.2)

$$p_0 + p = \frac{R}{M_{\text{mol}}}(\varrho_0 + \varrho)(T_0 + T) \approx \frac{R}{M_{\text{mol}}}(\varrho_0 T_0 + \varrho_0 T + T_0 \varrho), \tag{2.11}$$

where the (small quadratic) product between sound density and sound temperature is neglected in the last step. The static quantities vanish in (2.11), because they are a solution to the Boyle-Mariotte equation themselves and for the sound field related quantities

$$p = \frac{R}{M_{\text{mol}}}(\varrho_0 T + T_0 \varrho) \tag{2.12}$$

remains. This equation becomes a little clearer when dividing by the static pressure p_0 which gives

$$\frac{p}{p_0} = \frac{\varrho}{\varrho_0} + \frac{T}{T_0} \,. \tag{2.13}$$

If the resulting quotients are designated as 'relative quantities', (2.13) states that the relative sound pressure is the sum of the relative density and the relative sound temperature.

The second relationship between the sound field quantities is given by the adiabatic equation of state (2.9) which will be adapted for the comparatively small sound field quantities in what follows.

First, note that the adiabatic equation of state (2.9) states a non-linear relationship between pressure and density in the gas. On the other hand, only the smallest alterations around the operating point (p_0, ϱ_0) are of interest; thus the curved characteristics can be replaced by its tangent at the operating point. In other words, the characteristics can be linearized, because quadratic and higher order terms of a Taylor series expansion can be neglected.

To do this, the sound field quantities in eq.(2.10) are applied to the specific adiabatic equation for the aggregate values (2.9)

$$\frac{p_0 + p}{p_0} = 1 + \frac{p}{p_0} = \left(\frac{\varrho_0 + \varrho}{\varrho_0}\right)^{\kappa} = \left(1 + \frac{\varrho}{\varrho_0}\right)^{\kappa} .$$

The power series truncated after the linear term $f(x) = (1+x)^\kappa$ at $x = 0$ exists in $f(x) = 1 + \kappa x$, therefore

$$1 + \frac{p}{p_0} = 1 + \kappa \frac{\varrho}{\varrho_0} .$$

The linearized, adiabatic equation of state adapted for acoustic purposes then becomes

$$\frac{p}{p_0} = \kappa \frac{\varrho}{\varrho_0} . \tag{2.14}$$

Sound fields are nearly always described by means of their pressure distribution, because the sound pressure can be detected with microphones very easily, whereas the sound density can only be determined indirectly by the pressure. So, whenever possible, the sound pressure will be used in the following formulations. For that purpose, if a density occurs it has to be expressed by a sound pressure. This can be achieved by solving (2.14) for the density

$$\varrho = \frac{p}{c^2} , \tag{2.15}$$

with

$$c^2 = \kappa \frac{p_0}{\varrho_0} . \tag{2.16}$$

Obviously, sound pressure and sound density have the same time and spatial dependence. If the relative density is eliminated in (2.13) by using (2.14), the relative sound temperature becomes

$$\frac{T}{T_0} = \frac{p}{p_0} - \frac{\varrho}{\varrho_0} = \left(1 - \frac{1}{\kappa}\right) \frac{p}{p_0} .$$

All three relative quantities have the same wave form, apart from different scaling factors.

The next section will show that the constant c introduced in (2.16) has a special physical meaning: c denotes the speed at which the sound propagates in the gas. Verifying the dimensions (the units) of the quantities yields no contradiction to this statement, although it is not a proper proof:

$$\dim(c) = \sqrt{\frac{\dim(p)}{\dim(\varrho)}} = \sqrt{\frac{\mathrm{N\,m^3}}{\mathrm{m^2\,kg}}} = \sqrt{\frac{\mathrm{kg\,m}}{\mathrm{s^2}} \frac{\mathrm{m}}{\mathrm{kg}}} = \frac{\mathrm{m}}{\mathrm{s}}$$

The unit of c, $\dim(c)$, is indeed equal to that of a speed.

If the Boyle-Mariotte equation (2.2) (which is valid for the static quantities, too) is inserted into (2.16), the speed of sound c becomes

$$c = \sqrt{\kappa \frac{R}{M_{\mathrm{mol}}} T_0} . \tag{2.17}$$

It depends only on material constants and the absolute temperature, and is independent of the static pressure and the static density. To check (2.17), the parameters of air $M_{\mathrm{mol}} = 28.8\,10^{-3}\,\mathrm{kg}$ for $T_0 = 288\,\mathrm{K}$ (15°C) are inserted into the equation and the well-known value of $c = 341\,\mathrm{m/s}$ is obtained. For practical applications, it is nearly always sufficient to neglect temperature changes of up to 10°C and calculate with the rounded value of 340 m/s.

It should perhaps be noted that the assumption of isothermal compression for sound-related processes (which is actually not valid in the unbounded gas) would lead to a speed of sound

$$c_{\mathrm{iso}} = \sqrt{\frac{RT_0}{M_{\mathrm{mol}}}} = \frac{c_{\mathrm{adia}}}{\sqrt{\kappa}} \approx 0.85 c_{\mathrm{adia}}$$

which is too small. As a matter of fact, it was the discrepancy between c_{iso} and experimental values which showed that sound related compression processes are not isothermal but adiabatic. The experimentally determined speed of sound must therefore be equal to c_{adia}.

2.2 One-dimensional sound fields

2.2.1 Basic equations

The aim of the last section was to clarify the meaning of the physical variables of state sound pressure, sound density and sound temperature that arise in sound fields. The next section turns to the crucial question of acoustics: how is the phenomenon of (non-dispersive) wave propagation of sound in gases physically explained and described?

To derive the basic properties, the effects mentioned in the introduction, as well as the attenuation with increasing distance and reflections, are left out of consideration. What remains is the simple case of one-dimensional sound fields, depending only on a single coordinate. Such a one-dimensional wave guide can be realized, for example, by capturing the sound field in a tube with a rigid lining filled with just air. The air is then forced to propagate in one direction, i.e. the tube axis (which does not necessarily produce a sound field which is constant over the cross sectional area of the tube.) This will be shown in more detail in Chap. 6 which deals with sound absorption.

The main properties of sound fields can be deduced from the basic assumption that a perfect gas is an elastically shapeable and mass-adherent medium. A very simple and illuminating explanation for the propagation of waves is obtained if the air column in the one dimensional wave guide is segmented into a number of smaller sections (Fig. 2.2) and the segments alternate between elements with 'mass-characteristics' and elements with 'spring-characteristics'. Thus a so-called chain of elements model is obtained for the segmented air column. The excitation of the air column can, for example, be realized with a loudspeaker. The loudspeaker, assigned to the chain of elements, represents the displacement of the first mass in Fig. 2.2. If the first mass suddenly starts to move to the right, the first (air) spring is compressed; the spring puts a

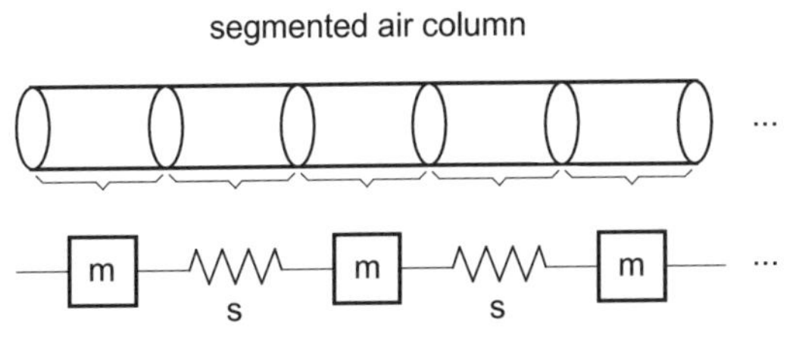

Fig. 2.2. Segmented air column, consisting of alternating elements of volume, representing mass substitutes and spring substitutes

force onto the second mass. Initially, the mass is not moving and due to its mass inertia, it also does not immediately start to move. The onset of motion actually begins with 'delayed' displacement. To illustrate the inertia law, the time dependence of a force which suddenly starts to excite, and the displacement of a coupled 'free' mass is shown in Fig. 2.3, where the mass is gradually put into motion. Therefore, the displacement of the second mass in the chain of elements starts with a delayed offset to the force of the spring. The mass compresses the next spring to the right and is slowed down. The 'delay' is cumulative, as the displacement is transmitted 'on down the line' or 'from mass to mass.' The process is replicated throughout the chain of elements and the pulse initiated at the left end of the chain migrates to the right with a finite speed.

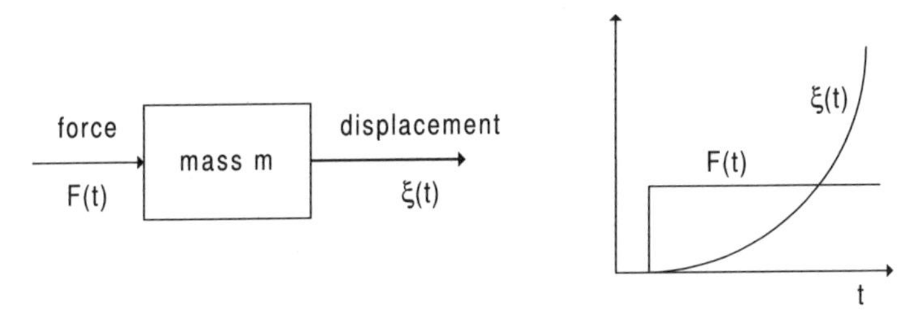

Fig. 2.3. Free mass and exemplary force-time characteristics with resulting displacement-time characteristics

Obviously, two different sorts of speed must be distinguished here. One is the 'migration speed' of the pulse through the wave guide. It is called propagation speed or wave speed, denoted in this book by c. The other one, which has to be distinguished from c, is the speed at which the local gas masses move around their equilibrium position, as the wave 'runs through' them. For better distinction the speed of the local gas elements is called 'velocity'. It is always denoted by v in this book.

The aforementioned physical aspects can now readily be formulated in equations. Two investigations are necessary: it has to be discussed how the air springs are compressed by the displacement at their boundaries to the left and to the right and the problem has to be solved how the air masses are accelerated by the forces of the springs acting on them. Small air volumes of length x are used for both investigations. The elements of length Δx (initially assumed to have a finite length for illustration purposes) will finally shrink to the infinitesimal length $\mathrm{d}x$, because the description of the physical facts is a lot easier with functions and their derivatives.

The inner compression of one gas element with mobile boundaries is derived by the fact that the mass between the boundaries is constant. If one element is compressed, the density increases. The mass of the element depicted in Fig. 2.4 is $S\Delta x\varrho_0$ if the medium is at rest (without sound), where S is the cross-sectional area of the column. If an elastic deformation takes place (in the presence of sound), the motion of the left boundary defined by $\xi(x)$ and the motion of the right boundary defined by $\xi(x+\Delta x)$ takes place, the mass is given by $S\left[\Delta x+\xi(x+\Delta x)-\xi(x)\right]\rho_tot$. The mass is equal to the mass at rest and using $\rho_tot=\varrho_0+\varrho$, we obtain

$$(\varrho_0+\varrho)\left[\Delta x+\xi(x+\Delta x)-\xi(x)\right]S=\varrho_0\Delta xS\,,$$

or, after the dividing the surface S and multiplying it out,

$$\varrho\Delta x+\varrho_0\left[\xi(x+\Delta x)-\xi(x)\right]+\varrho\left[\xi(x+\Delta x)-\xi(x)\right]=0\,. \tag{2.18}$$

The small quadratic term of sound density and displacement are insignificant, even at the highest most relevant sound levels. The sound density in question here can be described by

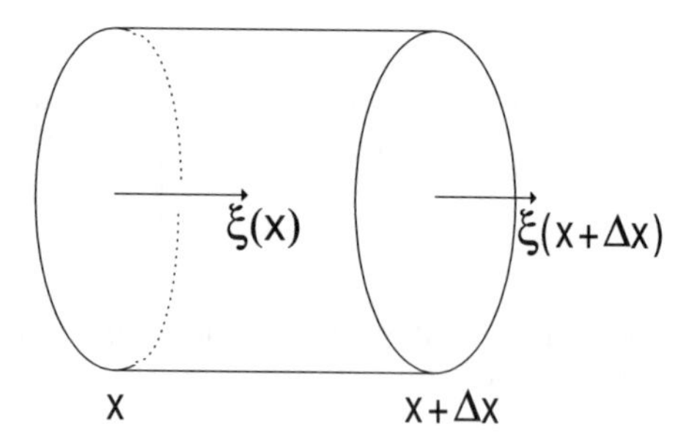

Fig. 2.4. Deformation of an element of the gas column leads to a change in internal density

$$\varrho=-\varrho_0\frac{\xi(x+\Delta x)-\xi(x)}{\Delta x}\,.$$

In the limiting case of infinitesimal small gas elements $\Delta x\to\mathrm{d}x$, the difference quotient becomes the differential quotient

$$\frac{\varrho}{\varrho_0} = -\frac{\partial \xi(x)}{\partial x}. \tag{2.19}$$

The sound density is directly related to the spatial derivative of the displacement. The latter is also called 'elongation' (or dilatation). This derivation is crucial for the following investigations. It states that the relative sound density is equal to the negative elongation.

It should be noted that this fact can also be interpreted as the spring equation. If the sound density in the penultimate step is replaced by the sound pressure $\varrho = p/c^2$ and multiplied by the cross sectional area S, we get

$$Sp = -S\varrho_0 c^2 \frac{\xi(x+\Delta x) - \xi(x)}{\Delta x}.$$

is obtained. The left side, Sp, represents the force F produced by deformation of the gas spring of length Δx. Hook's law can be applied to the case of springs with moving ends, validating

$$Sp = -s[\xi(x+\Delta x) - \xi(x)],$$

where s represents the stiffness of the spring. In the case of layers of elastic material, as in a gas element, with a cross sectional area S and length Δx

$$s = \frac{ES}{\Delta x} \tag{2.20}$$

is given, where E represents a material constant, the so-called elastic modulus. It should be indicated for the interpretation of (2.20) that producing a certain change in displacement at the ends requires an applied force which has to be larger. The larger the cross sectional area of the layer is, the smaller the thickness of the layer. The elastic modulus in gases is obviously related to the propagation speed by

$$E = \varrho_0 c^2 . \tag{2.21}$$

The second phenomenon pertaining to sound wave propagation that needs to be investigated is how gas particles are accelerated by the applied forces of the springs. The answer is found in Newton's law, which is applicable to the (small) volume element of the gas column as shown in Fig. 2.5. The acceleration $\partial^2\xi/\partial t^2$ of the enclosed mass is caused by the force 'pushing from the left' $Sp(x)$, from which the force 'pushing back from the right' $Sp(x+\Delta x)$ must be subtracted. The acceleration caused by the change in force is smaller, the smaller the mass m of the element is. Applying Newton's law, we obtain

$$\frac{\partial^2\xi}{\partial t^2} = \frac{S}{m}\left[p(x) - p(x+\Delta x)\right].$$

Alternatively, using m = volume $\times$ density = $\Delta x S \varrho_0$, we obtain

$$\frac{\partial^2\xi}{\partial t^2} = -\frac{1}{\varrho_0}\frac{p(x+\Delta x) - p(x)}{\Delta x}.$$

The element is finally compressed, and using

$$\lim_{\Delta x \to 0} = \frac{p(x+\Delta x) - p(x)}{\Delta x} = \frac{\partial p}{\partial x},$$

we arrive at the 'inertia law of acoustics',

$$\varrho_0 \frac{\partial^2 \xi}{\partial t^2} = -\frac{\partial p}{\partial x} \,. \tag{2.22}$$

Equations (2.19) and (2.22) form the basic equations in acoustics. They

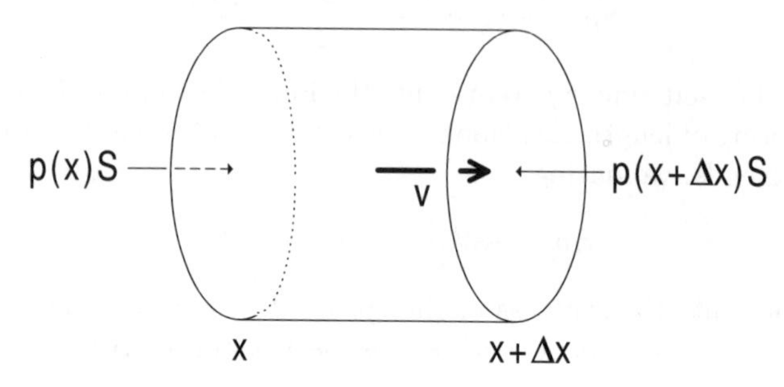

Fig. 2.5. Accelerated element of the gas column

are able to describe all (one-dimensional) sound incidences. The compression of the elastic continuum 'gas' caused by space-dependent displacement is described in (2.19); how the displacement is caused by compression, on the other hand, is described in (2.22). If both observations are combined they yield the explanation for wave propagation. 'Combining' the two observations in terms of equations means inserting one equation into the other. The displacement is consequently eliminated in (2.19) and (2.22). This can be achieved by a twofold differentiation of (2.19) by time

$$\frac{1}{\varrho_0}\frac{\partial^2 \varrho}{\partial t^2} = -\frac{\partial^3 \xi}{\partial x \partial t^2}$$

and differentiating (2.22) by space

$$\frac{\partial^3 \xi}{\partial x \partial t^2} = -\frac{1}{\varrho_0}\frac{\partial^2 p}{\partial x^2} \,.$$

Hence it follows

$$\frac{\partial^2 p}{\partial x^2} = \frac{\partial^2 \varrho}{\partial t^2} \,,$$

where the sound density can ultimately be replaced by the sound pressure $\varrho = p/c^2$ from (2.15) as stated earlier

$$\frac{\partial^2 p}{\partial x^2} = \frac{1}{c^2}\frac{\partial^2 p}{\partial t^2} . \tag{2.23}$$

Equation (2.23), which is referred to as the wave equation; all occurring sound incidences must satisfy (2.23).

As has already been shown, the wave equation is derived from both 'basic equations of acoustics', the law of compression eq.(2.19) and the law of inertia as applied to sound fields eq.(2.22), together with the 'material law' $\varrho = p/c^2$. ξ appears in both equations. It makes sense to express them in terms of sound velocity

$$v(x,t) = \frac{\partial \xi}{\partial t}. \tag{2.24}$$

The next section will explore the reason for this. For the simplest case of progressive waves, the signal forms of pressure and velocity are the same. Therefore, defining sound vibrations by their velocity, and not by their particle displacement, is the conventional approach in the field of acoustics. Sound velocity will from here on be used in this book as well. For this reason, the basic equations (2.19) and (2.22) will be noted here as a reminder, but hereafter only in terms of pressure and velocity. Thus, the universal law of compression is therefore

$$\frac{\partial v}{\partial x} = -\frac{1}{\varrho_0 c^2}\frac{\partial p}{\partial t} , \tag{2.25}$$

and the law of inertia exists in

$$\varrho_0 \frac{\partial v}{\partial t} = -\frac{\partial p}{\partial x} . \tag{2.26}$$

Both equations apply universally, even for sound fields where progressive waves are travelling in both directions. From here on, we will only refer to the notations (2.25) and (2.26). Also, the 0 index in the density constant component ϱ_0 will be left out, as there will be no cause for subsequent confusion of the terms hereafter. Sound density will only be treated in Chapter 2.

2.2.2 Progressive waves

In general, any function which is exclusively dependent on the argument $t-x/c$ or $t+x/c$ is a solution to the wave equation (2.23)

$$p(x,t) = f(t \mp x/c) . \tag{2.27}$$

$f(t)$ stands for a signal form possessing a structure which has been created by the emitter, that is, the sound source. c signifies the constant already defined in the previous section, where it was already alluded to as the definition for the sound propagation speed. The following considerations will serve as immediate proofs to this fact. First, it should be explained why (2.23) is described as the wave equation. The name is derived from a graphical representation of its solutions (2.27) as a function of space, as shown in Figure 2.6 for constant, frozen time (here for $f(t - x/c)$, that is, the negative value in the argument). The graph depicts a set of curves of the same parallel offset space-dependent functions which cross each other. As can be seen from the graph, the state of gas "'sound pressure"' migrates as at constant speed along the x axis. This migration of the local states is known as a "'wave"'.

The still unanswered question of the physical meaning of the constant c can now be explained. Imagine a certain value of the function f (in Fig. 2.6 the maximum of f is chosen) which is located at x at the time t and travels by the distance Δx during the time Δt

$$f(x, t) = f(x + \Delta x, t + \Delta t) \ .$$

This is the case if $(t - x/c)$ is the same in both cases, which is equivalent to

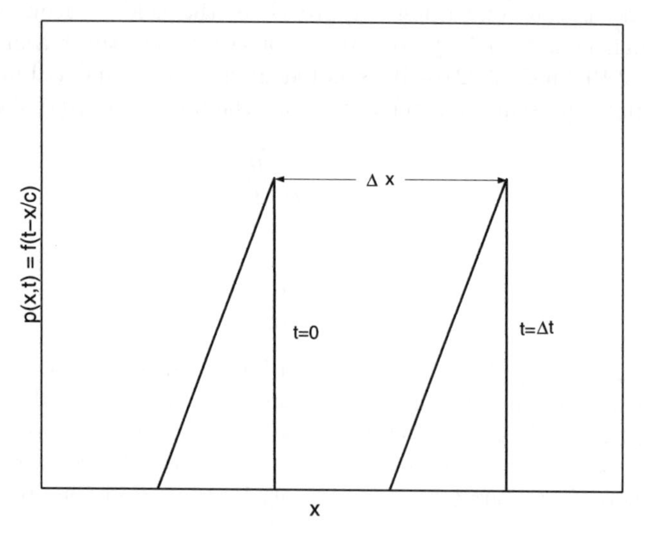

Fig. 2.6. Principal characteristics of $p = f(t - x/c)$ for two different times $t = 0$ and $t = t$

$$t - \frac{x}{c} = (t + \Delta t)\frac{x + \Delta x}{c}$$

and results in

$$\frac{\Delta x}{\Delta t} = c \ .$$

Because the speed is calculated by 'speed=distance/time to travel', c describes the 'transport speed of the function', i.e. the propagation speed of the wave. Obviously, it is independent of the characteristics of the function f; in particular, all frequencies travel at the same speed. The fact that the signal characteristics are not altered during propagation is an important feature of sound propagating in gases (compare the dispersive bending waves in beams and plates, see Chap. 4) which is one of the most important physical preconditions for acoustic communication (e.g. speech).

If only one wave occurs, travelling in a specific direction, it is called a plane propagating wave. Combinations of waves travelling in opposite directions contain standing waves (see also Sect. 2.2.4 on p. 34). For travelling waves with $p(x,t) = f(t - x/c)$ the inertia law of acoustics (2.22) yields

$$\varrho_0 v = -\int \frac{\partial p}{\partial x} \mathrm{d}t = -\int \frac{\partial f(t - x/c)}{\partial x} \mathrm{d}t = \frac{1}{c} \int \frac{\partial f(t - x/c)}{\partial t} \mathrm{d}t = \frac{p}{c} ,$$

which says that sound pressure and sound velocity have a constant ratio, independent of space and time, called the wave resistance or specific resistance of the medium:

$$\frac{p(x,t)}{v(x,t)} = \varrho_0 c . \tag{2.28}$$

Eq.(2.28) provides a simple answer to the remaining question of how the sound source leaves its "footprint" on the signal form of the sound pressure $f(t)$. The following assumptions are made for This model presupposes the following in respect to the one-dimensional wave guide model–as previously mentioned, a tube filled with air:

- There are no reflections (the tube is terminated in an extremely effective absorber known as a wave capturer
- The sound source is composed of a plane membrane (such as that of a loudspeaker), which oscillates at a previously designated membrane velocity of $v_M(t)$ at its resting position in $x = 0$

Since no reflections are present, the sound pressure solely consists of a progressive wave travelling in the x direction in the form of

$$p(x,t) = f(t - x/c) ,$$

and for the molecular sound velocity v in the wave guide, the following can be derived from eq.(2.28)

$$v(x,t) = p(x,t)/\varrho_0 c = f(t - x/c)/\varrho_0 c .$$

The velocity of the medium v must be equal to the velocity of the membrane at the position of the source $x = 0$, therefore

$$f(t)/\varrho_0 c = v_M(t) ,$$

according to which sound pressure

$$p(x,t) = \varrho_0 c v_M(t - x/c)$$

and sound velocity

$$v(x,t) = v_M(t - x/c)$$

in the wave guide are simply

- defined by the source signal 'membrane velocity $v_M(t)$' and
- by the existence of progressive waves in this situation.

Sound pressure can be defined using a more formal and universal approach as a phenomenon which meets the criteria of the wave equation. However, the exact structure can only be defined based on a given constraint at the position $x = 0$, $v(0, t) = v_M(t)$. This method is known as the 'solution of a boundary value problem.' The simplest example of such a problem was outlined above.

Definition of quantities

For travelling waves, (2.28) can be used to assess the order of magnitude of velocity and displacement. A relatively high level of 100 dB is related to a rms-value of the sound pressure of $p_{rms} = 2\,\mathrm{N/m^2}$. In a plane progressive wave $v_{rms} = p_{rms}/c$, where $\varrho_0 = 1.2\,\mathrm{kg/m^3}$, and $c = 340\,\mathrm{m/s}$ is $v_{rms} = 5\,10^{-3}\,\mathrm{m/s}$ $= 5\,\mathrm{mm/s}$. The local particle speed 'velocity' is therefore very, very small compared to $c = 340\,\mathrm{m/s}$. Even the displacement is not very large. It is calculated by

$$\xi_{rms} = \frac{v_{rms}}{\omega} , \tag{2.29}$$

assuming pure tones only and $v = \mathrm{d}\xi/\mathrm{d}t$. For 1000 Hz this would result in $\xi_{rms} = 10^{-6}\,\mathrm{m} = 1\,\mu\mathrm{m}$! In acoustics, the displacement often ranges in only one-thousandth of the diameter of an atom.

Compared to that, the accelerations occurring in acoustics can be considerably larger. Based on

$$b_{rms} = \omega v_{rms}, \tag{2.30}$$

an acceleration of $b_{rms} = 30\,\mathrm{m/s^2}$, a threefold ground acceleration, is obtained for a 100 dB-sound pressure level at $f = 1000\,\mathrm{Hz}$.

Harmonic time dependencies

Sound and vibration problems are, for practical reasons, often described using harmonic (sinusoidal) time-dependencies. Generally, the sound pressure of a wave travelling along the x-direction has the form

$$p(x, t) = p_0 \cos\omega(t - x/c) , \tag{2.31}$$

which can be written in abbreviated form using the so-called wave number

$$k = \omega/c$$

as

$$p(x, t) = p_0 cos(\omega t - kx) . \tag{2.32}$$

As is generally known, ω, which is

$$\omega = 2\pi f = \frac{2\pi}{T} \tag{2.33}$$

contains the time period T of a complete cycle. Likewise, the wave number k must contain the spatial period

$$k = \frac{\omega}{c} = \frac{2\pi}{\lambda} \,. \tag{2.34}$$

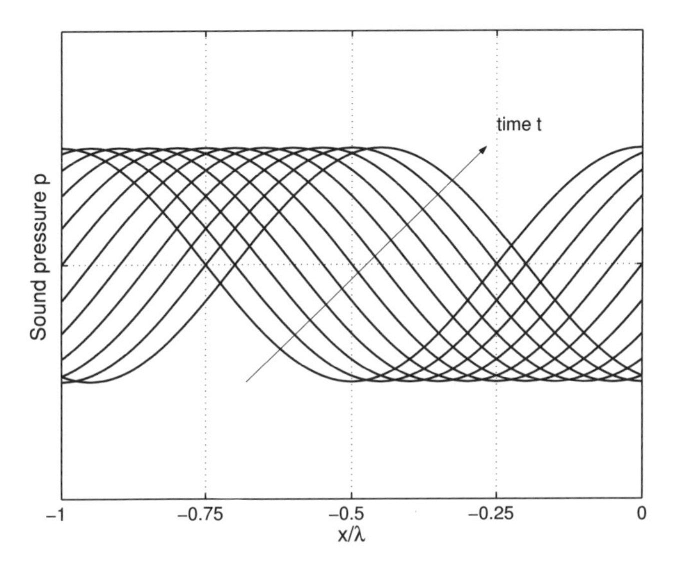

Fig. 2.7. spatial dependence of the sound pressure in a propagating wave at a constant time. Half a period is shown. The sound velocity v exists in the same space and time function due to $v = p/\varrho_0 c$.

The spatial period λ of a complete cycle is generally known as the wavelength. As can be seen clearly, this term is restricted to pure tones only. For the wavelength, using (2.33) and (2.34),

$$\lambda = \frac{c}{f} \tag{2.35}$$

is obtained. For non-dispersive sound waves in air the wavelength is roughly inversely proportional to frequency; it ranges from $\lambda = 17\,\mathrm{m}$ ($f = 20\,\mathrm{Hz}$) to $\lambda = 1.7\,\mathrm{cm}$ ($f = 20000\,\mathrm{Hz}$). This is a considerably large interval. It should not be surprising that the size of objects has to be expressed in terms of wavelength in acoustics (like in optics). Most objects and structures are acoustically 'invisible' in the low frequency range, where their size is small compared to the wavelength. At high frequencies they are acoustically effective, representing either sound absorbers, more or less complex reflectors, or diffusers, respectively. The aforementioned components in eq.(2.31) for progressive waves in pure tones are summarized again in Figure 2.7. It entails a cosine-formed space-dependent function, which moves from left to right at the propagation rate of c. According to eq.(2.28), pressure and particle speed possess the same space- and time-dependent signal characteristics.

2.2.3 Complex notation

From here on in this book, pure tone waves will be described using only complex amplitudes. More details on how to use, the purpose of and advantages of describing real-valued processes using complex numbers are outlined in Appendix B2. This appendix describes how to notate a sinusoidal wave travelling in the x-direction using space-dependent complex amplitudes

$$p(x) = p_0 \mathrm{e}^{-jkx} .$$

Waves travelling in the negative x-direction can by described by

$$p(x) = p_0 \mathrm{e}^{jkx}$$

If reflections are present or there are waves travelling in opposite directions, (as in the case of two sources or in a reflective room), a summation of both terms may occur. To describe the reverse transformation of the complex amplitudes to real-valued time and space domain functions, we use the time convention

$$p(x,t) = \mathrm{Re}\{p(x)\mathrm{e}^{j\omega t}\} \tag{2.36}$$

The time convention eq.(2.36) applies to the excitation of pure tones and for all physical quantities, as well as for all velocity components of any sound and reverberation field, for electrical voltage, currents, etc. Complex amplitudes are also referred to as 'pointers' or 'complex pointers.'

Subsequently, sound velocity, which results from a presumably known sound pressure spatial dependency, also often needs to be defined. As previously mentioned, this is calculated using the acoustical inertia law (2.26), which is written in complex notation as

$$v = \frac{j}{\omega \varrho_0} \frac{\partial p}{\partial x} \tag{2.37}$$

2.2.4 Standing waves and resonance phenomena

If a progressive wave is impeded by a barrier, it can be reflected at that location. If the one-dimensional wave guide has boundaries on both sides – at left, for example, bounded by a sound source and at the right, a reflector, as assumed in the following – standing waves and resonance phenomena will occur. As mentioned in the model of the one-dimensional wave guide (as well as for pure tones) the complex sound pressure amplitudes are composed of two terms

$$p(x) = p_0[\mathrm{e}^{-jkx} + r\mathrm{e}^{jkx}] . \tag{2.38}$$

As previously described, the first summand describes a wave moving in the $+x$ direction and the second summand, one moving in the $-x$ direction. p_0 describes the amplitude of the wave travelling toward the reflector. In the sound field ansatz (2.38) , we have already taken into consideration that the returning wave $-x$ direction may be attenuated relative to the approaching wave by a reflection coefficient r, if only a semi-reflection is at hand (as in the case of partial absorbers at the end of the tube in $x = 0$, refer to Chapter 6 for more information). The sound velocity pointing in the x direction pertaining to the sound pressure (2.38), using (2.37), becomes

$$v(x) = \frac{k}{\omega \varrho_0} p_0[\mathrm{e}^{-jkx} - r\mathrm{e}^{jkx}] = \frac{p_0}{\varrho_0 c}[\mathrm{e}^{-jkx} - r\mathrm{e}^{jkx}] . \tag{2.39}$$

For the sake of simplicity, let us initially assume that the reflector located at $x = 0$ is "'rigid"'. Therefore, the reflector must be made up of a large, immobile mass or non-elastic structure. Because the air particles, which span across the reflector at $x = 0$, are unable to penetrate the fixed barrier, their velocity, which is described by the sound velocity, must be zero:

$$v(x = 0) = 0 . \tag{2.40}$$

Therefore, the reflection coefficient r which characterizes the rigid reflector must be (2.39)

$$r = 1 , \tag{2.41}$$

leading to the spatial dependency for sound pressure of

$$p(x) = 2p_0 \cos kx \tag{2.42}$$

and for sound velocity of

$$v(x) = \frac{-2jp_0}{\varrho_0 c} \sin kx . \tag{2.43}$$

The time convention helps us determine the resulting space and time domain functions

$$p(x) = 2p_0 \cos kx \cos \omega t \tag{2.44}$$

and

$$v(x) = \frac{2p_0}{\varrho_0 c} \sin kx \sin \omega t . \tag{2.45}$$

Regardless of the relationship between the sound pressure amplitude p_0 and the sound source, which we have not yet explored here, the equations (2.44) and (2.45) serve as a universal description of a standing wave. Both space dependencies in terms of pressure and velocity within fixed time periods are represented in Figures 2.8 and 2.9. The sound field is characterized as stationary, because the space function is likewise stationary and is never shifted through time. The function is simply "dimmed in" or "dimmed out" depending on the local amplitude. Sound pressure and sound velocity do not comprise a fixed relationship independent of space and time as is the case with progressive waves. On the contrary, the domain functions of pressure and velocity for standing waves are of a varying, phase-shifted nature.

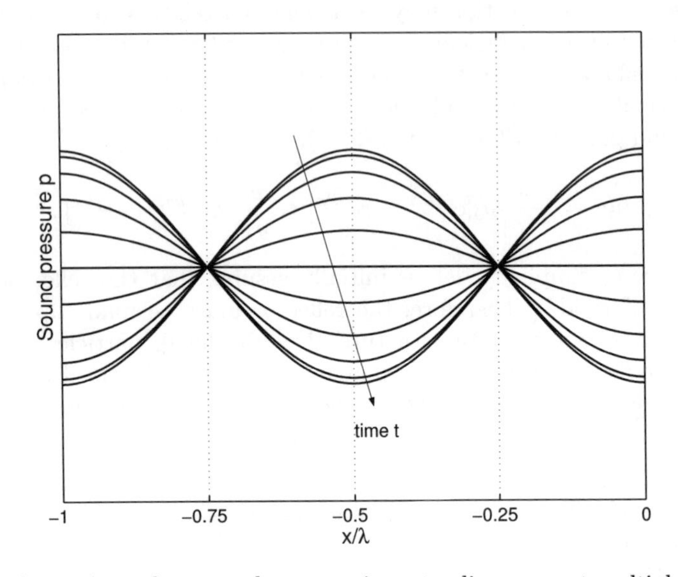

Fig. 2.8. Space-dependent sound pressure in a standing wave at multiple constant times. Half a period is shown.

As described above, resonance phenomena can be explained by the presence of multiple reflections at both ends of the tube. Furthermore, let us assume that the one-dimensional gas continuum is excited by a single in-phase oscillating surface at the location $x = -l$. The velocity of the surface v_0 is therefore independent of y. Of course, the velocity of the single oscillating surface must be equal to that of the sound field. According to eq.(2.43), the velocity is therefore

$$v_0 = v(-l) = \frac{2jp_0}{\varrho_0 c} \sin kl \; , \tag{2.46}$$

resulting in

$$p_0 = \frac{-j\,\varrho_0 c v_0}{2 \sin kl} \tag{2.47}$$

denoting the relationship between the field constant p_0 and the velocity v_0, the latter quantity which describes the sound source.

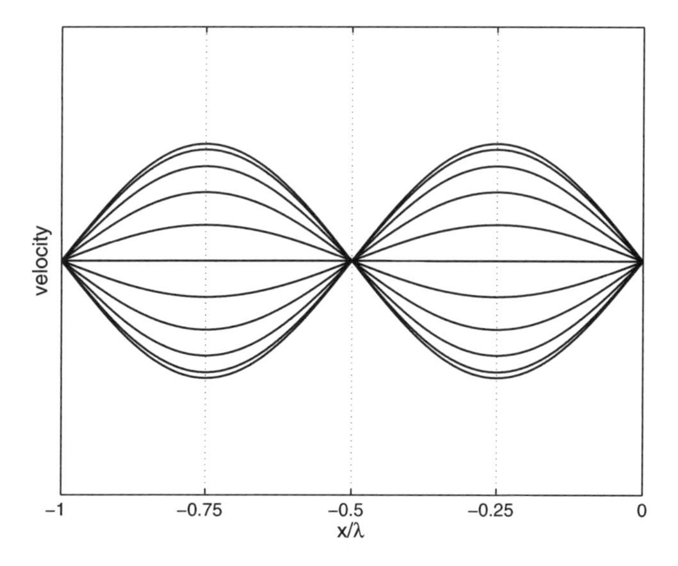

Fig. 2.9. Space-dependent sound velocity in a standing wave at multiple constant times. Half a period is shown.

The resonance frequencies of an oscillator are those frequencies at which a sound or reverberation field develops for any source, no matter how weak the signal, given in a loss-free scenario. In other words, resonance phenomena can also be referred to as "vibrations without excitation". For the air-filled tube bounded on both side with rigid-reflecting walls discussed here, resonances obviously occur at $sin(kl) = 0$, for $kl = n\pi$ $(n = 1, 2, 3, ...)$. Based on $k = \omega/c = 2\pi f/c = 2\pi/\lambda$, the resonance frequencies are

$$f = \frac{nc}{2l} \, . \tag{2.48}$$

For the wavelength λ corresponding to each resonance frequency,

$$l = n\frac{\lambda}{2} \, . \tag{2.49}$$

When resonance exists, the length of the resonator is divided up into multiples of half the wavelength.

The reasons behind this result are easy to understand. A wave propagating from the source travels twice the distance of the length of the tube $2l$, to and from the reflector. If it aligns itself in-phase with the next emitted wave, the sum of the waves then doubles. This in-phase alignment has already taken place an arbitrary amount of times in the steady state, meaning, that the field has already reached a resonance. In-phase alignment thus produces an infinite amount of completely identical space dependencies, whose sum grows without bound. In-phase alignment occurs only when the propagation path $2l$ is a multiple of the wavelength, thus $2l = n\lambda$.

The sound field analyzed above becomes infinitely large at the resonance frequencies, the reason being, that no loss is assumed to occur in the sound waves, neither along the medium nor at the reflections along the tube's walls. This pre-existing condition simplifies the analysis of the sound field, but is not realistic. In reality, the sound field consistently looses energy, whether it be through the friction at the tube's walls (viscous gas damping) or through the finite sound damping occurring between the walls and the termination of the tube, both factors which were left out of this hypothetical model. In actual experimental situations, the observer can easily hear the tone from outside of the tube.

In most actual rooms we use in everyday life, such as apartments, lecture halls, etc., the boundary surfaces (walls) behave neither as completely reflective nor completely absorbent at corresponding frequencies. Rather, walls of rooms possess reflection coefficients which can be anywhere between 0 and 1. In this case, the sound field is invariably composed of both progressive and standing waves, which can be easily illustrated using the one-dimensional continuum model. The approaching wave appearing in eq.(2.38) $p_0 e^{-jkx}$ can be theoretically decomposed into a fully-reflective component $rp_0 e^{-jkx}$ and a non-reflective component $(1-r)p_0 e^{-jkx}$

$$p_0 \mathrm{e}^{-jkx} = rp_0 \mathrm{e}^{-jkx} + (1-r)p_0 \mathrm{e}^{-jkx} \,. \tag{2.50}$$

With this, the total field referred to in eq.(2.38) can be expressed as

$$p(x) = rp_0(\mathrm{e}^{-jkx} + \mathrm{e}^{jkx}) + (1-r)p_0 \mathrm{e}^{-jkx} \,. \tag{2.51}$$

The first term with the coefficient r describes, as shown, a standing wave, and the second term with the coefficient $1-r$ describes a progressive wave. Except for the extreme cases $r = 0$ and $r = 1$, sound fields are always made up of both types of waves. If the walls of a room are neither completely reflective or completely anechoic (non-reflecting), a mixture of both standing and progressive waves is present. Progressive waves are described as the 'active field' and standing waves as the 'reactive field'. That is to say that all sound fields are comprised of both active and reactive components.

2.3 Three-dimensional sound fields

The one-dimensional wave propagation explained in the previous section can be easily transferred to the more general case of three-dimensional wave propagation. The three-dimensional extension of the principle of mass conservation (2.19) has to account for the fact that the volume element hosting the constant mass can now be constrained in all three space dimensions. Instead of (2.19), it is given simply by

$$\frac{\varrho}{\varrho_0} = -\frac{\partial \xi_x}{\partial x} - \frac{\partial \xi_y}{\partial y} - \frac{\partial \xi_z}{\partial z} \tag{2.52}$$

As previously mentioned, this book describes the sound field by the variables pressure and velocity. Equation (2.52) is therefore differentiated by time, and $\varrho = p/c^2$ is inserted:

$$\frac{1}{\varrho_0 c^2}\frac{\partial p}{\partial t} = -\frac{\partial v_x}{\partial x} - \frac{\partial v_y}{\partial y} - \frac{\partial v_z}{\partial z} \tag{2.53}$$

The three dimensional extension of the acoustic inertia law is even simpler. Considerations concerning forces can be applied to each component of the dimension separately for a thorough analysis. It should be noted for completeness that (2.26) refers to the x-component of the velocity, and the corresponding force balance equations for the two other dimensions can be added:

$$\varrho_0 \frac{\partial v_x}{\partial t} = -\frac{\partial p}{\partial x} \tag{2.54a}$$

$$\varrho_0 \frac{\partial v_y}{\partial t} = -\frac{\partial p}{\partial y} \tag{2.54b}$$

$$\varrho_0 \frac{\partial v_z}{\partial t} = -\frac{\partial p}{\partial z} \tag{2.54c}$$

For the derivation of the three-dimensional wave equation, the velocity in (2.53) and (2.54) (a,b,c) must be eliminated. If (a) is differentiated by x, (b) by y and (c) by z, and the result is inserted into (2.53) and differentiated by t, the wave equation obtained is

$$\frac{\partial^2 p}{\partial x^2} + \frac{\partial^2 p}{\partial y^2} + \frac{\partial^2 p}{\partial z^2} = \frac{1}{c^2}\frac{\partial^2 p}{\partial t^2}\,. \tag{2.55}$$

Equations (2.53) to (2.55) are often written with vector differential operators. Analogous to (2.53),

$$\mathrm{div}\mathbf{v} = -\frac{1}{\varrho_0 c^2}\frac{\partial p}{\partial t} \tag{2.56}$$

can be written, where div is the divergence. Analogous to (2.54–c),

$$\mathrm{grad}p = -\varrho_0 \frac{\partial \mathbf{v}}{\partial t}\,, \tag{2.57}$$

can be written, where grad is the gradient. The wave equation then becomes (Δ is the delta operator)

$$\Delta p = \frac{1}{c^2}\frac{\partial^2 p}{\partial t^2}\,. \tag{2.58}$$

The formulations in (2.56) to (2.58) can also be interpreted independently of a specific coordinate system. They can be directly 'translated' into a specific coordinate system (such as cylindrical or spherical coordinates), using, for example, a mathematical handbook. From that point of view, (2.52) to (2.55) appear to be the 'Cartesian release' of the more general relations (2.56) to

(2.58). The descriptions using vectorial differential operators cannot be found elsewhere in this book. They are only given for completeness.

In terms of mathematical field theory, a complete description of the sound field by means of a scalar spatial function p, whose gradient represents the vector field $\mathbf{v}$, is given by (2.56) through (2.58). The acoustic field theory, where the wave equation is solved under the presence of certain boundary conditions, is not treated in this book. The interested reader can refer to the book by P.M. Morse and U. Ingard "Theoretical Acoustics" (McGraw Hill, New York 1968).

It should be mentioned that it can directly (and perhaps a little formally) be inferred from (2.56) through (2.58) that all sound fields are 'irrotational' or conservative. Because rot grad = 0, it is in particular

$$\mathrm{rot}\mathbf{v} = 0 \ . \tag{2.59}$$

The attribute 'irrotational' is a peculiarity of the propagation of sound in gases which does not, for example, apply to solid structures.

2.4 Energy and power transport

The investigations of Sects. 2.1 and 2.2.1 have shown that the entity of wave propagation exists in the local compression of the medium (described by the pressure) and, consequently, in local vibrations of the gas elements. The 'disturbance pattern' (in comparison to the equilibrium position) migrates – for plane progressive waves – along one of the spatial axes.

This implies that energy is stored locally and momentarily in the medium. It consumes the same amount of energy to compress gases as does to accelerate the gas masses in motion. This effect can also be observed, when looking at the 'chain of elements' again, where the springs store potential energy and the masses store kinetic energy.

The kinetic energy of a mass m moving at speed v is given by

$$E_{\mathrm{kin}} = \frac{1}{2}mv^2, \tag{2.60}$$

as is generally known. The potential energy of a spring with the stiffness s, compressed by a force F is given by

$$E_{\mathrm{pot}} = \frac{1}{2}\frac{F^2}{s} \ . \tag{2.61}$$

The momentarily stored energy E_V of a gas element of volume ΔV (which is again 'small' and has the length Δx and cross-sectional area S) can be deduced by these two energies. The kinetic energy becomes

$$E_{\mathrm{kin}} = \frac{1}{2}\varrho_0 v^2 \Delta V \ .$$

The potential energy (of the spring), using $F = pS$ and $s = ES/\Delta x = \varrho_0 c^2 S/\Delta x$ with (2.20) and (2.21) becomes

$$E_{\text{pot}} = \frac{1}{2}\frac{p^2 S^2 \Delta x}{\varrho_0 c^2 S} = \frac{1}{2}\frac{p^2 \Delta V}{\varrho_0 c^2} .$$

The net energy stored in the element of volume results in

$$E_{\Delta V} = \frac{1}{2}\left\{\frac{p^2}{\varrho_0 c^2} + \varrho v^2\right\}\Delta V . \tag{2.62}$$

As each single point in the gas can store energy,

$$E = \frac{1}{2}\left\{\frac{p^2}{\varrho_0 c^2} + \varrho v^2\right\} \tag{2.63}$$

describes the energy density of the sound field. For small volumes ΔV the stored energy is simply given by

$$E_V = EV . \tag{2.64}$$

The state of energy in a gas has the same wave characteristics as the field quantities pressure and velocity. In particular,

$$p = f(t - x/c) \quad \text{and} \quad v = p/\varrho_0 c$$

is given for a plane progressive wave (see (2.27) and (2.28)) which results in an energy

$$E(x,t) = \frac{p^2}{\varrho_0 c^2} = \frac{1}{\varrho_0 c^2} f^2\left(t - \frac{x}{c}\right), \tag{2.65}$$

which has the same wave form as the square of the pressure, but also describes a transport process along the x-axis. The stored energy 'runs with the sound field' and is therefore also a wave. The energy distribution is displaced 'somewhere else' 'some time later'. Summarizing, one can imagine that in plane progressive waves, the source emits energy which migrates through the gas at the speed of sound. This energy is irrecoverably lost from the transmitter.

The energy transport of stationary (i.e. continuously driven) sources is more easily described in terms of a power quantity. To recapitulate the difference between the terms energy and power, the household light bulb should be mentioned. The power – usually measured in Watts – specifies the instantaneous consumption of light and heat. The bill paid to the supplier of the electricity is calculated by the product of the usage time multiplied by the consumed power. The consumed energy increases linearly with time, the power is the temporal change, i.e. the differentiation by time of the energy-time characteristics.

When investigating acoustic power flows, the cross-sectional area through which the power flows must be taken into account because the propagation of sound is a spatially distributed process. The sound power of the plane wave, for example, flowing through the area S, increases with increasing S. It is therefore wise to describe the power by the product

$$P = IS \; . \tag{2.66}$$

This newly defined quantity I is called intensity, which represents the acoustic sound power surface density. Intensity is generally a vector pointing in the direction of the propagating wave. We assumed one-dimensional sound fields for the purposes of deriving (2.66). Therefore, I points in the x-direction (using the notation of this chapter). It was also assumed that the intensity is constant over the area S.

Energy density and power density are related quantities. Their relationship results from the principle of energy conservation which, in this case, is applied to the (small) gas column from Fig. 2.4. The energy outflow at $x+\Delta x$ during the time interval Δt is $I(x+\Delta x)S\Delta t$, the energy inflow during this time interval is $I(x)S\Delta t$. The difference between the energy inflow and outflow must produce a difference $VE(t+\Delta t) - VE(t)$ of the energies stored at the times $t+\Delta t$ and t:

$$S\Delta x \left(E(t+\Delta t) - E(t)\right) = S\left(I(x) - I(x+\Delta x)\right)\Delta t$$

Dividing both sides by $S\Delta x\Delta t$ and using the limiting cases $\Delta x \to 0$ and $\Delta t \to 0$ yields

$$\frac{\partial I}{\partial x} = -\frac{\partial E}{\partial t} . \tag{2.67}$$

For power and intensity measurements in particular, the question arises as to how the intensity can be derived from both field quantities pressure and velocity. Using the energy density according to (2.65), (2.67) already gives the answer

$$\frac{\partial I}{\partial x} = -\frac{1}{2}\left(\frac{1}{\varrho_0 c^2}\frac{\partial p^2}{\partial t} + \varrho_0 \frac{\partial v^2}{\partial t}\right) = -\left(\frac{p}{\varrho_0 c^2}\frac{\partial p}{\partial t} + \varrho_0 v \frac{\partial v}{\partial t}\right) .$$

Here, $\partial p/\partial t$ is expressed by $\partial v/\partial x$ according to (2.25) and $\partial v/\partial t$ by $\partial p/\partial x$ according to (2.26):

$$\frac{\partial I}{\partial x} = p\frac{\partial v}{\partial x} + v\frac{\partial p}{\partial x} = \frac{\partial (pv)}{\partial x} .$$

A lucid result is obtained by integration:

$$I(t) = p(t)v(t) \tag{2.68}$$

The intensity is equal to the product of sound pressure multiplied by sound velocity. This is also valid for the more general, three-dimensional case, where (2.68) is replaced by

$$\mathbf{I} = p\mathbf{v} \; . \tag{2.69}$$

The power flowing through the area S is generally calculated by

$$P = \int \mathbf{I} d\mathbf{S} \tag{2.70}$$

where $d\mathbf{S}$ is the vectorial area element (normal to the area S everywhere).

For stationary sources, the time average of the power is of interest only, resulting in

$$\bar{I} = \int_0^T I(t) dt \tag{2.71}$$

and

$$\bar{P} = \bar{I} S \, . \tag{2.72}$$

For plane progressive waves

$$I(t) = \frac{p^2(t)}{\varrho_0 c} \tag{2.73}$$

and for pure tones $p = p_0 \cos \omega t$, we get

$$\bar{P} = \bar{I} = \frac{{p_0}^2}{2\varrho_0 c} = \frac{{p_{rms}}^2}{\varrho_0 c} \tag{2.74}$$

(p_{rms} is the root-mean-square). As it is clear from (2.74) the determination of the sound pressure alone is sufficient when measuring the intensity of plane progressive waves. For that reason, sound power can often be calculated are under free field conditions (e.g. in an anechoic chamber) at a large distance to the source. Under such conditions, it can be assumed that $p = \varrho_0 c v$, in fact, holds. To calculate the sound power, the (imaginary) surface enveloping the source is divided into N 'small' partitions S_i. The root-mean-square of the sound pressure is then calculated for each partition. The radiated sound power thus results in

$$\bar{P} = \sum_{i=1}^{N} \frac{{p_{\mathrm{rms},i}}^2}{\varrho_0 c} \, . \tag{2.75}$$

Finally, it should be noted that power and intensity can also be described by their corresponding levels. The required reference values P_0 and I_0 are defined in

$$L_{\mathrm{I}} = 10 \lg \frac{\bar{I}}{I_0} \tag{2.76}$$

and in

$$L_{\mathrm{w}} = 10 \lg \frac{\bar{P}}{P_0} \tag{2.77}$$

in such a way, that for the case of the plane progressive wave flowing through a surface of $S = 1\,\mathrm{m}^2$, the same values for the sound pressure level L, the intensity level L_{I} and the sound power level L_{w} are obtained. Using

$$L = 10\log\left(\frac{p_{rms}}{p_0}\right)^2 \quad \text{with} \quad p_0 = 2\,10^{-5}\,\mathrm{N/m^2}$$

$$I_0 = \frac{{p_0}^2}{\varrho_0 c} = 10^{-12}\,\mathrm{W/m^2} \tag{2.78}$$

and

$$P_0 = I_0 \times 1\,\mathrm{m^2} = 10^{-12}\,\mathrm{W} \tag{2.79}$$

are obtained (where $\varrho_0 c = 400\,\mathrm{kg/m^2s}$).

2.5 Intensity measurements

Measuring the sound power under free field conditions can be reduced to the determination of the sound pressure on a surface, as introduced before. This requires a special, anechoic test facility which may not be available. As a matter of fact, it is sometimes impossible or would be too expensive to put certain technical sound sources into anechoic environments. There is enough reason to use a sound power measurement technique which is independent of special environmental conditions.

Such a measurement technique must necessarily include the determination of the sound velocity. The basic idea of intensity measurements is to estimate the pressure gradient which is needed to determine the velocity by measuring the difference between the sound pressures acting on two microphone types. In place of the actual velocity

$$\varrho_0 \frac{\partial v}{\partial t} = -\frac{\partial p}{\partial x},$$

we use the measured velocity

$$\varrho_0 \frac{\partial v_{\mathrm{M}}}{\partial t} = \frac{p(x) - p(x+\Delta x)}{\Delta x} \tag{2.80}$$

to determine the intensity. x and $x + \Delta x$ denote the positions of the two intensity-probe microphones.

The direction in which the two microphones are positioned Δx is not necessarily the same as the actual (or suspected) direction of the sound propagation. The measurement technique described in this section invariably determines the vectorial component of the intensity pointing in the direction given by the axis of the two measurement positions.

Indeed, (2.80) approximates the 'actual sound velocity'. The sound intensity determined by the aid of (2.80) will include systematic errors which are also subject to further investigations here. First, the measurement techniques will be described in detail. Then, an error analysis specifying the procedure's limitations will follow.

As already mentioned, sound power measurements are especially useful for stationary ('permanently running') sources which are implied in the following. The intensity measurement technique utilizing (2.80) can either implement the time averaged mean value of the local intensity (where the required time domain signal can, for instance, be A-weighted as well) or the spectral analysis of the frequency components. The following sections will examine how intensity is measured in both the time domain and in the frequency domain.

2.5.1 Time domain

To measure intensity in the time domain, the pressure difference in (2.80) must be integrated by an analogue electrical circuit or a digital signal processor:

$$v_{\mathrm{M}}(t) = \frac{1}{\Delta x \varrho_0} \int [p(x) - p(x + \Delta x)]\, dt \tag{2.81}$$

The time dependence of the intensity results in the product of pressure and velocity. We use the space average of the two pressure signals because there are two signals for the sound pressure. These are obtained from the two adjacent positions

$$p_{\mathrm{M}}(t) = \frac{1}{2} [p(x) + p(x + \Delta x)] \ , \tag{2.82}$$

resulting in

$$I(t) = p_{\mathrm{M}}(t) v_{\mathrm{M}}(t) = \frac{1}{2\varrho_0 \Delta x} [p(x) + p(x + \Delta x)] \int [p(x) - p(x + \Delta x)]\, dt \ . \tag{2.83}$$

The time average is again obtained with the aid of an analogue or digital integrator

$$\bar{I} = \frac{1}{2\varrho_0 \Delta x T} \int_0^T [p(x) + p(x + \Delta x)] \int [p(x) - p(x + \Delta x)]\, dt dt, \tag{2.84}$$

where T is the averaging time.

2.5.2 Frequency domain

Harmonic time dependencies

Initially, we can assume that the exciting sound source emits a single tone of a given and known (or at least easily measurable) frequency. This assumption allows us to select a simple method for measuring intensity. In addition, this preliminary discussion provides a control for the investigations of the more general case of arbitrary time dependencies to be elaborated in the next section.

Here we can measure the amplitude spectra of the two pressure signals, denoted in the following by $p(x)$ and $p(x+\Delta x)$. According to the time conventions $p(x,t) = \mathrm{Re}\{p(x)e^{j\omega t}\}$ and $p(x+\Delta x,t) = \mathrm{Re}\{p(x+\Delta x)e^{j\omega t}\}$ apply. Based on these definitions, the complex amplitude of the sound velocity (2.80) becomes

$$v_{\mathrm{M}} = \frac{-j}{\omega\varrho_0\Delta x}\left[p(x) - p(x+\Delta x)\right]. \tag{2.85}$$

The pressure p_{M} is again determined by the average value of the two measured quantities

$$p_{\mathrm{M}} = \frac{1}{2}\left[p(x) + p(x+\Delta x)\right] . \tag{2.86}$$

The time average of the intensity (the effective intensity) is thus formed by

$$I_{\mathrm{M}} = \frac{1}{2}\mathrm{Re}\left\{p_{\mathrm{M}}{v_{\mathrm{M}}}^*\right\} = \frac{1}{4\omega\varrho_0\Delta x}\mathrm{Re}\left\{j\left[p(x)+p(x+\Delta x)\right]\left[p^*(x)-p^*(x+\Delta x)\right]\right\} \tag{2.87}$$

(* = complex conjugate). Because pp^* is a real quantity, we are left with

$$I_{\mathrm{M}} = \frac{1}{4\omega\varrho_0\Delta x}\mathrm{Re}\left\{-j\left[p(x)p^*(x+\Delta x) - p^*(x)p(x+\Delta x)\right]\right\}.$$

Applying $\mathrm{Re}\{-jz\} = \mathrm{Re}\{-j(x+jy)\} = y = \mathrm{Im}\{z\}$ yields

$$I_{\mathrm{M}} = \frac{1}{4\omega\varrho_0\Delta x}\mathrm{Im}\left\{p(x)p^*(x+\Delta x) - p^*(x)p(x+\Delta x)\right\},$$

or by using $\mathrm{Im}\{z - z^*\} = 2\mathrm{Im}\{z\}$, we ultimately arrive at

$$I_{\mathrm{M}} = \frac{1}{2\omega\varrho_0\Delta x}\mathrm{Im}\left\{p(x)p^*(x+\Delta x)\right\} . \tag{2.88}$$

As can be seen in the above, we only need to know the two amplitudes $|p(x)|$ and $|p(x+\Delta x)|$ and the phase difference between them to calculate I_{M}.

The argument $p(x)p^*(x+\Delta x)$ is also called the cross-spectral density, whose imaginary part yields the intensity. Once again, it should be noted that $p(x)$ and $p(x+\Delta x)$ denote complex amplitudes or complex amplitude spectra. The unit of their product is [N/m^2] and does not represent a power density function [power/Hz]. The intensity contained in a frequency band is derived by the summation of the included spectral components.

Arbitrary time dependencies

To measure the frequency domain for arbitrary source time-dependencies, one only has to decompose the time dependencies $p(x,t)$ and $p(x+\Delta x,t)$– occurring at the locations x and $x+\Delta x$ – into their frequency components. Chapter 13 provides a more detailed description of how to carry out the necessary steps for this type of spectral decomposition. At this stage, it suffices to provide an overall outline of this procedure in the sections which follow.

First, we can conclude that the required time signals can only be observed during a certain finite time period T – a truly infinite time length is impossible for the period of observation. The signals are therefore only known in the given interval $0 < t < T$. On the other hand, intensity measurements usually only make sense for stationary operating sources. It is therefore logical to assume that the signals beyond the interval $0 < t < T$ behave 'similarly' to the one within the interval. The simplest way to define this criteria is to assume that the signals perpetuate themselves periodically with the observation period T, that $p(x, t+T) = p(x, t)$ (and of course, $p(x+\Delta x, t+T) = p(x+\Delta x, t)$ both apply. The period T does not have to define an actual physical period, such as the rotation of a running motor, rather, it can simply be a time arbitrarily set for the purpose of the measurement. The fact that this mathematically 'strict' periodization excludes the possibility of significant error underscores an important characteristic of stationary signals: local intensity and global signal power by no means depend on the time interval extracted from the longer stationary signal to define the period T. Furthermore, an average can be obtained by taking random samples of time length T.

The advantage gained by the pre-existing condition of artificial periodization is that only certain discrete frequencies $n\omega_0$ (with $\omega_0 = 2\pi/T$) can occur within such signals. This drastically simplifies the quantitative analysis. As shown by eq.(13.38), for example, the time dependencies $p(x, t)$ and $p(x + \Delta x, t)$ can actually be expressed in Fourier series

$$p(x,t) = \sum_{n=-\infty}^{\infty} p_n(x) e^{jn\omega_0 t} \tag{2.89}$$

and

$$p(x+\Delta x,t) = \sum_{n=-\infty}^{\infty} p_n(x+\Delta x) e^{jn\omega_0 t}. \tag{2.90}$$

The complex-valued spectral amplitudes of each time dependency are expressed in the quantities $p_n(x)$ and $p_n(x + \Delta x)$. Eq.(13.37) shows how the complex amplitude series $p_n(x)$ and $p_n(x + \Delta x)$ are derived from the (periodic) time-dependencies $p(x, t)$ and $p(x+\Delta x, t)$. Because constant components do not occur in acoustics (constant pressures do not produce waves – if they did, they would not be able to heard!), the pre-existing condition is namely $p_0(x) = p_0(x+\Delta x) = 0$ for the following deliberations. As a reminder, 'conjugated symmetry' – $p_{-n} = p_n^*$ (* = complex conjugates) – applies, because by nature, the time dependencies $p(x, t)$ and $p(x+\Delta x, t)$ resulting from the summation in themselves must have real values. The summation does not have to occur without bound. Indeed, the frequencies they contain depend on the filter connected upstream, or on the measuring device itself, which of course, always possesses a low-pass or band-pass characteristic.

With the help of both aforementioned series decomposition eq.(2.84) converges to

$$\bar{I} = \frac{1}{2\varrho_0 \Delta x T} \int_0^T \sum_{n=-\infty}^{\infty} [p_n(x) + p_n(x+\Delta x)] e^{jn\omega_0 t}$$

$$\sum_{m=-\infty}^{\infty} \frac{1}{jm\omega_0} [p_m(x) - p_m(x+\Delta x)] e^{jm\omega_0 t} dt \,. \tag{2.91}$$

In the second sum, we use m for the summation index to avoid confusion with the first sum. After multiplying the entire equation out, it results in

$$\bar{I} = \frac{1}{2\varrho_0 \Delta x T} \int_0^T \sum_{n=-\infty}^{\infty} \sum_{m=-\infty}^{\infty} [p_n(x) + p_n(x+\Delta x)]$$

$$\frac{1}{jm\omega_0} [p_m(x) - p_m(x+\Delta x)] e^{j(n+m)\omega_0 t} dt \,. \tag{2.92}$$

Due to

$$\int_0^T e^{j(n+m)\omega_0 t} dt = 0$$

for $n + m \neq 0$ (for $n + m = 0$ the integral is equal to T), only the summands are left with $m = -n$, resulting in

$$\bar{I} = \frac{-1}{2\varrho_0 \Delta x} \sum_{n=-\infty}^{\infty} [p_n(x) + p_n(x+\Delta x)]$$

$$\frac{1}{jn\omega_0} [p_{-n}(x) - p_{-n}(x+\Delta x)] \,. \tag{2.93}$$

For one of the partial sums obtained by multiplying out the term inside the parentheses,

$$\sum_{n=-\infty}^{\infty} \frac{p_n(x) p_{-n}(x)}{jn\omega_0} = 0$$

applies, because each summand results in zero (with $n = N$ and $n = -N$) and also due to the precondition that was defined as $p_0(x) = 0$. Likewise,

$$\sum_{n=-\infty}^{\infty} \frac{p_n(x+\Delta x) p_{-n}(x+\Delta x)}{jn\omega_0} = 0$$

is true. For the effective intensity, this results in

$$\bar{I} = \frac{1}{2\varrho_0 \Delta x} \sum_{n=-\infty}^{\infty} \frac{1}{jn\omega_0} [p_n(x) p_{-n}(x+\Delta x) - p_n(x+\Delta x) p_{-n}(x)],$$

using eq.(2.93), or using the 'conjugate symmetry,'

$$\bar{I} = \frac{1}{2\varrho_0 \Delta x} \sum_{n=-\infty}^{\infty} \frac{1}{jn\omega_0} \left[p_n(x) p_n^*(x+\Delta x) - p_n(x+\Delta x) p_n^*(x) \right] .$$

The basic signal structure of the summands therefore can be defined as $(z - z^*)/j$, or simply $(z - z^*)/j = 2\,\mathrm{Im}\{z\}$. Thus follows

$$\bar{I} = \frac{1}{\varrho_0 \Delta x} \sum_{n=-\infty}^{\infty} \frac{1}{n\omega_0} \mathrm{Im}\{p_n(x) p_n^*(x+\Delta x)\} . \tag{2.94}$$

The expression $\mathrm{Im}\{p_n(x)p_n^*(x+\Delta x)\}/n\omega_0\varrho_0\Delta x$ constitutes the frequency response of the effective intensity. Obviously, every frequency component can be seen as a separate and distinct energy storage. The net intensity is therefore derived from the sum of the spectral intensity components.

The term $p_n(x)p_n^*(x+\Delta x)$ is typically referred to as spectral cross power. Spectral intensity is defined by its imaginary component.

2.5.3 Measurement error and limitations

High frequency error

The most obvious and immediately evident problem of intensity measurements exists in the fact that accumulation of differentials at the differentiation itself only provides sound conclusions for large wavelengths at their respective low frequencies. Even the simplest model indicates the magnitude of the resulting error. A plane progressive wave along the x-direction

$$p(x) = p_0 \mathrm{e}^{-jkx}$$

is assumed to be the sound field. The corresponding actual intensity I is given by

$$I = \frac{1}{2} \mathrm{Re}\left\{ p v^* \right\} = \frac{1}{2} \frac{{p_0}^2}{\varrho_0 c}$$

where $v = p/\varrho_0 c$ was used for plane progressive waves. In contrast, the measured intensity, according to (2.88), is

$$I_{\mathrm{M}} = \frac{{p_0}^2}{2\omega\varrho_0\Delta x} \mathrm{Im}\left\{ \mathrm{e}^{-jkx} \mathrm{e}^{jk(x+\Delta x)} \right\} = \frac{{p_0}^2}{2\omega\varrho_0\Delta x} \sin k\Delta ,$$

which results in

$$\frac{I_{\mathrm{M}}}{I} = \frac{\varrho_0 c}{\omega\varrho_0\Delta x} \sin k\Delta x = \frac{\sin k\Delta x}{k\Delta x} . \tag{2.95}$$

The actual and measured intensity are identical only at low frequencies, where $k\Delta x \ll 1$, because of $\sin k\Delta x \approx k\Delta x$. For values of $k\Delta x = 2\pi\Delta x/\lambda = 0.182\pi$

already, $\sin k\Delta x/k\Delta x = 0.8$; the error, using $10 \lg I_M/I = -1$, is thus 1 dB. The measurement error is therefore only smaller than 1 dB if approximately $x < \lambda/5$ applies. If the error does not exceed 1 dB, it is only necessary to measure up to $\lambda = 12.5$ cm and therefore up to $f = c/\lambda = 2700$ Hz for a spacing of only $x = 2.5$ cm

Low frequency error

The second error, concerning the lower frequency limit, occurs because the intensity probe, consisting of two microphones, seems to detect a 'phantom intensity' which is actually not present due to small errors in the phase relation between the microphones. To explain this effect, it must be clarified that there are sound fields which carry power (i.e. the time-averaged power is non-zero) as well as sound fields which do not carry power (i.e. the time-averaged power is zero). The first case is true for plane progressive waves, whereas the second case applies to standing waves. First, we will discuss the power transport for both fundamental wave types.

Power transport in progressive waves

As explained earlier, plane progressive waves with

$$p(x) = p_0 e^{-jkx}$$

and with

$$p(x,t) = \text{Re}\left\{p(x)e^{j\omega t}\right\} = p_0 \cos(\omega t - kx)$$

consist of a spatial dependency travelling with time (see Fig. 2.7). The phase difference between the two sound pressures in the distance Δx is $\Delta\varphi = k\Delta x = 2\pi\Delta x/\lambda$. The effective intensity is $I = p_0{}^2/2\varrho_0 c$.

Power transport in standing waves

As described above, standing waves have the sound pressure space-time domain function of

$$p(x,t) = \text{Re}\left\{p(x)e^{j\omega t}\right\} = 2p_0 \cos kx \cos \omega t\ , \tag{2.96}$$

and for sound velocity,

$$v(x,t) = \text{Re}\left\{v(x)e^{j\omega t}\right\} = \frac{2p_0}{\varrho_0 c} \sin kx \sin \omega t\ . \tag{2.97}$$

Both spatial dependencies are shown in Figure 2.8 over a constant time period.

The phase relationship between two microphone positions is either $\varphi = 0°$ or $\varphi = 180°$. Standing waves do not carry any intensity nor power in a time average, as can be seen from the pressure nodes shown in Fig. 2.8. At nodes with $p = 0$, the intensity is thus zero at all times ($I(t) = p(t)v(t) = 0$). Power never penetrates through surfaces where $p = 0$.

Due to the principle of energy balance, no power can flow through any surface over a time average. This can also be shown mathematically. From the complex pressure and velocity the intensity becomes

$$I(x,t) = \frac{2{p_0}^2}{\varrho_0 c} \sin kx \cos kx \sin \omega t \cos \omega t = \frac{{p_0}^2}{2\varrho_0 c} \sin 2kx \sin 2\omega t \,. \tag{2.98}$$

The time-averaged intensity is therefore zero at each position. The fact that standing waves can manage without consuming external energy can be explained by the exceptions assumed for them. No energy is lost during the (assumed) total reflection, for example. The sound wave can travel eternally between two reflectors without loosing energy, because the air was also assumed to be without losses. The assumption of no losses is, of course, more or less violated in practice.

Summarizing, it can be stated that power flow is bound to sound fields, where the phase of the sound pressure is different at two different positions. If, in contrast, the signals at two (arbitrarily chosen) positions are either identical or opposite in phase, the time-averaged power flow is zero. The aforementioned facts describe, more or less, the second problem of intensity measurements. In a reverberant sound field with low absorption at the walls, the sound field consists more or less of standing waves. If a small phase error between the two microphone signals arises in the measurement setup, a non-existent effective intensity is detected. Thus, the intensity measurement technique is not necessarily independent of the chosen test environment; rooms with long reverberation times are not suitable.

To estimate the measurement error due to phase errors, a sound field consisting of plane progressive and standing waves is initially assumed to be

$$p = p_{\mathrm{p}} \mathrm{e}^{-jkx} + p_{\mathrm{s}} \cos kx, \tag{2.99}$$

where p_{p} denotes the amplitude of the plane wave and p_{s} denotes the amplitude of the standing wave. In the following passages, both quantities are assumed to be real.

For sufficiently low frequencies, or assuming sufficiently large distances $k\Delta x \ll 1$, a very accurate estimate of the actual intensity is given by the measurement prescription (2.88), if no phase error occurs during the measurement. For simplicity, the first measurement position is chosen to be $x = 0$ and

$$I_{\mathrm{M}} = \frac{1}{2\omega \varrho_0 \Delta x} \mathrm{Im}\left\{p(0)p^*(\Delta x)\right\}$$

describes the intensity without experimental error. The measured intensity with a phase error is

$$I_{\mathrm{M}} = \frac{1}{2\omega\varrho_0\Delta x}\mathrm{Im}\left\{p(0)p^*(\Delta x)\mathrm{e}^{j\varphi}\right\} .$$

The sound pressure at the first position $p(0)$ can be considered real because only the phase difference between the measured signals is important, thus

$$\frac{I_{\mathrm{M}}}{I} = \frac{\mathrm{Im}\{p^*(\Delta x)\mathrm{e}^{j\varphi}\}}{\mathrm{Im}\{p^*(\Delta x)\}} = \frac{\mathrm{Im}\{p(\Delta x)\mathrm{e}^{-j\varphi}\}}{\mathrm{Im}\{p(\Delta x)\}} .$$

It is correct to assume that the phase error is a small quantity; microphone manufacturers, for example, specify $\varphi = 0.3°$ (!) as a tolerable phase error. Applying $\mathrm{e}^{-j\varphi} = 1 - j\varphi$ yields

$$\frac{I_{\mathrm{M}}}{I} = 1 - j\frac{\mathrm{Im}\{jp(\Delta x)\}}{\mathrm{Im}\{p(\Delta x)\}} = 1 - \varphi\frac{\mathrm{Re}\{p(\Delta x\}}{\mathrm{Im}\{p(\Delta x\}}, \tag{2.100}$$

where $\mathrm{Im}\{jz\} = \mathrm{Re}\{z\}$ is also used. According to (2.99) and using $kx \ll 1$, it is

$$p(\Delta x) = p_{\mathrm{p}}\mathrm{e}^{-jk\Delta x} + p_{\mathrm{s}}\cos(k\Delta x) \approx p_{\mathrm{p}} + p_{\mathrm{s}} - jp_{\mathrm{p}}k\Delta x$$

and consequently, (2.100) becomes

$$\frac{I_{\mathrm{M}}}{I} = 1 + \varphi\frac{p_{\mathrm{p}} + p_{\mathrm{s}}}{k\Delta x p_{\mathrm{p}}} = 1 + \frac{\varphi}{k\Delta x}\left(1 + \frac{p_{\mathrm{s}}}{p_{\mathrm{p}}}\right) . \tag{2.101}$$

In practice, $\varphi/k\Delta x$ is a small quantity even at low frequencies ω (for $\varphi = 0.3\pi/180$, at $f = 100\,\mathrm{Hz}$ and $\Delta x = 5\,\mathrm{cm}$ it is, for example, $\varphi/k\Delta x \approx 1/20$). The phase error only plays an important role if the amplitude of the standing wave field p_{s} is considerably larger than that of the plane wave, $p_{\mathrm{s}} \gg p_{\mathrm{p}}$. The ratio of the measured intensity I and the actual intensity I_{M} can be estimated under this conditions to be

$$\frac{I_{\mathrm{M}}}{I} = 1 + \frac{\varphi}{k\Delta x}\frac{p_{\mathrm{s}}}{p_{\mathrm{p}}} . \tag{2.102}$$

If a measurement error of 1 dB is still tolerable, then

$$\frac{\varphi}{k\Delta x}\frac{p_{\mathrm{s}}}{p_{\mathrm{p}}} < 0.2 \tag{2.103}$$

has to be fulfilled in the measurement. The the standing wave quota decreases in practice with increasing frequency. Therefore, (2.103) can also be interpreted as the determination of the lower band limit of the measured frequency range

$$f > \frac{\varphi}{2\pi}\frac{5c}{\Delta x}\frac{p_{\mathrm{s}}}{p_{\mathrm{p}}} . \tag{2.104}$$

Using $\varphi = 0.3\pi/180$ (this corresponds to $0.3°$) and $\Delta x = 0.05$ as an example, it follows that

$$f > 28\frac{p_\mathrm{s}}{p_\mathrm{p}}\,\mathrm{Hz}\,.$$

If $p_\mathrm{s} = 10p_\mathrm{p}$, a tolerance of 1 dB would be achieved above $f = 280\,\mathrm{Hz}$. As will be shown in Chap. 6, the walls of a test stand should have an absorption coefficient of $\alpha = 0.3$.

On one hand, equation (2.95) requires small microphone spacing intervals Δx in order to achieve small errors in the high frequency range. However, (2.104) requires, in contrast, a large Δx to account for low frequencies. For broadband measurements, the frequency range is usually split into two intervals, using two different microphone spacing increments.

2.5.4 Standards

The following standards should be applied to measuring intensity and performance:

- ISO 9614-1: Determining the sound level of noise sources in intensity measurements - Part 1: Measurements at discrete points (from 1995)
- EN ISO 9614-2: Determining the sound performance level of noise sources in intensity measurements - Part 2: Measurements with continuous sampling (from 1996)
- EN ISO 9614-3: Determining sound performance levels from noise sources in intensity measurements - Part 3: Scanning procedures of accuracy class 1 (from 2003)
- DIN EN 61043: Electrical acoustics; Sound intensity measuring devices; Measurements with microphone pairs (from 1994)

2.6 Wave propagation in a moving medium

This section is intended to explain the principle phenomena and effects attributed to the moving medium gas. It has already been established that motion is always relative to a point or coordinate system that is imagined to be stationary. The question pertaining to the fluid medium means – more precisely – the examination of its relationship to the sound source and receiver (ear or microphone). To this end, we will consider the following three commonplace situations:

1. In a windy, open-air environment, the sound source and sound receiver remain at a fixed location on the ground while the medium as a whole is passing over the object. Here one can think of both the sender and ear or microphone as stationary while the gas is flowing over it.
2. A common everyday experience is hearing an almost stationary sound source, that is moving at a particular driving speed U relative to the air, such as a siren on a police squad car or a fire truck. In this situation, the receiver can be construed as sitting in a stationary medium while the sound source is moving relative to both.

3. And finally, the receiver, (the vehicle driver's ear, for instance) can be moving to or away from the stationary sound source in the medium. In this situation, the stationary source can be construed as being fixed in the medium while the receiver is moving relative to both.

These three cases make up two groups which differ from one another by an important characteristic. In the first case, the distance between the sender and receiver stays the same, as well as the duration of the sound incidence between both remains constant over time. For this reason, the sound signal traverses the entire signal system without distorting the signal process. On the other hand, if the sender and receiver are moving relative to one another, the distance between them and the corresponding signal duration T itself is time-dependent, and therefore results in a distortion of the sound signal as it is transferred.

Figure 2.10 clearly illustrates this distinction. A signal transmitted with

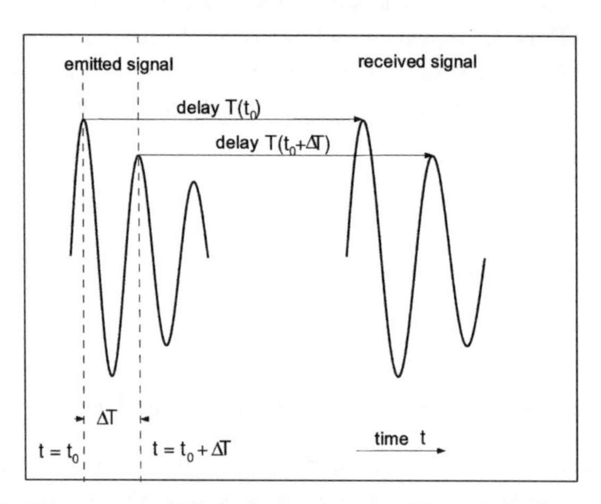

Fig. 2.10. Diagram outlining the basic principles of the Doppler effect

the duration $T(t_0)$ is emitted at the point in time t_0. The signal emitted at the point in time $t_0 + \Delta T$ requires $T(t_0 + \Delta T)$ to reach its receiver. Only when the time durations are the same, that is, when the source and receiver are relatively stationary to one another and $T(t_0) = T(t_0 + \Delta T)$, is the received signal an exact unaltered image of the original signal. In the case of pure sound sources, (frequency f_Q) the same frequency arrives at the receiver, thus for the receiver frequency f_E, $f_E = f_Q$ is true. Chapter 13 will show how this implicates an important characteristic for all time-invariant (and linear) transmitters: the frequencies of the stimulus (source) and the effect (receiver) are the same.

The relative motion of the sender and receiver to one another constitutes a nearly exact universal image for all time-dependent, that is, time-invariant, transmitters. In this case, both time durations $T(t_0)$ and $T(t_0 + \Delta T)$ differ because the distance between source and receiving point has changed during the time interval ΔT. Therefore, pure tones on the sender-side undergo a change in frequency during the transmission, resulting in a difference between the frequencies of sender and receiver, f_Q and f_E respectively. This effect is known, according to its first discoverer, as the Doppler effect.

The acoustical systems of stationary sources and receivers, or sources and receivers moving relative to one another – and/or relative to a stationary or moving medium – are analogous to a flowing river or a placid lake, where waves are moving along the surface (see Figure 2.11); the latter images depict air sound waves in gas.

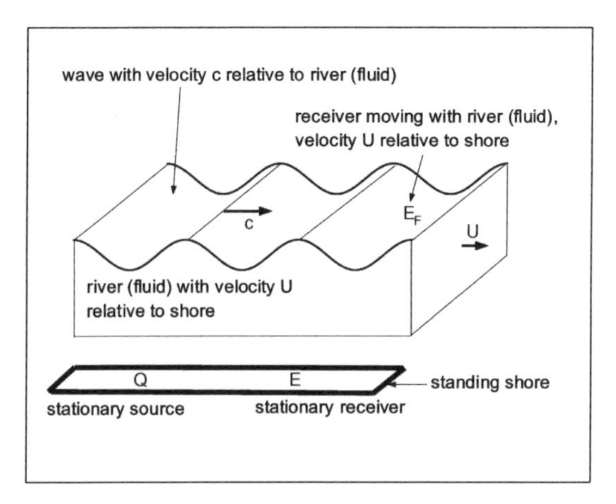

Fig. 2.11. Diagram outlining the basic principles of the Doppler effect for a receiver moving with the fluid

First let us consider the first case of a stationary source as observed by a likewise stationary onlooker, standing on the river bank (Figure 2.11). The stationary source can be imagined as some sort of an impact testing machine which is repeatedly striking the surface of the water with a period of T_Q. The source frequency which it emits is $f_Q = 1/T_Q$. In a stationary medium $U = 0$ the disturbances caused by the machine move along the wave guide at the velocity c, covering exactly $\Delta x = cT_Q$ distance during a time period. For a fluid medium, the fluid disturbances are taken up as well, left by the source 'in its wake' at a velocity of $c+U$, covering a distance of exactly $\Delta x = (c + U)T_Q$ during one time period. The distance between two disturbances can be defined by its local period and therefore, by its wavelength λ, which is equal to the distance covered by the disturbances in T_Q. The wavelength is therefore described by

$$\lambda = \frac{c + U}{f_Q} . \tag{2.105}$$

The wavelength has therefore increased in proportion to the relatively stationary medium. This is why it is completely irrelevant whether the wave surges are observed by someone standing (or photographing) from the river bank or by a receiver moving along with the medium. The same wavelength would even be observed by a ship swimming with the current as by someone standing on the bank.

As mentioned earlier, the frequencies of the source and the receiver are the same. This is also implicated by simple analogy of the running wave surges caused by the machine. During the time interval ΔT the disturbances $(c + u)\Delta T$ pass by the receiver. The number of disturbances passing by the observer N_E during the time interval ΔT is therefore

$$N_E = \frac{\Delta x}{\lambda} = \frac{(c + U)\Delta T}{\lambda} . \tag{2.106}$$

The ratio $N_E/\Delta T$ describes the frequency at the receiver of

$$f_E = \frac{N_E}{\Delta T} . \tag{2.107}$$

Eq.(2.105) results in $f_E = f_Q$.

A similar observation can be made of the Doppler shift from an observer moving with the fluid. In this scenario, the waves are moving past him at a velocity of c (see Figure 2.11), whereby the following applies to the number of wavelengths passing by him N_E

$$N_E = \frac{c\Delta T}{\lambda} . \tag{2.108}$$

After applying the wavelength according to eq.(2.105) the result is

$$N_E/\Delta T = \frac{c}{\lambda} = f_Q \frac{c}{c + U} . \tag{2.109}$$

The left side, once again, describes the receiving frequency. The right side can be simplified using the velocity ratio U/c by substituting it with the Mach number M

$$M = \frac{U}{c} . \tag{2.110}$$

The Doppler shift for a stationary source (frequency f_Q) with an observer moving with the medium at a frequency of f_E) can be expressed as

$$f_E = \frac{f_Q}{1 + M} . \tag{2.111}$$

As explained in the beginning, only relative motion is taken into account. Whether the fluid is moving with an embedded receiver and the source is stationary (on the river bank as depicted in the above example), or whether the fluid and receiver can be considered stationary and the source is moving away from these two in the opposite direction, is irrelevant. Only when fluid and receiver are stationary relative to one another can eq. (2.111) be used to describe the Doppler shift. It goes without saying that M can, in this case, be either positive or negative. Negative values of U, which describe the case when source and receiver are moving toward either, are allowed here and in the text which follows.

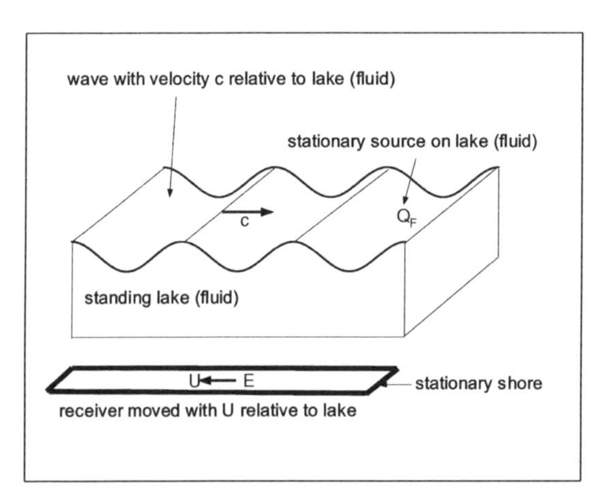

Fig. 2.12. Diagram outlining the basic principles of the Doppler effect for a stationary source and moving receiver in fluid

In contrast, the Doppler shift changes in the third scenario described above. In this case, the source is moving with the medium. Figure 2.12 provides the simplest way to describe this phenomena: a stationary source is floating in a likewise placid lake (fluid) while the receiver at the river bank is moving left at the velocity of U as shown in Figure 2.12. Of course, in this case, the wave motions on the water are not of interest to the observer on the river bank, and the wavelength can simply be expressed as

$$\lambda = \frac{c}{f_Q}\,. \tag{2.112}$$

The number of periods in the waves N_E that rush by the moving receiver during the time interval ΔT is

$$N_E = \frac{\Delta x}{\lambda} = \frac{(c-U)\Delta T}{\lambda}\,. \tag{2.113}$$

Because the observer tries to move in the opposite direction of the waves, the waves are passing him at a velocity of c-U. Therefore, in this situation where the number of wave periods N_E surge past the observer during the time interval ΔT the Doppler shift can be described as

$$f_E = \frac{N_E}{\Delta T} = \frac{(c-U)}{\lambda} = f_Q \frac{c-U}{c} = (1-M) f_Q \,. \tag{2.114}$$

Eq.(2.114) describes the specific Doppler shift scenario when the medium and source stationary relative to one another.

As can be seen from the examples above, one must be able to distinguish whether it is the sender or the receiver which is moving relative to the medium. Both cases result in different Doppler shifts. When the source is moving relative to the medium, we have to also consider a change in wavelength. The

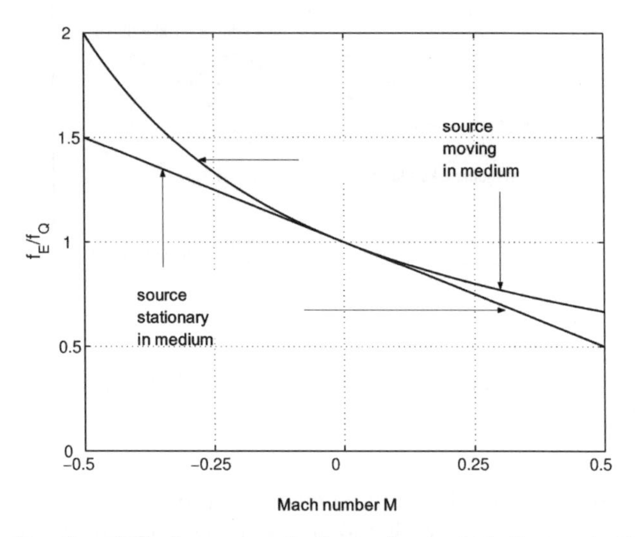

Fig. 2.13. Doppler shifts for sources both moving and stationary in the medium.

differences between the rules for (2.111) and (2.114) are, however, quite small for smaller Mach numbers, as shown in Figure 2.13. Furthermore, one must take into consideration that a Mach number of only 0.1 in air already corresponds to a velocity of $34\,m/s$, or more than $120\,km/h$; higher velocities than this seldom occur in acoustics.

Up until now, the preceding investigation has only served to explain the essentials of a one-dimensional wave guide. Doppler shift does not occur in three-dimensional sound propagation under windy conditions with a stationary source and likewise stationary observer system on the earth's surface. However, in this case, sound propagates in a directional wavelength, which is easily imaginable: the wavelength moving in the same direction as the wind will be larger than when it is moving away, and is not yet influenced by fluid velocity flanking it on the sides. In general, it can be shown that the sound field $p_M(x, y, z)$ for small Mach numbers M taken from the sound field in a

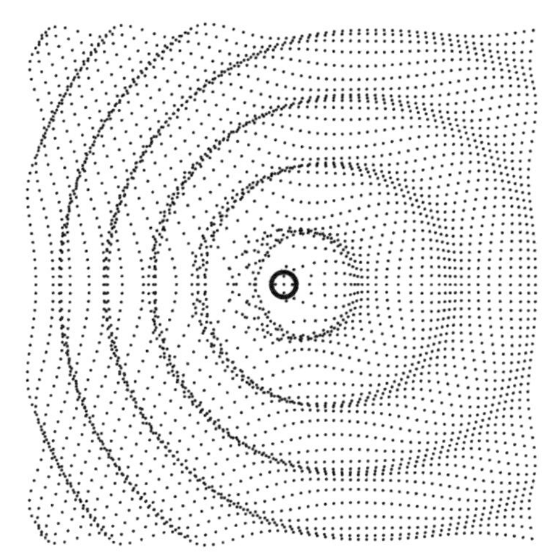

Fig. 2.14. Particle movements during sound propagation in fluid medium ($M = 0,33$) for a stationary source applied to a stationary coordinate system

stationary medium $p(x, y, z)$ can be arrived at with more precision as in the following

$$p_M(x, y, z) = e^{jkMx} p(x, y, z) . \tag{2.115}$$

Here we are assuming fluid motion at U in the direction of x. k describes the wavenumber in a stationary medium ($k = \omega/c = 2\pi/\lambda$). The effects already described can be discerned in Figure 2.14 depicting particle movements. To further illustrate the depiction in the graph, the wind is shown as blowing from left to right with a rather large Mach number of $M = 0,33$ (approx. $400\,km/h$).

2.7 Raised waves

Non-linearities have been left out of the previous sections in this chapter for good reason. Up to the highest perceivable levels of up to 140 dB, the field quantities sound pressure p, sound density ϱ, and sound temperature T are so small in comparison to the static dimensions p_0, ϱ_0 and T_0 that the quadratic expressions composing these dimensions can be left out.

The non-linear components only become significant when the levels are higher than what is humanly tolerable, such as the sound levels used to test satellite parts. 'Highest sound level acoustics' is obviously not a topic in this 'Lecture on Acoustical Engineering'. However, a few remarks can be made here as a completion to the study of acoustics as a whole.

The most important aspect here is easy to understand. As described in the last section, the sound propagation speed changes in fluid medium. Now the sound particle velocity also describes the local movement of the medium which occurs in a sound incidence. For this reason, the speed of the sound increases in regions that are moving at a high local particle velocity in the same direction as the wave propagation, and decreases in regions that are moving at a high local particle velocity, but in the opposite direction of the wave propagation. The particular velocity maxima are moving 'at supersonic speed' and therefore faster than the particle velocity minima, which are, in principle, moving at 'sub-sonic speed.' In this process, maxima are moving farther away from their preceding minima and approaching the next minima. This effect can be demonstrated by simulation processing. The results are illustrated in Figure 2.15. First, the linear wave can be expressed as

$$v_l(x,t) = v_0 \cos \omega (t - \frac{x}{c}) \tag{2.116}$$

and the non-linear wave, as

$$v(x,t) = v_0 \cos \omega (t - \frac{x}{c + v_l(x,t)}) = v_0 \cos (\omega t - \frac{2\pi x}{\lambda(1 + v_l(x,t)/c)}), \tag{2.117}$$

each a spatial function for $t = 0$ in the case of $v_0 = 0.025c$, corresponding to a sound level of approximately 163 dB. Much smaller particle velocity amplitudes, of course, disrupt this effect. The slope of the flanks increase, because, as mentioned before, the maxima are moving toward the minima which follow them, explaining the wave raising effect. As can be seen here, the wave is further raised with increasing distance to the source (here assumed to be at $x = 0$). At a certain 'critical' distance x_{cr}, the maximum will have surpassed the next minimum, causing a kind of 'breaking' effect, similar to ocean waves. The critical distance can be easily determined in this hypothetical model. At all points in time $\omega t = 2n\pi$ the maxima can be found at

$$\frac{x_{max}}{\lambda(1 + v_0/c)} = n \, ,$$

and the minima at

$$\frac{x_{min}}{\lambda(1 - v_0/c)} = n + 1/2 \, , .$$

$x_{max} = x_{min}$ is true for the 'breaking point,' resulting in

$$n = \frac{1}{4} \frac{1 - v_0/c}{v_0/c} \approx x_{cr}/\lambda \, . \tag{2.118}$$

According to this analysis, it is only a question of distance from the source until the raised wave effect occurs. In the scenario depicted in Figure 2.15 $x_{cr}/\lambda \approx 10$. In contrast, when reaching 100 dB, $v_0 \approx 7 \; 10^{-3} m/s$ and thus $v_0/c \approx 2 \; 10^{-5}$, or $x_{cr} \approx 10^4 \lambda$. Even at $1000 Hz$, the breaking point would therefore be about 3.5 km from the source. This example should make it clear how the internal attenuation inherent in the medium over long propagation

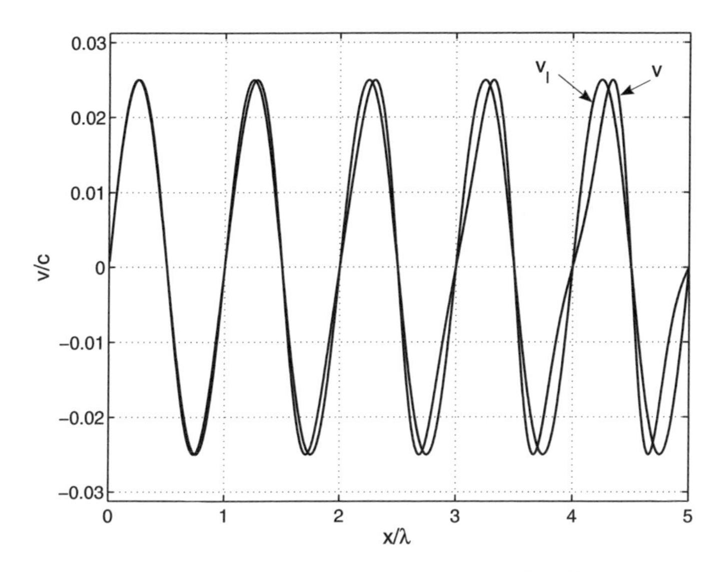

Fig. 2.15. Simulation of a raised wave with $v_0 = 0.025c$ for all points in time $\omega t = 2n\pi$

distances causes a flattening of the raised signal, accompanied by the usual effects of decreasing sound level with increasing propagation distance in any three-dimensional continuum.

It remains to be mentioned that the previously indicated (and neglected) physical non-linearities indeed lead to a change in sound velocity. The easiest way to illustrate this is to retain the quadratically small Taylor expansion term in the adiabatic equation of state. In $f(x) = (1+x)^\kappa \approx 1+\kappa x+\kappa(\kappa-1)x^2/2$, one can replace (2.14) with

$$\frac{p}{p_0} = \kappa\frac{\varrho}{\varrho_0} + \frac{\kappa(\kappa-1)}{2}(\frac{\varrho}{\varrho_0})^2 = \kappa\frac{\varrho}{\varrho_0}[1+\frac{\kappa-1}{2}\frac{\varrho}{\varrho_0}]\,. \tag{2.119}$$

Now when applying $c^2 = \kappa p_0/\varrho_0$, as we did before, the result is

$$p = c_N^2\varrho\,, \tag{2.120}$$

whereby c_N indicates a non-linear, space- and time-dependent sound velocity

$$c_N^2 = c^2[1+\frac{\kappa-1}{2}\frac{\varrho}{\varrho_0}]\,. \tag{2.121}$$

Similarly, it can be shown that the non-linear expression contained in the mass conservation principle according to eq.(2.18) can likewise be interpreted as a non-linear change in the sound velocity.

2.8 Summary

Sound consists of very small changes in pressure p, density ϱ, and temperature T of gases, which disperse in the form of waves in the medium at the speed of c. For pure tones, the waves have a wavelength of $\lambda = c/f$. Thermal conduction does not occur in sound phenomena, which undergo very rapid changes in time. For this reason, along with the Boyle-Mariotte equation, the adiabatic equation of state holds for sound phenomena. Based on the Boyle-Mariotte equation, acoustic quantities are given by

$$\frac{p}{p_0} = \frac{\varrho}{\varrho_0} + \frac{T}{T_0} \, .$$

The adiabatic equation expressed in terms of quantities of state is

$$\varrho = \frac{p}{c^2} \, .$$

Herein c signifies the sound propagation speed, which is

$$c = \sqrt{\kappa \frac{R}{M_{\mathrm{mol}}} T_0}$$

and is not only dependent on the material (type of gas), but also on the temperature. The particle velocity v describes the local air movements associated with the changes in density. For plane progressive waves (in a reflection-free wave guide), the ratio of pressure and particle velocity remains constant:

$$p = \varrho_0 c \, v$$

The constant $\varrho_0 c$ is known as the wave impedance or as the specific resistance of the medium.

Standing waves result from two progressive waves of equal amplitude but propagating in opposite directions. These waves either originate due to reflection or are the product of two sources. In the case of an incomplete reflection, the sound field is comprised of progressive ('active') and standing ('reactive') components.

Resonances can result from losses of sound energy in a gas-filled volume which are not caused by either inner losses or by losses due to outward transportation. These types of losses can be explained by the 'principle of in-phase alignment' in the case of one-dimensional wave guides.

The instantaneously stored sound energy travels along with sound in the medium. Sound energy consists of two components: energy of motion and compression energy. The transport of the sound energy is described by the intensity $I = P/S$. The intensity is the ratio of the power P permeating the surface area S to the surface area S itself. Overall, it can be shown that the intensity consists of the product of pressure and particle velocity

$$I = pv .$$

Acoustic power measurements, as in intensity field measurements, involve finding v, either by using a pair of microphones, or by taking measurements under free-field conditions in the far-field, where the intensity can be measured based on measurements of the sound pressure alone.

The Doppler effect results form situations with relative motion between sound source and receiver. The frequency shifts are caused by time-dependent time delays. The Doppler frequency slightly differs for moving sources or for moving receivers relative to the fluid.

2.9 Further reading

The book "Waves and Oscillations" by K. U. Ingard (Cambridge University Press, Cambridge 1990) provides an excellent and easily understandable description of the nature of waves. It describes acoustical waves in gases, fluids and solid state bodies. It also describes the behavior of other types of waves, such as electromagnetic and water surface waves. F. Fahy's book "Sound Intensity" (Elsevier, London and New York 1995) is recommended for further reading on the topic of sound intensity measurement techniques.

2.10 Practice exercises

Problem 1

Which of the following space and time functions satisfy the wave equation?

$$f_1(x,t) = ln(t + x/c)$$
$$f_2(x,t) = e^{(x-ct)}$$
$$f_3(x,t) = sh(ß(ct + x))$$
$$f_4(x,t) = cos(ax^2 + bx^3 - ct)$$

Problem 2

For experimental purposes, a cavity between double window panes is filled three different times with hydrogen (density 0.084 kg/m^3), oxygen (density 1.34 kg/m^3), and carbon dioxide (density 1.85 kg/m^3). To ensure that the glass panes do not cave in or expand, the pressure in the three gases is kept equal to the outside air pressure.

- How high is the speed of the sound in the above bivalent gases (air density $= 1.21\, kg/m^3$, speed of sound $= 340\, m/s$)?
- How large are the elasticity moduli $E = \varrho_0 c^2$ for the above gases and for air?
- How large are the wavelengths at the frequency of $1000\, Hz$?

Problem 3

An effective sound pressure value of $0.04\,N/m^2$ is determined in a progressive plane wave.How high is

- the particle velocity of the sound (at $\varrho_0 c = 400\,kg/sm^2$),
- the particle displacement at the frequencies $100\,Hz$ and $1000\,Hz$,
- the sound intensity,
- the sound power permeating a surface of $4\,m^2$ and
- sound pressure, sound intensity, and sound power levels for the surface $4\,m^2$?

Problem 4

The A-weighted sound pressure levels listed in the table are measured in an anechoic chamber on a cubic enveloping surface which encompasses a sound source. The 6 partitions of the enveloping surface are $2\,m^2$ each. How high is the A-weighted sound performance level of the sound source?

Partition	$L/dB(A)$
1	88
2	86
3	84
4	88
5	84
6	83

Problem 5

A sound source moving at the frequency of $1000\,Hz$ is moving at speeds of $50\,km/h$ (or $100\,km/h$ or $150\,km/h$) relative to a receiver. How high is the receiver frequency if

- a) the source is at rest in the medium?
- b) the receiver is at rest in the medium?

Consider for both scenarios two cases each: the first case, when the sender and receiver are moving toward each other, and the second case, when the sender and receiver are moving away from each other.

Problem 6

How high is the sound propagation speed at a temperature of $20^0\,C$, in 'stale' air (nitrogen) and in pure oxygen?

Problem 7

Reflections with low acoustic impedance are described by a reflection coefficient of $r = -1$. Such a reflector exists in situations such as a tube with a drastic increase in the cross-section at the opening (e.g. the lower end of a recorder).

- Give the equation for the space-dependent sound pressure in front of the reflector in the spatial range of $x < 0$ (origin is at the reflector)? Where do the nodes lie on the pressure-dependent continuum?
- Find the particular sound velocity. Where are the sound nodes located?
- Derive the resonance function and give the first three resonance frequencies for a tube length of $25\,cm$. A distance of the sound source l from the reflector in an expanded membrane moving at a speed of v_0 is given.

Problem 8

How large are the wavelengths in water (c=$1200\,m/s$) at $500\,Hz$, $1000\,Hz$, $2000\,Hz$ and $4000\,Hz$?

Problem 9

A half-infinite, one-dimensional wave guide (a tube filled with air with a wave capturer at each end) with a surface cross-section S is stimulated by a plane loudspeaker membrane at $x = 0$. Let the membrane particular velocity $v_M(t)$ be $v_M(t) = v_0 \sin \pi t/T$ in the time interval $0 < t < T$; beyond this interval for $t < 0$ and for $t > T$, $v_M(t) = 0$. Define

- the sound pressure for all points in time and space,
- the sound pressure for the particular sound velocity,
- the sound pressure for the energy density,
- the sound pressure for the intensity and
- the total energy radiating from the source.

How much energy does the source give off at a particular speed of $v_0 = 0,01\,m/s = 1\,cm/s$, at a wave guide diameter of $10\,cm$ with a circular cross-sectional surface, and the signal duration at $T = 0.01s$?

Problem 10

A highly significant error margin of $2\,dB$ ($3\,dB$) is tolerated for measurements using an intensity test probe.

- How large is the ratio of sensor distance Δx to wavelength λ at the highest permissable frequency reading?
- How high is the frequency reading for $\Delta x = 2.5\,cm$?

Problem 11

What is the highest acceptable phase tolerance for the microphones when the intensity sensors are $\Delta x = 5\,cm$ apart from each other and are to measure sound intensities with a standing wave component of $p_s/p_p = 10$ ($p_s/p_p = 100$) down to as low as $100\,Hz$ with an accuracy of $1\,dB$?

Problem 12

A vehicle with a siren, a police patrol car or fire truck, is driving by a microphone situated above the freeway with the siren on at no wind at a constant speed U. The microphone registers a frequency of $f_{E1} = 555.6\,Hz$ (the most significant frequency of a warning signal) just before the car drives by. After it drives by, the frequency is registered just at $f_{E2} = 454.6\,Hz$. How fast did the vehicle pass by? How high is the transmitting frequency of the siren?

3

Propagation and radiation of sound

As we know from experience (and as will be shown in the next sections), sound sources have a directivity. The sound level perceived by the observer is not dependent on the distance from the source alone. If the source is rotated, the level changes with angle.

On the other hand, it is known from several technical sound sources of interest that they radiate sound in all directions uniformly. Sound sources which are not too large like small machinery, ventilating system outlets emitting low frequency sound, ramming, hammering and banging and lots of other mainly broadband sound incidents have a negligible beam pattern in the sound field they produce. It can generally be shown that unilateral extruding sound sources show an omnidirectional radiation, if their size is small compared to the wavelength. Their directivity at sufficiently low frequencies is spherical. Finally, when estimating the radiation of sound sources, where the details of their directivity are unknown, one has to assume that their sound field is omnidirectional (which might actually not be the case).

Thus, the chapter on propagation and radiation begins with a discussion of omnidirectional sound radiation in free field where secondary influences such as weather conditions are neglected.

3.1 Omnidirectional sound radiation of point sources

Applying an energy principle considerably simplifies the investigation of omnidirectional sound sources. The acoustic power P penetrating an arbitrary surface which surrounds the source must be identical for every surface (assuming that propagation losses can be neglected for distances not too far from the source). This assumption is also valid for the source surface itself which is located directly on the source. Therefore, P has to be equal to the power which the source injects into the medium.

A spherical surface $S = 4\pi r^2$ is chosen for an omnidirectional radiation pattern, with the source located at the center (Fig. 3.1). At larger distances

M. Möser, *Engineering Acoustics*, DOI 10.1007/978-3-540-92723-5_3,

from the source the waveform of the radiated spherical waves resembles increasingly that of a plane wave, because the radius of curvature of the wavefronts decreases. Using the power relations derived in Sect. 2.4 for the far-field ($r \gg \lambda$ (see also Sec. 3.5.4)) we obtain

$$P = \frac{1}{\varrho c} p_{rms}^2 S = \frac{1}{\varrho c} p_{rms}^2 4\pi r^2 \, . \tag{3.1}$$

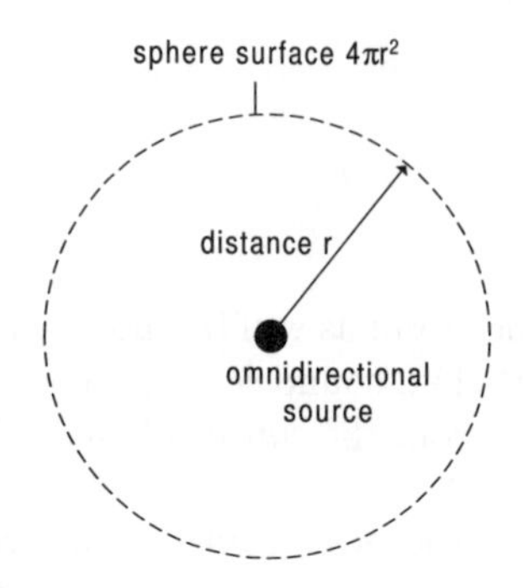

Fig. 3.1. Spherical surface around an omnidirectional point source used for determining radiated sound power

The mean square pressure is thus inversely proportional to the square of the distance. It is sensible to express (3.1) logarithmically, i.e. sound pressure level. Equation (3.1) is non-dimensionalized by the reference value $P_0 = {p_0}^2 S_0/\varrho c$ with $S_0 = 1\,\mathrm{m}^2$ (see (2.79) on p. 44) and the logarithm to the base 10 is taken ($\lg = \log_{10}$), resulting in

$$L_\mathrm{p} = L_\mathrm{w} - 20 \lg \frac{r}{\mathrm{m}} - 11\,\mathrm{dB}, \tag{3.2}$$

where L_p represents the sound pressure level at distance r. Based on the distance law (3.2), the level falls off at 6 dB per doubling of distance. If the noise source is located on a totally reflecting surface (like the ground), the power flows through a hemisphere only. In this case, instead of (3.2), we obtain

$$L_\mathrm{p} = L_\mathrm{w} - 20 \lg \frac{r}{\mathrm{m}} - 8\,\mathrm{dB}. \tag{3.3}$$

3.2 Omnidirectional sound radiation of line sources

Sometimes in practice, extended noise sources occur, which may consist of multiple, omnidirectional radiating (and incoherent) point sources. Examples are trains and busy streets. The power is calculated using a cylindrical surface (Fig. 3.2) and with l being the length of the line source, we obtain

$$P = \frac{p_{\mathrm{rms}}^2}{\varrho c} 2\pi r l, \tag{3.4}$$

resulting in a level of

$$L_{\mathrm{p}} = L_{\mathrm{w}} - 10 \lg \frac{l}{\mathrm{m}} - 10 \lg \frac{r}{\mathrm{m}} - 8\,\mathrm{dB} \tag{3.5}$$

or, if the source is again located on a reflecting surface

$$L_{\mathrm{p}} = L_{\mathrm{w}} - 10 \lg \frac{l}{\mathrm{m}} - 10 \lg \frac{r}{\mathrm{m}} - 5\,\mathrm{dB}\,. \tag{3.6}$$

Here, the sound pressure level only decreases at 3 dB per doubling of distance. Consequently, very long sources, like busy freeways, are still audible at large distances. For instance, the level measured at a distance of 1 km is only 16 dB lower than at a distance of 25 m. The value of $L_{\mathrm{eq}}(25\,\mathrm{m}) = 76\,\mathrm{dB(A)}$ for the equivalent A-weighted sound pressure level is certainly not underestimated, therefore 60 dB(A) remain in the distance of 1 km! Fortunately the presence of ground, plants and buildings alleviates the noise impact.

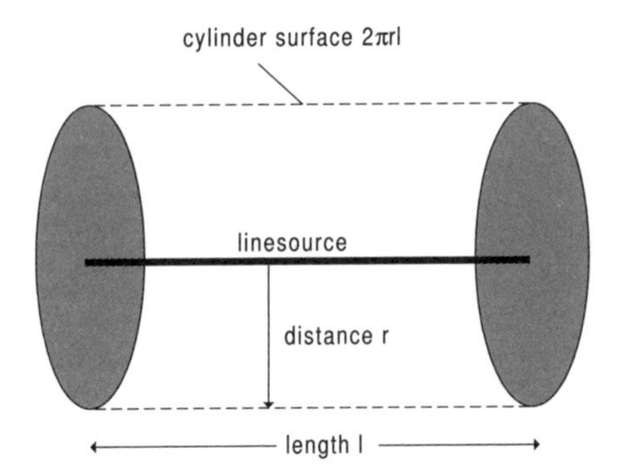

Fig. 3.2. Cylindrical surface around an omnidirectional line source used for determining radiated sound power

In the case of shorter line sources (e.g. public transportation trains), (3.6) is applied to shorter distances, whereas (3.3) applies to larger distances. In close vicinity to a source of finite length, the source characteristics resemble that of a long line source, whereas at larger distances, the individual sources shrink to a point. The transition between line and point source behavior occurs at a critical distance of $r_{\mathrm{cr}} = l/2$, as can be seen by setting (3.1) and (3.4) to be equal to one another. At distances $r < r_{\mathrm{cr}}$, the source behaves like a line source with the level decreasing by 3 dB per doubling of distance; at distances $r > r_{\mathrm{cr}}$, it acts like a point source with the level decreasing by 6 dB per doubling of distance. If the source power has to be measured in practice, it is usually measured close to the source in order to keep background noise low. Using (3.6), one can calculate the power and subsequently implement (3.3) to obtain a reasonable prognosis for distances $r > l/2$.

3.3 Volume velocity sources

As shown earlier, the root-mean-square sound pressure is inversely proportional to the distance, assuming the simplest idealistic case of omnidirectional radiation. For spherically-symmetric outgoing waves, the sound pressure therefore is of the form

$$p = \frac{A}{r} \mathrm{e}^{-jkr} , \tag{3.7}$$

where k is the wavenumber $k = \omega/c = 2\pi/\lambda$ and A the pressure amplitude. Although it is formulated based on plausibilities, (3.7) fulfils the wave equation (2.58), as can easily be shown.

To obtain such a 'mathematically ideal' field with perfect spherical symmetry, the sound source has to be constructed in a particular way. It consists of a 'pulsating sphere' which is a spherical surface $r = a$, pulsating radially with uniform local velocity v_a (see Fig. 3.3). The pulsating sphere is also called a simple source (of order zero) or 'monopole source', to indicate that the radiation is independent of the angle.

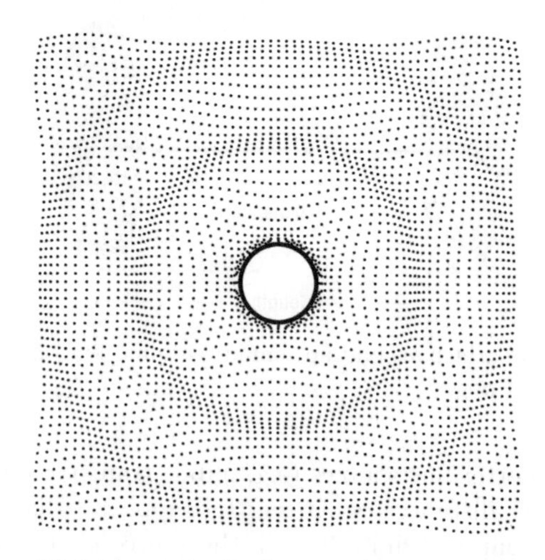

Fig. 3.3. Sound field (particle displacement) of a sphere pulsating with v_a

The pressure amplitude A, yet unknown in (3.7), can be calculated from the velocity v_a on the surface of the sphere $r = a$. Similar to (2.26)

$$\varrho \frac{\partial v}{\partial t} = -\frac{\partial p}{\partial r}$$

can be formulated (see also (2.57) in spherical coordinates) and assuming complex amplitudes as in (3.7)

$$v = \frac{j}{\omega\varrho}\frac{\partial p}{\partial r} = \frac{A}{r\omega\varrho}\left(k - \frac{j}{r}\right)\mathrm{e}^{-jkr} = \frac{A}{\varrho c}\left(1 - \frac{j}{kr}\right)\frac{\mathrm{e}^{-jkr}}{r} \tag{3.8}$$

is obtained. Because of $v = v_a$ for $r = a$ it follows that

$$A = \frac{\varrho c v_a a \mathrm{e}^{jka}}{1 - \frac{j}{ka}} \,. \tag{3.9}$$

Restricting (3.9) to small sources $ka = 2\pi a/\lambda \ll 1$, the 1 in the denominator can be neglected, which yields

$$A = jk\varrho c v_a a^2 = j\omega\varrho v_a a^2 \,. \tag{3.10}$$

The sound pressure in (3.7) becomes

$$p = j\omega\varrho v_a a^2 \frac{\mathrm{e}^{-jkr}}{r} \tag{3.11}$$

now described by source terms only and stating the fact that the energy of radially outgoing waves decreases with increasing distance.

The radiation of such an exact mathematically defined spherical monopole source would be of theoretical interest only if the derived knowledge would not be transferable to all volume velocity sources which are small compared to the wavelength. The main characteristics of such sources can be interpreted as a temporal change of their volume or the outflow of fluid mass. Some examples of expanding bodies are a loudspeaker cone in a closed box, whose dimensions are small compared to the wavelength, as well as explosions, the orifice of the exhaust pipe of a car, opening (or closing) valves (e.g. opening a bottle of sparkling wine). (3.11) can be applied to calculate all these small volume velocity sources,. The term, describing the source characteristics, is the volume velocity Q, generally calculated by

$$Q = \int_s v \mathrm{d}S \tag{3.12}$$

which provides the basis for determining the velocity of the source and the surface area S of the source. For example, let $v(t)$ be the velocity of a gas flowing through an exhaust pipe of cross-sectional area S. In that case, the volume velocity Q has to be distributed over the surface of the pulsating sphere as explained in the latter example $Q = 4a^2 v_a$. In general, volume velocity sources are described by

$$p = j\omega\varrho Q \frac{\mathrm{e}^{-jkr}}{4\pi r} \,. \tag{3.13}$$

In contrast to plane progressive waves with $p = \varrho c v$, the three-dimensional radiation of sound behaves like a time differentiation of the source velocity.

Because $j\omega$ represents time differentiation and e^{-jkr} represents a delay of $e^{-j\omega\tau}$ with the delay time $\tau = r/c$, (3.13) can be written in the time domain as

$$p = \frac{\varrho}{4\pi r} \frac{\mathrm{d}Q(t - r/c)}{\mathrm{d}t} . \qquad (3.14)$$

If low noise emission is required, the change of the volume velocity with respect to time has to be small. A sudden, jerky opening of valves is unfavorable in the sense of noise control; the process could be done more quietly by opening the valves gradually. An illustration is given in Fig. 3.4.

The sound pressure versus frequency characteristics $p \sim j\omega Q = bS$ are proportional to the acceleration b. This fact is more important in the determination of frequency response functions of loudspeakers. The results from this chapter will certainly be used in Chap. 11.

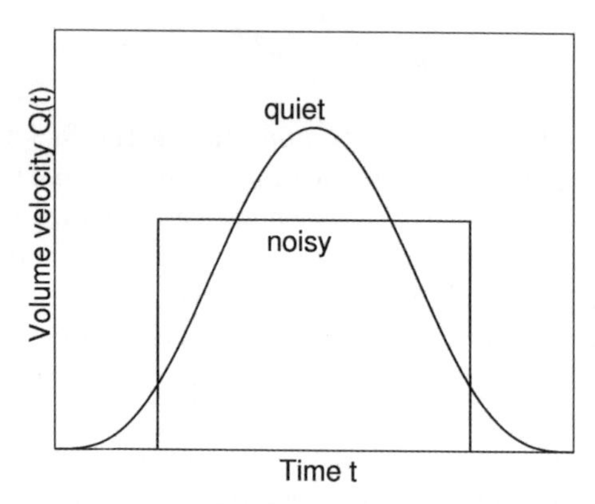

Fig. 3.4. 'Noisy' and 'quiet' change in volume velocity with equal amount $\int Q(t)\mathrm{d}t$

3.4 Sound field of two sources

There are a lot of good reasons to discuss the sound field of two (small) volume velocity sources. Arrangements of two equal sources, which are opposite in phase, can often be found in practice. At sufficiently low frequencies, each small and rigid oscillating surface which is not mounted into a chassis, as in a loudspeaker without a box or a baffle, can be interpreted as a dipole source.

If the surface 'pushes' the air to the right (Fig. 3.5), it 'sucks' air from the left at the same time. The compressed air on the right flows around the edge to the back of the surface and compensates the density difference (and thus the pressure difference). The medium produces a 'short circuit' of the mass. The fact that this effect can be described by a pair of equal sources opposite in phase leads to a non-uniform beam pattern and, at low frequencies, accounts for a substantial smaller radiation of sound than for single sources, as will

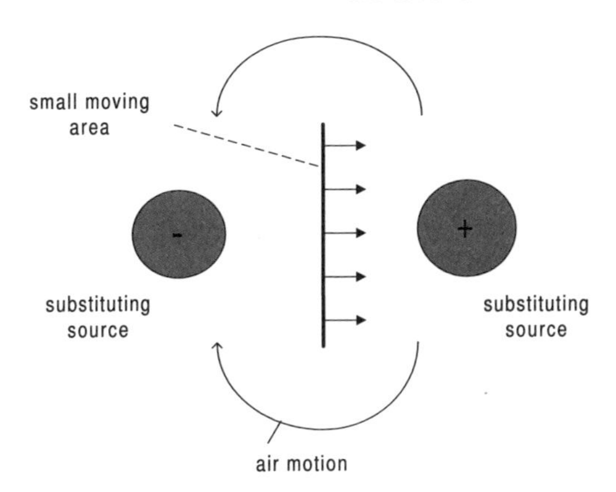

Fig. 3.5. A rigidly oscillating surface operates as a dipole source

be shown in the following discussions. 'Active noise control', a common topic of discussion these days, attempts (amongst other things) to superimpose a secondary sound field on that of an existing (primary) source which is opposite in phase. Even in this case, the simplest model consists of two equal but opposite sources.

Finally, it is currently a matter of interest in public address systems, for example, how sound field changes are measured when a second (equal and coherent) source is added.

Also, discussing the combination of two small sources is the simplest of the more general case being where a source is composed of an arbitrary number of multiple elements. The latter results in loudspeaker arrays. Moreover, oscillating surfaces in general (like plates, walls, ceilings, etc.) can be regarded as composed of a superposition of higher order sources.

So there is enough reason to discuss different combinations of two sources. Practical examples already mentioned will be pointed out when appropriate.

The model under investigation in the following paragraph is shown in Fig. 3.6 using a spherical coordinate system. The sources are aligned along the z-axis separated by a distance h; thus, a cylindrically symmetric sound field is produced which is independent of the circumference angle φ. The angle between the radius R, pointing from the origin towards the field point, and the z-axis is usually denoted by ϑ; the angles in what follows are defined with respect to that angle definition. The surface element $\mathrm{d}S$, for example, appearing in the (following) surface integral is given in spherical coordinates by

$$\mathrm{d}S = R^2 \sin\vartheta \mathrm{d}\vartheta \mathrm{d}\varphi \, . \tag{3.15}$$

The surface integral of a sphere is covered by the intervals $0 < \varphi < 2\pi$ and $0 < \vartheta < \pi$. In measuring process, it is also quite common to use the angle ϑ_{N} normal to the axis. These two angles are related by

$$\vartheta + \vartheta_{\mathrm{N}} = 90^\circ \, . \tag{3.16}$$

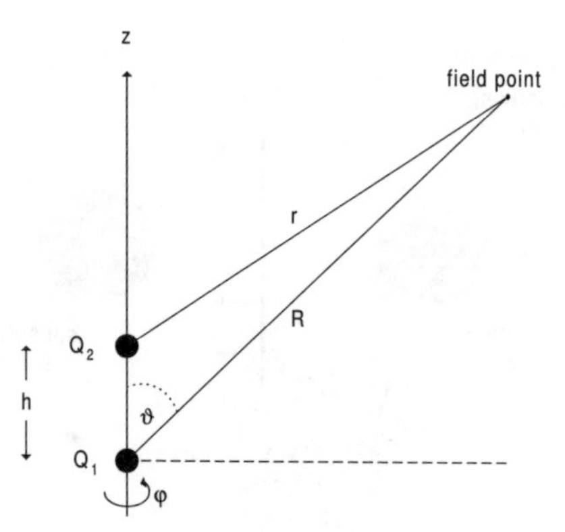

Fig. 3.6. Arrangement of two sources in a spherical coordinate system and definition of the used quantities

When predicting beam patterns, ϑ_N will be used in the following; when integrating intensities to obtain the power, it is easier to express ϑ_N by ϑ using (3.16), in which case universal equations for spherical coordinates can be applied.

The sound field of two sources in general, using (3.13), is given by

$$p = \frac{j\omega\varrho}{4\pi}\left\{Q_1\frac{\mathrm{e}^{-jkR}}{R} + Q_2\frac{\mathrm{e}^{-jkr}}{r}\right\} . \tag{3.17}$$

Due to the linearity of the wave equation, the sound field is simply the sum of the two components.

The main properties of the totals field can certainly be shown in a few pertinent examples. For this purpose, a color-coded scheme is used to illustrate the overall sound levels corresponding to each source combination for equally large sources at different distances a in Figures 3.7 through 3.10. 'Small,' 'medium,' and 'large' distances a – relative to the wavelength – are plugged into various distances expressed as $a/\lambda = 0.25$, $a/\lambda = 0.5$, $a/\lambda = 1$ and $a/\lambda = 2$. The location of the sources is denoted by the two pink-colored dots. Figures 3.11 through 3.14 depict the case of two identically large sources (magenta and green dots) opposite to one another at the same distance.

The tendencies illustrated in the aforementioned figures can be easily explained. At the shortest distance $h = 0.25\lambda$ between two equally large sources, the sources (almost) behave as if they were 'at the same place.' The total field is, in principle,6 dB brighter than the fields of each source alone. If the sources are pulled farther apart (or the frequencies diverge at the same distance accordingly), the first signs of interference become apparent. At $h = 0.5\lambda$, the individual pressures along the middle axis joining both sources virtually disappear at a sufficient distance from each source. In this case, the totals field is obviously much smaller than the components it is comprised of.

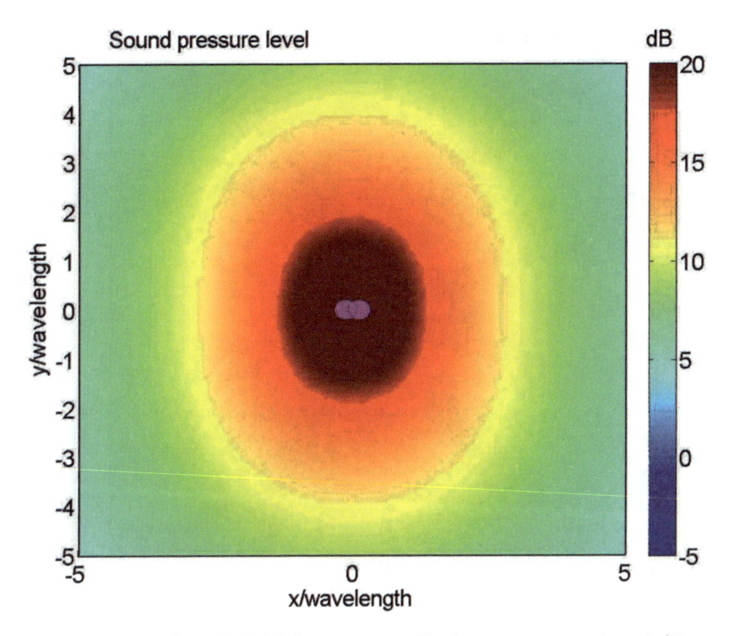

Fig. 3.7. Sound field for two equally large sources, $h = \lambda/4$

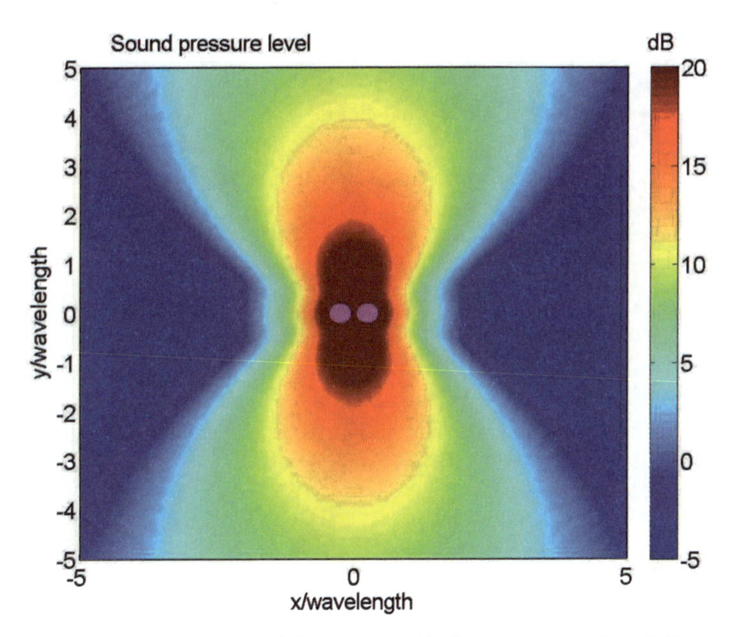

Fig. 3.8. Sound field for two equally large sources, $h = \lambda/2$

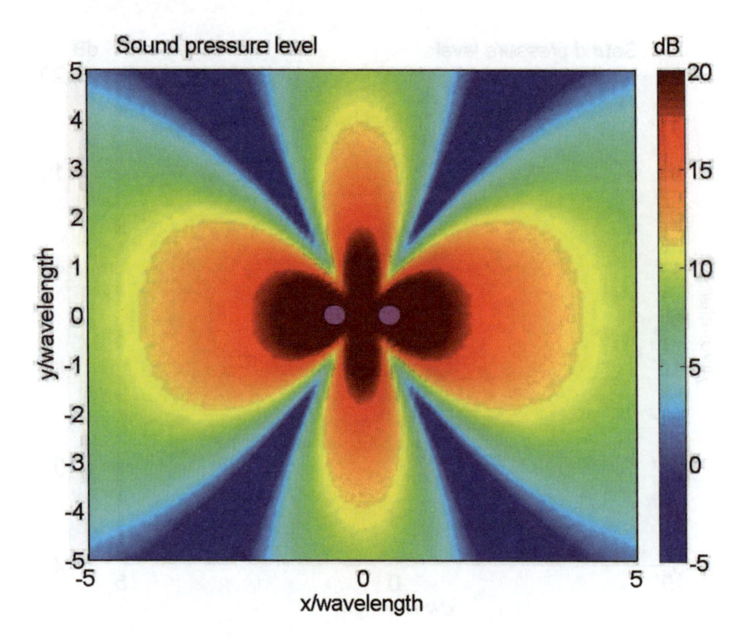

Fig. 3.9. Sound field for two equally large sources, $h = \lambda$

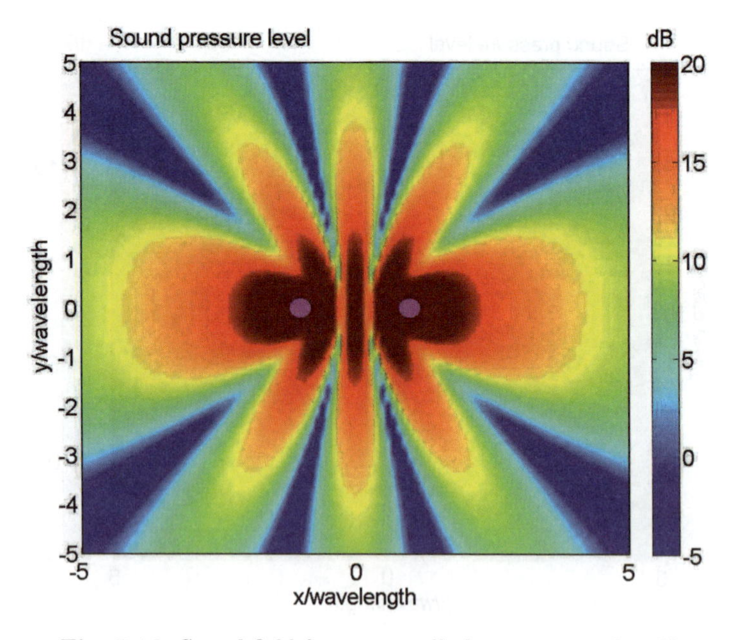

Fig. 3.10. Sound field for two equally large sources, $h = 2\lambda$

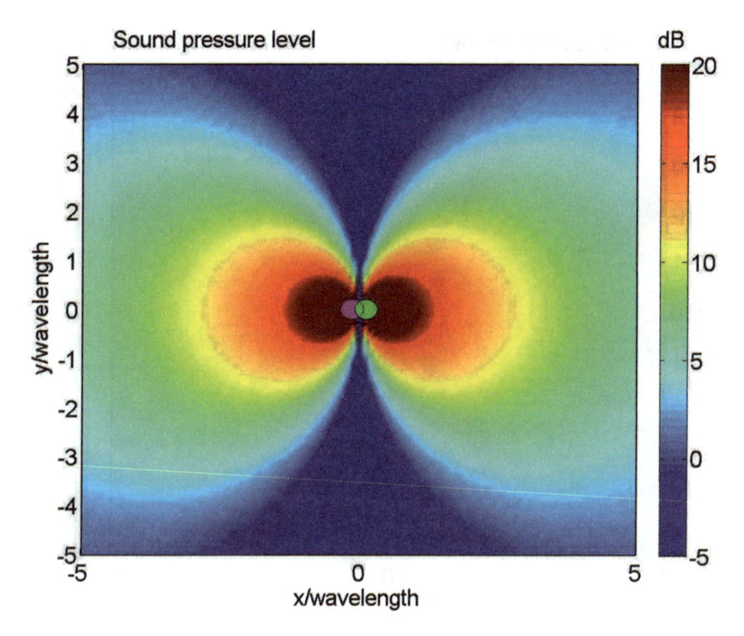

Fig. 3.11. Sound field for two opposite but equally large sources, $h = \lambda/4$

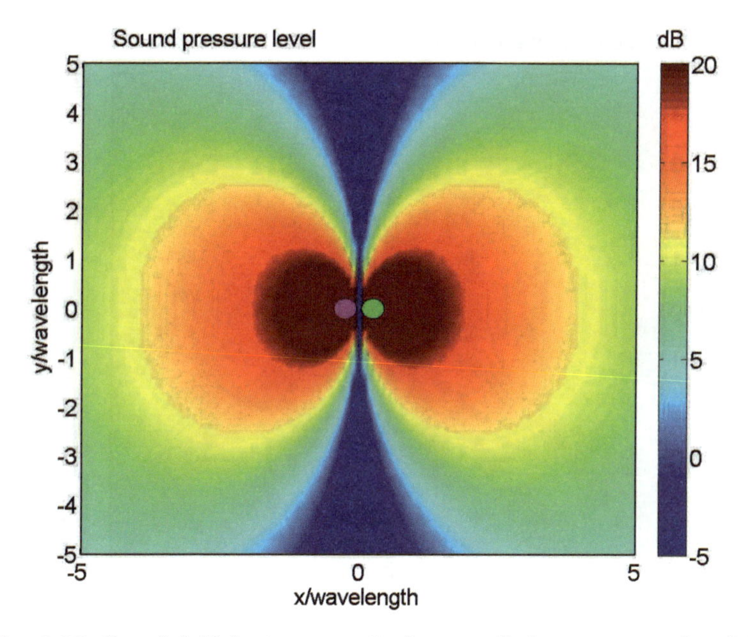

Fig. 3.12. Sound field for two opposite but equally large sources, $h = \lambda/2$

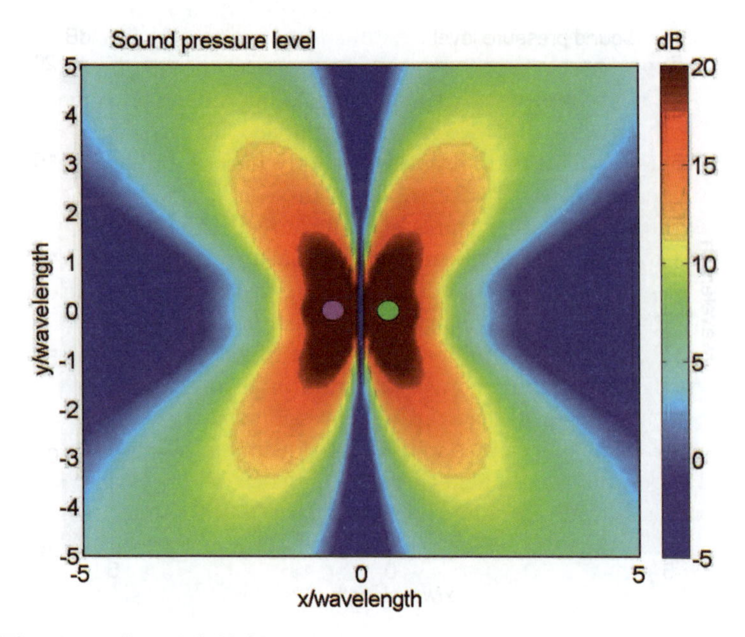

Fig. 3.13. Sound field for two opposite but equally large sources, $h = \lambda$

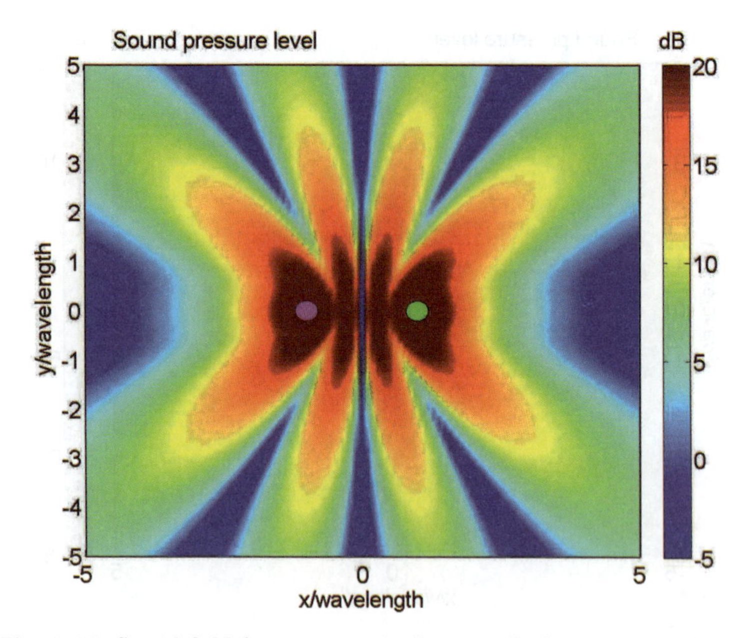

Fig. 3.14. Sound field for two opposite but equally large sources, $h = 2\lambda$

This phenomenon can also be described as 'destructive interference.' This case only results in a 'residual sound field' because the distance decreases at a slightly different rate of $1/R$ than of $1/r$. This difference continues to diminish with increasing distance from the sources. That means that the field as observed from the outside in grows steadily darker. Of course, the sound pressure on the mid-plane between both sources inevitably reaches twice the amount as the individual pressures of each source. If the frequency continues to increase to $h = \lambda$, a global interference pattern ensues, consisting of alternating 'bright' and 'dark' bands. This pattern becomes more prominent as the distance further increases, such as at a distance of $h = 2\lambda$.

The equally large opposing sources represented by the magenta and green dots behave in a similar manner. Upon an initial examination, very small distances between the sources lead to the impression that the sources are located in the same place in this case as well. The field sum is therefore globally 'close to zero'. The residual sound field therefore only occurs due to the fact that the sources cannot actually occupy the same space at the same time. Naturally, in this case, the partial pressures in the mid-plane between the sources will always add up to zero, likewise resulting in the residual sound being concentrated mainly on the adjoining axis in sources at a very small distance to one another for sources close to one another. Of course, at increasing frequencies (or at increasing distance between the sources) interference immediately crops up again, albeit with different aspects. Thus we see 'constructive' interference at the source axis for $h = \lambda/2$. This leads to an elimination of the opposite signs of the sources at precisely one-half wavelength apart in the dispersion of the sound components. For this reason, within such spatial dimensions and under these circumstances, the total field amounts to about double of each partial field. If the sources move even farther apart, an increasingly distinctive band of alternating bright and dark tones reappears, similar to the case of the equally large sources described above, only the location of the bands is reversed, as shown in Figures 3.10 and 3.14.

The principles and effects described above should be understood in light of their overall implications. Now the sound field varies greatly from field point to field point within close proximity to both sources $r < h$. At times, one source factors in greater than the other, or vice versa, depending on the distance. For this reason, it makes sense to construct a "far-field approximation" for eq.(3.17). As can be seen in the following, far-field conditions allow for a simple quantitative description.

Although equation (3.17) is correct, it is a little too complex. In close vicinity to the two sources $r < h$, the sound field changes rapidly between different field points. At one point the source at the closer distance dominates the sound field, at another point it might be the other source, due to the dependence on distance. For that reason, a 'far-field approximation' is defined for (3.17). As can quickly be shown in the following, this provides us with simple and easy results for far-field calculations.

The first simplification of (3.17) is based on the consideration that the amplitude decrease for increasing distances is roughly the same for $r \gg h$ and both sources. Therefore, we can presume $1/r \approx 1/R$, leading us to

$$p_{far} \approx \frac{j\omega\varrho}{4\pi R}\left\{Q_1 \mathrm{e}^{-jkR} + Q_2 \mathrm{e}^{-jkr}\right\} .$$

Despite the fact that $1/r \approx 1/R$ also implies $r \approx R$, the phase relations e^{-jkr} and e^{-jkR} are subject to closer investigation, the interference can be defined as 'constructive' or 'destructive' or 'somewhere in between' in terms of phase. This is simply recognized by the transformation

$$p_{far} \approx \frac{j\omega\varrho}{4\pi R}\mathrm{e}^{-jkR}\left\{Q_1 + Q_2 \mathrm{e}^{-jk(r-R)}\right\} . \tag{3.18}$$

The phase function appearing in the last brackets depends on the difference of the two distances relative to wavelength. Despite the fact that r and R can be nearly equal in order of magnitude (as assumed in the amplitude decrease), they can be different, however, by as much as one entire wavelength. It is this relatively small difference that determines the actual value of the phase function $\mathrm{e}^{-jk(r-R)}$.

To clarify this even more, r is expressed with the aid of a cosine theorem in terms of R and ϑ

$$r^2 = R^2 + h^2 - 2Rh\cos\vartheta,$$

or

$$r^2 - R^2 = (r-R)(r+R) = h^2 - 2Rh\cos\vartheta,$$

which can be solved for the sought difference

$$r - R = \frac{h^2}{r+R} - \frac{2Rh}{r+R}\cos\vartheta .$$

In the far-field $R \gg h$, a first order approximation can be determined where terms with $(h/R)^2$ and higher order terms are omitted:

$$r - R \approx -h\cos\vartheta \tag{3.20}$$

Equation (3.18) is thus approximated by

$$p_{far} \approx \frac{j\omega\varrho}{4\pi R}\mathrm{e}^{-jkR}\left\{Q_1 + Q_2 \mathrm{e}^{jkh\cos\vartheta}\right\} = p_1\left\{1 + \frac{Q_2}{Q_1}\mathrm{e}^{jkh\sin\vartheta_{\mathrm{N}}}\right\}, \tag{3.21}$$

where p_1 denotes the sound pressure of the first source alone (the case $Q_2 = 0$).

Sound fields are reasonably described by a 'global' quantity and a field distribution in all directions given by the sound pressure. The term in brackets in (3.21) describes the beam pattern of the source pair. Since only the difference from one angle to another is of interest, the scale can be set arbitrarily. For a measure of the 'source strength' it is unwise to concentrate on a single point or a certain direction; the radiated net power P is used as a global measure instead. It is calculated by

$$P = \frac{1}{2\varrho c} \int_0^{2\pi} \int_0^{\pi} |p_{far}|^2 R^2 \sin\vartheta \mathrm{d}\vartheta \tag{3.22}$$

(see also (3.15)). After inserting (3.21)

$$P = P_1 \left\{ 1 + \left(\frac{Q_2}{Q_1} \right)^2 + 2 \frac{Q_2}{Q_1} \frac{\sin(kh)}{kh} \right\} \tag{3.23}$$

where P_1 is the radiated power of Q_1 alone. The ratio Q_2/Q_1 in (3.23) is assumed to be real. (One difficulty in calculating (3.23) is the solution of

$$I = \int_0^{\pi} \cos(kh \cos\vartheta) sin\vartheta \mathrm{d}\vartheta \; .$$

The substitution $u = \cos\vartheta, \mathrm{d}u = -\sin\vartheta \mathrm{d}\vartheta$ leads to a simplification of the integral.)

Treating source attributes separately on the basis of low and high frequencies likewise simplifies the discussion of source behavior as follows.

Low frequencies $h/\lambda \ll 1$

At low frequencies, based on (3.21) and using $\mathrm{e}^{jx} \approx 1 + jx$ ($|x| \ll 1$), it can be stated that

$$p_{far} \approx p_1 \left\{ 1 + \frac{Q_2}{Q_1} (1 + jkh \sin\vartheta_{\mathrm{N}}) \right\} \; . \tag{3.24}$$

As long as the angle-dependent part of $1+Q_2/Q_1$ is non-zero, i.e. a dipole with $Q_2 = -Q_1$, the two source strengths can roughly be added: at low frequencies, the sources act as if they were put ‘in the same place’:

$$p_{far} \approx p_1 \left(1 + \frac{Q_2}{Q_1} \right)$$

The sound power using (3.23) and $\sin(kh)/kh \approx 1$ becomes

$$P \approx P_1 \left(1 + \frac{Q_2}{Q_1} \right)^2 \; .$$

In the case of identical sources $Q_2 = Q_1$, the sound pressure is twice the sound pressure of the single source

$$p_{far} \approx 2p_1 \tag{3.25}$$

and the sound power is fourfold the power of the single source

$$P \approx 4P_1 \,. \tag{3.26}$$

On the other hand, in the case of a dipole source $Q_2 = -Q_1$, based on (3.24) we obtain

$$p_{far}(\text{dipole}) \approx -jkhp_1 \sin \vartheta_{\mathrm{N}} \tag{3.27}$$

and based on (3.23),

$$P(\text{dipole}) \approx 2P_1 \left\{1 - \frac{\sin{(kh)}}{kh}\right\} \approx P_1 \frac{(kh)^2}{3} \tag{3.28}$$

(with $\sin(x)/x \approx 1 - x^2/6$ for $|x| \ll 1$). The dipole has an cardioid beam pattern at low frequencies, as shown on the left side of Fig. 3.15, the so-called 'figure-8 pattern'. The corresponding particle displacement is also shown in Fig. 3.15. The three-dimensional extension is developed by rotating the char-

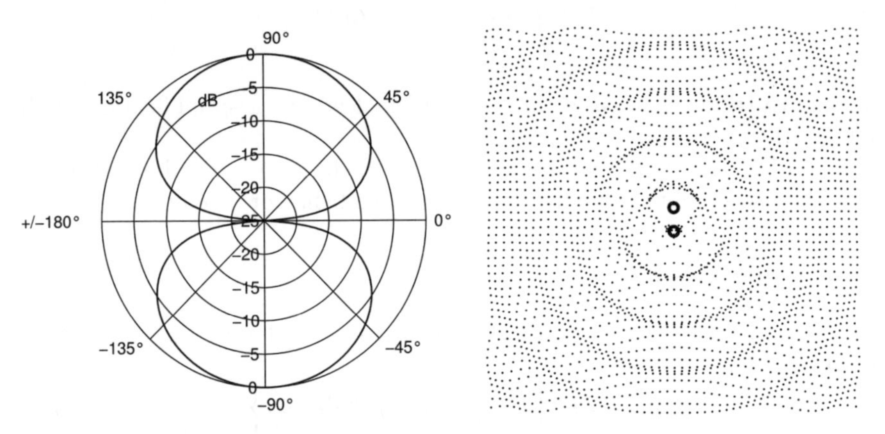

Fig. 3.15. Beam pattern (left) and particle displacement (right) of a dipole. Source distance $\lambda/2$

acteristics around the axis, where the sources are attached (the term 'double-sphere characteristics' would actually be more appropriate). The sound power radiated from the dipole at low frequencies is less than that of a single monopole source

$$L_{\mathrm{w}}(\text{dipole}) = L_{\mathrm{w}}(\text{single}) + 10 \lg \frac{(kh)^2}{3} \,. \tag{3.29}$$

If the sound power difference were expressed in numbers, it would be not very large anyway if unrealistic small distances or far too low frequencies are neglected. For $h/\lambda = 0.125$, it amounts to $10 \lg((kh)^2/3) = -6.8\,\mathrm{dB}$; the power produced by the dipole is about 6 dB smaller than that of the single sources. If the point sources are assumed to represent technical sound sources of finite expanse, they must only have small dimensions in order to reduce the power, even at the low frequencies actually favorable to power reductions. At 170 Hz with $\lambda = 2\,\mathrm{m}$, it is $\lambda/8 = 0.25\,\mathrm{m}$; the distance of the two sources (and therefore the source dimensions) must be smaller than 25 cm in order to achieve a power reduction of 6 dB. Larger distances would result in a smaller sound power level difference (at the same frequency).

The situation described above is one of the reasons that the expectations with respect to 'active noise control' should be exercised with caution – at least in the case of free-field radiation. Even if – for demonstration purposes – loudspeakers with opposite phase are used at low frequencies, their center points are usually spaced more than a quarter of a wavelength apart. The resulting noise reduction is – in terms of perceived changes – not very large. Moreover, this low frequency experiment often tempts the engineer to overdrive the loudspeakers, leading to inflated driving voltages, as low frequencies are also more difficult to perceive. As a consequence, the loudspeakers produce non-linear distortion at higher frequencies, where active control is ineffective anyway. Higher frequencies are more perceptible. However, active control performance may not be audible at all.

High frequencies $h \gg \lambda$

At high frequencies, the beam pattern rapidly changes with varying angle ϑ_{N} and alternates uniformly between the sound pressure maxima

$$|p_{far}|^2_{\max} = |p_1|^2 \left(1 + \left|\frac{Q_2}{Q_1}\right|\right)^2 \tag{3.30a}$$

and the sound pressure minima

$$|p_{far}|^2_{\min} = |p_1|^2 \left(1 - \left|\frac{Q_2}{Q_1}\right|\right)^2 . \tag{3.30a}$$

The reason is the interference of two fields whose magnitudes are added in the antinodes (maxima) and subtracted in the nodes (minima).

The net radiated power for $kh \gg 1$, according to (3.23), is

$$P \approx P_1 \left\{1 + \left(\frac{Q_2}{Q_1}\right)^2\right\} = P_1 + P_2 . \tag{3.31}$$

At high frequencies (in contrast to low frequencies), the sound powers of the individual sources are added to find the net power of the source pair. This fact is equivalent to the directivity changes already mentioned, where maxima and minima alternate and can be found in pairs. Therefore the mean square of the sound pressure is

$$\overline{p^2} = \frac{1}{2}\,|p_{far}|^2_{\max} + \frac{1}{2}\,|p_{far}|^2_{\min} = |p_1|^2 \left(1 + \left|\frac{Q_2}{Q_1}\right|\right)^2 ,$$

which also results in (3.31), because the mean square of the sound pressure and the radiated sound power are proportional quantities.

In the case of identical sources $Q_2 = Q_1$ – as well as for equal and opposite sources $Q_2 = -Q_1$ – the sound power is doubled. It thus follows that the addition of a secondary source with opposite phase for active noise control purposes may not only be ineffective, but may additionally represent a disadvantage. In achieving the minimum radiated sound power, it is therefore better to omit the 'active' source.

This also becomes clear, when the ratio of the two source strengths $V = Q_2/Q_1$ that leads to the minimum radiated power, is determined. Using V in (3.23) yields

$$\frac{P}{P_1} = 1 + V^2 + 2V\frac{\sin(kh)}{kh},$$

which, after differentiating, yields the optimum ratio 'with minimum net power'

$$V_{\text{opt}} = \left.\frac{Q_2}{Q_1}\right|_{\text{opt}} = -\frac{\sin(kh)}{kh}. \tag{3.32}$$

As is obvious from (3.32), the optimum source strength ratio is achieved for $Q_2 = -Q_1$. With increasing frequency, the optimum ratio gets smaller, and can even have the same sign as the 'primary source'. At high frequencies, V_{opt} tends to zero.

An illustration of these principles is given in Fig. 3.16a using the calculated radiated power from (3.23) for $Q_2 = Q_1$, for $Q_2 = -Q_1$ and for the optimum case (with respect to active noise control), where $Q_2 = -Q_1 \sin(kh)/kh$, according to (3.32). At lower frequencies, the components of the two sources are added, allowing an increase of 6 dB compared to the single source for $Q_2 = Q_1$, and a decrease of power for equal but opposite sources $Q_2 = -Q_1$. The phase relation becomes unimportant at high frequencies: the net power is always equal to the sum of the individual powers. The particle displacement for two different cases is shown in Fig. 3.16b for illustration.

The aforementioned principles can also be extended to a larger number of sources. The net power for N sources at low frequencies is given by

$$P_{tot} = \sum_{i=1}^{N} Q_i \frac{j\omega\varrho}{4\pi R} \mathrm{e}^{-jkR} \tag{3.33a}$$

(all source distances are small compared to the wavelength λ)

and at high frequencies,

$$P_{tot} = \sum_{i=1}^{N} P_i \tag{3.33a}$$

(all distances are large compared to λ).

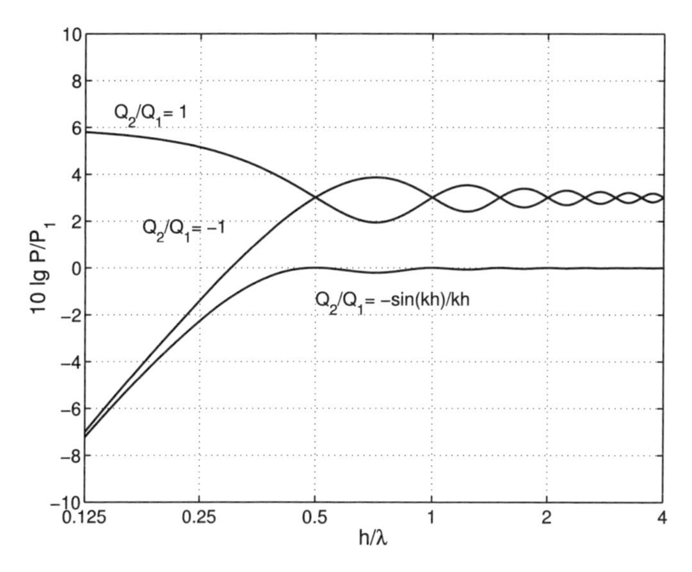

Fig. 3.16. (**a**) Frequency response functions of the power radiated by two sources

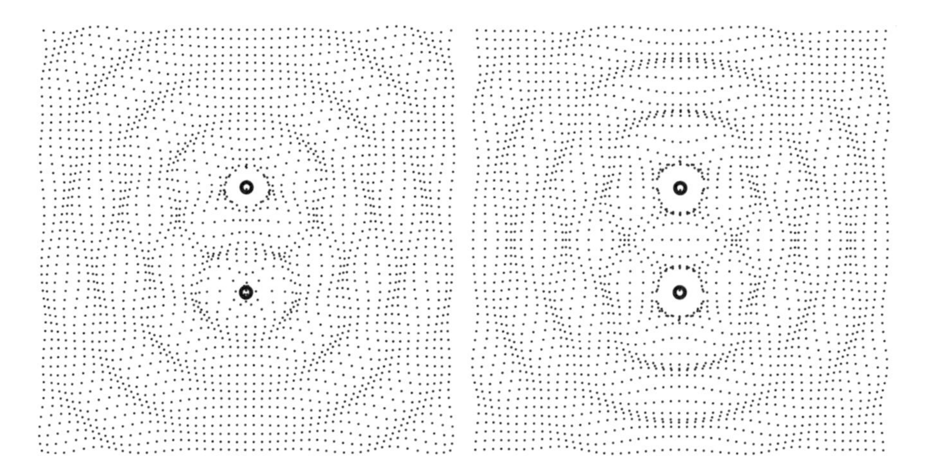

Fig. 3.16. (**b**) Particle displacement of two equal sources at a distance 2λ. Left: Sources in antiphase. Right: Sources in phase

3.5 Loudspeaker arrays

Consider a combination of arbitrary omnidirectional coherent sources which are aligned along one axis. The only practical realization of such one-dimensional chain of sources are loudspeaker arrays, as depicted in Fig. 3.17. For the sake of simplicity, the velocity characteristics on the source surface are described by a continuous function $v(z)$. The array has a width b which is always small compared to the wavelength.

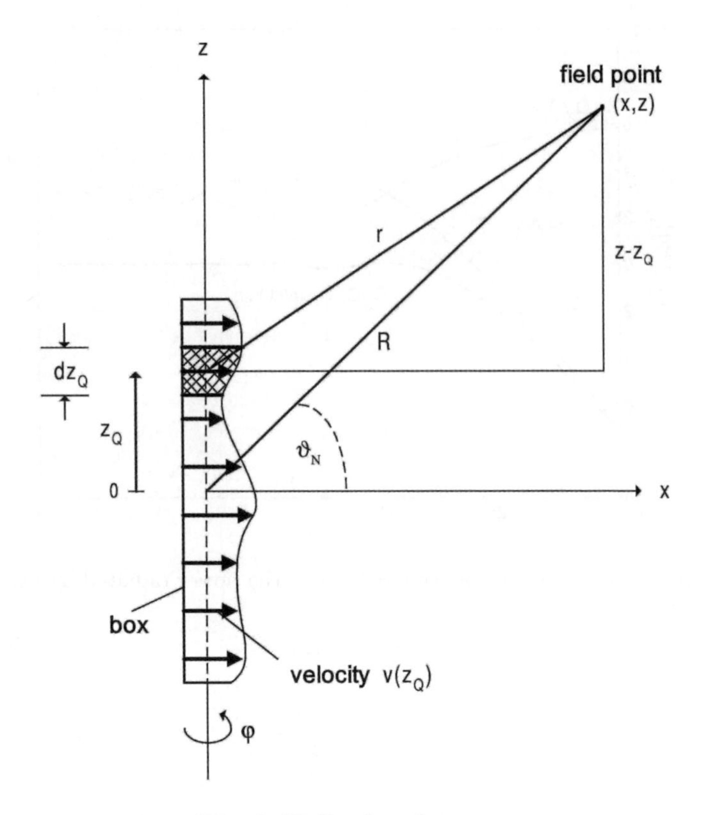

Fig. 3.17. Loudspeaker array

The contribution of the infinitesimal small source element to the sound pressure at the field point, also depicted in Fig. 3.17, is given by

$$\mathrm{d}p = \frac{j\omega\varrho\, b\, v(z_\mathrm{Q})}{4\pi r} \mathrm{e}^{-jkr} \mathrm{d}z_\mathrm{Q} \;, \tag{3.34}$$

and thus the total sound pressure is

$$p = \frac{j\omega\varrho\, b}{4\pi} \int_{-l/2}^{l/2} v(z_\mathrm{Q}) \frac{\mathrm{e}^{-jkr}}{r} \mathrm{d}z_\mathrm{Q} \;, \tag{3.35}$$

where l is the length of the array and r is the distance between the source and the field point (x, z)

$$r = \sqrt{(z - z_\mathrm{Q})^2 + x^2} \;.$$

It is assumed that each source element forms a volume velocity source. Therefore, the loudspeakers have to be built into an enclosure (a box), to prevent a short circuit of the mass (described in the last section). The sound field given by (3.35) is again cylindrically symmetric with respect to the φ-axis (Fig. 3.17).

At large distances from the source, (3.35) takes on a clearer form. To derive the far-field approximation using (3.35), the same procedure is used as outlined in the last section.

$$r - R = -z_Q \cos\vartheta = -z_Q \cos(90^\circ - \vartheta_N) = -z_Q \sin\vartheta_N \tag{3.36}$$

is written instead of (3.20) for the source element located at z_Q (using the center distance R as depicted in Fig. 3.17), where the terms with ${z_Q}^2$ (and higher order terms) have already been omitted. The frequency range at which the latter can be neglected will be explained in more detail in Sect. 3.5.4 'Far Field Conditions'. Moreover, distances $R \gg l$ are the focal point here, where the decrease of amplitude per distance $1/r \approx 1/R$ is roughly the same for all source elements. In the far field (3.35) becomes

$$p_{far} = \frac{j\omega\varrho b}{4\pi R} e^{-jkR} \int_{-l/2}^{l/2} v(z_Q) e^{jkz_Q \sin\vartheta_N} dz_Q \,. \tag{3.37}$$

The expression in front of the integral depicts sound waves whose amplitudes are inversely proportional to the distance and thus decrease with increasing distance. The integral describes the power outflow of the space-dependent source and the field distribution along the different radiation angles. It should also be noted that the integral represents the Fourier transform of the source velocity.

The relationship between beam pattern and source array geometry and how they are influenced by design criteria is more clearly shown using a concrete example. First, consider the simplest case, where all loudspeakers are in phase and driven with the same amplitude $v(z_Q) = v_0 = \text{const.}$

3.5.1 One-dimensional piston

For the one-dimensional piston described by $v(z_Q) = v_0$, using its net volume velocity of $Q = v_0 bl$, (3.37) results in

$$p_{far} = \frac{j\omega\varrho Q}{4\pi R} e^{-jkR} \frac{1}{l} \int_{-l/2}^{l/2} [\cos(kz_Q \sin\vartheta_N) + j\sin(kz_Q \sin\vartheta_N)]\, dz_Q \,.$$

Due to symmetry, the imaginary part of the integral can be dropped and

$$p_{far} = p_Q \frac{\sin\left(k\frac{l}{2}\sin\vartheta_N\right)}{k\frac{l}{2}\sin\vartheta_N} = p_Q \frac{\sin\left(\pi\frac{l}{\lambda}\sin\vartheta_N\right)}{\pi\frac{l}{\lambda}\sin\vartheta_N} \tag{3.38}$$

is obtained, where p_Q is the sound pressure of the 'compact' source (with equal volume velocity)

$$p_Q = \frac{j\omega\varrho Q}{4\pi R} e^{-jkR} . \tag{3.39}$$

It is easier to discuss the result in (3.38) by first discussing the general characteristics of the so called 'sinc-function' $\sin(\pi u)/\pi u$ appearing on the right. The beam pattern is obtained by substituting $u = l/\lambda \sin\vartheta_N$ and taking the interval $|u| \leq l/\lambda$ from the sinc-function. Varying ϑ_N by $-90° \leq \vartheta_N \leq 90°$ spans the interval $-l/\lambda \leq u \leq l/\lambda$, resulting in the beam pattern.

The 'radiation function' $G(u) = \sin(\pi u)/\pi u$ which can be used to produce all beam patterns using the appropriate intervals, is shown in Figs. 3.18 and 3.19, where Fig. 3.18 shows the function itself (for subsequent calculations). Fig. 3.19 depicts a representation of the corresponding level. The main attributes of $G(u)$ are:

- for $u = 0$ the function value is $G(0) = 1$
- $G(u)$ consists of alternating positive and negative half-periods of a sine wave under the envelope function $1/u$
- the level representation shows a structure which consists of a main lobe (at the origin $u = 0$) followed by side lobes (with the centers $u = \pm(n + 0.5)$, $n = 1, 2, 3, \ldots$).

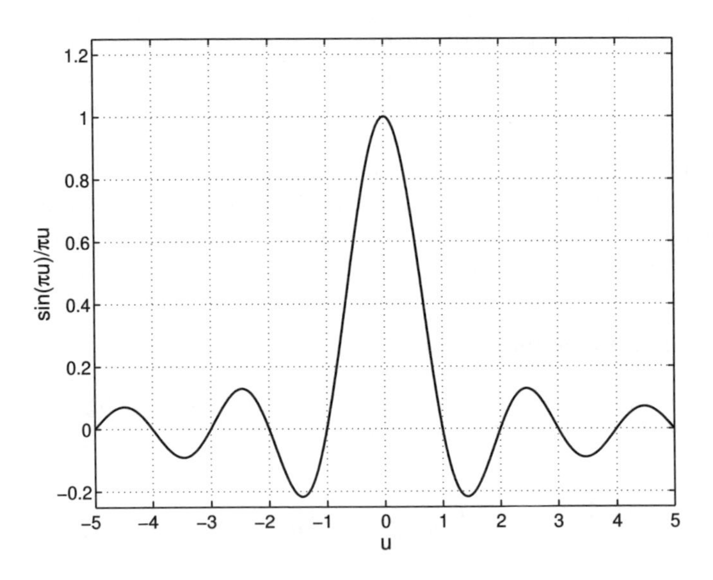

Fig. 3.18. Linear representation of the sinc-function

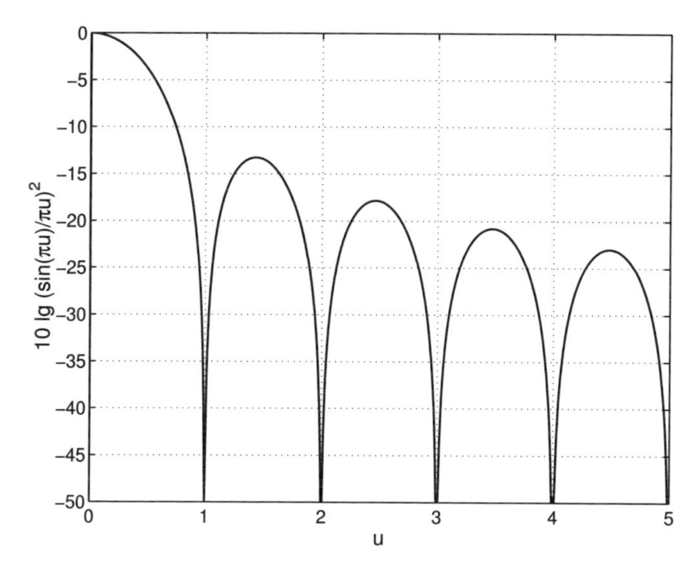

Fig. 3.19. Level (logarithmic) representation of the sinc-function

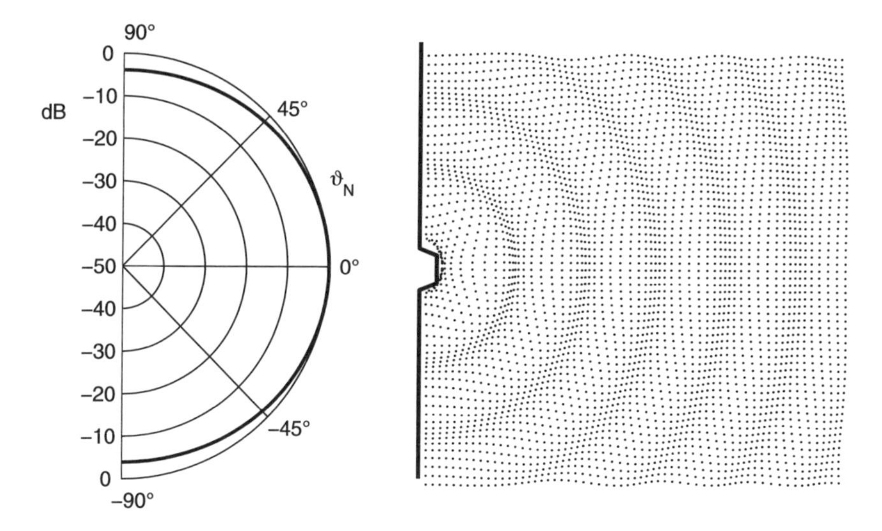

Fig. 3.20. (**a**) Beam pattern (left) and particle displacement (right) of a loudspeaker array for $l/\lambda = 0.5$

Some examples of the resulting beam patterns are given in Figs. 3.20 (**a**), (**b**), and (**c**) for different ratios of array length and wavelength. At low frequencies $l \ll \lambda$ ($l/\lambda = 0.5$, as shown in Fig. 3.20 (**a**)) a nearly omnidirectional radiation results using the interval $|u| < 0.5$ as shown in Fig. 3.19. A small

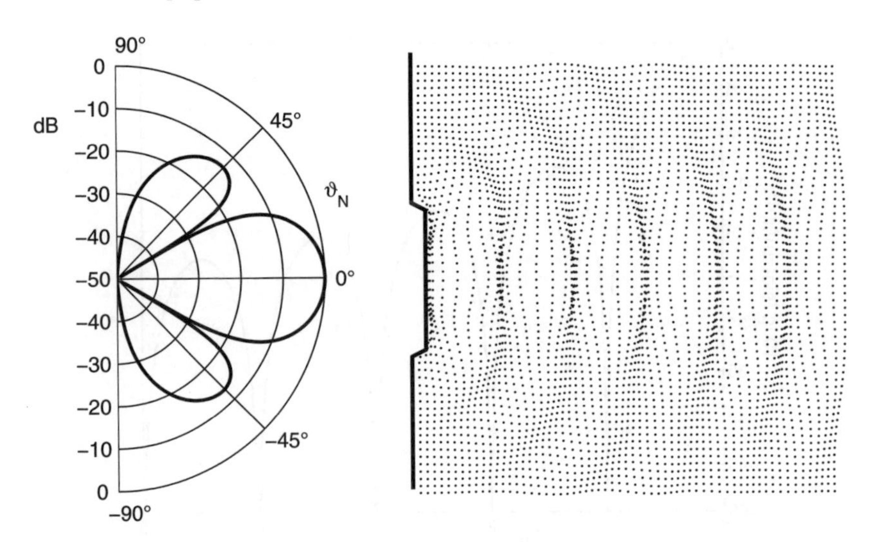

Fig. 3.20. (**a**) Beam pattern (left) and particle displacement (right) of a loudspeaker array for $l/\lambda = 2$

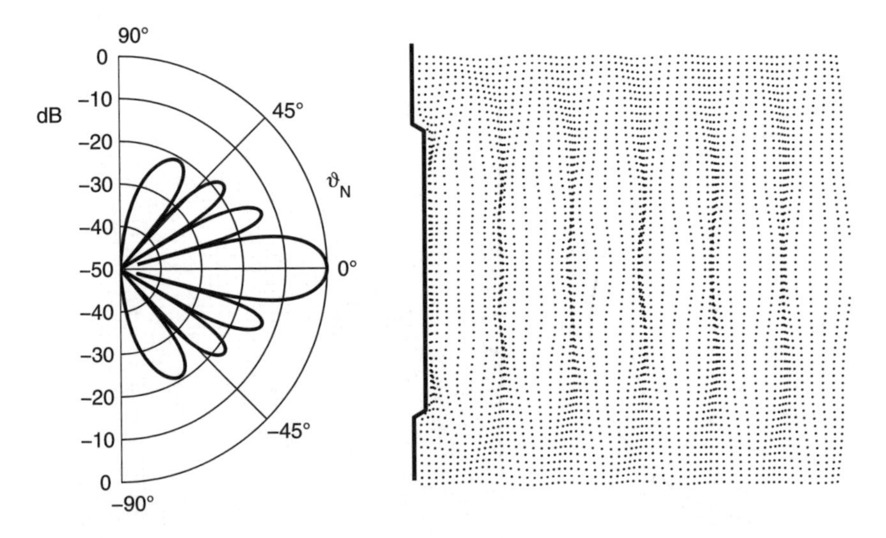

Fig. 3.19. (**c**) Beam pattern (left) and particle displacement (right) of a loudspeaker array for $l/\lambda = 4$

reduction of a few dB can only be detected at the edges $\vartheta_N \approx 90°$. In the case of the still considerably low 'mid' frequency range $l/\lambda = 2$ ($l = 1.5$ m would result in $\lambda = 0.75$ m and a frequency of $f = 227$ Hz) the interval $|u| < 2$ becomes relevant. The pattern shows a preference towards an angle of $\vartheta_N = 0°$, followed by a side lobe at $\vartheta_N = 45°$, which is 13.5 dB below the main lobe. The 'dip' is located at $l/\lambda \sin \vartheta_N = 1$, which is at $\sin \vartheta_N = 1/2$ or at $\vartheta_N = 30°$.

Finally, at high frequencies $l/\lambda = 4$ ($l = 1.5\,\mathrm{m}$ would result in $\lambda = 37\,\mathrm{cm}$ and a frequency of nearly 1000 Hz) the beam pattern shows a sharp grouping to the front, with a thin main lobe, followed by three side lobes on each side. The corresponding particle displacement is also shown in Figs. 3.20.

For practical purposes, the sound pressure of the main lobe which is $p(\vartheta_N = 0°) = p_0$ (see (3.38)) is mainly of interest. The radiated power in that case is only of marginal significance, hence we have left it out of the discussion at hand.

3.5.2 Formation of main and side lobes

Sometimes the formation of main and side lobes, as occurs in the one-dimensional piston, is undesirable. Take, for example, loudspeakers installed in an auditorium to positioned to address the audience seats while ensuring that the microphone in the room does not pick up any radiated sound in order to avoid feedback. In other implementations, certain areas are to be supplied with sound without interfering with other areas (such as during announcements at train stations). So, applications exist, where side lobes are disturbing and must be suppressed; the following will discuss a method to obtain this effect.

The underlying basic idea of side lobe suppression can be deduced from the simple relationship between time signals and their spectral composition. If the 'square-formed' step-function at the edges of the piston is transformed to the time domain, discontinuities, which would be audible as 'cracks', occur in time signal. These two signal discontinuities are responsible for the broadband characteristics of the square signal. As a matter of fact, the sinc function $sin(\pi u)/\pi u$ can be interpreted as the frequency spectrum of the time signal, where $u = fT$ (T = period of the signal). It is now fairly easy to reduce the higher frequencies which correspond to the side lobes: the discontinuity at the edges has to be transformed to a gradually smoother transition. A signal of the form $f(t) = \cos^2(\pi t/T)$ for $-T/2 < t < T/2$ certainly has a narrower bandwidth than a square wave signal. When transferred to the space representation of the speaker array, a velocity of $\cos^2$-form is expected to yield a suppression of the side lobes.

For this reason, the following discussion will examine the velocity gradient

$$v(z_Q) = 2v_0 \cos^2 \pi \frac{z_Q}{l} \ . \tag{3.40}$$

The factor of 2 causes a net volume velocity of

$$Q = b \int\limits_{-l/2}^{l_2} v(z_Q)\mathrm{d}z_Q = v_0 bl$$

at which is equal to that of the one-dimensional piston.

The radiated sound field in large distances can again be calculated using (3.37):

$$p_{far} = p_Q \frac{1}{l} \int\limits_{-l/2}^{l/2} 2\cos^2\left(\pi \frac{z_Q}{l}\right) e^{jkz_Q \sin\vartheta_N} dz_Q \,.$$

With the aid of

$$2\cos^2\alpha = 1 + \cos 2\alpha = 1 + \frac{1}{2}\left(e^{j2\alpha} + e^{-j2\alpha}\right)$$

this results in

$$p_{far} = p_Q \frac{1}{l} \int\limits_{-l/2}^{l/2} \left\{ e^{jkz_Q \sin\vartheta_N} + \frac{1}{2} e^{j\left(k\sin\vartheta_N + \frac{2\pi}{l}\right)z_Q} + \frac{1}{2} e^{j\left(k\sin\vartheta_N - \frac{2\pi}{l}\right)z_Q} \right\} dz_Q \,.$$

Using the symmetry properties again, the three integrals can easily be solved and

$$p_{far} = p_Q \left\{ \frac{\sin\left(\pi \frac{l}{\lambda}\sin\vartheta\right)}{\pi \frac{l}{\lambda}\sin\vartheta} + \frac{1}{2}\left[\frac{\sin\left(\pi\left(\frac{l}{\lambda}\sin\vartheta + 1\right)\right)}{\pi\left(\frac{l}{\lambda}\sin\vartheta + 1\right)} + \frac{\sin\left(\pi\left(\frac{l}{\lambda}\sin\vartheta - 1\right)\right)}{\pi\left(\frac{l}{\lambda}\sin\vartheta - 1\right)} \right] \right\} \tag{3.41}$$

is obtained for the sound pressure in the far field. It makes sense to first analyze the typical radiation function as was done in the last section:

$$G(u) = \frac{\sin(\pi u)}{\pi u} + \frac{1}{2}\frac{\sin(\pi(u+1))}{\pi(u+1)} + \frac{1}{2}\frac{\sin(\pi(u-1))}{\pi(u-1)} \tag{3.42}$$

The different beam patterns varying with frequency are simply obtained from intervals $l/\lambda \sin\vartheta_N$ taken from the function characteristics $G(u)$.

The principal characteristics of $G(u)$ are easily summarized. The three components – one un-shifted sinc function and two sinc functions, one shifted by 1 to the right and one shifted by 1 to the left and each multiplied by 1/2 – are depicted in Fig. 3.20. 'At first sight' the change in the sum as compared to the central sinc-function alone, which is related to the one-dimensional piston, becomes apparent:

- the width of the main lobe is doubled by the summation and
- the components in the area of the side lobes gradually add up to zero, the farther the distance of the side lobes is from the main lobe, the more the summation acts as a suppression of the side lobes.

These effects are again summarized in the graph of $G(u)$ in Fig. 3.21 (linear) and in Fig. 3.22 (level). The area under the side lobes in Fig. 3.21 is

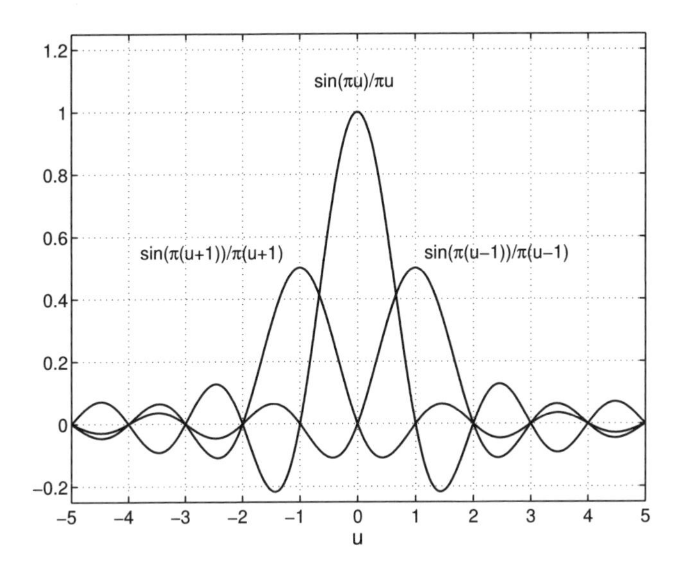

Fig. 3.20. The three components of the typical radiation function $G(u)$ in (3.42)

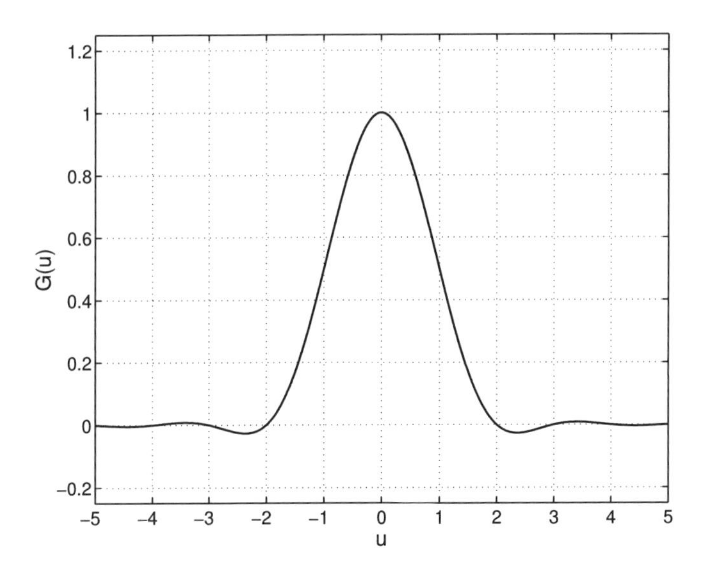

Fig. 3.21. Linear representation of $G(u)$

reduced significantly compared to the 'single' sinc function in Fig. 3.18. The consequences resulting for the levels can be seen, when comparing Fig. 3.22 with Fig. 3.19.

The description of the resulting beam patterns is actually superfluous: they consist of intervals of $G(u)$ drawn into a polar diagram. Nevertheless, examples are given in Figs. 3.23a, b and c using the same parameters as for

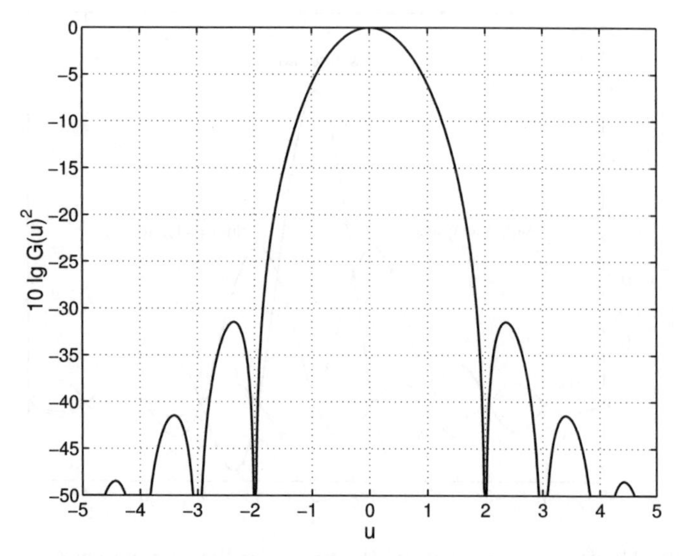

Fig. 3.22. Logarithmic (level) representation of $G(u)$

the piston in Fig. 3.20. In the mid and high frequency range the suppression of the side lobes can be observed clearly, while at the same time the main lobe is broadened. The corresponding particle displacement is also shown in Figs. 3.23a–c.

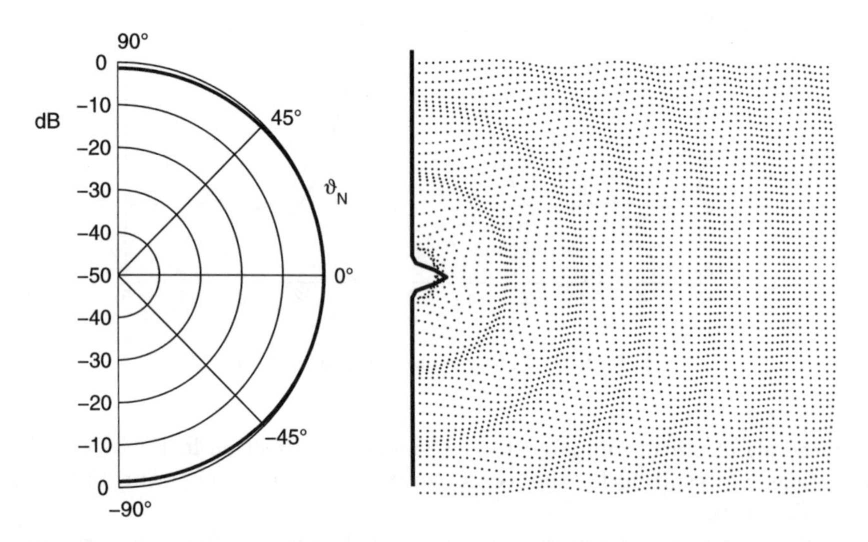

Fig. 3.23. (**a**) Beam pattern (left) and particle displacement (right) with formed lobes for $l/\lambda = 0.5$

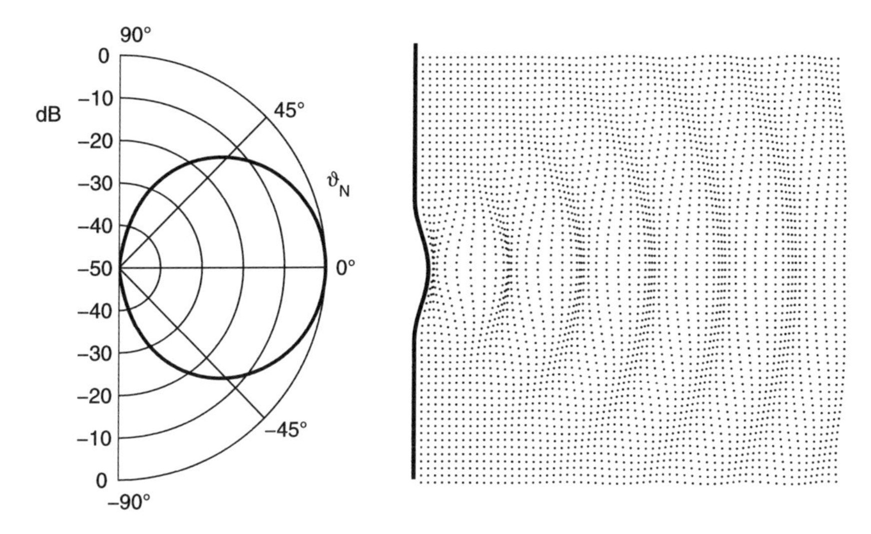

Fig. 3.23. (**b**) Beam pattern (left) and particle displacement (right) with formed lobes for $l/\lambda = 2$

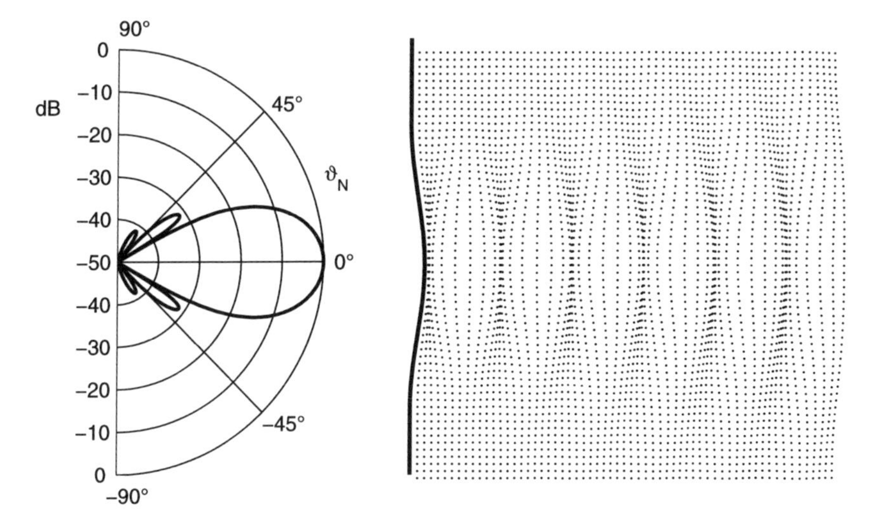

Fig. 3.23. (**c**) Beam pattern (left) and particle displacement with formed lobes for $l/\lambda = 4$

There are several other known spatial signal dependencies which result in a suppression of the side lobes. What they all have in common is that the reduction of the side lobes is always inherently linked to producing a broader main lobe. The differences between the individual signal characteristics play a minor role in the 'beam forming' of loudspeaker arrays. They are masked by the unavoidable tolerances and inaccuracies.

3.5.3 Electronic beam steering

It is of practical interest, however, if the main lobe can be steered into a specified direction with the aid of an electronic control device for each source element of the array. Electronic beam steering can be achieved as described in the following.

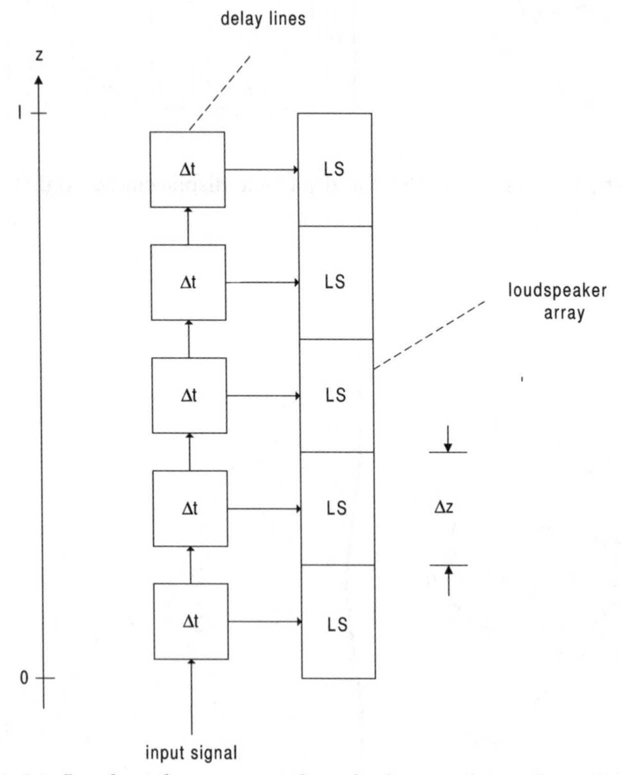

Fig. 3.24. Loudspeaker array with each element driven by a delay line

As a matter of fact, the setup which is needed to achieve the desired effect can be easily constructed. The voltages driving the loudspeakers need to be delayed in respect to one another by a time Δt, as shown in Fig. 3.24. The ith-loudspeaker (counting from the bottom) receives the signal $u(t - i\Delta t)$. The same applies for the velocity signals of the loudspeaker membranes. The velocity of a source located further up is a delayed version of the velocity of the first source, depending on its location. Altogether, the loudspeaker array, whose elements are driven by an array of equal time delays, acts as a wave guide. Ideally, given that the source elements are small, the velocity of the attached source in z can be described by

$$v(z,t) = f\left(t - \frac{z + l/2}{c_{\mathrm{s}}}\right) \tag{3.43}$$

where $f(t)$ is the velocity time characteristics at the lower far end of the array

$$v(-l/2, t) = f(t) \tag{3.44}$$

A spatial function is described in (3.43) which 'progresses with time'. The space-time characteristics represent that of a wave. The constant c_{s} in (3.43) is the propagation speed of the wave along the loudspeaker array; thus it is

$$c_{\mathrm{s}} = \Delta z / \Delta t, \tag{3.45}$$

where Δz is the distance between two elements (Fig. 3.24) and (as mentioned) Δt is the corresponding time delay.

Assuming that the speakers are solely driven by pure tones

$$f(t) = \mathrm{Re}\left\{v_1 \mathrm{e}^{j\omega t}\right\}$$

and using

$$f(z,t) = \mathrm{Re}\left\{\underline{v}(z)\mathrm{e}^{j\omega t}\right\}$$

we obtain the well-known wave characteristics

$$\underline{v}(z) = v_1 \mathrm{e}^{-j\frac{\omega}{c_{\mathrm{s}}}(z+l/2)} = v_0 \mathrm{e}^{-j\frac{\omega}{c_{\mathrm{s}}}z} \tag{3.46}$$

for the array velocity wave. As with any harmonic wave form, the ratio ω/c_{s} can be expressed in terms of the source wavenumber k_{s}

$$k_{\mathrm{s}} = \omega / c_{\mathrm{s}} \ . \tag{3.47}$$

This wavenumber k_{s} (and c_{s}) already contains the 'spatial period', the so called 'source wavelength' denoted by

$$\lambda_{\mathrm{s}} = \frac{c_{\mathrm{s}}}{f} = \frac{2\pi}{k_{\mathrm{s}}} \ . \tag{3.48}$$

It should be noted that so far, we have only discussed the properties of the composed source array, and not its radiation. c_{s}, k_{s} and λ_{s} are thus properties of the source which need to be distinguished from the properties of the medium c, k and λ (i.e. propagation speed, wave number and wavelength in air).

The radiation of the sound source defined above can be easily calculated with the aid of (3.37)

$$p_{far} = p_{\mathrm{Q}} \frac{1}{l} \int\limits_{-l/2}^{l/2} \mathrm{e}^{j(k \sin\vartheta_{\mathrm{N}} - k_{\mathrm{s}}) z_{\mathrm{Q}}} \mathrm{d}z_{\mathrm{Q}} = p_{\mathrm{Q}} \frac{\sin\left(\frac{kl}{2}\sin\vartheta_{\mathrm{N}} - \frac{k_{\mathrm{s}}l}{2}\right)}{\frac{kl}{2}\sin\vartheta_{\mathrm{N}} - \frac{k_{\mathrm{s}}l}{2}} \tag{3.49}$$

(again, $p_Q = j\omega\varrho b l v_0 e^{-jkR}/4\pi R$ is the pressure of the 'compact' source).

To analyze (3.49), the simplest method, as it was already outlined in the previous two sections, is to explain the 'typical' radiation function

$$G(u) = \frac{\sin(\pi(u - l/\lambda_s))}{\pi(u - l/\lambda_s)} . \tag{3.50}$$

The radiation is described by the interval $|u| < l/\lambda$. Equation (3.50) simply represents a sinc function shifted to the right by l/λ_s. An example is given in Fig. 3.25 for $l/\lambda_s = 2$. The crucial question is this: does the interval $|u| < l/\lambda$, which is 'visible' in the radiation pattern, include the maximum of the sinc function shifted to $u = l/\lambda_s$ or not?

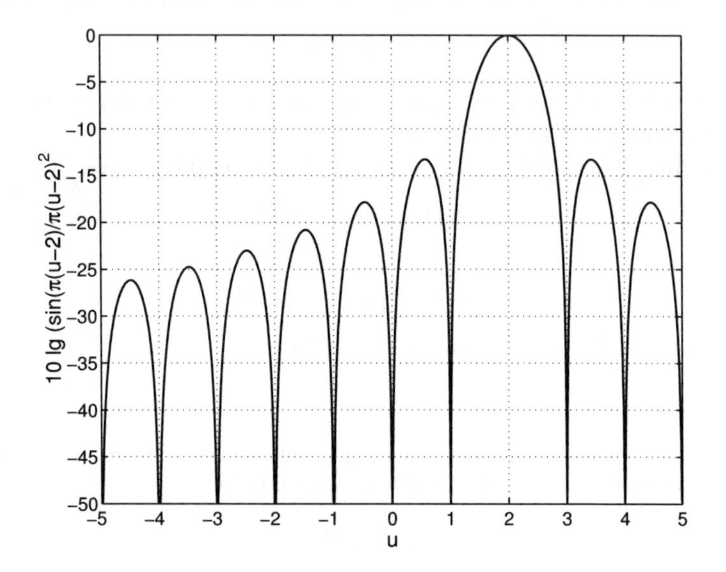

Fig. 3.25. $G(u)$ according to (3.50) for $l/\lambda_s = 2$

Source wavelength λ_s < air wavelength λ

If the source wavelength λ_s is smaller than the wavelength in air, the maximum value of the sinc function is located outside the visible interval $|u| < l/\lambda$. The radiation pattern is described here only by its side lobes; if λ_s is significantly less than λ, a very weak sound radiation results which is distributed over several equal-ranking side lobes – depending on the source wavelength l/λ .

The fact that short-wave sources produce a weak radiation will be explained in the last chapter of this book more detailed. The most interesting application of electronically steered loudspeaker arrays prove that only long wave arrays $\lambda_s > \lambda$ and thus $c_s > c$ can be implemented in practice.

Source wavelength $\lambda_s >$ air wavelength λ

The main lobe of the long wave source $\lambda_s > \lambda$ in the typical radiation function $G(u)$ shifted by $u = l/\lambda_s$ is always located in the interval visible to the beam pattern. The main radiation angle ϑ_M using

$$\frac{l}{\lambda} \sin \vartheta_M = \frac{l}{\lambda_s}$$

results in

$$\sin \vartheta_M = \frac{\lambda}{\lambda_s} = \frac{c}{c_s} . \tag{3.51}$$

Thus, the loudspeaker array defined in the beginning has the same main radiation angle direction for all frequencies. The width of the main lobe and the number of side lobes only depend on the source length expressed in air wavelengths. If $l/\lambda \ll 1$, the steered beam pattern is nearly omnidirectional (Fig. 3.26(**a**)); for the mid frequencies (Fig. 3.26(**b**)) and high frequencies (Fig. 3.26(**c**)) the beam pattern exists in larger portions of $G(u)$. The corresponding particle displacement is also shown in Figs. 3.26a–c.

Beam formation (as described in the previous section) and beam steering can of course be combined.

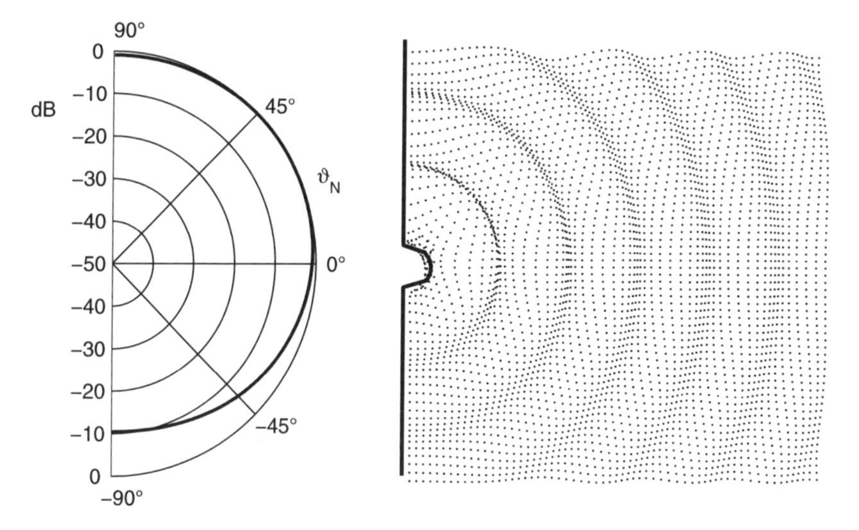

Fig. 3.26. (**a**) Steered beam pattern (left) and particle displacement for $\lambda_s/\lambda = 2$ and $l/\lambda = 0.5$

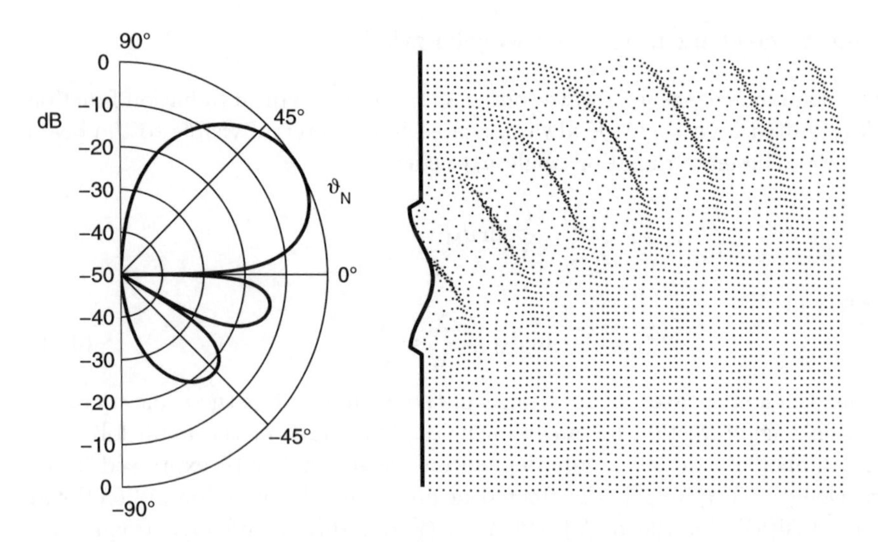

Fig. 3.26. (**b**) Steered beam pattern (left) and particle displacement for $\lambda_s/\lambda = 2$ and $l/\lambda = 2$

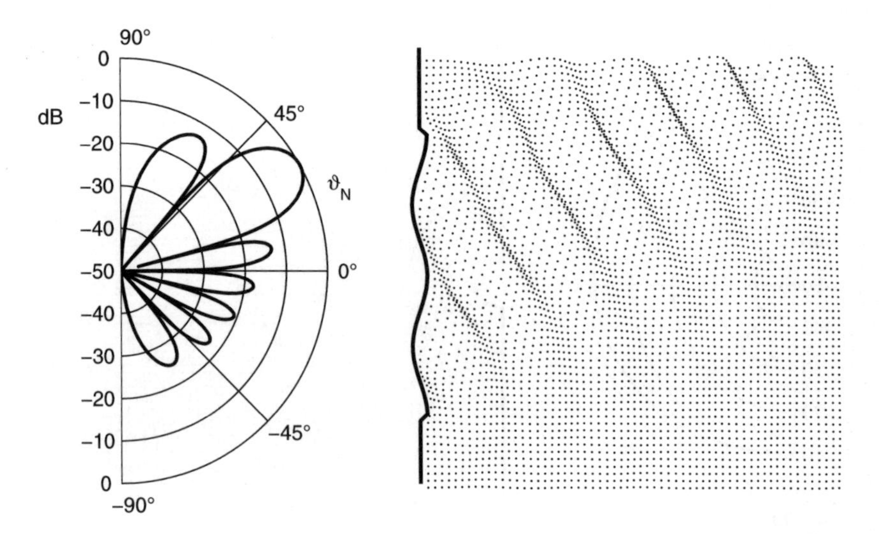

Fig. 3.26. (**c**) Steered beam pattern (left) and particle displacement (right) for $\lambda_s/\lambda = 2$ and $l/\lambda = 4$

3.5.4 Far-field conditions

The discussions outlined in the previous sections were performed in the 'far-field'. In order to maintain the train of thought in the theoretical discussion of sound propagation and radiation, we have deferred until now a more detail inquiry into the pre-existing conditions under which a far-field can exist. The answer to this question of far-field conditions is not only a matter of cognitive completeness but must be verified by experiment, as has the physical phenomena described in the previous sections. We proceed with a discussion of the measurement parameters that apply to the far-field.

Firstly, it has to be reiterated that an essential condition for far field approximation (3.37) was the assumption that all sources have about the same decrease of amplitude per decrease of distance. It directly follows from this assumption, that the distance R between the source elements and the field point has to be large compared to the array length l. The first far field condition thus is given by

$$R \gg l\,. \tag{3.52}$$

Secondly, it has to be reminded that the expression defining the phase $r-R$ (as a function of the location z_Q of a source element) was approximated by its linear term only. In order to explore all the conditions eliminating gross errors, it is necessary to approximate to the quadratic term and examine its effects. For the triangle formed by R, r and z_Q, as shown in Fig. 3.17,

$$r(z_\mathrm{Q}) \approx \sqrt{R^2 + z_\mathrm{Q}^2 - 2Rz_\mathrm{Q}\cos\vartheta}$$

applies. The customary Taylor series (truncated at the quadratic term including z_Q^2) is

$$r \approx r(0) + z_\mathrm{Q}\left.\frac{\mathrm{d}r}{\mathrm{d}z_\mathrm{Q}}\right|_{z_\mathrm{Q}=0} + \frac{z_\mathrm{Q}^2}{2}\left.\frac{\mathrm{d}^2 r}{\mathrm{d}z_\mathrm{Q}^2}\right|_{z_\mathrm{Q}=0}.$$

The coefficients, calculated by the derivatives, are $r(0) = R$,

$$\frac{\mathrm{d}r}{\mathrm{d}z_\mathrm{Q}} = \frac{z_\mathrm{Q} - R\cos\vartheta}{r}, \quad \text{hence} \quad \left.\frac{\mathrm{d}r}{\mathrm{d}z_\mathrm{Q}}\right|_{z_\mathrm{Q}=0} = -\cos\vartheta\,,$$

and

$$\frac{\mathrm{d}^2 r}{\mathrm{d}z_\mathrm{Q}^2} = \frac{r - (z_\mathrm{Q} - R\cos\vartheta)\frac{\mathrm{d}r}{\mathrm{d}z_\mathrm{Q}}}{r^2}, \quad \text{hence} \quad \left.\frac{\mathrm{d}^2 r}{\mathrm{d}z_\mathrm{Q}^2}\right|_{z_\mathrm{Q}=0} = \frac{R - R\cos^2\vartheta}{R^2} = \frac{\sin^2\vartheta}{R}\,,$$

resulting in

$$r \approx R - z_\mathrm{Q}\cos\vartheta + \frac{z_\mathrm{Q}^2\sin^2\vartheta}{2R}\,.$$

In a second order approximation the phase function is given by

$$\mathrm{e}^{-jkr} = \mathrm{e}^{-jkR}\mathrm{e}^{jkz_\mathrm{Q}\cos\vartheta}\mathrm{e}^{-j\frac{kz_\mathrm{Q}^2\sin^2\vartheta}{2R}}\,.$$

To approximate the exponential function with z_Q^2 as the argument by 1 (as implied for the far-field approximation (3.37)), the condition

$$\frac{kz_Q^2 \sin^2 \vartheta}{2R} \ll \pi/4$$

has to be met for all ϑ and z_Q. Because the maximum of z_Q is $l/2$ and the maximum of $\sin\vartheta$ is 1, the condition is met when

$$\frac{2\pi}{\lambda}\frac{l^2}{4}\frac{1}{2R} \ll \pi/4,$$

or in simplified form,

$$\frac{l}{\lambda} \ll \frac{R}{l}\,. \tag{3.53}$$

Equation (3.53) defines the second condition which has to be met to make the far-field approximation (3.37) usable. Allowing a phase error of $\pi/4$, '$\ll$' in (3.53) can be replaced by '$<$'.

Thirdly, the term 'far-field' is based on the pre-existing condition that the impedance $z = p/v_R$ (v_R = radial component of the velocity) must be the same as the impedance of the plane progressive wave ϱc. Thus, 'far-field' also implies that the intensity can be determined solely based on pressure measurements. By using

$$v_\mathrm{R} = \frac{j}{\omega\varrho}\frac{\partial p}{\partial R}$$

in (3.37),

$$\frac{p}{v_\mathrm{R}} = \frac{\varrho c}{1 + \frac{j}{kR}}$$

is obtained. The third condition for the far-field requires, using

$$R \gg \lambda\,, \tag{3.54}$$

that the distance R has to be large compared to the wavelength. Because $kR \approx 6.3$ for $R = \lambda$, the deviation from ϱc is small, if '$\gg$' is replaced by '$>$' in (3.54), too.

Summarizing again, it can be stated that the field point R is in the far-field if all three conditions

$$R \gg l \tag{3.52}$$

$$\frac{R}{l} \gg \frac{l}{\lambda} \tag{3.53}$$

$$R \gg \lambda \tag{3.54}$$

are met. Usually, a tolerable error is acceptable in (3.53) and (3.54) if the pre-existing condition 'much larger' is replaced by the less rigorous constraint 'larger'.

The meaning of (3.52) through (3.54) for the admitted measurement range quickly becomes clear, if once assumes a given source in a constant distance

R. The latter is specified in such a way that according to (3.52) $R \gg l$ is fulfilled. For instance, let $R = 5\,\mathrm{m}$ and $l = 1\,\mathrm{m}$. Equation (3.53) then states that with increasing frequency, the far-field conditions are not fulfilled by a certain limiting frequency. In the given example $R = 5\,\mathrm{m}$ and $l = 1\,\mathrm{m}$, (3.53) is only valid for $\lambda > 20\,\mathrm{cm}$, thus for frequencies below 1700 Hz. Equation (3.54), on the other hand, states that the wavelength has to be smaller than 5 m; the frequency range for the far-field therefore starts $f = 68\,\mathrm{Hz}$ and higher. Generally (3.53) and (3.54) define the band limits, within which far-field conditions are met. Equation (3.52) is related to a geometrical condition.

The far-field conditions (3.52) to (3.54) can easily be transferred to the sources in the following sections, where l corresponds to the largest of the source dimensions then.

3.6 Sound radiation from plane surfaces

One is often interested in the sound radiation of large, vibrating surfaces, like walls, ceilings and windows in buildings and plates, which could represent, for example, the chassis of a machinery or vibrating parts of vehicles, aircrafts and trains. There is a large number of examples of sources which consist of prolated plane surfaces, providing reason enough to expand the discussion of one-dimensional sources to two-dimensional sources.

The method used is exactly the same as in section 3.5: the vibrating surface is split into infinitesimally small volume velocity sources, whose sound pressures are summed through integration at the field point. One major difference to the one-dimensional, narrow array has to be kept in mind. The sound field due to a single source element is reflected at the larger surface of the plane. This becomes clear, when imagining a vibrating element of the surface in an otherwise rigid, motionless surface, where the spherical waves of the infinitesimal source are totally reflected. The reflection of the sound field of one individual source element at all the others was neglected from our previous discussion of loudspeaker arrays. This omission was justified, however, because the array width b was always assumed to be small compared to the wavelength and thus consisting of a non-reflecting array.

When dealing with sound radiation of extended surfaces, the reflection at the source surface itself has to be taken into account. Because the reflected field of surfaces of finite expansion depends on its nature and its dimensions as well as on the location of the small volume velocity source element of interest, the discussion of the radiating surface of finite expanse in the 'free-field' would be far too complicated. If infinite surfaces are considered, these dependencies disappear. Independent of its location, each source element experiences the same total reflection at the infinite surface. It will be assumed in the following that the velocity $v_z(x, y)$ pointing in the z-direction is known and given on the whole surface $z = 0$. This does not, on the other hand, exclude the discussion of vibrating surfaces of finite expanse. They are only regarded as a part of an otherwise motionless rigid, infinite baffle with $v_z = 0$.

The contribution of one source element in the baffle to the sound field can be determined with the aid of a simple theoretical experiment. Imagine a small volume velocity source placed in front of a reflector $z = 0$ in a distance z. The reflected sound field can be regarded as originating from a mirror source at the point $-z$ behind the baffle. The total sound field is therefore produced by sources at z and $-z$. If the original source is moved back to the baffle, it can be seen that the reflection causes a doubling of the source or a doubling of the sound pressure respectively. Yet the contribution $\mathrm{d}p$ of the infinitesimally small volume velocity source with the volume velocity $v(x_Q, y_Q)\mathrm{d}x_Q\mathrm{d}y_Q$ is given analogue to (3.34)

$$\mathrm{d}p = \frac{j\omega\varrho v(x_Q, y_Q)}{2\pi r} \mathrm{e}^{-jkr} \mathrm{d}x_Q \mathrm{d}y_Q \, , \tag{3.55}$$

where r describes the distance between the source center and the field point

$$r = \sqrt{(x - x_Q)^2 + (y - y_Q)^2 + z^2} \, . \tag{3.56}$$

The total sound pressure due to all the source elements is thus

$$p(x, y, z) = \frac{j\omega\varrho}{2\pi} \int\limits_{-\infty}^{\infty} \int\limits_{-\infty}^{\infty} v(x_Q, y_Q) \frac{\mathrm{e}^{-jkr}}{r} \mathrm{d}x_Q \mathrm{d}y_Q \, . \tag{3.57}$$

Equation (3.57) is known as the 'Rayleigh integral'. As already mentioned, it refers to the velocities given in the whole $z = 0$-plane. For vibrating surfaces of finite expanse, the Rayleigh-integral assumes that these are a part of a rigid, motionless surface. Equation (3.57) can therefore only be applied proviso and with restrictions to the radiation of vibrating surfaces such as, for example, train wheels, free and un-boxed vibrating loudspeakers, etc. The Rayleigh integral is still a useful approximation for the actual sound field in such cases, where the dimensions of the radiating surface in the high frequency range are already large compared to the wavelength. At low frequencies, the short circuit of the mass between the front and the rear of the surface ($z > 0$ and $z < 0$) plays a major role in the radiation, and this short circuit is prevented by the baffle implicitly contained in (3.57). The Rayleigh integral will therefore lead to wrong predictions for free sources 'without baffle' at low frequencies.

The Rayleigh integral can be controlled analytically only in rare cases (an example with a closed solution of (3.57) at least for the center axis around $z = 0$ is given in what follows). In contrast, a simple and lucid far-field approximation can again be deduced from (3.57). The spherical coordinates used to describe the field point are given as is generally known as

$$\begin{aligned} x &= R \sin\vartheta \cos\varphi \\ y &= R \sin\vartheta \sin\varphi \\ z &= R \cos\vartheta \end{aligned}$$

(R: distance of the point (x, y, z) from the origin,
ϑ: angle between the z-axis and the ray between origin and field point
φ: angle between the x-axis and the ray projected to the $z = 0$-plane).

First of all, the assumption of a surface of finite extent is required for the far-field approximation, indicated here by an integral with finite limits

$$p(x, y, z) = \frac{j\omega\varrho}{2\pi} \int_{-l_y/2}^{l_y/2} \int_{-l_x/2}^{l_x/2} v(x_Q, y_Q) \frac{e^{-jkr}}{r} dx_Q dy_Q \, . \tag{3.58}$$

As discussed in Sect. 3.5.4, the far-field conditions are assumed as

$$R \gg l, \quad R \gg \lambda \quad \text{and} \quad R/l \gg l/\lambda$$

($l = \max(l_x, l_y)$). Again, $1/r \approx 1/R$ (R = distance of the center points) is assumed in the far-field and can be written in front of the integral. For r

$$\begin{aligned} r^2 &= (x - x_Q)^2 + (y - y_Q)^2 + z^2 = x^2 + y^2 + z^2 + x_Q^2 + y_Q^2 - 2(xx_Q + yy_Q) \\ &\approx R^2 - 2(xx_Q + yy_Q) \, , \end{aligned}$$

is obtained, because x_Q^2 and y_Q^2 can be neglected under far field conditions. Using spherical coordinates, we obtain

$$r^2 - R^2 = (r - R)(r + R) = -2R(x_Q \sin\vartheta \cos\varphi + y_Q \sin\vartheta \sin\varphi) \, ,$$

or (with $r + R = 2R$ neglecting small quadratic terms),

$$r - R = -(x_Q \sin\vartheta \cos\varphi + y_Q \sin\vartheta \sin\varphi) \, .$$

The far-field approximation for the radiation of plane surfaces therefore becomes

$$\begin{aligned} p_{far}(R, \vartheta, \varphi) = {} & \frac{j\omega\varrho}{2\pi R} e^{-jkR} \\ & \int_{-l_y/2}^{l_y/2} \int_{-l_x/2}^{l_x/2} v(x_Q, y_Q) e^{jk(x_Q \sin\vartheta \cos\varphi + y_Q \sin\vartheta \sin\varphi)} dx_Q dy_Q \end{aligned} \tag{3.59}$$

(the double integral on the right represents the twofold Fourier transform of the source velocity).

For most source models of interest, (3.59) can easily be solved and can be reduced to products of beam patterns already discussed for loudspeaker arrays. For a square piston, for example, with $v = v_0$ for $|x| \le l_x/2$, $|y| \le l_y/2$ and $v = 0$ everywhere else (with $Q = v_0 l_x l_y$),

$$p_{far} = \frac{j\omega\varrho Q}{4\pi R} e^{-jkR} \frac{\sin\left(\pi \frac{l_x}{\lambda} \sin\vartheta \cos\varphi\right)}{\pi \frac{l_x}{\lambda} \sin\vartheta \cos\varphi} \frac{\sin\left(\pi \frac{l_y}{\lambda} \sin\vartheta \cos\varphi\right)}{\pi \frac{l_y}{\lambda} \sin\vartheta \cos\varphi}$$

is obtained. Similar results are obtained when assuming sources which have a wave-formed shape.

It particularly follows from (3.59) at low frequencies $kl_x \ll 1$ and $kl_y \ll 1$ that

$$p_{far} \simeq \frac{j\omega\varrho}{4\pi R} \mathrm{e}^{-jkR} \int_{-l_y/2}^{l_y/2} \int_{-l_x/2}^{l_x/2} v(x_Q, y_Q) \mathrm{d}x_Q \mathrm{d}y_Q \, . \tag{3.60}$$

In a first order approximation, the sound field is therefore proportional to the net volume velocity of the source. For sources with a wave-formed shape, minor details possibly determine the net volume velocity and thus the radiation, as discussed earlier.

Finally, it should be mentioned that by definition the impedance ϱc is present in the far-field. Hence, the intensity is given by

$$I = \frac{1}{2\varrho c} |p_{far}|^2 \, . \tag{3.61}$$

The sound power can therefore be calculated by integrating over a semi-sphere:

$$P = \frac{1}{2\varrho c} \int_0^{\pi/2} \int_0^{2\pi} |p_{far}|^2 R^2 \sin\vartheta \mathrm{d}\varphi \mathrm{d}\vartheta \, . \tag{3.62}$$

3.6.1 Sound field on the axis of a circular piston

In arriving at the far-field approximation in the previous sections, great importance was attached to the conditions applying for it. It is certainly of interest to assess what effects can be expected if the far-field conditions are not met. The following discussions will answer this question by means of an example. The sound pressure on the center axis of a circular piston (velocity $v_0 = \text{const.}$ at $r < b$) is calculated from the Rayleigh integral (3.57) (see Fig. 3.27).

Expressed in polar coordinates

$$\begin{aligned} x_Q &= R_Q \cos\varphi_Q \\ y_Q &= R_Q \sin\varphi_Q \\ \mathrm{d}x_Q \mathrm{d}y_Q &= \mathrm{d}S = R_Q \mathrm{d}R_Q \mathrm{d}\varphi_Q \end{aligned}$$

equation (3.57) becomes

$$p = \frac{j\omega\varrho v_0}{2\pi} \int_0^{2\pi} \int_0^{b} \frac{\mathrm{e}^{-jkr}}{r} R_Q \mathrm{d}R_Q \mathrm{d}\varphi_Q \, , \tag{3.63}$$

where $r = \sqrt{R_Q^2 + z^2}$ represents the distance between the source element R_Q and the field point on the z-axis. The radius of the piston is denoted by b. Equation (3.63) is synonymous with

$$p = j\omega\varrho v_0 \int_0^b \frac{e^{-jk\sqrt{R_Q^2+z^2}}}{\sqrt{R_Q^2+z^2}} R_Q \mathrm{d}R_Q \, . \tag{3.64}$$

With aid of the substitution

$$u = \sqrt{R_Q^2+z^2}$$
$$\mathrm{d}u = \frac{R_Q \mathrm{d}R_Q}{\sqrt{R_Q^2+z^2}}$$
$$R_Q = 0 \; : \; u = z$$
$$R_Q = b \; : \; u = \sqrt{b^2+z^2}$$

equation (3.63) can easily be solved, as in

$$p = j\omega\varrho v_0 \int_z^{\sqrt{b^2+z^2}} e^{-jku}\mathrm{d}u = \varrho c v_0 \left(e^{-jkz} - e^{-jk\sqrt{b^2+z^2}}\right) \, ,$$

or

$$p = \varrho c v_0 e^{-j2\pi z/\lambda} \left\{1 - e^{-j2\pi\left(\sqrt{(b/\lambda)^2+(z/\lambda)^2} - z/\lambda\right)}\right\} \, . \tag{3.65}$$

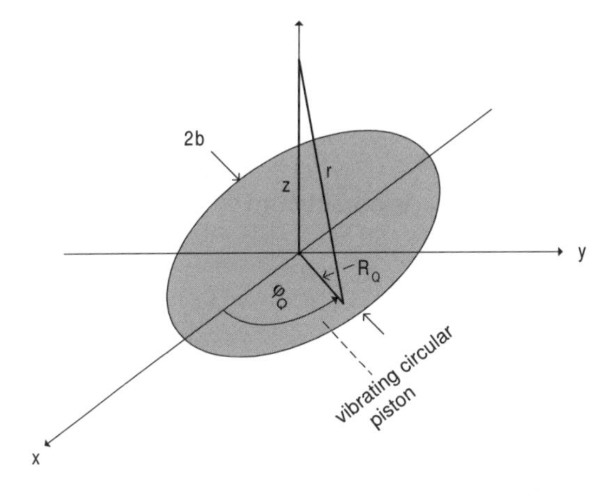

Fig. 3.27. Location of the circular piston in a polar coordinate system with a description of the quantities

Obviously, the sound pressure can have zeros on the z-axis. The location of the zeros $p(z_0) = 0$ is obtained by

$$\sqrt{(b/\lambda)^2 + (z_0/\lambda)^2} - z_0/\lambda = n \, .$$

It follows that

$$(n + z_0/\lambda)^2 = n^2 + (z_0/\lambda)^2 + 2nz_0/\lambda = (b/\lambda)^2 + (z_0/\lambda)^2$$

or

$$z_0/\lambda = \frac{(b/\lambda)^2 - n^2}{2n} \,. \tag{3.66}$$

In (3.66) n passes through the values $n = 1, 2, 3, \ldots$ as long as positive values for z_0/λ are obtained. For example,

- for $b/\lambda = 1$ a single zero $z_0/\lambda = 0 (n = 1)$ at the center of the piston itself is obtained and
- $b/\lambda = 4$ results in the zeros $z_0/\lambda = 7.5$ $(n = 1)$, $z_0/\lambda = 3$ $(n = 2)$, $z_0/\lambda = 1.1667$ $(n = 3)$ and $z_0/\lambda = 0$ $(n = 4)$

(some examples of the axial level distribution are shown in Fig. 3.28). The zero with the largest distance from the source for $n = 1$ is approximately given by

$$z_{max}/\lambda \approx \frac{1}{2}(b/\lambda)^2 \,. \tag{3.67}$$

In the range $z < z_{max}$ axial nodes $p = 0$ are present, with a number of (approximately) b/λ.

Now a sound field structure with pressure nodes and antinodes in the direction of the measured distance (the z-direction, in that case) is a contradiction to the assumption of a far-field. As (3.60) shows, the far-field can be considered as the range of distances between source elements, where the only dependency on R is the amplitude decrease by $1/R$ (and therefore the level falls off at 6 dB per doubling of distance). According to (3.60) the structure of alternating minima and maxima along the distance axis is impossible in the far-field.

Thus, the range $z < z_{max}$ is outside of the far-field. Only for

$$z \gg z_{max} = \frac{1}{2}\frac{b^2}{\lambda}$$

or for

$$\frac{b}{\lambda} \ll \frac{z}{b} \tag{3.68}$$

'far-field conditions' apply. Equation (3.68) is identical to (3.53) which was derived earlier.

On the other hand, based on the preceding discussions, the predictable effects can be shown if too small of a measurement distance z is selected, violating (3.68). The measured circumferential level can show sound pressure minima which are actually present for a particular distance but do not appear at other, larger distances. The measured beam pattern is therefore untypical for other distances and thus quite meaningless. Only beam patterns measured

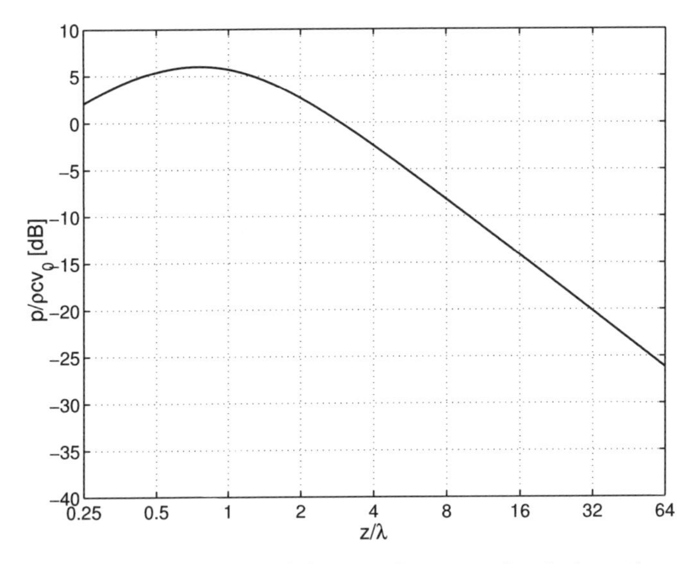

Fig. 3.28. (**a**) Space dependence of the sound pressure level along the center axis z in front of the circular piston for $b/\lambda = 0.5$

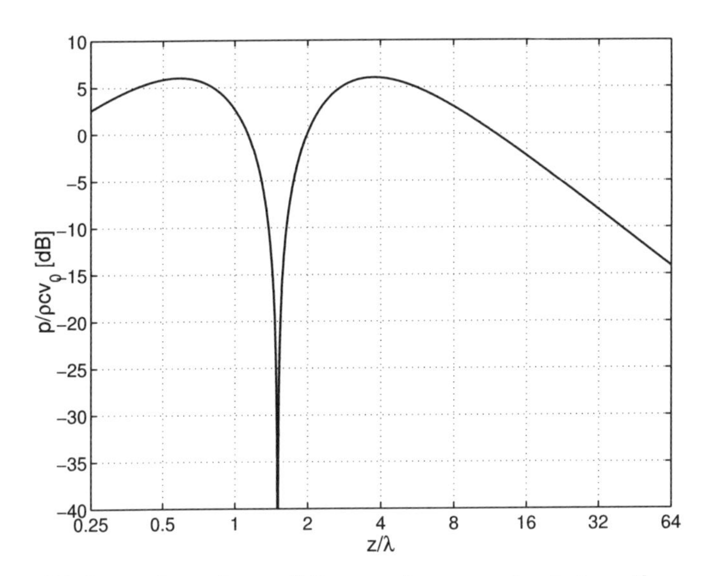

Fig. 3.28. (**b**) Space dependence of the sound pressure level along the center axis z in front of the circular piston for $b/\lambda = 2$

in the far-field no longer fluctuate, and this can be understood as being the aim of the far-field definitions.

Finally, the far-field approximation related to (3.64) is derived. If $z \gg b$ is implied, then

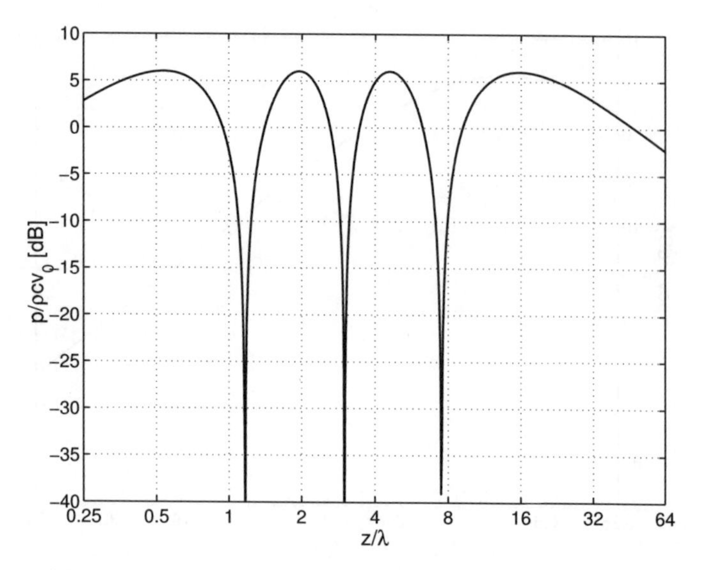

Fig. 3.28. (c) Space dependence of the sound pressure level along the center axis z in front of the circular piston for $b/\lambda = 4$

$$\sqrt{z^2+b^2} = z\sqrt{1+(b/z)^2} \approx z\left(1+\frac{1}{2}\frac{b^2}{z^2}\right) = z + \frac{1}{2}\frac{b^2}{z^2}$$

is valid and the far field approximation becomes

$$p_{far} \approx \varrho c v_0 \mathrm{e}^{-j2\pi z/\lambda}\left\{1-\mathrm{e}^{-j\pi\frac{b^2}{\lambda z}}\right\} .$$

If, according to (3.68), $b^2 \ll \lambda z$, using $\mathrm{e}^{-jx} \approx 1 - jx$ can be applied to determine the far-field pressure

$$p_{far} = \frac{j\varrho c v_0 b^2\pi}{\lambda z}\mathrm{e}^{-j2\pi z/\lambda} = \frac{j\omega\varrho\pi b^2 v_0}{2\pi z}\mathrm{e}^{-j2\pi z/\lambda} . \tag{3.69}$$

Equation (3.69) is identical to the result in (3.59) for $\vartheta = 0$ (which denotes the z-axis).

3.7 Summary

The sound pressure level decreases by $6\,dB$ with the doubling of the distance for point sources, and by $3\,dB$ for line sources. Small volume velocity sources create a sound field with spheres as wave fronts. Combined sources result in interferences which create beam patterns. The latter may change according to the distance from the source. Only in the far-field which is defined by the three conditions $r >> l$, $r >> \lambda$ and $r/l >> l/\lambda$ can the beam pattern remain uninfluenced by the distance to the source. In the case of a dipole, the beam pattern consists in a (rotating) eight. In-phase vibrating sources with large surface areas, such as loudspeaker arrays, etc., create beam patterns, which, depending on the size and wavelength, consist of a main lobe followed by side lobes. This structure can be specifically altered by weighting the elements. By introducing time delays to loudspeaker control signals, the main lobe is shifted to some desired angle. Such wave-formed radiators produce sound fields along their edges only, given that the source's wavelength λ_s is shorter than that of the surrounding medium λ, or $\lambda_s < \lambda$. For sources with long wavelengths, $\lambda_s > \lambda$, the entire surface of the source is part and parcel to the entire acoustical process. The directivity of the main lobe is characterized by a plane wave radiated into an oblique direction which can be derived from $sin\, \vartheta = \lambda/\lambda_s$.

3.8 Further reading

To deepen the contents of this chapter, it is particularly recommended to read Chap. 5 in the book (in German) by Meyer and Neumann "Physikalische und Technische Akustik" (Vieweg Verlag, Braunschweig 1967) and Chap. 7 on sound radiation from structures in the book by Lothar Cremer and Manfred Heckl "Structure Borne Sound", translated by B.A.T. Petersson (Springer, Berlin and New York 2004).

3.9 Practice exercises

Problem 1

A sound pressure level of 50 $dB(A)$ has already been registered at an emission control center coming from a neighboring factory. A pump is planned for installation 50 m away from the emission control center. The pump can only emit up to $L_P = 53.3\, dB(A)$, so that the overall sound level does not exceed 55 $dB(A)$. How high can the A-weighted sound power level of the pump be in this case? See Problem 1 from the Practice Exercises in Chapter 1.

Problem 2

A small valve emits a volume flow $Q(t)$ according to the sketch in Figure 3.29. Find the time-dependent sound pressure square at 10 m distance for $Q_0 = 1\, m^3/s$ and for the signal slope times of $T_F = 0.01\, s$, $0.0316\, s$ and $0.1\, s$. How high is the sound pressure level here?

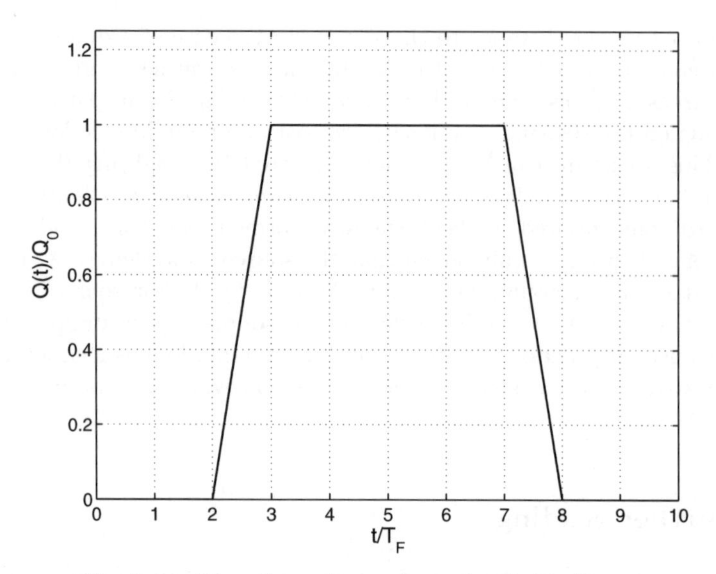

Fig. 3.29. Time-dependent volume flow in Problem 2

Problem 3

A suspended loudspeaker can produce a considerable decibel level of $100\,dB$ heard from $2\,m$ away. How great is the emitted sound power and power level? How great is the efficiency rate if the electrical power input is $50\,W$?

Problem 4

A loudspeaker array has a space-dependent particular velocity as shown in the diagram in Figure 3.30(l=array length). Find the beam patterns for all frequencies.

Problem 5

Far-field measurements are to be taken of a sound source extending to $1\,m$ ($50\,cm$, $2\,m$). To satisfy the geometric requirements $R >> l$, the measurements are taken at a distance of $R = 5\,l$, or $R = 5\,m$ $(2,5\,m,\ 10\,m)$.

a) If in the remaining far-field conditions, the condition '>>' ('much larger than') is replaced with 'five times greater than': in which frequency range can the far-field measurements be taken?
b) Otherwise, if the '>>' ('much larger than') condition is replaced by 'twice as large as': in which frequency range can the far-field measurements be taken?

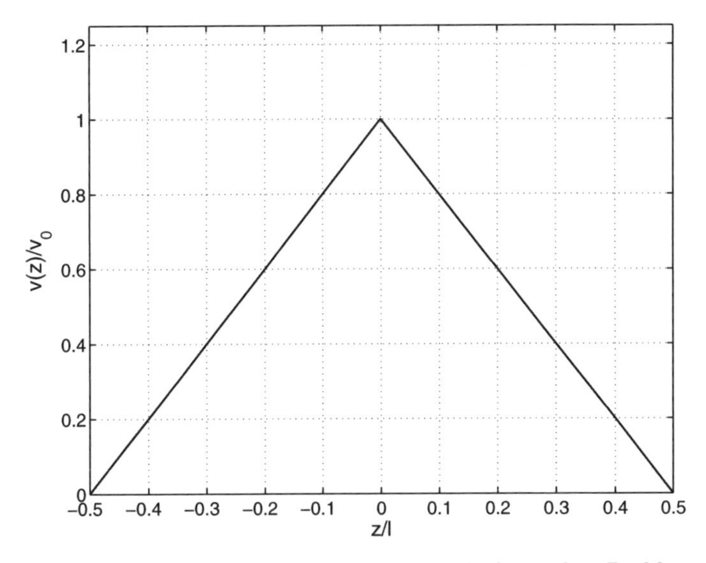

Fig. 3.30. Space-dependent particular speed of sound in Problem 4

Problem 6

A sound pressure level of $84\,dB(A)$ is measured $10\,m$ away from a finite line source (train) which is $100\,m$ long. How high is the sound pressure level measured at distances of $20\,m$, $200\,m$ and $400\,m$?

Problem 7

Four sound sources are arranged on coordinate axes as shown in the following diagram.

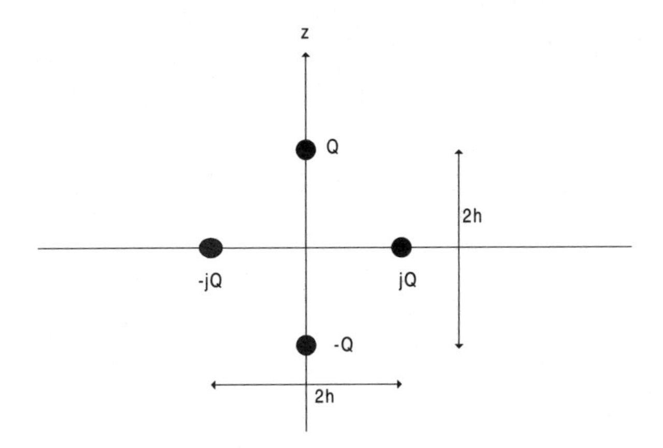

Fig. 3.31. Arrangement of four sound sources with relative phase shifts

The sources' volume flows have the same absolute value; however, they vary relative to one another clockwise in phase by 90°, as shown in the diagram. Determine the spatial sound pressure. State the particle motion in the film using the parameters $2h/\lambda = 0.25$; 0.5; 1 and 2. Take the following circumstances into account:

Assume a line source, where the field can be defined by

$$p = Ae^{-jkr}/\sqrt{r}$$

(where r is the distance between the source point and field point). In contrast to point sources, the interference pattern is unaffected. The only effect is a slower reduction in amplitude from the source to the field.

Rather than analyzing the particle velocity by differentiation, the particle velocity can be approximated by calculating the difference of pressure using the proportionalities

$$v_x \sim p(x + \lambda/100, y) - p(x, y)$$

and

$$v_y \sim p(x, y + \lambda/100) - p(x, y) .$$

Also take into consideration that the dimension of the source (and thereby the velocity scale) is completely arbitrary.

Problem 8

The displacement of a piston ξ (with surface area S) of a small loudspeaker in a box is brought about my electronic means:

- $\xi = 0$ for $t < 0$,
- for $0 < t < T_D$ is $\xi = \xi_0 \sin^2\left(\frac{\pi}{2}\frac{t}{T_D}\right)$ and
- for $t > T_D$ is $\xi = \xi_0$.

What is the time- and space-dependent sound pressure in the entire space?

Problem 9

Determine the sound pressure emanating from the circular piston with radius b in a reverberant wall in far-field.

Problem 10

The objective is to investigate the sound radiation of the plate and bending vibration of the form

$$v(x, y) = v_0 \sin(n\pi x/l_x) \sin(m\pi y/l_y)$$

(Plate dimensions l_x and l_y, resonator in reverberant wall, source area in $0 < x < l_x$ and $0 < y < l_y$), where both wavelengths

$$\lambda_x = \frac{2l_x}{n}$$

and

$$\lambda_y = \frac{2l_y}{m}$$

are small relative to the sound source wavelengths in air λ ($\lambda_x << \lambda$ and $\lambda_y << \lambda$). How great is the sound pressure approximately in the far-field, when the cross-sectional measurements l_x and l_y are likewise very small in relation to the sound wavelengths in air?

Problem 11

Where do the pressure nodes lie on the axis just before a circular piston with a diameters $b/\lambda = 3.5$; 4.5 and 5.5?

Problem 12

A pair of sound emitters consists of two sources with volume flows Q_1 and $Q_2 = -(1 + jkh)Q_1$, whereby h describes the distance of the sources to one another. Find the directivity pattern of the source pair for low frequencies $kh << 1$.

4

Structure-borne sound

4.1 Introduction

Vibrations and waves in solid structures, like, for example, vibrations in plates and beams, walls, ships and buildings, etc., are summarized by the term 'structure-borne sound'. Structure-borne sound has major importance with respect to solving noise control problems: the air-borne sound into (or from) the aforementioned solid structures is caused by motion in the structure's surface. In many cases it is the structure-borne sound that is responsible for the resulting sound in air (or sound in liquids). Even the transmission through walls, ceilings and windows, etc. is essentially a structural problem.

A vital and fundamental difference exists between sound waves in air and sound waves in structures. A gas (or a liquid) reacts with a change in pressure if its volume is changed. A mere change of the geometrical shape of the gas mass has no influence on the pressure at all (apart from losses due to friction). The boundaries between elements of volume in a gas therefore only transmit forces normal to their surface.

As illustrated by means of a simple example, a bendable thin beam (such as a ruler),- solid structures not only try to resist a compression of the volume they fill, but also a deformation of their shape. The boundaries in solid structures therefore transmit tangential forces, or shear tension, as well. By means of the example, the bendable beam, the existence of forces normal to its axes can easily be observed: these shear forces keep the beam in its bendable shape. It would otherwise be impossible for it to stay in this shape.

Instead of only the normal component of the tension, which appears in gases, three components of forces have to be taken into account at the boundaries when dealing with elements of volume in solid structures (Fig. 4.1). Just as one uses the force exerted onto the surface (pressure) for the purposes of definition in the case of airborne sound, one uses tension to formulate the the force laws under the circumstance of structure-borne sound. The tension is equal to the ratio of force to surface area. Furthermore, a distinction has to be

M. Möser, *Engineering Acoustics*, DOI 10.1007/978-3-540-92723-5_4,

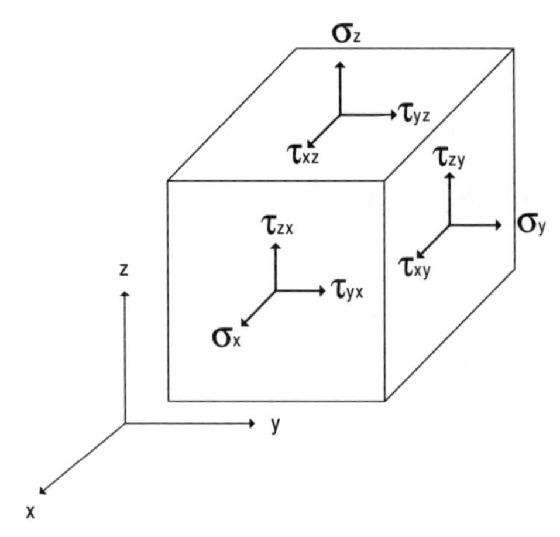

Fig. 4.1. Normal tension and shear tension on the element of volume in a solid structure

made between normal tension (normal to the imaginary boundary) and shear tension (tangential to the boundary).

All external tension components result in an elastic deformation of the structure which reacts with a resilient oscillation to its equilibrium position when the external tension is removed. The observed oscillation can be explained by a continuous conversion of potential energy which is stored in a change of shape and volume, into kinetic energy of the involved masses and vice versa. This 'reciprocal transformation' of the stored energy is not only taking place continually with time, but is also spatially distributed in such a way that vibrations occur as waveforms. In beams, for instance, the three-axial state of tension leads to different sorts of waves for each direction of motion. In beams

- bending waves, where the displacement is normal to the beam axis and therefore also normal to the propagation direction of the wave (Fig. 4.2a),
- the likewise transverse torsional waves, produced by torsion of the beam cross-section and
- the longitudinal waves caused by strain, where the displacement mainly appears along the beam axis (Fig. 4.2b)

occur.

The circumstances become even more complicated by the fact that bending waves with displacement can occur in both of the directions pointing normal to the beam axis. Only in the case of circular cross-sections (or quadratic cross-sections) do both types of bending waves show the same behavior. As can easily be observed in a beam with a flat, prolated cross-section (a ruler), the bending stiffness generally depends on the direction of the tension.

In addition, there are many more waveforms in beams and plates if the finite cross-sectional dimensions are taken into account. In the same manner as in air-borne sound, cross-sectional distributions of the sound (so-called modes) develop, each having an individual wave form (an example is given in Fig. 4.3). The waves mentioned above are only a special case of modes. It is quite obvious that the full breadth of structure-borne waves which can occur in beams and plates cannot be dealt with here. The interested reader should especially refer to the book "Structure Borne Sound" by Lothar Cremer and Manfred Heckl (translated by B.A.T. Petersson, Springer, Berlin and New York 2004) for a detailed description of the wave forms. Here, it is sufficient to limit the explanation to the bare necessities.

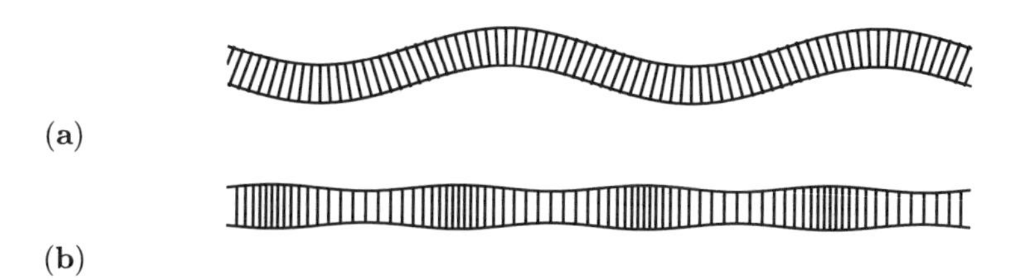

Fig. 4.2. (**a**) Bending waves and (**b**) longitudinal waves in beams

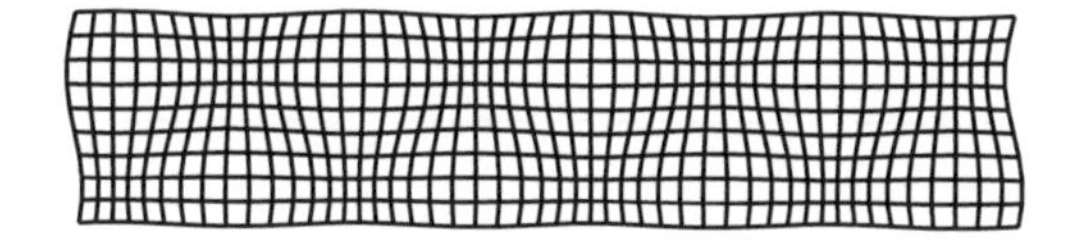

Fig. 4.3. Higher order wave (mode) in a thick plate

In engineering acoustics, the propagation of bending waves is mainly of interest because this wave form – also occurring in laterally extended plates – is the most essential one for a simple reason: the plate displacements are normal to the plate (or beam) surface and will preferably result in a radiation of air-borne sound rather than longitudinal strain or torsional waves with their motion mainly tangential to the surface. In addition to that, a second reason for the dominating role of bending waves can be deduced by intuition. Bending provides a much smaller resistance to the external exciting force (in most practically relevant cases) as does the reaction ot strain, for example. It can therefore be assumed that bending waves are more easily excitable and are thus the dominating type of vibration.

The discussion of bending waves starts with the simplest case of beams. The obtained results will be transferred to plates which are more relevant in practice.

4.2 Bending waves in beams

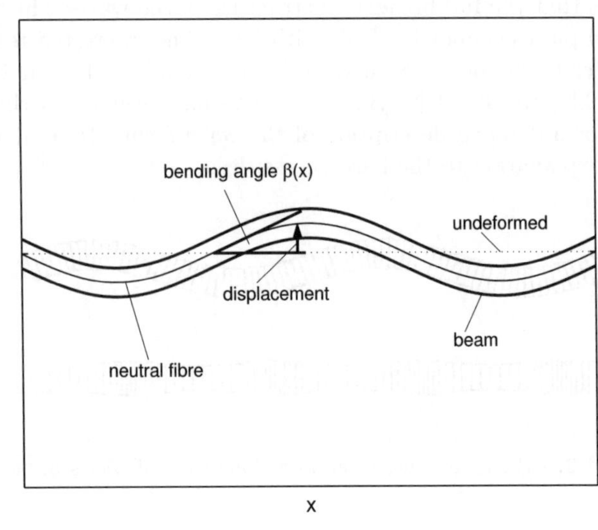

(a)

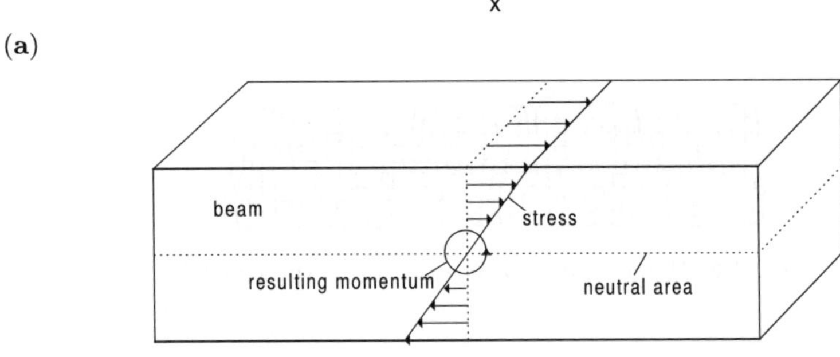

(b)

Fig. 4.4. Elastic bending of a beam. (**a**) Displacement and bending angle. (**b**) tension and resulting momentum

Static bending of beams and plates has understandably been a considerably important topic in the statics of structures and has been elucidated in past studies. One can therefore refer to the results of static bending science and readily use it here. The kinetic quantities are described by the beam displacement $\xi(x)$ and the bending angle $\beta(x)$ (Fig. 4.4a). Only small bending angles are of interest here, to which

$$\frac{\partial \xi}{\partial x} = \tan \beta \approx \beta \tag{4.1}$$

applies. Apart from these, the normal tension and the shear tension in the beam cross-section are also important. If it is assumed, as is conventional in static bending science, that the shear tension increases linearly from a (shear tension free) neutral beam fibre (Fig. 4.4a and 4.4b). The shear tension ε_x can be summarized in a resulting moment M, which acts on the neutral fiber

$$M = \int_S \varepsilon_x y \mathrm{d}S \,.$$

The stronger the beam is flexed by bending the larger this bending moment becomes. A reasonable assumption is

$$M \sim \frac{1}{r_\mathrm{k}} \,,$$

where r_k is the radius of the flexed circle at the corresponding beam point. For small displacements

$$\frac{1}{r_\mathrm{k}} = \frac{\partial^2 \xi}{\partial x^2}$$

applies, as is generally known. Using the proportional constant B, we obtain

$$M = -B \frac{\partial^2 \xi}{\partial x^2} \tag{4.2}$$

where the sign of M is specified in such a way that M and ξ have equal signs for waves of the form $\xi = \xi_0 cos(kx - \omega t)$. The constant B is accordingly called the bending stiffness. It not only contains the specific stiffness of the material E, but it is also dependent on the geometry of the cross-section. It was graphically shown above that the latter is included in the bending stiffness. Bending science shows that (Fig. 4.5a)

$$B = E \int_S y^2 \mathrm{d}y \mathrm{d}z = EI \tag{4.3}$$

is given. The specific stiffness E is called elastic modulus, also known as Young's modulus. It may be determined from the stiffness s of a material block which can be regarded as a spring, having the cross-sectional area S and the thickness h (Fig. 4.5b). If the block is loaded with the force F, then it is compressed by Δx according to Hook's law:

$$F = s\Delta x = \frac{ES}{h} \Delta x \tag{4.4}$$

The elastic modulus $E = sh/S$ therefore may be determined from the stiffness of a probe and its dimensions.

The quantity I, describing the geometry of the cross-section, is called the second moment of area. In this book, only rectangular cross-sections are of interest. Using the beam thickness h (pointing towards the direction of the exciting force) and the width b, I is obtained for the

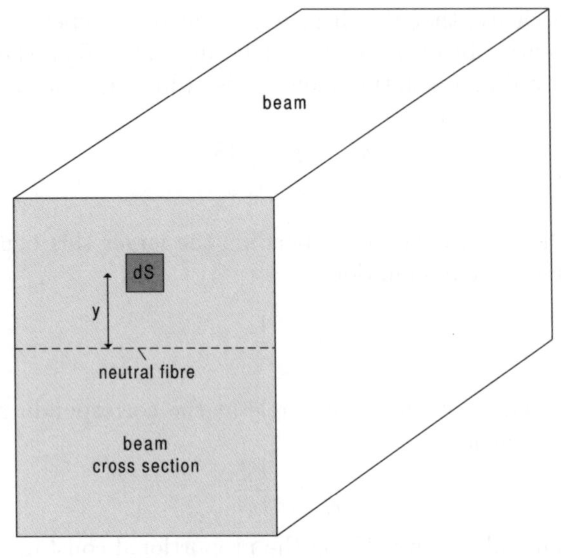

(a)

(b)

Fig. 4.5. (**a**) Definition of the second moment of area (**b**) Measurement setup for determination of the elastic modulus E

Rectangular cross-section

$$I = \frac{bh^3}{12} \; .$$

A second example, using the radius a, is mentioned for the
Circular cross-section

$$I = \frac{\pi}{4} a^4 \; .$$

Other values for the second moment of area can be found, for example in Dubbel's book "Taschenbuch für den Maschinenbau" (Springer, Berlin 2001), which is also a very useful book for acousticians.

Similar to summarizing the normal tension in a bending moment, a shear force can be assigned to the shear tension τ, pointing in the direction of the displacement:

$$F = \int_S \tau_{xy} \mathrm{d}y\mathrm{d}z \ . \tag{4.5}$$

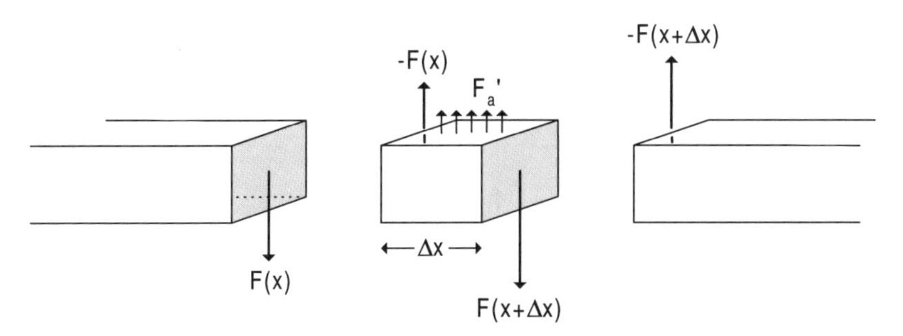

Fig. 4.6. Free body diagram of a beam element

As static bending science shows, the shear force, acting tangentially on the cross-section and pointing downwards, results in

$$F = -\frac{\partial M}{\partial x} \tag{4.6}$$

when using the bending moment. Due to 'actio=reactio', the transversal force pair $F(x + \Delta x)$ and $-F(x)$ now acts on the free body of the beam element (Fig. 4.6) with the length Δx. If an external force $F_a = F'_a \Delta x$ is taken into account, pointing in the direction of the displacement (e.g. the beam is excited by a hammer-blow) and uniformly distributed along the length Δx, Newton's law requires that

$$\Delta x \varrho S \ddot{\xi} = F(x) - F(x + \Delta x) + \Delta x F'_a \ .$$

Or, after proceeding to the limiting case $\Delta x \to 0$,

$$m' \ddot{\xi} + \frac{\partial F}{\partial x} = F'_a \ , \tag{4.7}$$

where m' is the beam mass and F'_a is the external force per corresponding unit length. (4.6) and (4.2) specify

$$\frac{\partial F}{\partial x} = -\frac{\partial^2 M}{\partial x^2} = B \frac{\partial^4 \xi}{\partial x^4} \ ,$$

and therefore the equation

$$m' \ddot{\xi} + B \frac{\partial^4 \xi}{\partial x^4} = F'_a$$

is finally obtained from (4.7) for bending waves. It is similar only to the wave equation in gases inasmuch as the second time derivative appears. The principal differences between bending waves and air waves become apparent, when pure tones

$$\xi(x,t) = \mathrm{Re}\left\{\xi(x)\mathrm{e}^{j\omega t}\right\}$$

are considered. For complex amplitudes, the wave equation for bending waves becomes

$$\frac{\partial^4 \xi}{\partial x^4} - \frac{m'}{B}\omega^2 \xi = \frac{F_a'}{B} \, ,$$

for the displacement, or

$$\frac{\partial^4 v}{\partial x^4} - \frac{m'\omega^2}{B} v = \frac{j\omega F_a'}{B} \tag{4.8}$$

for the velocity $v = j\omega\xi$, respectively. Using (4.1), (4.2) and (4.6), the more seldom needed angular velocity $w = \mathrm{d}\beta/\mathrm{d}t$, the bending moment M and the shear force F are given by

$$w = \frac{\partial v}{\partial x} \, , \tag{4.9}$$

$$M = -\frac{B}{j\omega}\frac{\partial^2 v}{\partial x^2} \quad \text{and} \tag{4.10}$$

$$F = \frac{B}{j\omega}\frac{\partial^3 v}{\partial x^3} \, . \tag{4.11}$$

Angular velocity, moment and shear force can be calculated from the velocity accordingly.

4.3 Propagation of bending waves

The main characteristics of bending waves can be explained using the basic ansatz for waves

$$v = v_0 \mathrm{e}^{-jk_\mathrm{B}x}$$

in the homogeneous differential equation

$$\frac{\partial^4 v}{\partial x^4} - \frac{m'}{B}\omega^2 v = 0 \, , \tag{4.12}$$

which is valid outside of local external forces. Inserting the ansatz into (4.12) yields the bending wavenumber

$$k_\mathrm{B}^4 = \frac{m'}{B}\omega^2 \, . \tag{4.13}$$

Using

$$k_\mathrm{B} = \frac{2\pi}{\lambda_\mathrm{B}} = \frac{\omega}{c_\mathrm{B}}$$

the bending wavelength becomes

$$\lambda_{\mathrm{B}} = 2\pi \sqrt[4]{\frac{B}{m'}} \frac{1}{\sqrt{\omega}} \tag{4.14}$$

and the propagation speed of bending waves becomes

$$c_{\mathrm{B}} = \sqrt[4]{\frac{B}{m'}} \sqrt{\omega} \, . \tag{4.15}$$

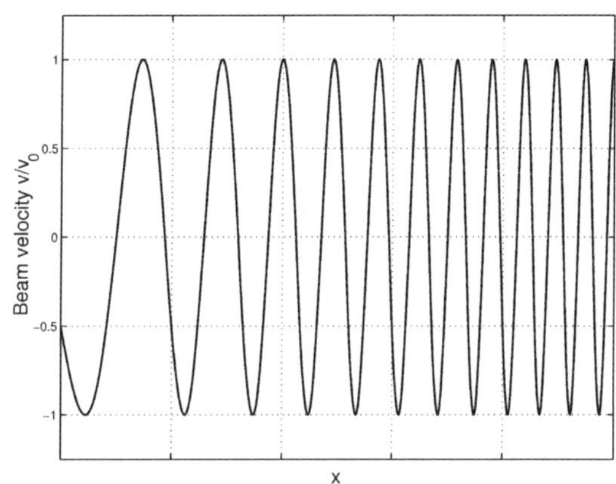

(a)

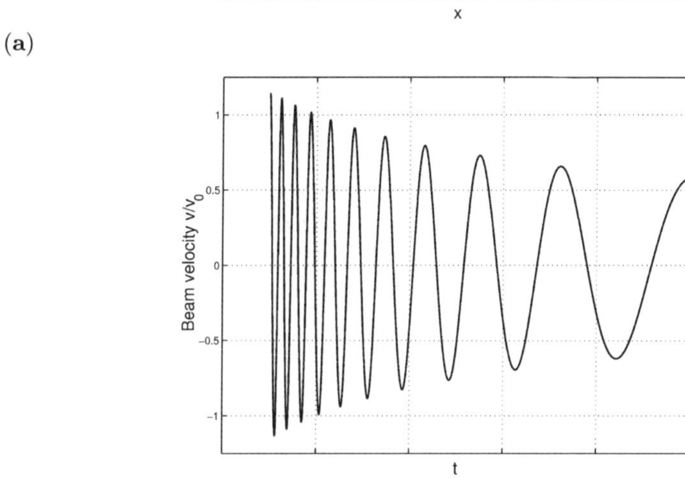

(b)

Fig. 4.7. Impulse response of the beam velocity. (**a**) Space characteristics for constant time. (**b**) Time characteristics for constant position

Thus, a bending wavelength λ_{B} is obtained which only decreases with the square-root of increasing frequency and has a frequency-dependent propagation speed. These facts illustrate the fundamental difference between air waves and bending waves. If, instead of pure tones, we consider time dependencies consisting of several spectral components, the frequency-dependent propagation speed causes the spectral components to 'run away from each other' the larger the distance is that they traverse and the more their frequencies diverge. This also means that the spectral composition at two different positions on the beam is different and it follows that, at two different positions, totally different time characteristics of the beam velocity are found. The time signal is distorted along the propagation direction of the bending wave.

This effect is called dispersion. It can be recognized, for example, in the 'impulse response' (i.e. the beam velocity, if the beam at $x = 0$ is excited at $t = 0$ by a short hammer blow). An example of this effect is shown in Fig. 4.7 above. The higher frequency components of a broadband excitation arrive before the lower ones, the instantaneous frequency at the arrival point decreases gradually. The space characteristics for a constant time behave accordingly: higher frequencies have travelled a larger distance than the lower components (Fig. 4.7b). The curves shown in Figure 4.7 are depicted in the graphic representation of the solution to Problem 8 in Chapter 13.

4.4 Beam resonances

Beam segments are affected by resonance vibrations much the same way as finite gas volume segments confined on all sides are, if virtually no wave energy is lost at the terminations. Resonance frequencies and types of vibrations that occur in beam segments depend on how the beam is suspended or mounted at the ends. A distinction is made between beams bilaterally supported, mounted or suspended beams:

- The supported bearing consists of a beam resting at one point. This impedes displacement, so that $v = 0$. The bearing absorbs the bending force, but not the bending momentum. Thus the beam bending momentum diminishes to $M = 0$. Beams can be also supported by fastening them to any type of moveable hinge.
- In the case of a beam mounted at its end, both force and momentum are absorbed by its bearing. The frequency oscillator is now not able to move or rotate at it its point of bearing. Therefore, the velocity and angular velocities of the beam are $v = 0$ and $w = 0$, respectively.
- In the case of a beam that is freely moving at its end, neither its directional nor its rotational motion is hindered. That is, neither force nor momentum is absorbed. For this reason, $M = 0$ and $F = 0$ are true for suspended beams.

A mathematical consideration of beam resonances generally necessitates an ansatz for beam velocity consisting of four linear-independent solutions of the bending wave function (4.8). Because – as mentioned before – it is necessary to find the non-stimulated vibrations, we will consider the homogenous bending wave equation ($F'_a = 0$ in eq.(4.8)). The purpose of the ansatz is to satisfy the constraints at the ends of the beam, resulting, of course, in a system of four equations. Symmetrical constructions simplify the analysis a bit. Such constructions consist of beam segments characterized by the same constraints at both ends. Here, one can make a separate distinction between point- and axis-symmetric wave forms. As shown in the following, this allows us to only use two equations to describe this situation. However, both have to be solved twice. The beam ends exist in this scenario at the points $x = -l/2$ and $x = +l/2$.

For axis-symmetric wave forms, the sound velocity is

$$v_g = A_1 \cos k_B x + A_2 \, ch \, k_B x \tag{4.16}$$

and for point-symmetric wave forms the sound velocity is

$$v_u = A_1 \sin k_B x + A_2 \, sh \, k_B x \, . \tag{4.17}$$

A_1 and A_2 are constants, *ch* and *sh* indicate the hyperbolic cosine and sine. In the following, we will consider the three previously mentioned cases of bilaterally supported, mounted, or suspended beams.

4.4.1 Supported beams

Axis-symmetric wave forms of bilaterally supported beams

The bearing constraints $v(l/2) = 0$ and $M(l/2) = 0$ for a beam supported at the end $x = l/2$ are

$$A_1 \cos k_B l/2 + A_2 \, ch \, k_B l/2 = 0 \tag{4.18}$$

and, because the momentum is proportional to the second-order spatial derivative of the velocity according to eq.(4.10):

$$A_1 \cos k_B l/2 - A_2 \, ch \, k_B l/2 = 0 \, . \tag{4.19}$$

By subtracting both equations, we obtain $A_2 = 0$. In order to provide the pre-existing condition of non-stimulated vibration that is not zero, the constant A_1 can also not be zero, $A_1 \neq 0$. Therefore, the resonance condition exists in $\cos k_B l/2 = 0$. This results in $k_B l/2 = \pi/2 + n\pi$ or

$$k_B l = (2n + 1)\pi \tag{4.20}$$

($n = 0, 1, 2, ...$). For the resonance frequencies we use $k_B = \sqrt[4]{m'\omega^2/B}$ to obtain

$$f = (2n + 1)^2 \frac{\pi}{2l^2} \sqrt{\frac{B}{m'}} \, . \tag{4.21}$$

The shape of the wave form function in the resonance is referred to as a mode. For the modes, we use the ansatz (4.16) where $k_B l/2 = \pi/2 + n\pi$ and $A_2 = 0$ to obtain

$$v_{g,mod} = \cos\left((2n+1)\pi\frac{x}{l}\right). \tag{4.22}$$

Wave modes are scalable. For this reason, we can arbitrarily set $A_1 = 1$.

Point-symmetric wave forms of bilaterally supported beams

The parameters $v(l/2) = 0$ and $M(l/2) = 0$ for the beam supported at the end $x = l/2$ result in

$$A_1 \sin k_B l/2 + A_2\, sh\, k_B l/2 = 0 \tag{4.23}$$

and – since, as in the above, the momentum is proportional to the second-order spatial derivative of the velocity according to eq.(4.9):

$$A_1 \sin k_B l/2 - A_2\, sh\, k_B l/2 = 0\,. \tag{4.24}$$

Again, we obtain $A_2 = 0$ by subtracting both equations. In order to define the pre-existing condition of a non-stimulated vibration that is not zero, the constant A_1 must also have a value other then zero, $A_1 \neq 0$. Therefore, the resonance constraint exists in $\sin k_B l/2 = 0$. This results in $k_B l/2 = n\pi$ or

$$k_B l = 2n\pi \tag{4.25}$$

$(n = 1, 2, ...)$. This leads to the definition of the resonance frequencies in

$$f = (2n)^2 \frac{\pi}{2l^2}\sqrt{\frac{B}{m'}}\,. \tag{4.26}$$

According to the ansatz (4.17) where $k_B l/2 = n\pi$ and $A_2 = 0$, the modes exist in

$$v_{u,mod} = \sin\left(2n\pi\frac{x}{l}\right). \tag{4.27}$$

Here, too, we can scale the modes to $A_1 = 1$ for the sake of simplicity.

Summary of bilaterally supported beams

The $k_B l$ pertaining to the resonances are alternately even- and odd-numbered multiples of π according to equations (4.20) and (4.25). In sum:

$$k_B l = m\pi \tag{4.28}$$

Due to $k_B = \sqrt[4]{m'\omega^2/B}$, the resonance frequencies altogether result in

$$f = m^2 \frac{\pi}{2l^2}\sqrt{\frac{B}{m'}} \tag{4.29}$$

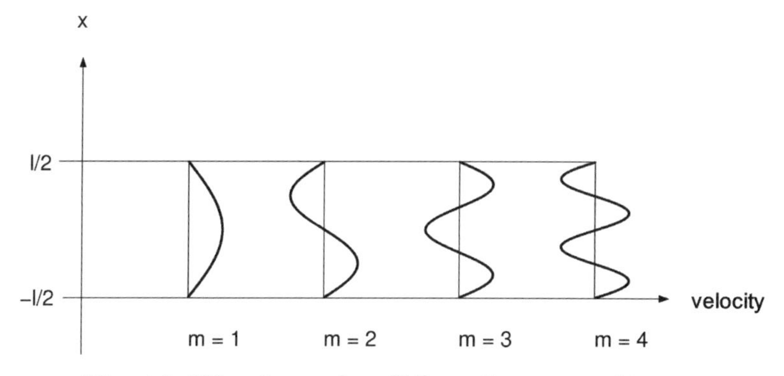

Fig. 4.8. Vibration modes of bilaterally supported beams

where $m = 1, 2, 3, ...$, and whereby odd-numbered m indicate axis-symmetric modes with forms as given in eq.(4.22) and even-numbered m indicate point-symmetric modes with forms given in eq.(4.27). As seen above, the distance between two resonances increases quadratically. The number of resonance incidences in a given finite frequency band decrease as the center frequency of the band increases. Unlike one-dimensional gaseous wave guides, the frequency density decreases for bilaterally supported beams. In addition, the bending resonance frequencies are inversely proportional to the square of the length of the wave guide l. This is also different from sound travelling through air, where the resonance frequencies are inversely proportional to the length itself.

Figure 4.8 provides a visual overview of the first four vibration modes.

4.4.2 Bilaterally mounted beams

Axis-symmetric wave Forms for bilaterally mounted beams

The criteria $v(l/2) = 0$ and $w(l/2) = 0$ for a beam mounted at its end $x = l/2$ result in

$$A_1 \cos k_B l/2 + A_2 \, ch \, k_B l/2 = 0 \tag{4.30}$$

and (because the angular velocity is proportional to the spatial derivative of the velocity according to eq.(4.9)):

$$A_1 \sin k_B l/2 - A_2 \, sh \, k_B l/2 = 0 \, . \tag{4.31}$$

Non-zero solutions A_1 and A_2 only exist when the determinant of the homogenous system of both equations vanishes. Only then do non-stimulated vibrations – resonances – occur. For this reason, the equation which defines the resonance frequencies is

$$\cos k_B l/2 \, sh \, k_B l/2 + \sin k_B l/2 \, ch \, k_B l/2 = 0 \, , \tag{4.32}$$

or in

$$tg\,k_B l/2 = -th\,k_B l/2 \tag{4.33}$$

(tg=tangent, th = hyperbolic tangent). (4.33) is a transcendent equation for the values $k_B l/2$, which characterize the resonances. It can easily be depicted graphically or solved for its approximation. This fact will be discussed in more detail in the mathematical considerations for the odd-numbered vibration forms.

Point-symmetric vibration forms for bilaterally mounted beams

The criteria $v(l/2) = 0$ and $w(l/2) = 0$ apply for odd-numbered wave forms in a beam mounted at its end at $x = l/2$

$$A_1 \sin k_B l/2 + A_2\,sh\,k_B l/2 = 0 \tag{4.34}$$

and (because the angular velocity is proportional to the spatial derivative of the velocity according to eq.(4.9)):

$$A_1 \cos k_B l/2 + A_2\,ch\,k_B l/2 = 0\,. \tag{4.35}$$

Non-zero solutions A_1 and A_2 are again only possible if the determinant of the homogenous system of both equations vanishes. Only then do non-stimulated vibrations – resonances – occur. For this reason, the equation which defines the resonance frequencies is

$$\sin k_B l/2\;ch\,k_B l/2 - \cos k_B l/2\;sh\,k_B l/2 = 0\,, \tag{4.36}$$

or

$$tg\,k_B l/2 = th\,k_B l/2\,. \tag{4.37}$$

Summary of bilaterally mounted beams

When applying equations (4.33) and (4.37), the resonances can be derived from the point where the tangent function intersects with the positive and the negative hyperbolic tangents, alternately (refer to Figure 4.9). At the first point of intersection which has the smallest argument $k_B l/2$, the tangent hyperbole already is close to the value 1. We can use the resonance criterium

$$tg\,k_B l/2 = \pm 1 \tag{4.38}$$

to achieve quite a close approximation. The equation results in

$$k_B l/2 = 3\pi/4;\; 5\pi/4;\; 7\pi/4;\; ...$$

(as can be seen in Figure 4.9) or

$$k_B l = (2m+1)\pi/2 \tag{4.39}$$

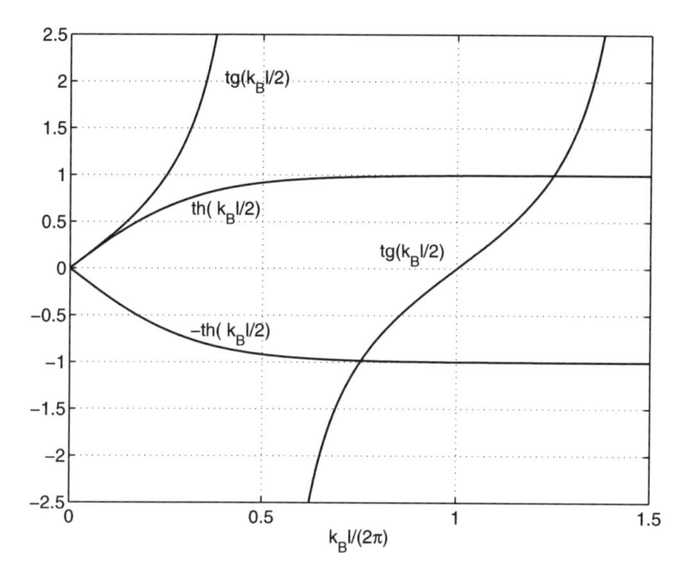

Fig. 4.9. Graphic solution of the eigenvalue equations (4.33) and (4.37)

$(m = 1, 2, ...)$. For the resonance frequencies, it follows that

$$f = (m + \frac{1}{2})^2 \frac{\pi}{2l^2} \sqrt{\frac{B}{m'}} \; . \tag{4.40}$$

For all practical considerations, the application of the above approximation is adequate. An even more precise assessment of the first point of intersection between the tangent and hyperbolic tangent for $m = 1$ produces the more accurate value $k_B l = 1.506\pi$ instead of $k_B l = 1.5\pi$. The margin of error for the lowest resonance frequency is therefore less than 1 percent.

The axis-symmetric modal forms can be obtained by solving the eq.(4.31) for A_2 and subsequently applying eq.(4.16) to the ansatz:

$$v_{g,mod} = \cos k_B x + \frac{\sin k_B l/2}{sh\; k_B l/2}\, ch\; k_B x \; . \tag{4.41}$$

Since these modes consist of axis-symmetric vibrations in resonance, we proceed by inserting the values $k_B l/2 = 3\pi/4; 7\pi/4; 11\pi/4....$ Again, it has been scaled to $A_1 = 1$.

The odd-numbered modal forms can be obtained by solving eq.(4.35) for A_2 and subsequently plugging eq. (4.17) into the ansatz:

$$v_{u,mod} = \sin k_B x - \frac{\cos k_B l/2}{ch\; k_B l/2}\, sh\; k_B x \; . \tag{4.42}$$

Since these modes are point-symmetric vibrations in resonance, we use the values $k_B l/2 = 5\pi/4; 9\pi/4; 13\pi/4....$ Here too, it was scaled to $A_1 = 1$.

The first four modal forms are depicted in Figure 4.10.

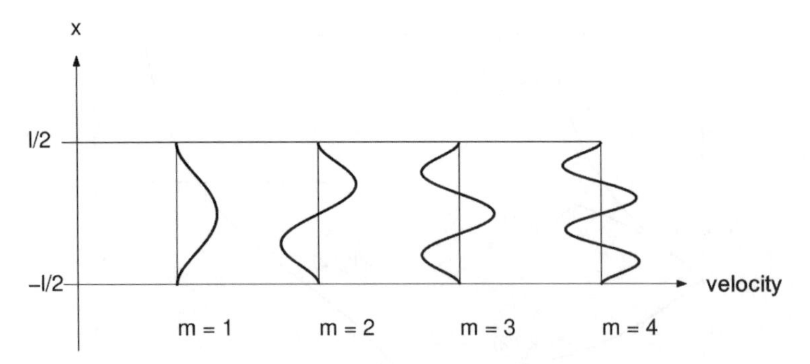

Fig. 4.10. Vibration modes of bilaterally mounted beams

4.4.3 Bilaterally suspended beams

The resonances of bilaterally suspended beams can be deduced using the same criteria as for bilaterally mounted beams, if the formulae in the equations (4.16) and (4.17) are applied for the bending momentum rather than the vibrational speed, that is, if (4.16) and (4.17) are replaced by

$$M_g = A_1 \cos k_B x + A_2\, ch\, k_B x \tag{4.43}$$

and, for the point-symmetric forms of the beam velocity,

$$M_u = A_1 \sin k_B x + A_2\, sh\, k_B x. \tag{4.44}$$

The constraint $F(l/2) = 0$ is synonymous with the first spatial derivative being eliminated at the onset of the momentum. Along with the parameter $M(l/2) = 0$, this results in the equations (4.30) and (4.31) for the axis-symmetric case, and (4.34) and (4.35) for the point-symmetric case. For this reason, the resonance frequencies of bilaterally suspended beams are equal to those of bilaterally mounted beams. This also means that (4.39) (and of course (4.40)) apply to the resonance frequencies for bilaterally suspended beams.

The vibration forms (as represented in the vibrational speed) are, of course, not identical to those for mounted beams. The suspended vibration modes are derived from the two-fold integration of equation (4.41) and result in

$$v_{g,mod} = \cos k_B x - \frac{\sin k_B l/2}{sh\, k_B l/2}\, ch\, k_B x, \tag{4.45}$$

(where $k_B l/2 = 3\pi/4; 7\pi/4; 11\pi/4...$) and by integrating equation (4.42) twice,

$$v_{u,mod} = \sin k_B x + \frac{\cos k_B l/2}{ch\, k_B l/2}\, sh\, k_B x \tag{4.46}$$

(where $k_B l/2 = 5\pi/4; 9\pi/4; 13\pi/4...$). Figure 4.11 shows the first four modes here as well.

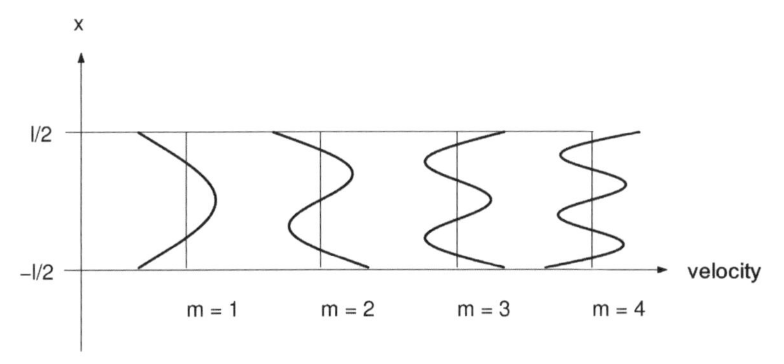

Fig. 4.11. Vibration modes of bilaterally suspended beams

4.5 Bending waves in plates

As can be shown with similar investigations as for the one-dimensional beam, the bending wave equation for homogeneous plates is given by

$$\frac{\partial^4 v}{\partial x^4} + 2\frac{\partial^4 v}{\partial x^2 \partial y^2} + \frac{\partial^4 v}{\partial y^4} - \frac{m''}{B'}\omega^2 v = \frac{j\omega p}{B'} \tag{4.47}$$

where p denotes an external force of area (e.g. a pressure) acting on the plate. This time m'' represents the mass per unit area

$$m'' = \varrho h \tag{4.48}$$

(ϱ is the plate density and h the thickness) which is obtained using the bending stiffness of the beam (for a rectangular cross-section) per unit length in the same way as the bending stiffness of the plate

$$B' = \frac{E}{1-\mu^2}\frac{h^3}{12} . \tag{4.49}$$

The denominator containing Poisson's ratio μ accounts for the fact that plate elements of volume are somewhat stiffer than beam elements. Whereas the material of the beam can elude laterally to a small extent (normal to the beam and the direction of displacement), this possibility is not given in the plate. Depending on the material, it is approximately $\mu \approx 0.3$, so that $\mu^2 \ll 1$ is negligible in so far that material parameters can be specified accurately.

In the case of one-dimensional wave propagation of the plate motion, excited, for instance, by a line force or an impinging sound wave, the bending wave equation becomes

$$\frac{1}{k_\mathrm{B}^4}\frac{\mathrm{d}^4 v}{\mathrm{d}x^4} - v = \frac{jp}{m''\omega} , \tag{4.50}$$

with

$$k_{\mathrm{B}}^4 = \frac{m''}{B'}\omega^2 \tag{4.51}$$

(these equations could also easily be derived by dividing the beam equation (4.8) by the beam width). Therefore the same dependencies for free bending waves

$$v = v_0 \mathrm{e}^{-jk_{\mathrm{B}}x} ,$$

which occur at a distance far enough from exciting forces, are obtained, as in the case of beams. Bending wavelength and propagation speed are

$$\lambda_{\mathrm{B}} = \sqrt[4]{\frac{B'}{m''}}\frac{2\pi}{\sqrt{\omega}} \tag{4.52}$$

$$c_{\mathrm{B}} = \sqrt[4]{\frac{B'}{m''}}\sqrt{\omega} . \tag{4.53}$$

The propagation speed of bending waves in plates is also frequency-dependent and the transmission of the displacement is dispersive. The wavelength is likewise inversely proportional to the square-root of the frequency. As a matter of principle, a radial symmetric sound field produced by a point force has the same wave form as plane waves in plates.

Due to energy balance, the amplitude has to be inversely proportional to the square-root of the distance r from the source. For not-too-short distances, it is

$$v = \frac{A}{\sqrt{r}}\mathrm{e}^{-jk_{\mathrm{B}}r} .$$

In the course of the discussion of bending waves in beams, the basic physical principles were of interest, whereas this time the focus shall be the relationships and the order of magnitudes which can be found in practice. It should first be stated that the use of the bending stiffness is not very handy in practical calculations. It is thus replaced by the ratio m''/B', which consists of more useful material parameters, while at the same time $\mu^2 \ll 1$ is neglected:

$$\frac{m''}{B'} = \frac{\varrho h 12}{E h^3} = \frac{12}{c_{\mathrm{L}}^2 h^2} \tag{4.54}$$

where

$$c_{\mathrm{L}} = \sqrt{\frac{E}{\varrho}}$$

is the longitudinal wave propagation speed in beams, consisting of the same material, as can be shown. c_{L}, thickness h and mass of area m'' are most commonly used for the acoustic description of plates. The influence of the material (expressed as c_{L}), thickness h and frequency f cannot be overlooked directly in (4.52) and (4.53), because the parameters B' and m'' not only depend on the material but also on the thickness. The dependencies can clearly be shown by inserting (4.54) into (4.52) and (4.53). We thus obtain the handy equations

Table 4.1. Commonly used material parameters

	Density ϱ (kg/m^3)	c_L (m/s)	η
Aluminium	2700	5200	$\approx 10^{-4}$
Steel	7800	5000	$\approx 10^{-4}$
Gold	19300	2000	$3\,10^{-4}$
Lead	11300	1250	$10^{-3} \ldots 10^{-1}$
Copper	8900	3700	$2\,10^{-3}$
Brass	8500	3200	10^{-3}
Concrete, light	600	1700	10^{-2}
Concrete, dense	2300	3400	$5\,10^{-3}$
Bricks (+mortar)	2000	2500–3000	10^{-2}
Plywood	600	3000	10^{-2}
Oak	700–1000	1500–3500	10^{-2}
Spruce	400–700	1200–2500	10^{-2}
Plaster	1200	2400	$8\,10^{-3}$
Beaverboard	600–700	2700	10^{-2}
Perspex	1150	2200	$3\,10^{-2}$
Sand, light	1500	100–200	10^{-1}
Sand, dense	1700	200–500	10^{-2}
Glass	2500	4900	$2\,10^{-3}$

$$\lambda_B \approx 1.35\sqrt{\frac{hc_L}{f}}$$

and for the propagation speed

$$c_B \approx 1.35\sqrt{hc_L f}\ .$$

to describe bending wavelength, respectively.

The parameters for the materials most commonly used in practice are given in Table 4.1. The loss factors correspond to pure internal damping, thus radiation losses and the damping at the connections of some compounds (e.g. bolt connections) have to be added when appropriate.

The longitudinal propagation speeds of different materials do not differ very much, as can be seen in Table 4.1: the range of 2000 m/s to 5000 m/s is roughly covered, whereas the range of thicknesses used in acoustics is a lot larger. Metal sheets of 0.5 mm thickness in vehicles are of the same interest as concrete walls of 0.5 m thickness, which represents a considerable ratio of 1:1000. The range of wavelengths is accordingly large. For 1000 Hz, for instance,

$$\lambda_B(0.5\,\text{mm metal sheet}) = 7\,\text{cm}$$
$$\lambda_B(25\,\text{cm light concrete}) = 90\,\text{cm}$$
$$\text{for comparison:}\quad \lambda_0(\text{air}) = 34\,\text{cm}$$

is obtained. As it becomes obvious by the examples, shorter wavelengths than in air occur in thin plates and longer wavelengths in thick plates.

The differences between 'long-wave' and 'short-wave' components for the transmission loss are of major importance, as it will be explained in Chap. 8 of this book. Generally, the property 'shorter wavelength than air' (or 'longer wavelength than air') can be assigned to a frequency interval. The two wavelengths λ_0 (for air) and λ_B (for the bending waves)

$$\lambda_0 = c/f \quad \text{and} \quad \lambda_B \approx 1.35\sqrt{\frac{hc_L}{f}}$$

are equal at a typical 'critical' frequency f_{cr}. Taking the square of the wavelengths and setting both equations to be equal shows that

$$f_{cr} = \frac{c^2}{1.82hc_L}$$

is obtained for the critical frequency f_{cr}. This is also called 'limiting frequency' or 'coincidence frequency'. For frequencies

- $f < f_{cr}$ below the critical frequency, the bending waves are shorter than the air waves, for frequencies
- $f > f_{cr}$ above the critical frequency, the bending waves are longer than the air waves.

Some examples of the critical frequency are given in the following table:

Plate consisting of	f_{cr}(Hz)
0.5 mm metal sheet	25000
4 mm glass	3000
5 cm plaster	530
25 cm heavy concrete	75

Here, it is clear that thick and massive walls and ceilings have a limiting frequency at the lower edge of the frequency range of interest, whereas the critical frequencies of windows and metal sheets are located at the upper edge.

Thin walls built, for example, into flats or offices, have a limiting frequency right in the middle of the frequency range of interest, thus they can suffer from poor transmission loss.

4.5.1 Plate vibrations

As is the case with beams, the vibrations and modal forms of plates are also dependent on the material their bearings consist of. With the exception of the scenario described below, where a plate is supported on all sides by its bearing, handling problems with plate resonance is by no way an easy task.

Usually, finding workable solutions entails complex and lengthy quantitative considerations, which have filled pages in lengthy treatises especially devoted to discussing this problem. Readers interested in more in-depth information on this topic are encouraged to take a look at Leissa, A.W.: 'Vibration of Plates', Office of Technology Utilization, National Aeronautics and Space Administration, Washington 1969 and Blevins, R. D.: 'Formulas for natural frequency and mode shape', Van Nostrand Reinhold, New York 1979. The latter work also contains sections discussing the vibration of beams, in particular, combined with various factors affecting the left and right ends.

For plates supported on all four edges $x = 0$, $y = 0$, $x = l_x$ and $y = l_y$, modal forms exist in

$$v = \sin(n_x \pi x / l_x) \sin(n_y \pi y / l_y) \tag{4.55}$$

($n_x = 1, 2, 3, ...$ and $n_y = 1, 2, 3, ...$). The plate dimensions are defined by l_x and l_y, the plate surface is therefore $S = l_x l_y$. The plate modes (1,1), (1,2) and (2,2) are shown in Figures 4.12, 4.13 and 4.14.

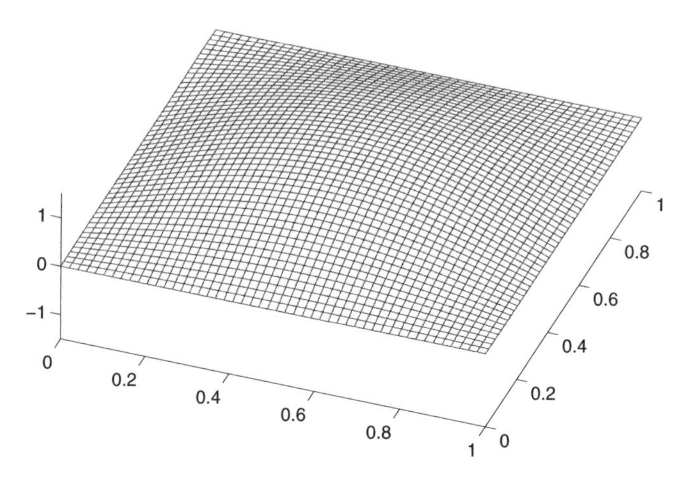

Fig. 4.12. Vibration mode $n_x = 1$ and $n_y = 1$

The resonance frequencies are obtained by plugging the modal forms into the bending wave equation (homogenous with $p = 0$) (4.47):

$$k_B^4 = \frac{m''}{B'}\omega^2 = (\frac{n_x \pi}{l_x})^4 + 2(\frac{n_x \pi}{l_x})^2(\frac{n_y \pi}{l_y})^2 + (\frac{n_y \pi}{l_y})^4 = [(\frac{n_x \pi}{l_x})^2 + (\frac{n_y \pi}{l_y})^2]^2 , \tag{4.56}$$

therefore

$$f = \frac{\pi}{2}[(\frac{n_x}{l_x})^2 + (\frac{n_y}{l_y})^2]\sqrt{\frac{B'}{m''}} . \tag{4.57}$$

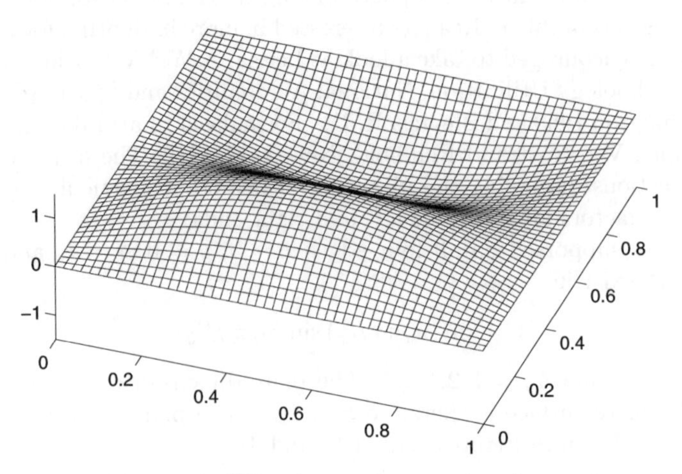

Fig. 4.13. Vibration mode $n_x = 1$ and $n_y = 2$

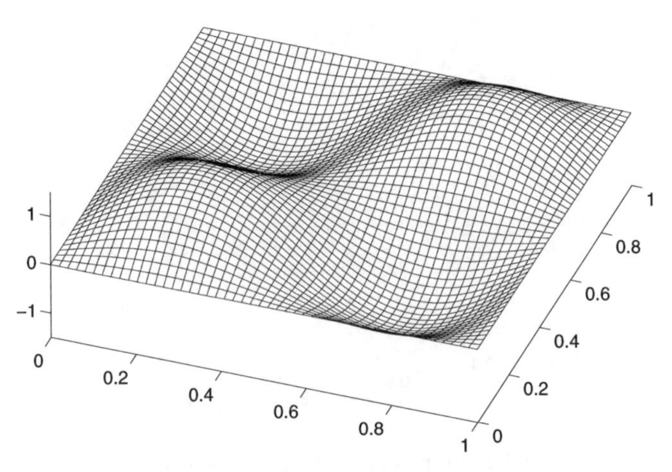

Fig. 4.14. Vibration mode $n_x = 2$ and $n_y = 2$

The resonance frequencies can be easily represented in graph form by taking the root of the last equation, thereby obtaining a grid point for each frequency root. These grid points exist in a constant eigenfrequency grid with grid edge lengths (Figure 4.15)

$$\sqrt{\frac{\pi}{2}}\sqrt[4]{\frac{B'}{m''}}\frac{1}{l_x}$$

and

$$\sqrt{\frac{\pi}{2}}\sqrt[4]{\frac{B'}{m''}}\frac{1}{l_y}\,.$$

Every grid node pertains to the root of each resonance frequency.

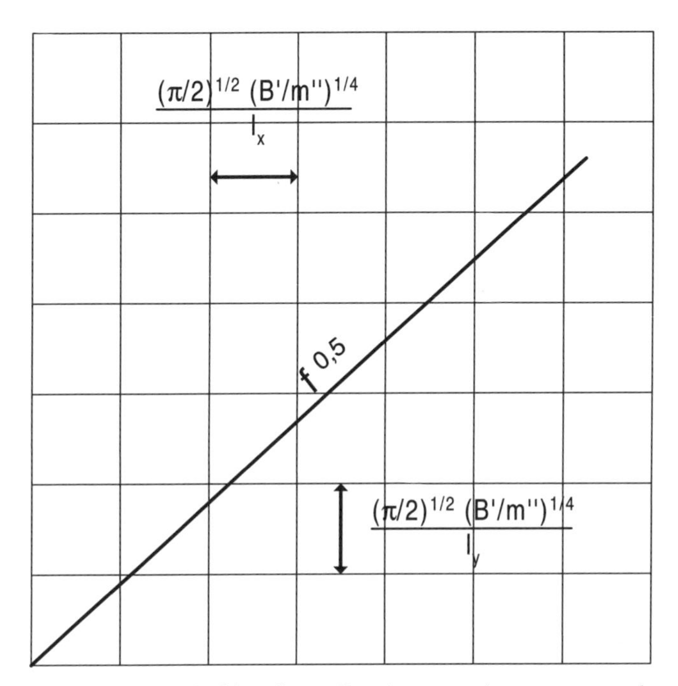

Fig. 4.15. Resonance grid of bending vibrations on plates supported at all edges

This graphical representation allows a simple estimation of the total resonance frequencies ΔN present in a single frequency band Δf. To do this, one must determine the number N of existing resonances within the interval 0 to f, which is roughly equal to the ratio of one quarter of a circle with a radius of $\sqrt{f}$ to the surface area of a grid element, that is

$$N = \frac{\frac{\pi}{4}f}{\frac{\pi}{2}\sqrt{\frac{B'}{m''}}\frac{1}{l_x l_y}} = \frac{1}{2}fS\sqrt{\frac{m''}{B'}}\,. \tag{4.58}$$

The differential quotient $\Delta N/\Delta f$ can be roughly substituted by

$$\frac{\Delta N}{\Delta f} \approx \frac{dN}{df} = \frac{1}{2}S\sqrt{\frac{m''}{B'}}\,,$$

thereby resulting in

$$\Delta N = \frac{1}{2} S \sqrt{\frac{m''}{B'}} \Delta f \,. \tag{4.59}$$

Furthermore, for practical calculations, it makes more sense to express the relationship of m'' and B' in terms of the longitudinal wave speed c_L and the thickness of the plate h according to eq.(4.54). This subsequently yields

$$\Delta N = \frac{1,73S}{hc_L} \Delta f \,. \tag{4.60}$$

Different than beams, the resonance density of plates is a frequency-independent constant. For instance, for sheet metal ($c_L = 5000\, m/s$) that is $1\, mm$ thick and which has a surface area $1\, m^2$, one obtains $\Delta N \approx 0.35\, \Delta f / Hz$. Thus, independent of the band's mid-frequency, the bandwidth $\Delta f = 100 Hz$ will always contain $\Delta N = 35$ resonances. A resonance occurs at roughly every $3\, Hz$. If the surface of the plate is reduced to a tenth of its original dimensions at $S = 0,1\, m^2$, there will only be about 3.5 resonances in a $100\, Hz$–bandwidth, and the average resonance intervals will be approx. every $30\, Hz$.

For other types of bearings, the resonance frequencies detune at the edges of the plate (as is the case for suspended, unmounted plates). However, even in such cases, the number of resonances ΔN in a sufficiently large band Δf remains in large part unchanged. This is why one can also use eq. (4.60) to estimate plate resonances regardless of the particular type of bearing arrangement.

4.6 Summary

The most essential structure-borne sound waves constitute bending waves occurring on plate and beam structures. As opposed to wave propagation in gases and liquids, flexural wavelengths depend on the frequency, growing proportionally to the square root of f. For mechanical sound signals consisting of multiple frequencies, each spectral component is transported at different speeds, leading to a dispersion effect in the signal shape. Due to this dispersion, bending wavelengths are inversely proportional to the square root of the frequency. As a result, bending wavelengths λ_B below a certain critical frequency f_{cr} are smaller than sound wavelengths in air λ. Conversely, $\lambda_B > \lambda$ is true for all f above f_{cr}. This fact plays an important role in both bending wave phenomena as well as sound insulation in plate-like structures such as walls, ceilings, windows, etc. (see Chapters 3 and 8). The threshold frequency is inversely proportional to the thickness of the building material $f_{cr} \sim 1/h$. The critical frequency is high for thin building materials and low for thick materials.

Resonance occurs in beam and plate structures of finite lengths, given that only a small amount of energy is lost beyond the material's boundaries. Resonance frequencies depend on the material's bearings. In beams, the distance between resonance frequencies increases with increasing frequency, that is, the density of resonances decreases. In plate structures, the resonance density is a constant, independent of frequency. The vibration patterns of the resonances are referred to as modes.

4.7 Further reading

Undoubtedly the work "Structure Borne Sound" by Lothar Cremer and Manfred Heckl (translated by B.A.T. Petersson, Springer, Berlin and New York 2004) is highly recommended for those dealing with the subject mentioned in the title. The book by Blevins, R. D.: "Formulas for natural frequency and mode shape" (Van Nostrand Reinhold, New York 1979) is also recommended for further reading in resonance processes and modal patterns.

4.8 Practice exercises

Problem 1

Find the coincidental critical frequencies for

- Plaster plates, $8\,cm$ thick ($c_L = 2000\,m/s$),
- window panes, $4\,mm$ thick and for
- a door panel made out of oak wood ($c_L = 3000\,m/s$), $25\,mm$ thick.

Problem 2

Find the first 5 resonance frequencies of aluminum beams with a thickness of $5\,mm$, and $50\,cm$ and one $100\,cm$ long. Consider two cases: bilaterally supported and bilaterally mounted.

Problem 3

Find the four lowest resonance frequencies

- of a window pane $4\,mm$ thick, measuring $50\,cm$ by $100\,cm$,
- a $10\,cm$–thick plaster wall ($c_L = 2000\,m/s$), measuring $3\,m$ by $3\,m$ and
- a $2\,mm$–enforced steel plate, measuring $20\,cm$ by $25\,cm$.

Problem 4

Find the resonance frequencies and the modal patterns of the mounted beam to the left and the supported beam to the right.

Problem 5

A force acting normal to a beam of infinite length (or a beam with a 'wave capturer' on both ends) induces bending vibrations, which resonate harmoniously with time, in the middle of the beam at $x = 0$. Graph the space-dependent beam velocity and beam displacement at the times $t/T = n/20$ (T=duration of a period, $n = 0, 1, 2, 3...20$) in the interval $-1 < x/\lambda_B < 1$. Give an account of the vibrations for the velocity $v(x = 0, t)$ and the beam's displacement $\xi(x = 0, t)$ at the point of the beam just below the point of induction $x = 0$. When do $v(x = 0, t)$ and $\xi(x = 0, t)$ finally reach their maxima in a period $0 < t < T$?

Problem 6

Find the sound velocity of an infinitely long beam induced to bending wave vibrations at its middle ($x = 0$) by a time-dependent sinusoidal concentrated force at an amplitude F_0, as a function of a field point x on the beam.

5

Elastic isolation

The most common solution to problems due to vibration impact into buildings or the ground, is to isolate the machinery, engines or other aggregates from their bearing foundations by spring elements. Applications for this technique, called elastic decoupling, are:

- mounting machinery on single springs to decouple them from the building (Fig.5.1)
- sub-ballast mats for train or subway rails in close vicinity of houses to reduce the vibration impact (Fig. 5.2) and
- the floating concrete floor, nowadays almost always used in buildings (Fig. 5.3, see also Chap. 8 on transmission loss)

The aforementioned examples span a wide range of applications and technical solutions.

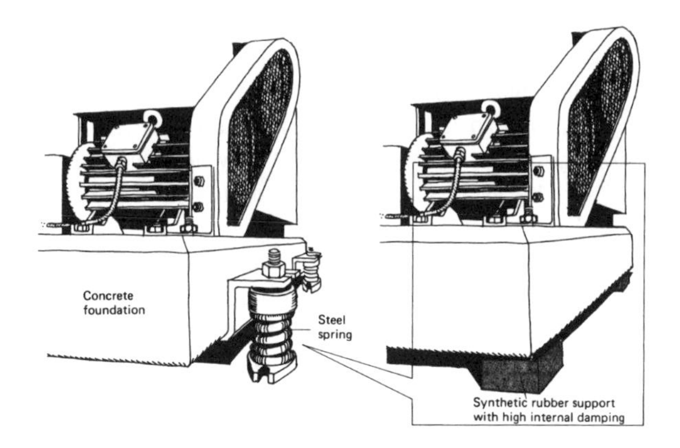

Fig. 5.1. Machine bearing on single springs

M. Möser, *Engineering Acoustics*, DOI 10.1007/978-3-540-92723-5_5,

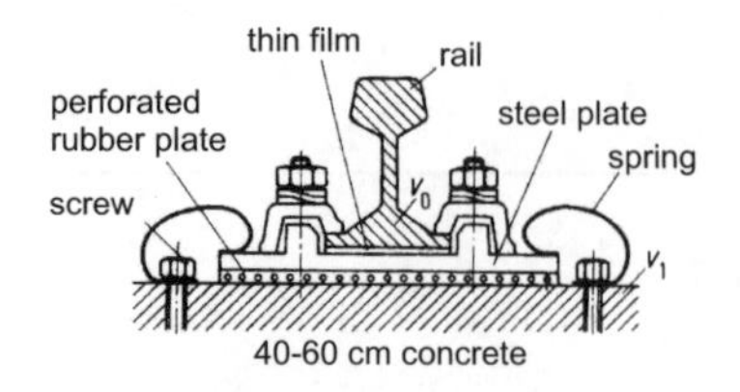

Fig. 5.2. Elastic decoupling of the rails against the track sub-structure on a thick plate

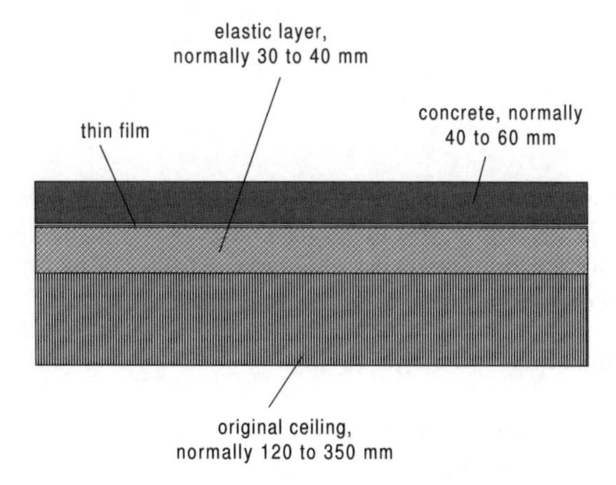

Fig. 5.3. Structure of a floating concrete floor for noise impact level reduction

It is, for instance, reasonable (instead of using a single point machine bearing as in Fig. 5.1) to mount an aggregate or machinery on a solid foundation (a couple of centimeters of concrete) and then decouple it on the whole surface against the foundation using a soft elastic layer. Technical appliances (like cooling aggregates) are often composed of several sub-structures jointed by cables and tubes, instead of representing a 'compact structure'. Not only for this reason, but also to increase the mass, mounting the machinery on a heavy foundation seems to be reasonable. The whole diversity of applications is also indicated by the variety of spring elements and elastic plates. A small survey of spring elements is presented in Fig. 5.4.

The discussions in this chapter will attempt to answer questions pertaining to elastic bearings, not only with respect to the basic principles, but also with respect to practical applications. The following section provides information about the basic principles by introducing the simplest possible model. The physical effects originally left out of the discussion will be investigated in order to gain a more realistic insight as to the expected level reduction. Finally some questions of practical interest will be answered such as how elastic bearings are designed and what are the conditions under which they make sense at all?

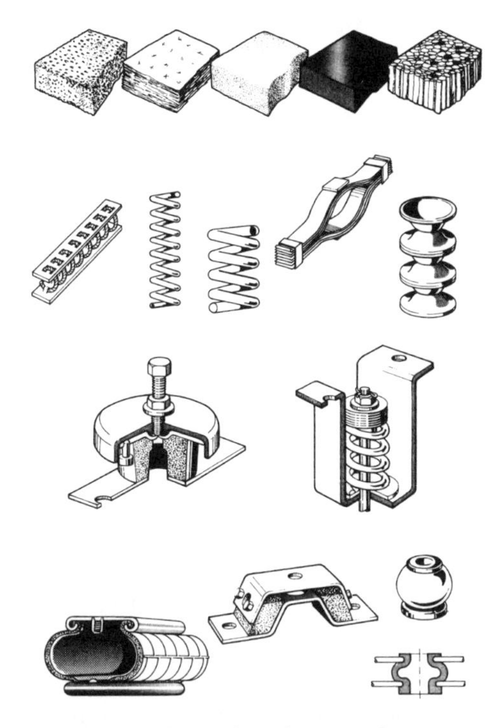

Fig. 5.4. Examples of spring elements

5.1 Elastic bearings on rigid foundations

The simplest model used to describe elastic decoupling consists of modelling the machinery (the engine, the aggregate, the train,...) as an inert mass which is excited by an alternating force F, bringing it to oscillation. It rests upon a rigid, stationary foundation to which it is connected by a spring (Fig. 5.5). The actual relevance of the foundation compliance under consideration will first be explained in Sect. 5.3. The inherent inner damping of the spring is accounted for by assuming the presence of a viscous damper.

Three external forces acting on the mass are

- the exciting force F,
- the resetting spring force F_s in opposite direction to F and
- the likewise resetting damping force F_r.

According to Newton's law, the sum of the aforementioned forces results in an accelerated motion of the mass

$$m\ddot{x} = F - F_\mathrm{s} - F_\mathrm{r} \,, \tag{5.1}$$

where x represents the displacement of the mass ($\dot{x}$: the velocity, $\ddot{x}$: the acceleration) counting into the direction of the force F. The resulting resetting forces F_s and F_r are defined by

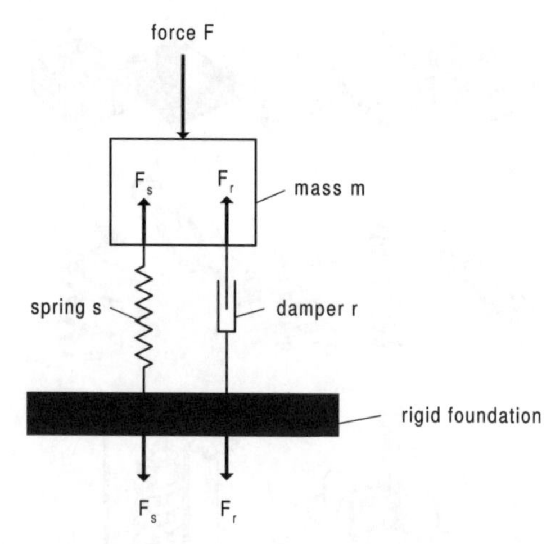

Fig. 5.5. Model for the calculation of the insertion loss of elastic bearings on a rigid foundation

- Hook's law (s = stiffness of the spring), result in

$$F_{\mathrm{s}} = sx \tag{5.2}$$

- and, assuming a damping force proportional to the velocity, in

$$F_{\mathrm{r}} = r\dot{x}, \tag{5.3}$$

  where r is the damping coefficient.

The equation of the oscillating mass is thus given by

$$m\ddot{x} + r\dot{x} + sx = F\ . \tag{5.4}$$

Assuming pure tones

$$x(t) = \mathrm{Re}\left\{x\mathrm{e}^{j\omega t}\right\}$$

(underlining the complex pointer x is omitted for simplicity)

$$-m\omega^2 x + j\omega r x + sx = F\ , \tag{5.5}$$

is obtained or of course

$$x = \frac{F}{s - m\omega^2 + j\omega r}\ . \tag{5.6}$$

When assessing the benefit of the elastic bearing, the force F_{F} acting on the foundation is mainly of interest. It is composed of the spring force and the damping force and is given by

$$F_{\mathrm{F}} = F_{\mathrm{s}} + F_{\mathrm{r}} \tag{5.7}$$

or, using (5.2) and (5.3) and assuming pure tones again,

$$F_F = (s + j\omega r)x \tag{5.8}$$

or, using (5.6),

$$F_F = \frac{s + j\omega r}{s - m\omega^2 + j\omega r} F \,. \tag{5.9}$$

To evaluate the success of the 'elastic bearing' compared to the 'rigid foundation' the ratio V is used as a measure

$$V = \frac{\text{fondation force, rigid}}{\text{foundation force, elastic}} = \frac{F_F(s \to \infty)}{F_F(s)} \,, \tag{5.10}$$

which according to (5.9) yields

$$V = \frac{s - m\omega^2 + j\omega r}{s + j\omega r} \,. \tag{5.11}$$

Finally, the insertion loss (i.e. the level difference of the foundation force 'without' minus 'with' elastic bearing) is defined as

$$R_E = 10 \lg |V|^2 \,. \tag{5.12}$$

R_E is a decibel measure for the success of the elastic bearing.

Obviously, the resonance frequency

$$\omega_0 = \sqrt{\frac{s}{m}} \tag{5.13}$$

plays an important role when interpreting the ratio V (and thus the insertion loss). In the lossless case $r = 0$ the mass displacement x can take infinite values according to (5.6) at the resonance $\omega = \omega_0$. Also, the ratio V clearly behaves totally differently at low frequencies $\omega \ll \omega_0$ than at high frequencies $\omega \gg \omega_0$.

(5.11) can be reformed by dividing the numerator and denominator by s, yielding a ratio of frequencies:

$$V = \frac{1 - \frac{\omega^2}{\omega_0^2} + j\eta\frac{\omega}{\omega_0}}{1 + j\eta\frac{\omega}{\omega_0}} \,. \tag{5.14}$$

The damping coefficient r is herein expressed as a dimensionless loss factor

$$\eta = \frac{r\omega_0}{s} \,. \tag{5.15}$$

As will be explained in one of the following sections, the loss factor can easily be determined by measurements. It was thus reasonable to replace the quantity 'damping coefficient r', which is somewhat difficult to assess, by a quantity which is well measurable. The loss factor η for common springs or elastic layers (apart from a few exceptions) takes on values in the range $0.01 < \eta < 1$.

Generally, four frequency ranges need to be considered when discussing the validity of (5.14):

1. **Low frequencies** $\omega \ll \omega_0$
   The elastic bearing is still ineffective: according to (5.14) $V \approx 1$ and thus $R_\mathrm{E} \approx 0\,\mathrm{dB}$.

2. **Mid to high frequencies ($\omega \gg \omega_0$, but also $\omega \ll \omega_0/\eta$)**
   In this frequency range it is

$$V \approx -\frac{\omega^2}{\omega_0^2}$$

   and thus

$$R_\mathrm{E} = 10\lg\frac{\omega^4}{\omega_0^4} = 40\frac{\omega}{\omega_0}\,. \tag{5.16}$$

   The insertion loss rapidly increases with increasing frequency at 12 dB per octave and can take on considerable values (e.g. $R_\mathrm{E} = 36\,\mathrm{dB}$ three octaves above the resonance frequency).

3. **Highest frequencies**
   The presence of damping reduces this steep gradient at the highest frequencies $\omega \gg \omega_0/\eta$ (and $\omega \gg \omega_0$). Here, it is only

$$V = j\frac{\omega}{\omega_0}\frac{1}{\eta}\,,$$

   leading to

$$R_\mathrm{E} = 20\lg\left(\frac{\omega}{\omega_0}\frac{1}{\eta}\right) = 20\lg\left(\frac{\omega}{\omega_0}\right) - 20\lg\eta \tag{5.17}$$

   for the insertion loss. R_E increases by only 6 dB/octave and depends on the loss factor. The larger η is, the smaller R_E is.

4. **Resonance range** $\omega \approx \omega_0$
   In close vicinity to the resonance frequency, the elastic bearing performs poorly compared to rigid coupling to the foundation. For $\omega = \omega_0$ it is

$$V = \frac{j\eta}{1+j\eta}\,,$$

   which, for small loss factors $\eta \ll 1$, also implies

$$R_\mathrm{E} \simeq 20\lg\eta\,. \tag{5.18}$$

   At resonance, the insertion loss is therefore negative. Therefore, the deterioration implicit in the insertion loss is greater the smaller the loss factor η is.

A summary of the aforementioned details can be found in the frequency response function of R_E shown in Fig. 5.6. Here, R_E is calculated for the various loss factors η according to (5.14) (and (5.12)) over the frequency ratio ω/ω_0. The tendencies described earlier can be observed:

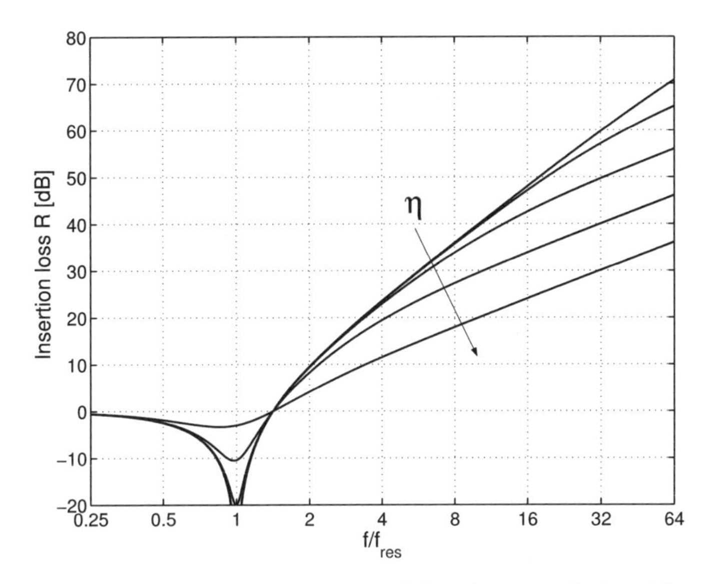

Fig. 5.6. Theoretical insertion loss on a rigid foundation, calculated for $\eta = 0.01$, 0.0316, 0.1 and 0.316

- no effect below the resonance frequency,
- deterioration at the resonance, moderated by the increasing loss factor,
- steep ascent of R_E at 12 dB/octave in a frequency range which gets narrower with increasing η, and, finally,
- deviation to an ascent of only 6 dB/octave, where an increasing loss factor decreases the insertion loss.

It is evident that an increasing loss factor alleviates the disadvantages in the vicinity of the resonance, on the one hand, yet on the other hand, it limits the advantages at the high frequencies. The latter high frequency deterioration due to the loss factor of the spring is in practice only of minor interest: such large sound reductions, as calculated here, are seldom obtained in practice. The main reason for that is the fact that real foundations have a finite compliance. This effect and its influence on the insertion loss will be investigated in more detail in the next section.

As already mentioned, the limitation due to damping found at higher frequencies is irrelevant in everyday practice. Often larger loss factors are preferred in applications where the machine (the engine, the aggregate, etc.) operates at a frequency far above the resonance. It has to be kept in mind during the process of turning on or off the machine, it passes through the resonance frequency domain. The displacement, which occurs at resonance, is specified, using (5.6) and (5.15), by

$$x(\omega = \omega_0) = \frac{F}{j\omega_0 r} = \frac{F}{j\eta s} \, . \tag{5.19}$$

The displacement by all means needs to have a limit; otherwise the aggregate can 'bounce' and possibly damage the plant.

It should finally be noted that the extent to which damping influences the insertion loss is determined by the assumed effect of viscous damping, as represented by 5.3. Other assumptions about the type of damping (e.g. so-called relaxation damping) can also be made. In addition, it is often prudent, whenever the cause of viscosity is unknown, to analyze the damping by way of complex spring stiffness. If this is the case, one can formulate 5.5) as

$$-m\omega^2 x + s(1 + j\eta)x = F \tag{5.20}$$

instead. As it is obvious, the frequency dependence of the damping force is dropped. This leads to a different insertion loss than that given in Fig. 5.6, especially at the highest frequencies. It should be clear by now that theoretical results strongly depend on the assumptions made beforehand. If another, perhaps even more optimistic evaluation of elastic bearings can be found somewhere else (e.g. in company catalogues), it may in fact be based on different assumptions, but most certainly not on metaphysical miracles.

5.2 Designing elastic bearings

From the acoustician's point of view, the practical design of spring elements is a simple task: the larger the ratio of operating frequency(-ies) ω to the resonance frequency ω_0 is, the greater the success of the solution will be. For this reason, an attempt is made to tune the resonance frequency as low as possible, preferably to 0 Hz.

Obviously, this is impossible. The machine (or whatever vibrating source it may be) needs to hover in order to achieve this, due to $s = 0$. As a matter of fact, the practical design of springs or elastic layers is dominated by 'non-acoustic' conditions. Such conditions may be:

a) Static spring loads

The springs have to be designed in such a way that they are able to carry their bearing weight statically. For high-grade springs or elastic layers, the manufacturer usually states the load limits of his product range. Some examples are given in Table 5.1.

The material described above is intended for use in bearings covering a surface area (similar to that in Fig. 5.7). As is generally known, the static pressure p_{stat} in this arrangement results in

$$p_{\mathrm{stat}} = \frac{Mg}{S} \tag{5.21}$$

Table 5.1. Description of the product range SYLOMER (according to technical data of the manufacturer)

Product type	G	L	P
Thickness (mm)	12/25	12/25	12/25
Density (kg/m^3)	150	300	510
Load limit (N/mm^2)	0.01	0.05	0.2
Loss factor	0.23	0.2	0.16
E modulus (N/mm^2)	0.18	0.35	2.2
	to 0.36	to 1.1	to 3.6

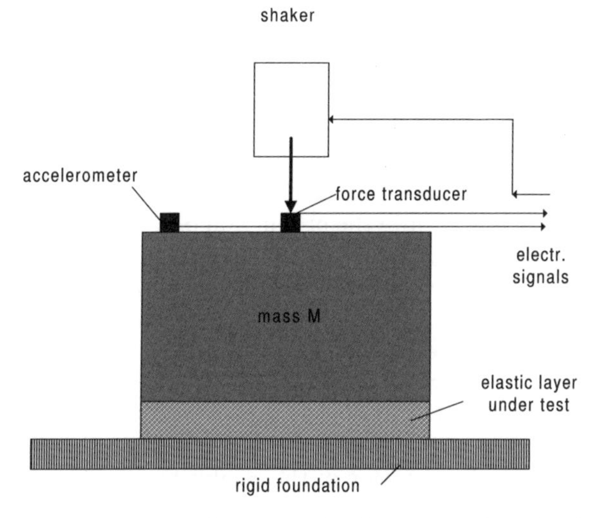

Fig. 5.7. Measurement setup to determine the elastic modulus E and the loss factor η of an elastic layer with surface area S

(M = total mass, S = bearing surface, g = ground acceleration $10\,\mathrm{m/s^2}$). If, for instance, $M = 1000\,\mathrm{kg}$ and $S = 1\,\mathrm{m^2}$ it is therefore $p_{\mathrm{stat}} = 10000\,\mathrm{kg\,m/(s^2m^2)}$ $= 104\,\mathrm{N/m^2} = 10^{-2}\,\mathrm{N/mm^2}$.

Thus, the type G had to be chosen from the product range in Table 5.1. It can also be verified by the resonance frequency if the so-defined bearing is acoustically reasonable. First, the stiffness of the elastic layer must be calculated using the elastic modulus E and the thickness d

$$s = \frac{ES}{d}\,, \tag{5.22}$$

and thus arriving at the resonance frequency of

$$f_0 = \frac{1}{2\pi}\sqrt{\frac{s}{M}} = \frac{1}{2\pi}\sqrt{\frac{ES}{Md}}\,. \tag{5.23}$$

In the aforementioned example ($M = 1000\,\text{kg}$, $S = 1\,\text{m}^2$, type G from Table 5.1, with d=25 mm and $E = 0.2\,\text{N/mm}^2$), the resonance frequency would approximately be $f_0 = 14\,\text{Hz}$.

It only makes sense to use this specific isolation if the operating frequencies f are at least one octave above the resonance frequency. The lowest possible operating frequency can often be determined from the rotation frequency of the machine, for example. Otherwise, its frequency components have to be determined by measurements.

The resonance frequency can be shifted toward lower frequencies, if necessary, by adding multiple elastic layers. This is equivalent to increasing the thickness. However, the stability of the whole structure puts a limit on this: the layer thickness can only be a small percentage of the smallest dimension.

b) Operating conditions

Some devices are only required to perform very small motions. This is the case, for example, for medical lasers or magnetic resonance scanners, but also for certain printing machinery or in the production of semi-conductors. Of course, the allowed motion certainly have to be taken into account. Most problems of that kind can be solved by mounting the device on a large add-on mass and elastically bearing both together.

Railway vehicles are likewise required to meet the highest safety standards. Train rails on sub-ballast mats, for example, should not 'statically' sink by a certain designated amount.

We are also often confronted with the task of scrutinizing limited available data for industrial implementation or the establishment of general safeguards. The elastic modulus (or Young's modulus) E can be determined by way of static experiments, where the sample surface S (of thickness d) is uniformly loaded with a mass M, and the resulting static displacement x_{stat} is measured. As is generally known, the force equation is given by

$$s x_{\text{stat}} = Mg \,, \tag{5.24}$$

or, using (5.22),

$$E = \frac{Mgd}{x_{\text{stat}} S} = p_{\text{stat}} \frac{d}{x_{\text{stat}}} \,, \tag{5.25}$$

where p_{stat} again denotes the static pressure (for the measurement setup).

If the static load is unknown or if the manufacturer's data has only been subject to a cursory review. Most materials allow for a change of their thickness by a maximum of 5% to 10% under static load, whereby it is always safest to assume the 5%. To fulfill this criterium, the required elastic modulus E, according to (5.25), is given by

$$E_{\text{erf}} = 20 p_{\text{stat}} \tag{5.26}$$

(where p_{stat} now refers to the *actual application* and not the experimental setup).

The material must have at least this elastic modulus E_{erf} in order to remain stable over a long period. Equation (5.26) represents a quite realistic dependence between the static load p_{stat} and the material parameter E, as shown in the examples in Table 5.1. The elastic moduli of the materials are roughly 10 to 20 times larger than the static load limits.

It should finally be mentioned that the elastic modulus of a sample can also be determined by measuring the resonance frequency with a setup as shown in Fig. 5.7. According to (5.13) it is

$$s = \frac{ES}{d} = M\omega_0^2 \,. \tag{5.27}$$

5.3 Influence of foundations with a compliance

Before the influence of the foundation compliance on the sound reduction of the elastic bearing can be discussed, the compliance itself has to be described by a technical quantity. It is typically described in terms of foundation impedance, which will be explained in the following.

5.3.1 Foundation impedance

Foundation impedance z_{F} is defined as the ratio of a force F_{F}, exciting the foundation at a fixed point, to the velocity v_{F}, occurring at the same point

$$z_{\text{F}} = \frac{F_{\text{F}}}{v_{\text{F}}} \,. \tag{5.28}$$

z_{F} is inversely proportional to the 'mobility' of the foundation, as can easily be shown by solving (5.28) for v_{F}

$$v_{\text{F}} = \frac{F_{\text{F}}}{z_{\text{F}}} \,.$$

Given a constant exciting force F_{F}, 'small motion' is obtained for large magnitudes of z_{F}, whereas small magnitudes of z_{F} lead to large foundation velocities. In this chapter, displacements will be calculated. Therefore, the dependence between foundation velocity v_{F} and foundation displacement x_{F} (for pure tones) is given by

$$v_{\text{F}} = j\omega x_{\text{F}} \,, \tag{5.29}$$

resulting in the subsequent dependence

$$F_{\text{F}} = j\omega z_{\text{F}} x_{\text{F}}. \tag{5.30}$$

Generally, the complex impedance may result in a complicated frequency response function which can be specified by measurement if necessary. If, for instance, the foundation represents a simple resonator itself (which might be the case for certain cornerstone foundations in buildings, for instance) the equation of motion (similar to (5.5)) is given by

$$\left[jm_{\mathrm{F}}\omega + \frac{s_{\mathrm{F}}}{j\omega} + r_{\mathrm{F}}\right] v_{\mathrm{F}} = F_{\mathrm{F}} , \tag{5.31}$$

and yields a foundation impedance

$$z_{\mathrm{F}} = jm_{\mathrm{F}}\omega + \frac{s_{\mathrm{F}}}{j\omega} + r_{\mathrm{F}} . \tag{5.32}$$

Examining such (or even more complicated) frequency response functions of the foundation impedance with respect to the design of elastic bearings could prove to be a valuable task for practical problems. To elaborate on the basic principles, it may be far more reasonable to restrict the discussion to certain 'impedance types'. Therefore, only impedances with mass characteristics $z_{\mathrm{F}} = j\omega m_{\mathrm{F}}$ and with spring characteristics $z_{\mathrm{F}} = s_{\mathrm{F}}/j\omega$ will be investigated in the following.

5.3.2 The effect of foundation impedance

First, the equation of motion has to be set up to be able to describe the effect of finite foundation impedance, as was done in Sect. 5.1, where this time we assume a model including a mobile foundation (Fig. 5.8).

As usual, the inertia force has to be compensated by the sum of the exciting force and the resetting spring and damping forces. Equation (5.1) for the mass displacement x therefore remains the same and is given by

$$m\ddot{x} = F - F_{\mathrm{s}} - F_{\mathrm{r}} . \tag{5.33}$$

But this time, the spring force is proportional to the *difference* of the mass displacement x and foundation displacement x_{F}

$$F_{\mathrm{s}} = s(x - x_{\mathrm{F}}) , \tag{5.34}$$

and in the same way for the damping force

$$F_{\mathrm{r}} = r(\dot{x} - \dot{x_{\mathrm{F}}}) . \tag{5.35}$$

The equation of motion (5.1) thus results in

$$m\ddot{x} + r(\dot{x} - \dot{x_{\mathrm{F}}}) + s(x - x_{\mathrm{F}}) = F . \tag{5.36}$$

The force F_{F} acting on the foundation is of primary interest here and it is given by

$$F_{\mathrm{F}} = F_{\mathrm{s}} + F_{\mathrm{r}} = s(x - x_{\mathrm{F}}) + r(\dot{x} - \dot{x}_F) \,. \tag{5.37}$$

Using complex amplitudes, (5.36) and (5.37) yield

$$-m\omega^2 x + (s + j\omega r)(x - x_{\mathrm{F}}) = F \tag{5.38}$$

and

$$F_{\mathrm{F}} = (s + j\omega r)(x - x_{\mathrm{F}}) \,, \tag{5.39}$$

where the foundation compliance according to (5.30) is additionally described by

$$F_{\mathrm{F}} = j\omega z_{\mathrm{F}} x_{\mathrm{F}} \,. \tag{5.40}$$

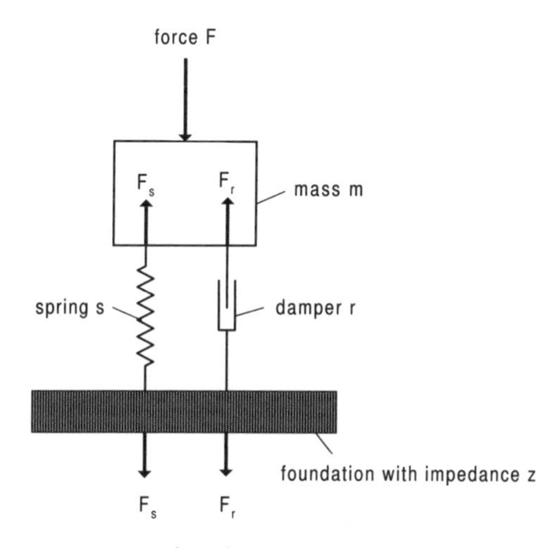

Fig. 5.8. Model arrangement for the calculation of the insertion loss of elastic bearings with compliant foundation

Equations (5.38) to (5.40) mathematically form a set of linear equations with the three unknown quantities x, x_{F} and F_{F}. They can be solved using the standard procedures. The result for F_{F} is especially of major interest. Without excluding other solving procedures, the author suggests the following procedure as being simple and thus reasonable:

1. Add $m\omega^2 x_{\mathrm{F}}$ on both sides of (5.38), obtaining the result

$$\left[-m\omega^2 + s + j\omega r\right](x - x_{\mathrm{F}}) = F + m\omega^2 x_{\mathrm{F}}$$

2. Express x_{F} by F_{F} using (5.40)

$$\left[-m\omega^2 + s + j\omega r\right](x - x_{\mathrm{F}}) = F + \frac{m\omega^2}{j\omega z_{\mathrm{F}}} F_{\mathrm{F}} = F - \frac{jm\omega}{z_{\mathrm{F}}} F_{\mathrm{F}}$$

3. According to (5.39) this results in

$$F_{\mathrm{F}} = \frac{s + j\omega r}{s - m\omega^2 + j\omega r} \left[F - \frac{jm\omega}{z_{\mathrm{F}}} F_{\mathrm{F}} \right]$$

Solving this for F_{F} finally yields

$$F_{\mathrm{F}} = \frac{s + j\omega r}{(s + j\omega r)\left(1 + \frac{jm\omega}{z_{\mathrm{F}}}\right) - m\omega^2} F \tag{5.41}$$

Again, the advantage gained by the elastic bearing and thus the ratio V is of interest here. This is given by

$$V = \frac{F_{\mathrm{F}}(s \to \infty)}{F_{\mathrm{F}}(s)}$$

and the insertion loss is again given by

$$R_{\mathrm{E}} = 10 \lg |V|^2 \, .$$

From (5.41) it follows that

$$F_{\mathrm{F}}(s \to \infty) = \frac{F}{1 + \frac{j\omega m}{z_{\mathrm{F}}}} \, ,$$

resulting in the ratio V of

$$V = 1 - \frac{\frac{m\omega^2}{s + j\omega r}}{1 + \frac{j\omega m}{z_{\mathrm{F}}}} \, , \tag{5.42}$$

or, if, for the sake of clarity, the resonance frequency (of the rigid foundation) ω_0 and the loss factor η are inserted with the aid of (5.13) and (5.15) ,

$$\omega_0^2 = s/m \quad \text{and} \quad \eta = \frac{r\omega_0}{s} \, ,$$

$$V = 1 - \frac{\omega^2}{\omega_0^2} \frac{1}{\left(1 + j\frac{\omega m}{z_{\mathrm{F}}}\right)} \frac{1}{\left(1 + j\eta\frac{\omega}{\omega_0}\right)} \tag{5.43}$$

is obtained.

This somewhat lengthy calculation (which, by the way, fortunately yields (5.14) for $z_{\mathrm{F}} \to \infty$ as a check) shows some remarkable results:

a) The foundation impedance with mass Characteristics, $z_{\mathrm{F}} = j\omega m_{\mathrm{F}}$

is

$$V = 1 - \frac{\omega^2}{\omega_0^2\left(1+\frac{m}{m_\mathrm{F}}\right)} \frac{1}{1+j\eta\frac{\omega}{\omega_0}} \,. \tag{5.44}$$

The finite foundation impedance acts as if detuning the resonance frequency towards higher frequencies, producing the same results as in the case of rigid foundations

$$V = 1 - \frac{\omega^2}{\omega_\mathrm{res}^2} \frac{1}{1+j\eta\frac{\omega}{\omega_0}} \,, \tag{5.45}$$

where the resonance frequency depends on the mass (i.e. the device) *and* the foundation:

$$\omega_\mathrm{res}^2 = s\left(\frac{1}{m}+\frac{1}{m_\mathrm{F}}\right) \,. \tag{5.46}$$

Here, the interpretations for the different frequency ranges, as given in Sect. 5.1, are also valid.

b) The foundation impedance with spring characteristics, $z_\mathrm{F} = s_\mathrm{F}/j\omega$

is

$$V = 1 - \frac{\omega^2}{\omega_0^2\left(1-\frac{\omega^2 m}{s_\mathrm{F}}\right)} \frac{1}{1+j\eta\frac{\omega}{\omega_0}} \,. \tag{5.47}$$

Naturally, a second resonance effect occurs in this case, because the mass (i.e. the mechanical device) and the foundation already form a resonator with the mass foundation resonance frequency

$$\omega_\mathrm{mF}^2 = \frac{s_\mathrm{F}}{m} \,, \tag{5.48}$$

therefore resulting in

$$V = 1 - \frac{\omega^2}{\omega_0^2\left(1-\frac{\omega^2}{\omega_\mathrm{mF}^2}\right)} \frac{1}{1+j\eta\frac{\omega}{\omega_0}} \,. \tag{5.49}$$

For practical applications, it is a safe assumption that the elastic bearing is much softer than the foundation, i.e.

$$s \ll s_\mathrm{F}$$

and thus

$$\omega_\mathrm{mF} \gg \omega_0 \,.$$

The interesting conclusion which can be drawn from (5.49) is the fact that the ratio V becomes independent of the frequency for high frequencies $\omega \gg m_\mathrm{F}$ (and small spring losses $\eta \approx 0$):

$$V \simeq 1 + \frac{\omega_{\mathrm{mF}}^2}{\omega_0^2} = 1 + \frac{s_{\mathrm{F}}}{s} \approx \frac{s_{\mathrm{F}}}{s} . \tag{5.50}$$

The insertion loss is given solely by the ratio s_{F} and s:

$$R_{\mathrm{E}} \approx 20 \lg \frac{s_{\mathrm{F}}}{s} . \tag{5.51}$$

An elastic bearing consisting of a spring, or an elastic layer, which has a stiffness of 10% of the foundation's stiffness, therefore has an insertion loss of 20 dB. Theoretically, the insertion loss, according to (5.51), could even decrease with increasing frequency for sufficiently large η.

The low frequency range is easily explained. At the lowest frequencies, it is again

$$R_{\mathrm{E}} \approx 0\,\mathrm{dB} , \tag{5.52}$$

followed by the resonance area with a negative R_{E}.

Strictly speaking, the resonance frequency is now given by

$$\omega_{\mathrm{A}}^2 = \omega_0^2 \left(1 - \frac{\omega_{\mathrm{A}}^2}{\omega_{\mathrm{mF}}^2}\right) \tag{5.53}$$

and is therefore

$$\frac{1}{\omega_{\mathrm{A}}^2} = \frac{1}{\omega_0^2} + \frac{1}{\omega_{\mathrm{mF}}^2} . \tag{5.54}$$

Often it is $\omega_{\mathrm{mF}}^2 \gg \omega_0^2$, so that, using $\omega_{\mathrm{A}} \approx \omega_0$, the detuning only plays a minor role.

In contrast, R_{E} tends to infinity at the 'mass-foundation resonance' $\omega \approx \omega_{\mathrm{mF}}$. The reason for that is simple: without the elastic bearing, mass and foundation are at resonance; the force $F(s \to \infty)$ acting on the foundation becomes infinite, because no damping was assumed in the spring of the foundation. As the foundation force F_{F} now stays finite due to the elastic bearing, they both seem to cause an 'infinitely large improvement'.

The characteristics in Fig. 5.9 account for a foundation damping in the calculation of V in (5.49) by additionally assuming a complex spring stiffness

$$s_{\mathrm{F}} \to s_{\mathrm{F}}(1 + j\eta_{\mathrm{F}}) \tag{5.55}$$

and therefore also a complex resonance frequency

$$\omega_{\mathrm{mF}}^2 \to \omega_{\mathrm{mF}}^2(1 + j\eta_{\mathrm{F}}) \tag{5.56}$$

which were used to calculate $R_{\mathrm{E}} = 10 \lg |V|^2$.

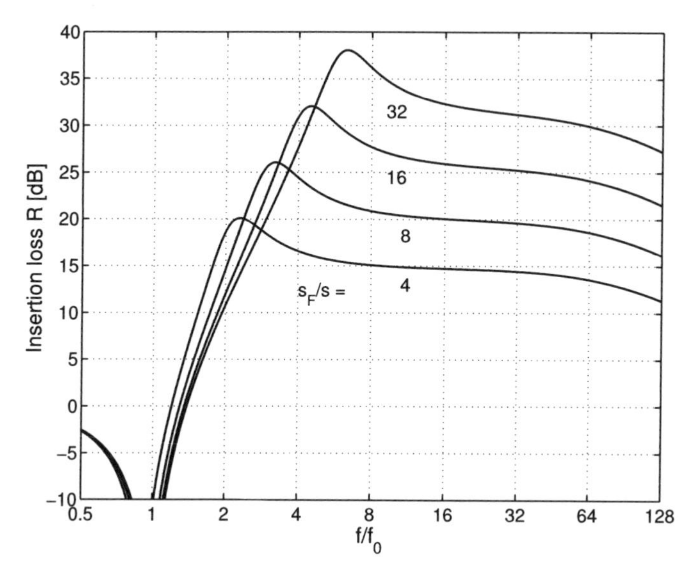

Fig. 5.9. Insertion loss for a foundation with spring characteristics, calculated for $\eta = 0.01$ and $\eta_{\text{foundation}} = 0.5$

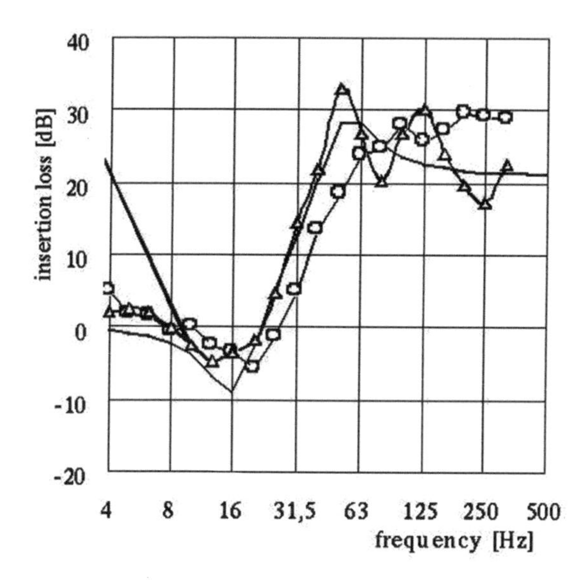

Fig. 5.10. Insertion loss of the sub-ballast mat Sylodyn CN235. Experiment: average over several positions and train types; southbound trains (*triangles*), northbound trains (*circles*). Calculation for dynamic stiffness of $s'' = 0,022\,\mathrm{N/mm^3}$ (*solid line*)

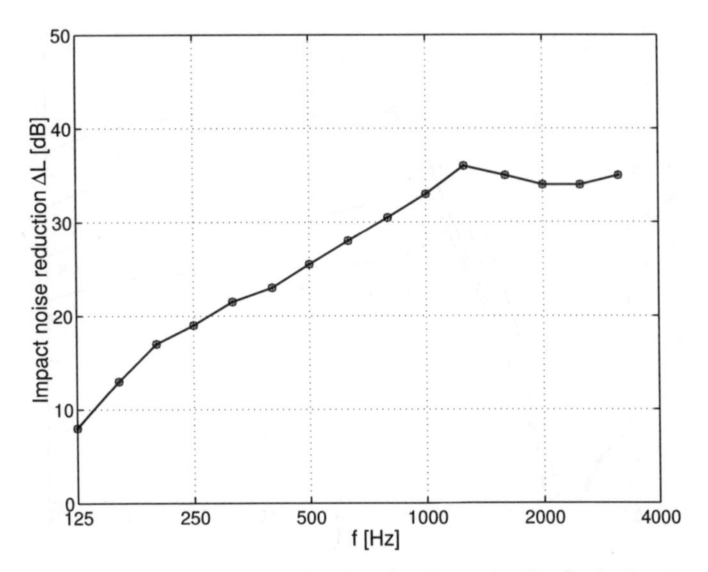

Fig. 5.11. Measured impact noise reduction (i.e. insertion loss) of a floating concrete floor. The ceiling structure consists of 50 mm cement floor on 0.2 mm PE foil on 35 mm solid foam plate on 120 mm reinforced concrete

To summarize, it can be stated that the foundation impedance has a considerable influence on the insertion loss. More exact statements on the effect of an elastic bearing require detailed knowledge of z_F. More generally, it can be stated that the actual characteristics at higher frequencies are roughly located between a frequency-independent line and a straight line with a slope of 12 dB/octave.

Experimental values, as shown in Fig. 5.10 and 5.11, behave accordingly. The slope of 12 dB/octave, valid only for a rigid foundation, is never actually achieved.

The example in Fig. 5.10 indicates that it represents a foundation with spring characteristics. The frequency response in Fig. 5.11 does not indicate a particular frequency response of a characteristic impedance – but a 'floating' concrete floor (a cement floor, resting upon an elastic layer, placed between the raw ceiling and the cement, see Fig. 5.3) does achieve quite a large insertion loss.

As final practical piece of advice, it suffices to say that either the details of any given problem have to be analyzed more accurately, or one should be aware of far too unrealistic promises about the effect of elastic bearings.

5.4 Determining the transfer path

Even for sufficiently heavy or rigid foundations the application of elastic bearings is not reasonable in any situation. This is the case, for example, whenever the sound field as observed from any given vantage point, is not produced by a force acting on the foundation of the machine.

A typical situation which can often be found in buildings is sketched in Fig. 5.12. An appliance mounted on one floor that produces unwanted high noise levels in another room. Will an additional elastic decoupling of the device be useful in that case?

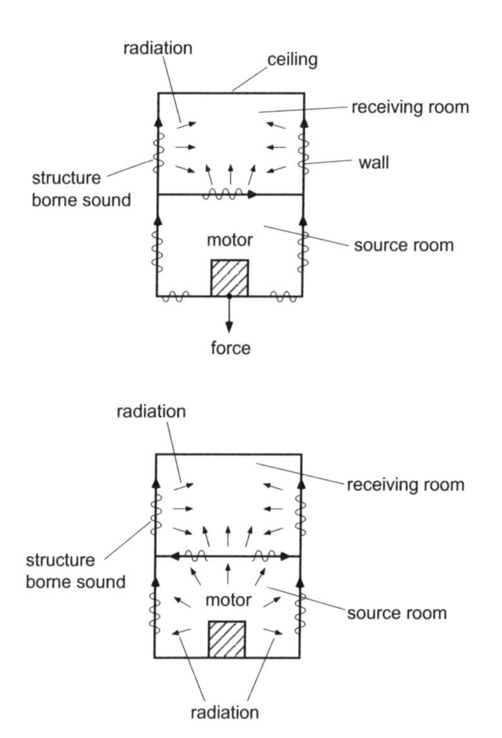

Fig. 5.12. The two transfer paths under consideration

To answer this question, it has to be kept in mind that the sound transmission (as sketched in Fig.5.12) can take two different paths:

1. Structure-borne vibrations are excited in the ceiling by an induced force and are transmitted to the flanking components, i.e. walls and ceilings. The wall motions of the 'receiving room walls' radiate air-borne sound. For simplicity this transfer path is called the 'structure-borne path'.
2. Simultaneously most of the devices directly radiate air-borne sound into the 'source room' where they are located. This air-borne sound also acts

on the adjoining walls as an exciting force (actually a spatially distributed pressure). Vibrations of the walls and ceilings are thereby produced, which are transmitted through the building and radiate into the receiving room. This transfer path is called the 'air-borne path'.

It should be obvious that an elastic bearing can only reduce the noise in the receiving room if the sound field transmitted along the structure-borne path is notably larger than that which is transmitted along the air-borne path. If necessary, the question of which of the two paths dominates a particular acoustic situation has to be verified by measurements.

The simplest test is to turn the machine off and put a loudspeaker into the source room in order to produce an artificial sound field. The measured level differences ΔL_{M} (i.e. the level in the receiving room minus the level in the source room while operating the machine) and ΔL_{L} (i.e. the level in the receiving room minus the level in the source room while operating the loudspeaker) can be used to deduce the transfer path. If ΔL_{M} is considerably larger than ΔL_{L} the transmission of structure-borne sound predominates that of air-borne sound. Elastic decoupling of the machine against the foundation makes sense in that case. The level reduction which can be expected by this solution is a maximum of $\Delta L_{\mathrm{M}} - \Delta L_{\mathrm{L}}$, because the air-borne path dominates the transmission beyond this point.

If, in contrast, ΔL_{M} is roughly equal to ΔL_{L}, either both transfer paths have about the same importance, or the air-borne sound dominates. An additional measurement is needed to distinguish these two cases. Artificial forces are induced into the foundation of the machine using shakers or suitable hammers. The source characteristics are taken from the velocity level on the foundation. The level differences ΔL_{M} (i.e. the level in the receiving room minus the level in the source room while operating the machine) and ΔL_{S} (i.e. the level in the receiving room minus the level in the source room while operating the shaker) are measured. If ΔL_{M} is considerably larger than ΔL_{S}, the air-borne path has precedence over the structure-borne path. Elastic isolation is absolutely useless in this case. Instead, the transmission loss between the two rooms needs to be improved (e.g. by using soft bending shells, see Chap. 8).

5.5 Determining the loss factor

To determine the loss factor the frequency response function of the ratio of the displacement x and the induced force F have to be measured. The experimental setup, requiring the use of an appropriate capturer, is depicted in Fig. 5.7. According to (5.6), the ratio is expected to be

$$\frac{x}{F} = \frac{1}{s} \frac{1}{1 - \frac{\omega^2}{\omega_0^2} + j\eta \frac{\omega}{\omega_0}} \tag{5.57}$$

(with ω_0 based on (5.13) and η based on (5.15)). The presence of a broad resonance peak offers a significant advantage when measuring large loss factors.

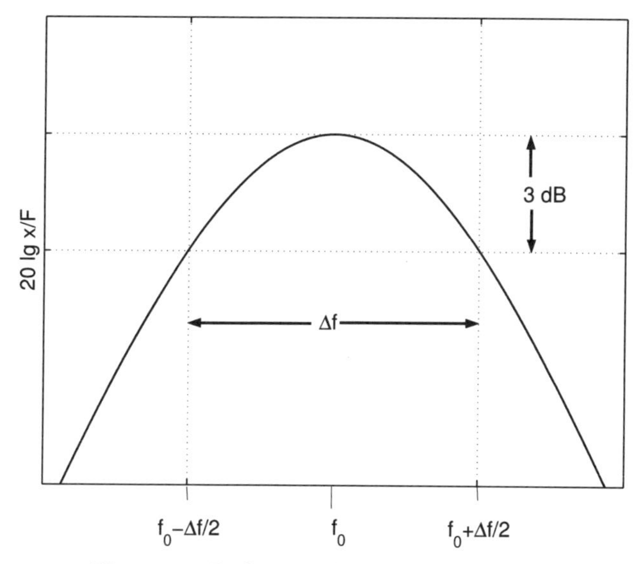

Fig. 5.13. Definition of half-bandwidth Δf

The so-called half-bandwidth $\Delta\omega$ is used (see also Fig. 5.13) as a scale for the width of the peak: on the left and right side of the actual maximum of the magnitude, two frequencies $\omega = \omega_0 + \Delta\omega/2$ and $\omega = \omega_0 - \Delta\omega/2$ exist, where the magnitude $|x/F|^2$ is half that of the maximum itself (as is generally known, half the magnitude corresponds to a level difference of 3 dB to the maximum). The frequency distance between these two points is referred to as half-bandwidth.

The connection between half-bandwidth and the loss-factor according to the aforementioned definition, is described by

$$\frac{1}{\left[1-\left(\frac{\omega_0\pm\Delta\omega/2}{\omega_0}\right)^2\right]^2+\left[\eta\frac{\omega_0\pm\Delta\omega/2}{\omega_0}\right]^2}=\frac{1}{2}\frac{1}{\eta^2}\,. \tag{5.58}$$

For a first order approximation using $\Delta\omega \ll \omega_0$,

$$\eta\frac{\omega_0\pm\Delta\omega/2}{\omega_0}\approx\eta$$

can be assumed, which keeps the error in η small. This results in

$$\left[1-\left(\frac{\omega_0\pm\Delta\omega/2}{\omega_0}\right)^2\right]+\eta^2\approx 2\eta^2\,,$$

or

$$1-\left(\frac{\omega_0\pm\Delta\omega/2}{\omega_0}\right)^2=\pm\eta\,.$$

Using

$$\left(\frac{\omega_0 \pm \Delta\omega/2}{\omega_0}\right)^2 = 1 \pm \frac{\Delta\omega}{\omega_0}\frac{1}{4}\left(\frac{\Delta\omega}{\omega_0}\right)^2 \approx 1 + \frac{\Delta\omega}{\omega_0},$$

where again $\Delta\omega/\omega_0 \ll 1$ is assumed, it follows that

$$\eta = \frac{\Delta\omega}{\omega_0} = \frac{\Delta f}{f_0}, \tag{5.59}$$

where the sign ($\eta > 0$) was chosen in a physically reasonable sense. Equation (5.59) indicates how the loss factor is calculated by the measured half-bandwidth Δf.

In carrying out measurements by digital means, using an FFT spectrum-analyzer, it should be kept in mind that

- a typically rectangular window function with a main lobe as narrow as possible is used and that
- at least six (or preferably more than ten) spectral lines are located within the half-bandwidth. A sufficiently high resolution can be obtained by using the FFT-zoom function, if necessary.

5.6 Dynamic mass

Machines, appliances, lathes, etc., cannot always be considered 'compact masses.' On the contrary, they can be subject to elastic deformations with resonance properties. The "moving," dynamic mass, which is significant for resonance frequencies of bearings, can therefore be far smaller than the total static mass "at rest", as shown in the following.

Take a railway or subway car as an example of an elastic structure consisting of spring parts. In order to ensure the comfort of the passengers, the actual passenger car rests on spring suspensions mounted to the wheel axles. To reduce the entry of vibrations into the ground, the track system is additionally insulated with elastic rail bearings or a sub-ballast mat. The entire structure is basically comprised of two masses and two springs, as depicted in Figure (5.14). In our example, the upper mass m_1 represents the passenger rail car, the spring s_1 consists of the steel spring suspensions between it and the wheel axles, the mass m_2 encompasses the wheel axles themselves, the tracks and the track bed. Finally, s_2 represents the sub-ballast mat, the bottom of which is to be considered rigid. The rolling impact occurs in the contact between the wheel and railway, thereby exerting the stimulating force on m_2. The friction forces will be neglected for the purposes of simplification. The equation of motion for Mass 1 suffices for the description of this scenario. The equation is

$$m_1\ddot{x}_1 = s_1(x_1 - x_2), \tag{5.60}$$

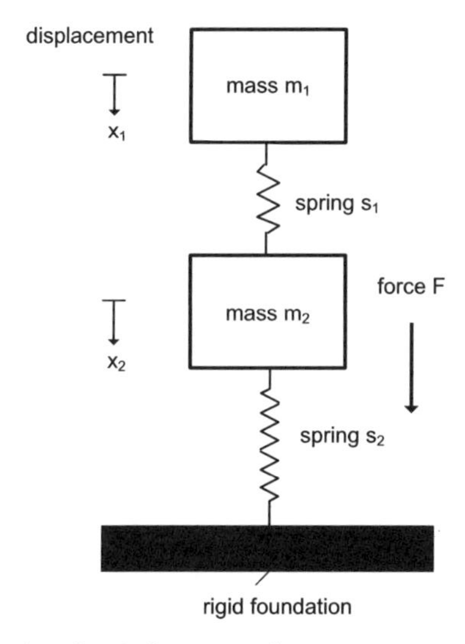

Fig. 5.14. Model for the elastic bearing of structures consisting of spring parts

or, for pure tones at the frequency ω and complex amplitudes

$$m_1\ddot{x}_1 = s_1(x_1 - x_2), \tag{5.61}$$

from which

$$x_1 = \frac{x_2}{1 - \omega^2/\omega_1^2} \tag{5.62}$$

with

$$\omega_1{}^2 = \frac{s_1}{m_1} \tag{5.63}$$

can be derived. The frequency ω_1, appearing in equation (5.63), exists in the resonance frequency which the structural component m_1 and s_1 would have if it were resting on a rigid foundation. This resonance, however, is usually set very low for the sake of the passenger's comfort. Therefore, it can at least be assumed that $\omega >> \omega_1$ is true for a frequency range large enough to ensure the transmission of the vibrations into the ground. This also implies $x_1 << x_2$: That is, the amplitude of the vibrations emitted by the upper mass m_1 can principally be disregarded in favor of the vibration amplitude of the second mass m_2. The upper mass m_1 is practically motionless – in any case, producing the desired effect for the passengers. It suffices to say that the systematically decoupled mass m_1 is virtually irrelevant in respect to the acoustical effect of the elastic bearing s_2. This can also be shown by the equation of motion for mass m_2 derived from

$$m_2\ddot{x}_2 = x_2(s_1 - s_2) + F, \tag{5.64}$$

or in the frequency domain

$$x_2 = \frac{F}{s_1 + s_2 - m_2\omega^2}. \tag{5.65}$$

The force exerted on the foundation is therefore

$$F_F = s_2 x_2 = \frac{F}{\left(1 - \omega^2/\omega_{12}^2\right)} \frac{s_2}{(s_1 + s_2)}, \tag{5.66}$$

with

$$\omega_{12} = \frac{s_1 + s_2}{m_2}. \tag{5.67}$$

In conclusion, the insertion loss

$$R = 10lg(F/F_F)^2 = 10lg(1 - \omega^2/\omega_{12}^2)^2 \tag{5.68}$$

is solely based on the resonance frequency resulting from mass m_2 and combined spring element $s_1 + s_2$. The mass m_1 does not exist from a dynamic standpoint, because m_2 is decoupled from the system by s_1. This has a detrimental effect on the elastic bearing s_2, as the spring layer has to be dimensioned based on the 'static' total mass $m_1 + m_2$. As mentioned previously, the ratio of static displacement to spring thickness cannot exceed a certain value. Unfortunately, this 'dynamically moving' mass m_2, which can be much smaller by comparison, is the only one that is significant for the dynamic effect of the insulation.

5.7 Conclusion

As usual, the previous remarks concerning elastic bearings are incomplete. The most important effects which are not discussed here are named in the following.

1. Vibrations of solid structures are not restricted to one degree of freedom (translational motion along one axis), as was tacitly assumed here. Objects can certainly perform one translational and one rotational vibration in each of the three room axes. To avoid motion parallel to the foundation and 'rocking oscillations', a low center of gravity is favorable, which can be achieved, for instance, by an additional mass (e.g. mounting of engines on concrete).
2. Machinery, devices, turning-lathes, etc. cannot always be regarded as 'compact masses'. In contrast, they can suffer from elastic deformations with resonance phenomena themselves. The 'moved' dynamic mass, valid for the resonance frequency of the bearing, can therefore be considerably smaller than the static mass.
3. Last, but not least, actual spring elements can form a waveguide themselves, where standing waves may occur at higher frequencies, resulting in large dips in the insertion loss at their resonances.

5.8 Summary

Elastic decoupling using springs or soft, elastic insulation between structure-borne sound sources and foundations can greatly reduce the permeation of structural noise into building foundations. Below the resonance frequency, there is an insertion loss of the mass-spring system of about $0\,dB$. In the resonance frequency, the insertion loss can have a negative value, depending on the loss captured in by the spring. The mitigating effects only occur above the resonance frequency. Here, the insertion loss increases with $12\,dB$ per octave. The main objective from the perspective of noise control is therefore maintaining low resonance by using the softest springs possible. A rule of thumb is to use springs which have a static depression of 5 to 10 percent of the spring's length or the layers thickness.

The structural foundation can have a considerable influence on the amount of insertion loss. If the foundation has a mass character, the resonance frequency shifts upward. If it has a spring characteristic, the insertion loss attains a frequency-independent constant value once it exceeds the resonance frequency. This value is defined by the stiffness ratio of the foundation to the spring. The effects of elastic decoupling can also heavily depend on whether the dynamically reverberating mass is actually much smaller than the remaining mass of the structure-borne sound source. Before implementing elastic decoupling, it is highly recommended to verify that the structure-borne sound is entering the foundation along the main transmission path to the area to be insulated, or whether it is entering it along another route.

5.9 Further reading

The "Handbook of Acoustical Measurements and Noise Control" (edited by C. M. Harris, Acoustical Society of America, ASA, New York 1998) provides additional information on elastic decoupling in Chapter 29. Furthermore, the "Handbook" provides a valuable research tool to answer a wide variety of questions in acoustics.

5.10 Practice exercises

Problem 1

A nuclear magnetic resonance tomograph has a mass of $1000\,kg$. It is resting on four square legs, each measuring $30\,cm$ by $30\,cm$. A flat bearing is to be mounted on an elastic decoupling system, whereby the static depression of the elastic bearing is 0.05 times the thickness of the layer. How large does the bearing's elasticity module have to be? How thick does the layer have to be in order to guarantee a resonance frequency of $14\,Hz$?

Problem 2

How stiff does the spring have to be in the foundation structure of the NMRT in Problem 1 to obtain a reduction in vibration emissions of $6\,dB$ ($10\,dB$, $20\,dB$) in the building?

Problem 3

During the operation of the NMRT, as described in the problems above, the following octave levels emitted from both a transmission room and waiting area (receiver room):

f/Hz	Source level $/dB$	Receiver level$/dB$
500	65.3	32.0
1000	64.4	31.4
2000	63.5	30.4

The NMRT is subsequently switched off. The following levels in the source and receiver rooms are measured using a loudspeaker:

f/Hz	Source level(L)$/dB$	Receiver level(L)$/dB$
500	85.2	45.3
1000	86.4	45.2
2000	83.8	42.4

Does it make sense to mount the foundation to an elastic decoupler? If yes, how much of a decibel reduction can be expected? How high is the non-weighted emission level in the receiver room after introducing elastic decouplers?

Problem 4

Suppose a machine of mass M is decoupled from a foundation. If the foundation is likewise a component of a mass, how much does the resonance frequency have to be detuned as opposed to the case of a rigid foundation if the foundation mass is double, four times, and eight times the mass to be decoupled?

Problem 5

The 'resonance frequency $\omega_{0\eta}$ of a damped resonator' is the frequency at which the frequency response $|x/F|^2$ (see equation(5.6)) reaches its maximum. How high is the damped resonance frequency? How much is the critical damping where the resonance frequency reaches zero? Express the damping constant r in terms of the loss factor η.

6

Sound absorbers

When designing the acoustics of rooms, the problem of influencing the sound reflections at the room boundaries often arises. In factory buildings, for example, it is necessary to prevent noise emitted by machinery from travelling to areas further away via reflections; one is interested in obtaining a sufficiently high absorption at the walls. On the other hand, the sound for audiences in auditoria like studios or concert halls must be enhanced by indirect reflections, while simultaneously reducing reverberation caused by multiple reflections which would otherwise make intelligibility more difficult.

Such design objectives can be achieved by implementing absorbent structures in the construction of walls and ceilings. These must contain standardized reflection properties suitable for the purpose. The present chapter discusses such building structures and explains their impact on the airspace surrounding them.

We will first begin this chapter by introducing the basic effects of normal impingement of a plane sound wave onto either a reflective or an absorbent wall structure or noise barrier. The acoustic properties of a wall are often measured in experiments based on this specific case. We will first proceed in determining the properties of the acoustic wall by testing the theory by experiment as is customary in empirical investigation. properties.

6.1 Sound propagation in the impedance tube

In order to determine the reflection and absorption of a wall structure specimen under the previously defined condition of normal sound incidence, it is first necessary to produce a plane wave in a laboratory. Simply exposing a wall area to sound would present certain difficulties in measuring the outcome, as a plane wave can only be generated in small areas of space. Furthermore, the results would be difficult to reproduce, as they would depend on the positioning and the direction of the sender.

M. Möser, *Engineering Acoustics*, DOI 10.1007/978-3-540-92723-5_6,

In contrast, unambiguous and easy to reproduce environmental conditions can be achieved by capturing the sound in a one-dimensional continuum. This can be done in a rigid tube, where the sound is internally guided and forced to propagate along the tube axis. Such a tube, used to determine the acoustic properties of a termination equipped with the sample, is called an impedance tube, or Kundt's tube (named after the acoustician Kundt implemented it to prove the wave properties of sound). As long as the tube diameter is small compared to the wavelength, it produces a plane sound wave, propagating along the tube axis.

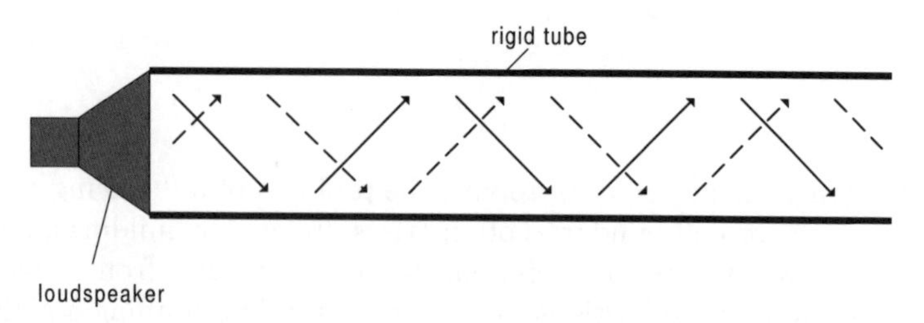

Fig. 6.1. Principle of propagating sound-rays in the tube

As the impedance tube is a widely used experimental device, this chapter on 'sound absorbers' starts with a discussion of its basic principles using a simplified two-dimensional tube model, consisting of two rigid parallel plates. In the following, we will consider how the results of the simplified two-dimensional discussion can be applied to a more realistic three-dimensional scenario.

Our initial subject of examination will be the rigid parallel plates at $y = 0$ and $y = h$. Transversal wave propagate to and from between these plates due to continuous reflections running in a 'zigzag course' (as shown in Fig. 6.1). This simple but appropriate association can be summarized by the following equation:

$$p = p_0 \mathrm{e}^{-jk_x x} \left(\mathrm{e}^{-jk_y y} + r\mathrm{e}^{jk_y y}\right) \tag{6.1}$$

The wave numbers in transverse direction k_y and in the direction of propagation k_x are still unknown. The nature of k_x and k_y is the focus of the following discussion, as these two quantities describe the principles of sound propagation in a two-dimensional tube: k_y determines the transverse distribution, which is normal to the propagation direction, and k_x defines how this cross-distribution propagates along the tube. Reflections are not permitted along the tube axis x because the basic principles of the sound transmitter 'tube' are of interest. Thus, the tube has to be regarded as being either 'infinite long' or 'terminated by an anechoic end'.

An expression can be easily derived for the transverse wave number k_y,due to the rigid boundaries at $y = 0$ and $y = h$. These two boundary conditions require that the velocity normal to the boundaries $y = 0$ and $y = h$ be equal to zero, therefore

$$\left.\frac{\partial p}{\partial y}\right|_{y=0} = \left.\frac{\partial p}{\partial y}\right|_{y=h} = 0$$

is given at the boundaries. The first boundary condition at $y = 0$ yields $r = 1$. This transforms eq.(6.1) into

$$p = 2p_0 \mathrm{e}^{-jk_x x} \cos k_y y \ .$$

The second boundary condition at $y = h$ requires

$$\sin k_y h = 0 \ .$$

This so-called 'eigenvalue equation' of the tube has the solutions

$$k_y = \frac{n\pi}{h} \ ; \quad n = 0, 1, 2, \dots \ . \tag{6.2}$$

There are therefore only certain given wave numbers k_y permissible for the transverse distributions of the pressure in the tube. These are called 'eigenvalues' of the tube. Each eigenvalue has its own unique given pressure dependency $f_n(y)$ which runs in the y–-direction

$$f_n(y) = \cos k_y y = \cos\left(\frac{n\pi y}{h}\right) \ ,$$

This pressure dependency is referred to as the "'eigenfunction"' or "'mode"' of the tube.The word "'mode"' indicates the "'state."' The modes contain all pressure states which can exist based on the boundary conditions of the tube.

Which of the n and which corresponding k_y occur, cannot be determined at this point. Therefore, one must initially permit all possibilities (allowing for all their corresponding oblique wave directions). For this reason, we will first introduce the reader to the general pressure ansatz, based on (6.1):

$$p = \sum_{n=0}^{\infty} p_n \cos\left(\frac{n\pi y}{h}\right) \mathrm{e}^{-jk_x x} \tag{6.3}$$

Cosinosoidal pressure distributions may develop along the y-axis are called modes (i.e. 'state'). They are characterized by pressure maxima (antinodes) at the edges (due to the boundary conditions). The mode shapes for $n = 0$, 1, 2 and 3 are shown in Fig. 6.2. They are simply segments of the cosine function transferred to the tube cross-section so that vanishing pressure derivatives (i.e. pressure maxima) result at the edges.

As already mentioned, the principle nature of the propagation of individual modes is mainly of interest, which are described by a corresponding wave number k_x which depends on the mode index n. The wave number is simply obtained by using the two-dimensional wave equation

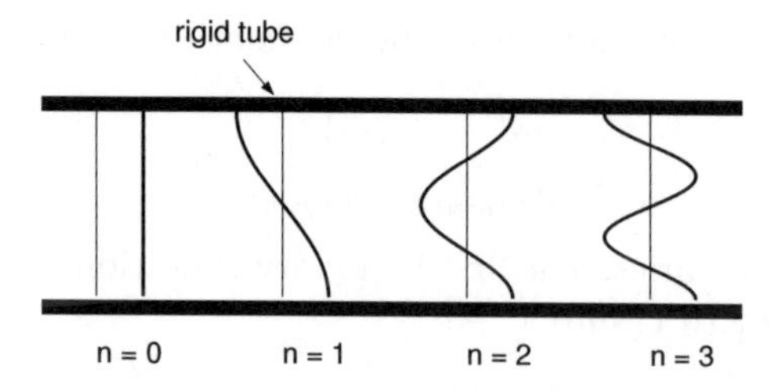

Fig. 6.2. Transverse sound pressure distribution (two-dimensional tube)

$$\frac{\partial^2 p}{\partial x^2} + \frac{\partial^2 p}{\partial y^2} + k^2 p = 0,$$

which, with $k = \omega/c$, requires

$$k_x^2 = k^2 - \left(\frac{n\pi}{h}\right)^2$$

in (6.3). When evaluating the square root of k_x, it is assumed that the wave number k_x either describes a sound propagation along the x-axis or a near-field which decays exponentially farther away from the source:

$$k_x = \begin{cases} +\sqrt{k^2 - \left(\frac{n\pi}{h}\right)^2} & ; |k| > \frac{n\pi}{h}, \\ -j\sqrt{\left(\frac{n\pi}{h}\right)^2 - k^2} & ; |k| < \frac{n\pi}{h} \end{cases} \tag{6.4}$$

A sound field which increases exponentially further away from the sender is physically not realistic. To be exact, it should be mentioned that positive and real values are explicitly assumed under the square roots in (6.4).

Obviously, a limiting frequency

$$f_n = n\frac{c}{2h} \tag{6.5}$$

corresponds to a mode with the index n. Wave propagation with a corresponding wave number k_x occurs only for the corresponding transverse distribution. Only at frequencies $f > f_n$ above the modal limiting frequency f_n. This fact is expressed by a real wave number k_x.

In contrast, no sound radiation occurs at frequencies below the modal limiting frequency $f < f_n$. A quickly decaying sound field develops, vanishing at a greater distance from the sender. This is expressed by a purely imaginary wave number k_x.

The fact that modes are unable to propagate below a certain frequency is called the 'modal cut-off' with the 'cut-off'-frequency f_n. Below the lowest limiting frequency

$$f_1 = c/2h$$

(where the tube diameter h is equal to half the wavelength, $\lambda_1/2 = h$) only a sound field which consists of plane waves $n = 0$ can be detected in somewhat close proximity to the sender. If the tube is only implemented below that frequency limit, only plane waves can occur, regardless of the loudspeaker's shape and its local velocity distribution. This effect is used to measure absorber samples.

Even though determining the pressure coefficients p_n is not necessarily required for subsequent investigations, we will nevertheless carry out the sound field calculations for the sake of thoroughness. The considerations required here are indeed somewhat fundamental and so we will still mention them in the following sections.

The key to the answer of the aforementioned problem is simply given by the fact that the local characteristics of the sound field always depend on the environment (in this case, the rigid plates), whereas the actual distribution of the sound field also depends on the source. The same is true here. For simplicity, let us assume a flat, prolated membrane at $x = 0$ which has a velocity distribution $v_0(y)$. The velocity pointing in the x-direction, which can be calculated based on the ansatz (6.3), has to be equal to the membrane velocity $v_0(y)$ given at $x = 0$. This leads to

$$v_0(y) = \frac{j}{\omega\varrho}\frac{\partial p}{\partial x}\bigg|_{x=0} = \frac{1}{\varrho c}\sum_{n=0}^{\infty} p_n \frac{k_x}{k}\cos\left(\frac{n\pi y}{h}\right) .$$

This equation can easily be solved to determine p_n. Initially, we arbitrarily select a pressure amplitude with the index m. Then we solve the above equation for p_m as follows. First, both sides are multiplied by $\cos(m\pi y/h)$ and integrated over y:

$$\frac{1}{\varrho c}\sum_{n=0}^{\infty} p_n \frac{k_x}{k}\frac{2}{h}\int_0^h \cos\left(\frac{n\pi y}{h}\right)\cos\left(\frac{m\pi y}{h}\right)\mathrm{d}y = \frac{2}{h}\int_0^h v_0(y)\cos\left(\frac{m\pi y}{h}\right)\mathrm{d}y$$

Due to

$$\frac{2}{h}\int_0^h \cos\left(\frac{n\pi y}{h}\right)\cos\left(\frac{m\pi y}{h}\right)\mathrm{d}y = \begin{cases} 0, & n \neq m \\ 1, & n = m \neq 0 \, , \\ 2, & n = m = 0 \end{cases}$$

only one term of the sum with $n = m$ remains. This term can actually be solved for p_m. For this particular pressure amplitude it is

$$p_m = \frac{2k}{k_x}\frac{\varrho c}{h}\int_0^h v_0(y)\cos\left(\frac{m\pi y}{h}\right)\mathrm{d}y \; ; \quad m \neq 0$$

$$p_0 = \frac{\varrho c}{h}\int_0^h v_0(y)\mathrm{d}y$$

($k_x = k$ for $m = 0$). Since m is arbitrary, it is irrelevant *which* m is used. Thus, *all* p_m can be calculated from the equation above. The method described is therefore referred to in mathematical terms as 'eigenfunction expansion of the loudspeaker velocity'.

Finally, it should be explicitly emphasized that the *modal composition* is absolutely *not* determined by the modal cut-off; it is only a matter of the source characteristics. The idea that modes cannot exist below the limiting frequency is wrong: they only occur in the near-field, but this does not necessarily mean that they do not exist. This error can be a grave one as Chap. 9 on silencers elaborates. Such a mistaken notion implies that some silencers would possess 'infinitely high' attenuation – and that is not the case.

6.1.1 Tubes with rectangular cross sections

The simplest scenario for three-dimensional wave dispersion is in the tubes with rectangular cross sections (four walls) . Again, pressure antinodes must be present at the edges $y = 0, a$ and $z = 0, b$. Therefore, two-dimensional pressure modes exist in

$$p = \sum_{n=0}^{\infty} \sum_{m=0}^{\infty} \cos\left(\frac{n\pi y}{a}\right) \cos\left(\frac{m\pi y}{b}\right) \mathrm{e}^{-jk_x x} \tag{6.6}$$

where a and b are the diameter measurements of the rectangular cross section. The modes are simply products of transverse distributions. As is evident based on the wave equation, the cut-off frequencies are given by

$$f_{nm} = \frac{c}{2}\sqrt{\frac{n^2}{a^2} + \frac{m^2}{b^2}}\,, \tag{6.7}$$

where the lowest frequency (f_{01} or f_{10}, respectively) is given by the larger of the two dimensions.

So basically, nothing has changed: the sound field can be composed of modes in two dimensions as well as in three dimensions. Each mode has a cut-on effect. The modal amplitudes are calculated from the source velocity.

6.1.2 Tubes with circular cross sections

And these facts also apply to cylindrical tubes with circular cross-sections. Such tubes are almost always used in measurements. The sound field is also composed of transverse distributions, where each of these has a modal cut-on frequency. For the sound pressure

$$p = \sum_{n=0}^{\infty} \sum_{m=1}^{\infty} p_{nm} \cos(n\varphi)\, J_n\left(x_{nm}\frac{r}{b}\right) \mathrm{e}^{-jk_z z}\,. \tag{6.8}$$

Here, r, φ, z represent the coordinates of the circular cylinder and the z-coordinate represents the axis of the rigid tube. The tube's lining is defined by

$r = b$. For the purposes of simplicity, only the pressure distribution symmetrical to the x-axis $cos n\varphi$ are accounted for. Otherwise, in the most general case, asymmetry can be expressed by adding the term $\sin(n\varphi)$. $J_n(x)$ denotes the Bessel function of order n (with $n = 0, 1, 2, \cdots$) as shown in Figure 6.3 . The

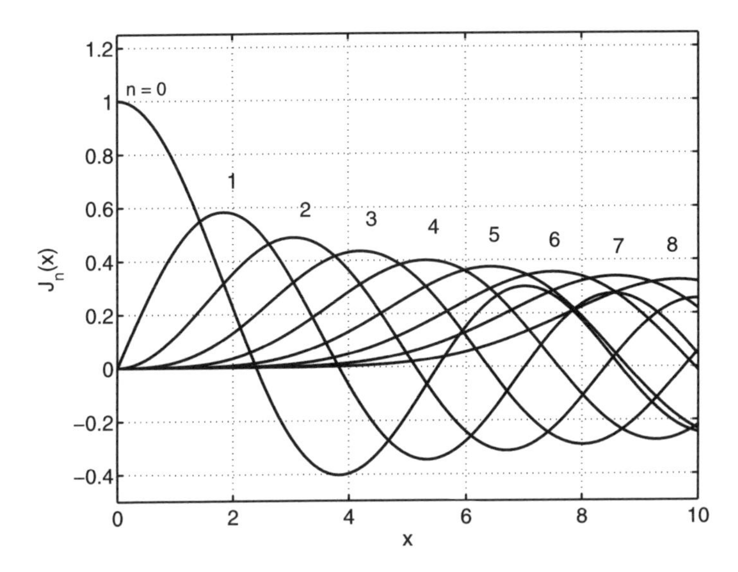

Fig. 6.3. Bessel functions $J_n(x)$ of the orders 0 to 8

factors x_{nm} indicate the roots of the (first) derivative of the Bessel functions. They are approximately represented in Figure 6.3. For example, the values for the first zeros of the derivative of $J_0(x)$ are $x_{01} = 0$, $x_{02} \approx 3.8$, and $x_{03} \approx 6.7$ (the more precise values to the third decimal place can be found in the table below.) Some modes $J_n(x_{nm}r/b)\cos(n\varphi)$ are listed in Figure 6.4. Based on the wave equation, the following applies to wave numbers k_z which describe the sound propagation of the modes along the tube's axis:

$$k_z^2 = k^2 - (\frac{x_{nm}}{a})^2 . \tag{6.9}$$

If the right side of the equation is positive, the modes pertain to a propagating wave with a real wave number. On the other hand, a negative term on the right indicates the existence of a near-field with an imaginary wave number. Therefore, the cut-on frequencies

$$f_{nm} = x_{nm}\frac{c}{2\pi a} , \tag{6.10}$$

pertain to the modes. The numerical values listed in ascending order in Table 6.2 can be substituted for x_{nm}.

The lowest cut-on frequency is given by

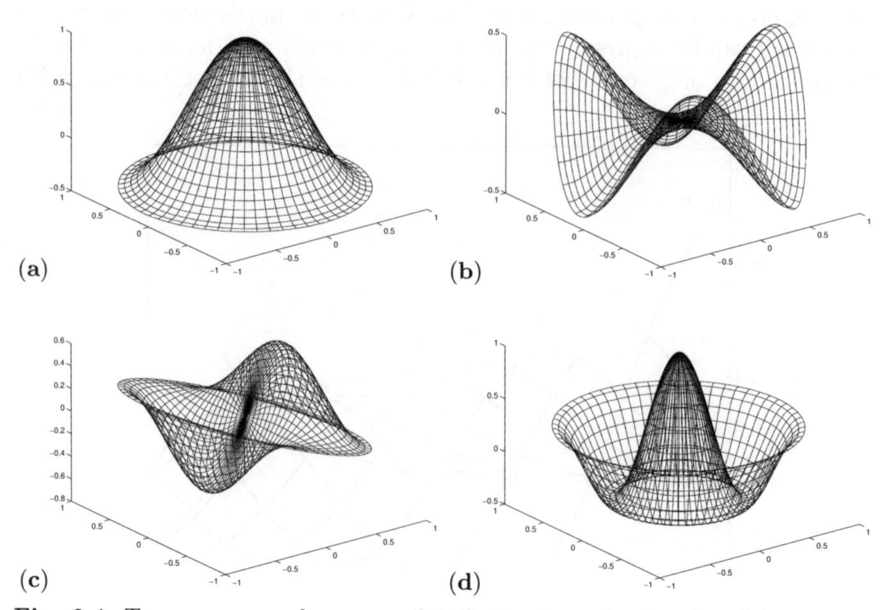

Fig. 6.4. Transverse sound pressure distribution in a circular duct (**a**) x_n=3.832 (**b**) x_n=4.201 (**c**) x_n=5.331 (**d**) x_n=7.016

Table 6.1. Cut-on frequencies

x_{nm}	0	1,841	3,054	3,832	4,201	5,331	6,706	7,016

Table 6.2. Roots x_{nm} of the Bessel functions in sequential order from smallest to largest. The cut-on frequencies of circular cylindrical tubes can be derived from these factors inserted into eq. (6.10).

$$f_1 = 0.59 \frac{c}{2a} \tag{6.11}$$

which is roughly equal to that of a rectangular duct ($f_1 = 0.5c/b$, b=width) with the same cross-sectional area ($b = \sqrt{\pi}a$).

The sound field exists as a sum of modes also in circular cylindrical tubes. Every mode has a cut-on effect. The cut-on effect of modal amplitudes are calculated from the source here as well.

6.2 Measurements in the impedance tube

As the previous section has shown, the impedance tube is a mode filter; it can be used below the lowest cut-off frequency to produce a plane wave. It represents a measurement device which can be used to characterize partially absorbing and partially reflecting structures at normal sound incidence.

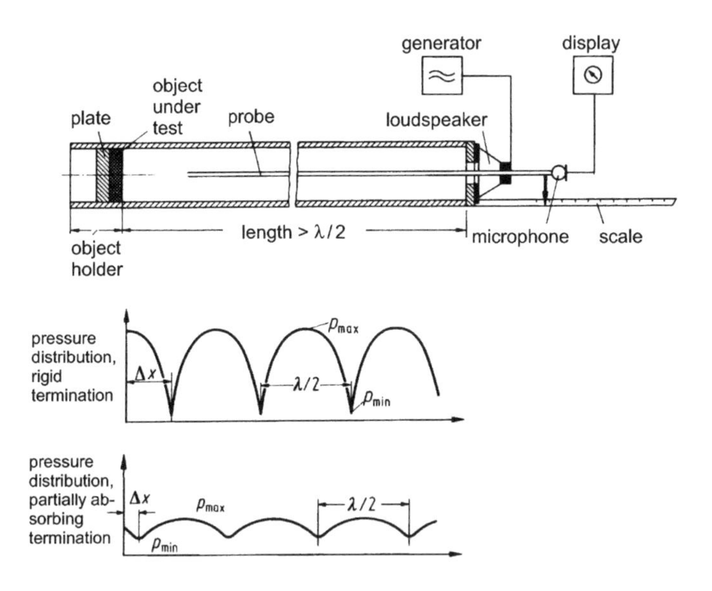

Fig. 6.5. Experimental setup for the determination of absorption coefficient and impedance in the tube

The tube is terminated by the sample under investigation (e.g. a sheet of fibers on a rigid surface) at one end (Fig. 6.5). The resulting sound field in the tube, when excited using pure tones, consists of one component travelling toward the sample, as well as one reflected component

$$p = p_0 \left\{ e^{-jkx} + r e^{jkx} \right\} , \tag{6.12}$$

where r is the pressure reflection coefficient, which is equal to the ratio of the sound pressures p_-/p_+ of the wave travelling in the positive direction

$$p_+ = p_0 e^{-jkx} \tag{6.13}$$

and the wave travelling in the negative direction

$$p_- = p_0 r e^{jkx} \tag{6.14}$$

at the position $x = 0$ of the sample. Generally, the reflection can also include a phase shift between the two wave components which results in a reflection coefficient

$$r = R e^{j\varphi} \tag{6.15}$$

(R is real), which is a complex number. In (6.12), a wave field is expressed by the sum of two waves travelling in opposite directions, where the reflected component p_- (due to $R \leq 1$) can have a smaller amplitude. This includes the case of incomplete reflection. To illustrate the space dependence of the sound field, the positive wave p_+ is decomposed into a totally reflected part plus a remainder:

$$p_+ = p_0 r \mathrm{e}^{-jkx} + p_0(1-r)\mathrm{e}^{-jkx} \, .$$

The total sound field

$$p = p_0 r \left(\mathrm{e}^{-jkx} + \mathrm{e}^{jkx}\right) + p_0(1-r)\mathrm{e}^{-jkx} = 2p_0 r \cos(kx) + p_0(1-r)\mathrm{e}^{-jkx}$$

is therefore composed by the sum

$$p = p_\mathrm{s} + p_\mathrm{f}$$

of a standing wave

$$p_\mathrm{s} = 2p_0 r \cos(kx) \tag{6.17a}$$

and a wave travelling in x-direction

$$p_\mathrm{f} = p_0(1-r)\mathrm{e}^{jkx} \, . \tag{6.17a}$$

The combination of standing and progressive waves is seen in the ripple of the locally measured rms-value (root mean square). When there is total reflection $r = 1$, the sound field is dominated by the standing wave only, with the rms-value $\tilde{p}$

$$r = 1: \quad \tilde{p}^2 = 2p_0^2 \cos^2(kx)$$

(see also Fig. 6.5) which represents a space dependence with large fluctuations. Without any reflection $r = 0$, the wave field is only given by the progressive wave

$$r = 0: \quad \tilde{p}^2 = \frac{p_0^2}{2}$$

with a constant local rms-value. A ripple $\tilde{p}_\mathrm{min}/\tilde{p}_\mathrm{max}$ develops in the spatial dependency between these two extrema ripple. This is explained by the combination of standing and progressing waves. Obviously, the ripple $\tilde{p}_\mathrm{min}/\tilde{p}_\mathrm{max}$ of the space dependence of $\tilde{p}(x)$ is a direct measure of the reflection coefficient which can thus be determined from the measurement of this pressure ratio.

6.2.1 Mini-max procedure

The measurement procedure can be deduced easily. First, the space dependence of the squared rms-value is calculated from (6.12)

$$\tilde{p}^2 = \frac{1}{2}|p|^2 = \frac{1}{2}pp^* = \frac{1}{2}p_0^2 \left(\mathrm{e}^{-jkx} + R\mathrm{e}^{j(kx+\varphi)}\right)\left(\mathrm{e}^{jkx} + R\mathrm{e}^{-j(kx+\varphi)}\right)$$
$$= \frac{1}{2}p_0^2 \left[1 + R^2 + 2R\cos(2kx+\varphi)\right] \ . \tag{6.18}$$

The maximum values obviously occur at $2kx + \varphi = 0, \pm 2\pi, \pm 4\pi, \ldots$. The maxima are then given by

$$\tilde{p}^2_{\text{max}} = \frac{1}{2}p_0^2 \left(1 + R^2 + 2R\right) = \frac{1}{2}p_0^2(1+R)^2 \ . \tag{6.19a}$$

The minima occur at $2kx+\varphi = \pm\pi, \pm 3\pi, \ldots$. The sound pressure minima are given by

$$\tilde{p}^2_{\text{min}} = \frac{1}{2}p_0^2 \left(1 + R^2 - 2R\right) = \frac{1}{2}p_0^2(1-R)^2 \ . \tag{6.19a}$$

Thus, the ratio of the extrema μ becomes

$$\mu = \frac{\tilde{p}_{\text{min}}}{\tilde{p}_{\text{max}}} = \frac{1-R}{1+R} \ , \tag{6.20}$$

or

$$R = \frac{1-\mu}{1+\mu} \ . \tag{6.21}$$

Equation (6.21) directly states the measurement procedure for determining the reflection coefficient. By moving a microphone probe along the tube-axis (Fig. 6.5) the minima and maxima of the rms-value can easily be detected. Since only the ratio is of interest, it is not necessary to calibrate the microphone.

Instead of the reflection coefficient R, the loss factor β is often specified for the characterization of samples. It is defined by the ratio of the sound power flowing through the sample surface $\underline{P}_\beta$ and the incident sound power $\underline{P}_+$

$$\beta = \frac{\underline{P}_\beta}{\underline{P}_+} \ .$$

In general, the sound power P, which is lost from the tube, is composed of a loss component $\underline{P}_\alpha$ which actually results from sound energy being transformed into heat, as well as a component $\underline{P}_\tau$ whose value may also contain sound energy transmitted to the exterior (e.g. an open or 'nearly open' tube).

$$\underline{P}_\beta = \underline{P}_\alpha + \underline{P}_\tau \ .$$

Similar to the loss factor β, the definitions are

- the absorption coefficient $\alpha = \underline{P}_\alpha / \underline{P}_+$ and
- the transmission coefficient $\tau = \underline{P}_\tau / \underline{P}_+$

Obviously, it is

$$\beta = \alpha + \tau .$$

As a matter of fact, only the net loss β, and not the causes α and τ, can be determined by the aforementioned measurements procedures in the impedance tube. For most applications, however, the interpretation is quite simple:

- For absorbing samples with a rigid termination, $\beta = \alpha$; these are the samples of practical interest.
- For thin and light terminations without an absorbing sheet, $\beta = \tau$.

The latter are seldom found in practice; they are mentioned here for completeness.

The relationship between the reflection coefficient R and the loss factor β can be deduced from the energy balance equation

$$\underline{P}_+ = \underline{P}_\beta + \underline{P}_- ,$$

where $\underline{P}_-$ denotes the reflected sound power. For plane progressive waves it is

$$\underline{P}_- = R^2 \underline{P}_+$$

and, using the definition of β, it follows that

$$\underline{P}_+ = \beta \underline{P}_+ + R^2 \underline{P}_+$$

or

$$\beta = 1 - R^2 . \tag{6.22}$$

After inserting (6.21), the relationship between β and the ripple parameter μ is finally defined:

$$\beta = \frac{2}{1 + \frac{1}{2}\left(\mu + \frac{1}{\mu}\right)} . \tag{6.23}$$

As shown earlier, the loss factor (and the magnitude of the reflection coefficient) can be calculated from the ripple of the local space dependency of the sound pressure rms-value. To measure the phase φ of the reflection coefficient, the position of the relative local extrema has to be determined. Since, in practice, the minima are usually easier to locate than the maxima, the location x_{min} of the first minimum, found in front of the sample, is used, for which

$$2kx_{\mathrm{min}} + \varphi = \pm\pi$$

or

$$\varphi = \pi \left(\frac{|x_{\mathrm{min}}|}{\lambda/4} \pm 1 \right) \tag{6.24}$$

is given, (note that $x_{\mathrm{min}} < 0$ due to the chosen origin of the coordinate system). The complex reflection coefficient $r = Re^{j}\varphi$ is therefore also known.

When using the measurement technique described above to measure the absorption coefficient, it should be kept in mind that the closest minimum to the sample is used to determine μ. The reason is due to the unavoidable losses which are caused by the damping of the sound wave along the direction of propagation. A small amount of internal damping in air is often the physical cause, not to mention the energy which is also lost to the exterior. The tube wall, of course, has a high, but obviously finite transmission loss. Thus, there is always some sound energy that escapes through the walls (for example, this effect is prevalent in large, self-made wooden rectangular measurement tubes). Both losses reduce the apparent reflection coefficient, the larger the distance of the sound radiated by the sender. Therefore, in any given sample, a larger absorption coefficient is measured than actually exists, and the resulting observed absorption coefficient is larger, the farther the measurement point is from the sample surface.

For a tube with a rigid termination, the apparent reflection coefficient at the point x would therefore be measured at

$$R = 1 - \Delta R|x|$$

a reduced value at x by a loss factor of ΔRx in the tube. It seems to be realistic to assume small damping effects per tube length ΔR. According to (6.19) and (6.19), these minor damping effects are specified by

$$\begin{aligned} p_{\text{max}}^2 &= \frac{1}{2} p_0^2 \left(1 + 1 - \Delta R|x|\right)^2 \approx 2p_0^2 \\ p_{\text{min}}^2 &= \frac{1}{2} p_0^2 \left(1 - (1 - \Delta R|x|)\right)^2 \approx \frac{1}{2} p_0^2 \Delta R^2 x^2 \, . \end{aligned}$$

In practice, the damping in the tube can only be identified at the minima. The minimum rms-values are located on a straight line along x

$$p_{\text{min}} = p_0 \Delta R x / \sqrt{2} \, .$$

The maxima remain virtually independent of the losses.

6.3 Wall impedance

The quantities discussed in the previous section and the corresponding measurement technique addresses the question of the acoustic impact of an 'actually present wall structure'. Using the descriptions given so far, it is impossible to draw conclusions for specific wall structures with respect to their effect.

A quantity describing the specific behavior of a reflecting device is the wall impedance z. It is simply defined by the ratio of the sound pressure and the sound velocity at the wall surface $x = 0$:

$$z = \frac{p(0)}{v(0)} \, . \tag{6.25}$$

It will be discussed in Sect. 6.5 as to how a specific arrangement can be described by its wall impedance. In the present section, we consider the relationship between the 'old and new quantities'.

The relationship between the quantity z which describes the structure and the active quantity β (or α or τ, respectively) can be clarified easily. Choosing the coordinate system so that that the origin is located at the wall surface,

$$p = p_0 \left(\mathrm{e}^{-jkx} + r\mathrm{e}^{jkx} \right) \tag{6.26}$$

and

$$v = \frac{j}{\omega\varrho} \frac{\partial p}{\partial x} = \frac{p_0}{\varrho c} \left(\mathrm{e}^{-jkx} - r\mathrm{e}^{jkx} \right) \tag{6.27}$$

are given in the domain $x < 0$ in front of the wall. The wall impedance is thus related to the reflection coefficient by

$$\frac{z}{\varrho c} = \frac{1+r}{1-r} \,. \tag{6.28}$$

As already mentioned, in the case of absorbent structures, the absorption coefficient is almost always used instead of the reflection coefficient. Therefore, Therefore, the relationship between the loss factor β and the wall impedance z can now be specified by solving (6.28) for r

$$r = \frac{\frac{z}{\varrho c} - 1}{\frac{z}{\varrho c} + 1}$$

and β is calculated by

$$\beta = 1 - |r|^2 = \frac{4\mathrm{Re}\left\{z/\varrho c\right\}}{\left[\mathrm{Re}\left\{z/\varrho c\right\} + 1\right]^2 + \left[\mathrm{Im}\left\{z/\varrho c\right\}\right]^2} \,. \tag{6.29}$$

It should be obvious now why (6.29) is called the 'matching law'. A large loss factor is clearly achieved for the matching case $z = \varrho c$, where $\beta = 1$. This case can be realized by either using a complete absorbent arrangement $\alpha = 1$ (and $\tau = 0$) or by using a simple (non-reflecting) and infinitely elongated tube with $\tau = 1$ ($\alpha = 0$). As is clear that an imaginary part of the impedance is always 'detrimental to absorption'. β has a maximum if the imaginary part of z is equal to zero $\mathrm{Im}\{z\} = 0$. When discussing wall impedances in the following sections, the complex frequency response function in the complex impedance plane will be represented graphically. Each frequency yields a certain complex number, corresponding to a point in the $z/\varrho c$-plane. If the frequency is changed, the point travels along a curve called a phasor curve. If the lines of constant loss factor in the complex wall impedance plane are known, the frequency response function of β can be deduced from the array of curves $\beta =$ const. and the phasor curve.

The lines $\beta =$ const. in the wall impedance plane are obtained by the following simple discussion. The real and imaginary part of $z/\varrho c$ are represented in short form by x and y in (6.29) for the sake of brevity:

$$x = \mathrm{Re}\{z/\varrho c\}, \quad y = \mathrm{Im}\{z/\varrho c\} \tag{6.30}$$

and (6.29) result in

$$(x+1)^2 - \frac{4}{\beta}x + y^2 = 0 \,. \tag{6.31}$$

If this is compared to the general equation of a circle

$$(x - x_c)^2 + (y - y_c)^2 = a^2$$

(x_c, y_c: coordinates of the center point, a = radius of the circle) it can be seen that the transformed equation (6.31)

$$\left(x - \left(\frac{2}{\beta} - 1\right)\right)^2 + y^2 = \frac{4}{\beta}\left(\frac{1}{\beta} - 1\right)$$

describes circles. Lines with β = const. are therefore circles with their center points located on the real axis and the coordinates of the center point

$$x_c = \frac{2}{\beta} - 1 \tag{6.32}$$

and the radius

$$a = \sqrt{\frac{4}{\beta}\left(\frac{1}{\beta} - 1\right)} \,. \tag{6.33}$$

Some lines β = const. for $\beta = 0.5, 0.55, 0.6 \ldots 0.9$ and 0.95 are shown in

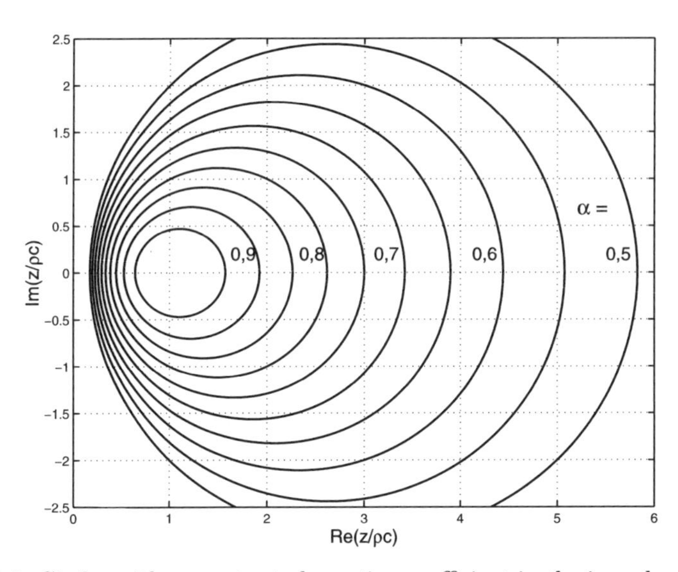

Fig. 6.6. Circles with a constant absorption coefficient in the impedance plane

Fig. 6.6 (here, it it assumed that $\alpha = \beta$, as it will also be assumed in what follows). Obviously, the circles surround each other; the center point moves to the right with decreasing β, while the radius increases. The curves with $\beta =$ const. are also called 'appollonic circles'. In the case of $\beta = 1$ the circle 'degenerates' to a point at $z/\varrho c = 1$, resulting in $\beta = 0$ at the imaginary axis.

6.4 Theory of locally reacting absorbers

The terms for the description of sound-reflecting and sound-absorbing structures were already explained in the previous sections. The physical structures and the absorption coefficient they can possess will be investigated in this section.

The absorbent material in particular, which is used for the purpose of sound absorption, plays an important role herein. Typically, porous and fibrous material is implemented. It is composed of many fibers or cells (for example glass- or mineral-wool, cocos fiber, felt, wood-shaving or porous cellular foam). The main property of a plate, which is composed of such a material, is the resistance r_s which it opposes to the air flowing through it. This results in a pressure difference

$$p_1 - p_2 = r_s U = \Xi \mathrm{d} U \tag{6.36}$$

between the front and the back, which is proportional to the resistance r_s and the speed U of the uniform flow. The flow resistance for the same material is certainly larger, the larger the thickness d of the plate is. We need a constant to describe a material independent of the material's dimensions in terms of its specific flow resistance (or 'flow resistivity'). This is defined as

$$\Xi = \frac{r_s}{d} \,.$$

According to (6.36), the physical unit of Ξ is

$$\dim(\Xi) = \frac{\mathrm{Ns}}{\mathrm{m}^4} = 10^{-3} \frac{\mathrm{Rayl}}{\mathrm{cm}} \,, \tag{6.37}$$

which is often given in Rayl/cm according to the unit conversion already shown, whereas the flow resistance is given in Rayl ($1\,\mathrm{Rayl} = 10\,\mathrm{N\,s/m^3}$). The range which is of technical interest for the flow resistivity is about $5\,\mathrm{Rayl/cm} < \Xi < 100\,\mathrm{Rayl/cm}$. The actual value primarily depends on the 'density' of the fibers in the material, but also depends on parameters like the position of the fibers relative to the flow.

The physical principle behind the pressure difference along the thickness of the material is friction, which the air particles experience while moving along the absorber skeleton. This friction comes about due to the viscosity of the air in very thin ducts which becomes irrelevant in larger duct cross-sections.

Thus, (6.36) can also be interpreted as a force balance equation, where the right side represents the damping force (per unit area) opposite to the external force (per unit area).

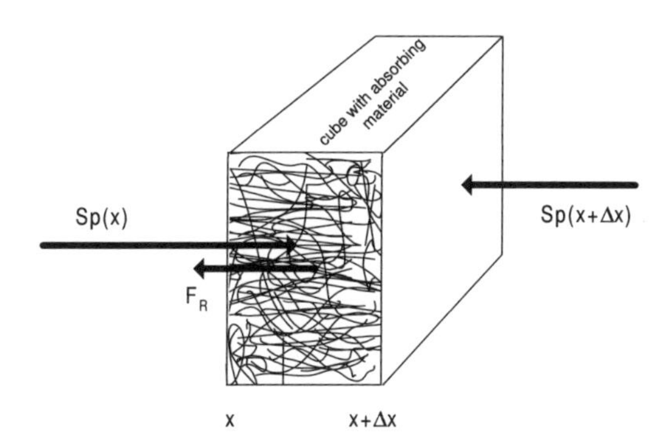

Fig. 6.7. Balance of forces in a cube of absorbing material filled with gas

It is of course the viscous damping in the pores and small ducts, which enables porous and fibrous materials to achieve substantial sound absorption in air. Kinetic energy is transformed to heat and this portion of energy is inevitably extracted from the sound field (the thermal conduction from the air to the fibers only plays an additional role at low frequencies.) It is already obvious by now that using porous is effective if they are used in areas, where the amplitudes of the particle velocity are large. If, on the other hand absorbers are located in areas of small velocity amplitudes (e.g. thin absorber sheets in front of a rigid wall), only a small absorption can be expected. It is this simple principle that partially explains the effect of most of the absorber structures and their design rules, which will be described in further detail later on.

In deriving the basic equations for the sound propagation in a porous medium, we consider an element of volume $S\Delta x$ (Fig. 6.7) located in a continuous medium. For the case of absorbent material, when using the force balance equation (2.22) for the fiber-free gas which has already been derived in Sect. 2.2.1 (see p. 28), an additional damping force needs to be brought into consideration. If it is assumed that (6.36) also applies to oscillatory flow, the damping force is $\Xi \Delta x S v$. It acts in the opposite direction of the velocity v. It is therefore

$$\varrho \Delta x S \frac{\partial v}{\partial t} = S\left[p(x) - p(x+\Delta x)\right] - \Xi \Delta x S v \,. \tag{6.38}$$

In the limits $\Delta x \to 0$, this results in

$$\varrho \frac{\partial v}{\partial t} = -\frac{\partial p}{\partial x} - \Xi v \,. \tag{6.39a}$$

As a matter of fact, there is a little inaccuracy in (6.39). The force balance equation (6.36) explicitly pertains to the damping force of the speed in air *in front of* (and behind) a thin sheet of porous material. The damping term Ξv thus denotes an 'external' velocity v_e, which is not exactly the same as the internal velocity v_i in the porous material. Since the air is squeezed in between the fibers, v_i has to be slightly larger than v_e to ensure flow continuity. In contrast, the inertia force per unit volume $\varrho\partial v/\partial t$ refers to the actual (average) air motion between the fibers, in other words, to the 'internal' velocity v_i. The corrected balance equation (6.39) is therefore given by

$$\varrho\frac{\partial v_\mathrm{i}}{\partial t} = -\frac{\partial p}{\partial x} - \Xi v_\mathrm{e} \,. \tag{6.39a}$$

A uniform description should either use v_i or v_e. Since we are mainly interested in the coupling of the absorber to the external sound field in air, we specify the external velocity v_e for our description. This has the advantage that these can all be calculated based on 'external' velocities; for the conditions at the boundaries between air and porous medium, it is only required that the (external) velocities be equal on both sides of the boundary.

Based on the assumption suggested in the 'Rayleigh model' of the absorber material that the fibers of the skeleton are stretched and in parallel, the 'internal' and 'external' velocity are related by the porosity σ ($\sigma < 1$)

$$\sigma = \frac{\text{volume of air in the absorber}}{\text{total volume of the absorber}} \,.$$

Using this assumption, the porosity is equal to the ratio of the total surface of the air ducts at the boundary toward the unbounded air and the total surface area of the absorber. Based on mass conservation it is $v_\mathrm{e} = \sigma v_\mathrm{i}$.

If it is taken into account that some ducts in the absorber are 'blind', using a structure coefficient κ ($\kappa > 1$), we obtain a smaller external velocity compared to the internal velocity

$$v_\mathrm{e} = \frac{\sigma}{\kappa} v_\mathrm{i} \,.$$

Equation (6.39) therefore leads to

$$\frac{\kappa\varrho}{\sigma}\frac{\partial v_\mathrm{e}}{\partial t} = -\frac{\partial p}{\partial x} - \Xi v_\mathrm{e} \,. \tag{6.39}$$

and unless otherwise specified, $\sigma = \kappa = 1$ has always been used in the examples discussed here in order to be able to begin with the basics.

For a complete description of the processes, a description of the compression in the absorber material is required. If the skeleton of the fibers is assumed to be rigid, it is similar to the case without absorbent material

$$\frac{\partial v_\mathrm{i}}{\partial x} = \frac{1}{\sigma}\frac{\partial v_\mathrm{e}}{\partial x} = -\frac{1}{\varrho c^2}\frac{\partial p}{\partial t} \,. \tag{6.40}$$

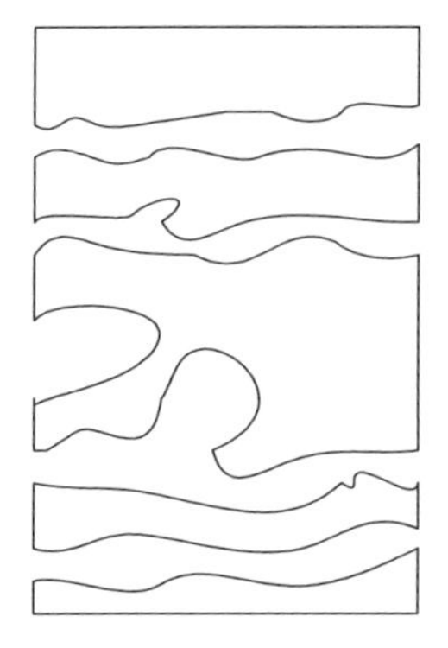

Fig. 6.8. Pockets and dead ends in the absorber material (principle)

Here, $v_e = \sigma v_i$ is also used because the spring characteristics of small volumes depend on the enclosed amount of air, but not on its distribution (within the structure).

For pure tones and complex amplitudes, the field equations (6.39) and (6.40) become

$$v_e = -\frac{1}{j\omega\varrho\kappa/\sigma + \Xi}\frac{\partial p}{\partial x} \tag{6.41}$$

and

$$\frac{\partial v_e}{\partial x} = -\frac{j\omega\sigma}{\varrho c^2}p \, . \tag{6.42}$$

Combining (6.41) and (6.42) results in the wave equation of the porous medium

$$\frac{\partial^2 p}{\partial x^2} + k^2\left(\kappa - j\frac{\Xi\sigma}{\omega\varrho}\right)p = 0 \, . \tag{6.43}$$

As usual $k = \omega/c$ is the wave number in air. The solutions to the wave equation in the absorber material

$$p = p_0 \mathrm{e}^{\pm jk_a x} \, , \tag{6.44}$$

using the complex wave number of the absorber

$$k_a = k\sqrt{\kappa}\sqrt{1 - j\frac{\Xi\sigma}{\omega\varrho\kappa}} \, , \tag{6.45}$$

represent attenuated waves. If k_a is split into its real and imaginary part

$$k_a = k_r - jk_i$$

(where k_r and k_i are positive real numbers) the pressure of (6.44)

$$p = p_0 \mathrm{e}^{\pm jk_r x}\mathrm{e}^{\pm k_i x} \tag{6.46}$$

is now attenuated along the propagation direction in both cases. The sound propagation speed c_a

$$c_a = \frac{\omega}{k_r}$$

and the level distribution along the propagation direction (now in the positive x-direction)

$$D(x) = -20 \lg e^{-k_i x} = 8.7 k_i x$$

are used as characteristic quantities of the waves. The level therefore decreases along the propagation direction. To present an illustration, the level along a certain thickness d of the material

$$D(d) = 8.7 k_i d$$

can, for instance, be cited as a characteristic value.

The most important facts expressed in the complex wave number k_a can easily be observed at frequency ranges below and above the folding frequency

$$\omega_f = \frac{\Xi \sigma}{\varrho \kappa} \tag{6.47}$$

if they are treated separately. The folding frequency 'typically' lies in an interval of about 500 Hz $< f_f <$ 5000 Hz for the range of 5 Rayl/cm $< \Xi <$ 50 Rayl/cm.

For $\omega \ll \omega_f$ it is

$$k_a \approx k\sqrt{\kappa}\sqrt{-j\frac{\omega_f}{\omega}} = \frac{1-j}{\sqrt{2}} k\sqrt{\kappa}\sqrt{\frac{\omega_f}{\omega}}$$

and

$$c_a = c\sqrt{\frac{2}{\kappa}}\sqrt{\frac{\omega}{\omega_f}} \, . \tag{6.48a}$$

The propagation speed is frequency-dependent. Therefore, the wave propagation is dispersive.

For $\omega \gg \omega_f$ it is

$$k_a \approx k\sqrt{\kappa}\left(1 - j\frac{1}{2}\frac{\omega_f}{\omega}\right) = k\sqrt{\kappa} - j\frac{\sigma}{2\sqrt{\kappa}}\frac{\Xi}{\varrho c} \, .$$

Here, the wave propagation is non-dispersive:

$$c_a = \frac{c}{\sqrt{\kappa}} \, . \tag{6.48a}$$

The absorber attenuation $D(d)$ for $\omega \gg \omega_f$ is given by

$$D(d) = \frac{4.35\sigma}{\sqrt{\kappa}}\frac{\Xi d}{\varrho c} \, . \tag{6.49}$$

The attenuation $D(d)$ is constant for $f > f_{\mathrm{f}}$. It can assume considerably large values. As will be demonstrated in the following sections, typically porous sheets are applied in practice. These have a damping coefficient $\Xi d/\varrho c$ which varies between the interval $0.25 < \Xi d/\varrho c < 8$. This results in an attenuation of up to 35 dB along the sheet thickness for all frequencies above $f > f_{\mathrm{f}}$.

The wall structures under consideration in the following sections sometimes require calculation of the velocity in the absorbent material according to (6.41) using the solution of the pressure ansatz. It is then more convenient to express the denominator in (6.41) in terms of the wave number k_{a} according to (6.45)

$$j\frac{\omega\kappa}{\sigma} + \Xi = j\frac{\omega\varrho\kappa}{\sigma}\left(1 - j\frac{\sigma\Xi}{\omega\varrho\kappa}\right) = j\frac{\omega\varrho\kappa}{\sigma}\frac{k_{\mathrm{a}}^2}{\kappa k^2} ,$$

and (6.41) resulting in

$$v_{\mathrm{e}} = \frac{j\sigma k}{\varrho c k_{\mathrm{a}}^2}\frac{\partial p}{\partial x} . \tag{6.50}$$

6.5 Specific absorbent structures

6.5.1 The 'infinitely thick' porous sheet

As shown above, it is easy to obtain large internal attenuation in a porous material. On the other hand, high absorption of a sound wave incident on the absorber surface cannot necessarily be achieved because this is a matter of matching between the two media and not of locally effective level reduction. It is by no means true that, if the level rapidly decreases along the propagation direction in the material, the contact surface to the unbounded air becomes non-reflecting. After all, the sound field must be able to penetrate the absorber in the first place in order to guarantee a high absorption. If faulty or insufficient matching results in the sound field already being reflected at the absorber's surface, an extremely high internal level reduction is useless. Adjusting the wall impedance which corresponds to the impinging sound wave is key for effective absorption.

A structure of academic importance is the porous hemisphere, which is simple and easy to understand. The 'infinitely thick' sheet of porous material can also be replaced by a sheet of finite thickness if the thickness and the attenuation are sufficiently large, making a reflection at the back irrelevant. The porous hemisphere can thus be regarded as a limiting case.

The discussion of the hemisphere is fairly simple, because no reflection occurs in the absorber. Therefore, we only need to take into account a single wave propagating in the x-direction

$$p = p_0 \mathrm{e}^{-jk_{\mathrm{a}}x} .$$

With the aid of (6.50), the characteristic impedance of the porous medium is obtained:

$$z_{\mathrm{a}} = \frac{p}{v_{\mathrm{e}}} = \frac{\varrho c}{\sigma}\frac{k_{\mathrm{a}}}{k}$$

At the boundary $x = 0$, separating air and absorber, the sound pressure and the ('external') velocity have to be equal

$$\frac{p_{air}(0)}{v_{air}(0)} = \frac{p(0)}{v_{\mathrm{e}}(0)}$$

thus the wall impedance z_∞ which is effective for the given sound field in air is equal to the characteristic impedance z_{a} in the absorbent medium

$$z_\infty = z_{\mathrm{a}} = \varrho c \frac{\sqrt{\kappa}}{\sigma}\sqrt{1 - j\frac{\Xi\sigma}{\omega\varrho\kappa}} = \varrho c \frac{\sqrt{\kappa}}{\sigma}\sqrt{1 - j\frac{\omega_{\mathrm{f}}}{\omega}} \,. \tag{6.51}$$

Apart from the porosity σ and the structure coefficient κ, the effect of the porous hemisphere is determined solely by the ratio of frequency to folding frequency.

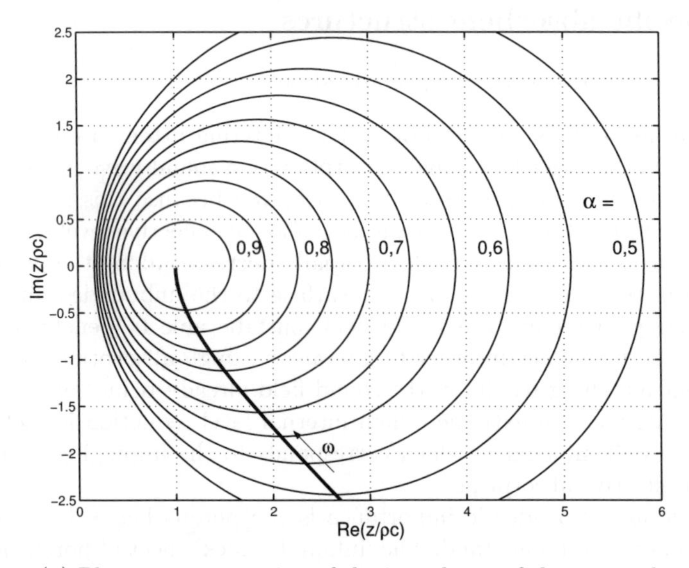

Fig. 6.9. (**a**) Phasor representation of the impedance of the porous hemisphere

The phasor curve can be easily drawn.

a) At low frequencies $\omega \ll \omega_{\mathrm{f}}$ it is

$$z_\infty \approx \varrho c \frac{\sqrt{\kappa}}{\sigma}\sqrt{\frac{\omega_{\mathrm{f}}}{\omega}}\mathrm{e}^{-j\pi/4} \,.$$

The phasor representation in the complex plane consists of a straight line inclined at a $-45°$ angle with the real axis (see Fig. 6.9a).

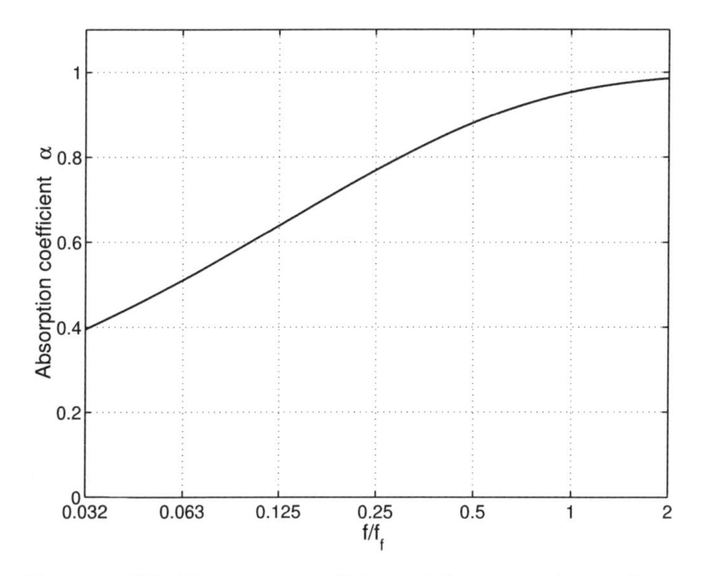

Fig. 6.9. (**b**) Absorption coefficient of the porous hemisphere

b) At high frequencies $\omega \gg \omega_\mathrm{f}$ it is

$$z_\infty \approx \varrho c \frac{\sqrt{\kappa}}{\sigma} \left(1 - j \frac{1}{2} \frac{\omega_\mathrm{f}}{\omega}\right)$$

The straight phasor representation line begins to curve and approaches a point on the real axis which corresponds to the maximum absorption. For $\sigma = \kappa = 1$ the high-frequency impedance is nearly matched. It is approximately ϱc.

The absorption at $\omega = \omega_\mathrm{f}$ is actually already high. For $\omega = \omega_\mathrm{f}$, using $1 - j = \sqrt{2} e^{-j\pi/4}$ and therefore

$$\sqrt{1-j} = \sqrt{\sqrt{2}\mathrm{e}^{-j\pi/4}} = \sqrt[4]{2}\mathrm{e}^{-j\pi/8} \approx 1.2\mathrm{e}^{-j\pi/8} ,$$

the corresponding impedance results in

$$z_\infty(\omega = \omega_\mathrm{f}) \approx 1.2 \varrho c \frac{\sqrt{\kappa}}{\sigma} \mathrm{e}^{-j\pi/8} ,$$

which is a point very close to the real axis (see Fig. 6.9a) with a very large absorption coefficient. The exact calculation using the matching law (6.29) gives an absorption coefficient $\alpha(\omega = \omega_\mathrm{f}) = 0.93$ (calculated using $\sigma = \kappa = 1$). Even for $\omega/\omega_\mathrm{f} = 0.1$, the absorption coefficient is larger than 0.6 (see also Fig. 6.9b). Such high absorption coefficients (at such low frequencies) cannot be achieved using porous sheets of finite thickness, as will be shown in what follows.

6.5.2 The porous sheet of finite thickness

The simplest construction used to absorb incident sound is a sheet of porous material, mounted on a rigid wall (Fig. 6.10a). The frequency response function of the absorption coefficient α is roughly estimated by a simple illustration. Since the reflection at the rigid wall implies a velocity node at the wall (Fig. 6.10a), the absorber layer is likewise affected by an area of small velocity if the layer thickness is small compared to the wavelength.

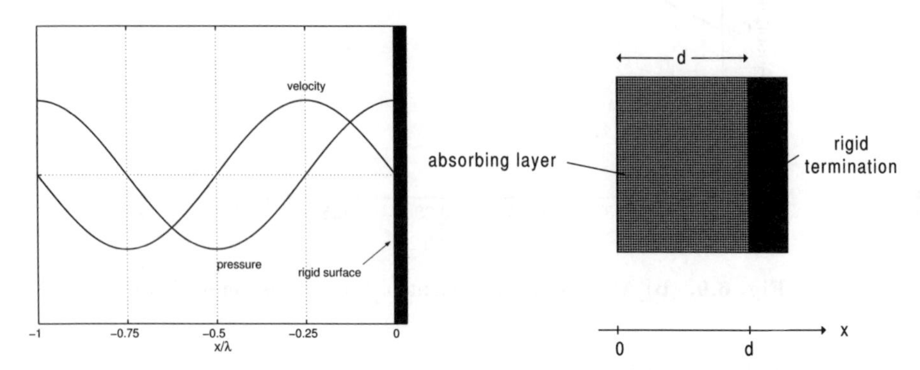

Fig. 6.10. (**a**) Left: Space characteristics of sound pressure and sound velocity in front of a rigid reflector. Right: Absorbent sheet in front of a rigid wall

At low frequencies, α is very small, because an absorber transforms kinetic energy into heat. Only if the first velocity maximum, which is located a quarter-wavelength in front of the wall, migrates into the absorber, are there areas of large velocity amplitude inside the absorbent material and the absorption coefficient increases. If the wavelength becomes slightly smaller, the porous sheet is a slightly more inefficient. The minimum is located at about $d = \lambda/2$. After that, α increases again up to $d = 3\lambda/4$, and so on. Behind the first maximum, weak alternating characteristics are obtained which gradually approach the value of α of the porous sheet of infinite thickness.

For the calculation of the absorption coefficient or the wall impedance, respectively, an approach is needed which requires that the sound field is composed of opposite progressive waves with a reflection coefficient $r = 1$ at the back rigid wall

$$p = p_0 \left\{ \mathrm{e}^{-jk_\mathrm{a}(x-d)} + \mathrm{e}^{jk_\mathrm{a}(x-d)} \right\} \tag{6.52}$$

$$v = \frac{k\sigma}{\varrho c k_\mathrm{a}} p_0 \left\{ \mathrm{e}^{-jk_\mathrm{a}(x-d)} - \mathrm{e}^{jk_\mathrm{a}(x-d)} \right\} , \tag{6.53}$$

where, additionally, the sound velocity was determined by the pressure, using (6.50). The sound velocity already satisfies the boundary condition $v(x = d) = 0$.

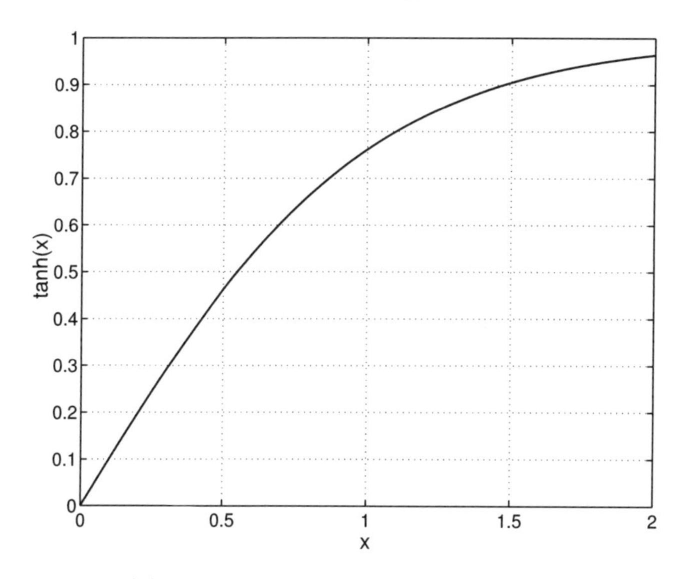

Fig. 6.10. (**b**) Characteristics of the hyperbolic tangent $\tanh(x)$

The effective impedance for the external sound field in air is again (due to the identity of pressure and velocity on both sides of the boundary $x = 0$ to the air) given by

$$z = \frac{p(0)}{v(0)} = -j\frac{\varrho c}{\sigma}\frac{k_\mathrm{a}}{k}\cot(k_\mathrm{a}d) = -jz_\infty \cot(k_\mathrm{a}d) \; . \tag{6.54}$$

We will begin our initial discussion of the characteristics of a porous sheet of finite thickness shall starting with the case of low frequencies $|k_\mathrm{a}d| \ll 1$. In a first-order approximation, using $\cot(k_\mathrm{a}d) \approx 1/k_\mathrm{a}d$, these are given by

$$z \approx -j\frac{\varrho c}{\sigma}\frac{1}{kd} = -j\frac{\varrho c^2}{\sigma d}\frac{1}{\omega} \; . \tag{6.55}$$

In (6.55) the pure impedance of a spring, which is caused by the air enclosed in the skeleton of the absorber, is described by the spring stiffness $\varrho c^2/\sigma d$. The first order approximation results in no absorption at all due to $\alpha = 0$. Only a second order approximation can determine (tiny) small absorption coefficients $\alpha \neq 0$. At low frequencies the phasor representation starts at the negative imaginary axis, which is crossed as the phasor representation moves toward the origin.

At higher frequencies, $\cot(k_\mathrm{a}d)$ is more consequently expressed in terms of exponential functions, using

$$\cot(k_\mathrm{a}d) = \frac{\cos(k_\mathrm{a}d)}{\sin(k_\mathrm{a}d)} = j\frac{\mathrm{e}^{jk_\mathrm{a}d} + \mathrm{e}^{-jk_\mathrm{a}d}}{\mathrm{e}^{jk_\mathrm{a}d} - \mathrm{e}^{-jk_\mathrm{a}d}} = j\frac{1 + \mathrm{e}^{-j2k_\mathrm{a}d}}{1 - \mathrm{e}^{-j2k_\mathrm{a}d}} \; ,$$

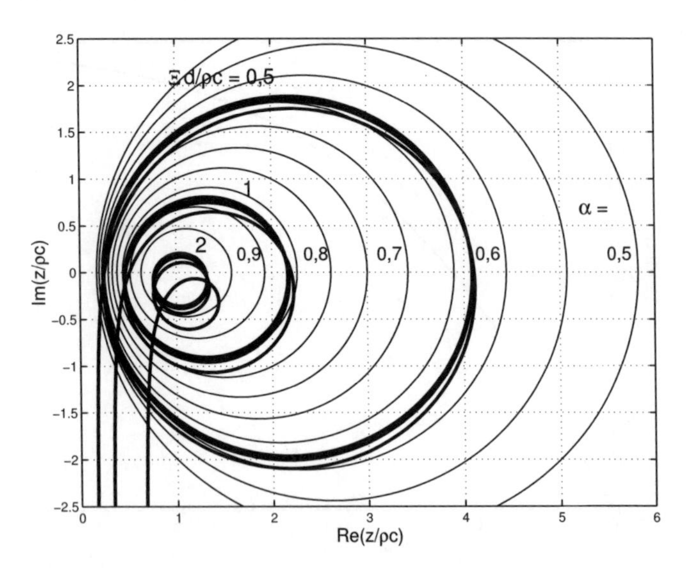

Fig. 6.10. (**c**) Phasor representations of the impedance of a porous layer in front of a rigid wall

and (6.54) results in

$$z = z_\infty \frac{1 + \mathrm{e}^{-j2k_\mathrm{a}d}}{1 - \mathrm{e}^{-j2k_\mathrm{a}d}} \,. \tag{6.56}$$

One can better imagine the phasor representation by assuming that the frequency range is above the folding frequency $\omega > \omega_\mathrm{f}$. Here,

$$k_\mathrm{a}d = kd\frac{k_\mathrm{a}}{k} = kd\left(1 - j\frac{1}{2}\frac{\Xi}{\omega\varrho}\right) = kd - j\frac{1}{2}\frac{\Xi d}{\varrho c}$$

can be used and (6.56) results in

$$z = z_\infty \frac{1 + \mathrm{e}^{-j2kd}\mathrm{e}^{-\Xi d/\varrho c}}{1 - \mathrm{e}^{-j2kd}\mathrm{e}^{-\Xi d/\varrho c}} \,. \tag{6.57}$$

It is of interest to consider points where the fraction on the right side is real. We distinguish here between two different cases $\mathrm{e}^{-j2kd} = +1$ and $\mathrm{e}^{-j2kd} = -1$:

a) At frequencies where

$$\frac{d}{\lambda} = \frac{1}{4} + \frac{n}{2}\,, \quad n = 0, 1, 2, \ldots$$

is given, the impedance results in

$$z = z_\infty \frac{1 - \mathrm{e}^{-\Xi d/\varrho c}}{1 + \mathrm{e}^{-\Xi d/\varrho c}} = z_\infty \tanh\left(\frac{1}{2}\frac{\Xi d}{\varrho c}\right), \tag{6.58}$$

due to $2kd = 4\pi d/\lambda = \pi + 2\pi n$. As shown in the previous section, $z_\infty = \varrho c$ can be assumed for $\omega > \omega_\mathrm{f}$. Thus, (6.58) denotes points on the real axis, where the distance to the origin is reduced by a factor $\tanh(\Xi d/2\varrho c)$ compared to ϱc (Fig. 6.10b recalls the characteristics of the hyperbolic tangent $\tanh(x)$).

b) At frequencies where

$$\frac{d}{\lambda} = \frac{n}{2}\,, \qquad n = 1, 2, 3, \ldots$$

is given, the impedance results in

$$z = z_\infty \frac{1 + \mathrm{e}^{-\Xi d/\varrho c}}{1 - \mathrm{e}^{-\Xi d/\varrho c}} = \frac{z_\infty}{\tanh\left(\frac{1}{2}\frac{\Xi d}{\varrho c}\right)} \tag{6.59}$$

due to $2kd = 4\pi d/\lambda = 2\pi n$. In (6.59) points on the real axis are denoted, where the distance to the origin compared to ϱc is increased by the factor $1/\tanh(\Xi d/2\varrho c)$

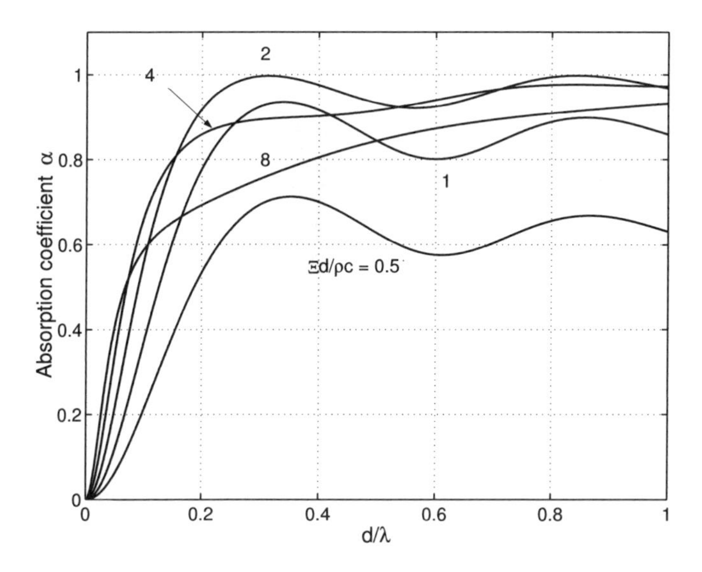

Fig. 6.10. (**d**) Absorption coefficient of a porous sheet in front of a rigid wall

As can be seen in Fig. 6.10c, the points will be alternately crossed, depending on 'case a' or 'case b', respectively, with increasing frequency, where (apart from small differences not considered above) phasor representations similar to a circle are overlaid repeatedly. Using the previous considerations, and from Fig. 6.10c and 6.10d, it follows that

- a smaller resistance $\Xi d/\varrho c$ results in a lower folding frequency, but produces an absorption coefficient, alternating with frequency and

- a larger resistance $\Xi d/\varrho c$ results in a smoother absorption coefficient $\alpha \approx 1$ above the folding frequency, but the folding frequency is very high.

As Fig. 6.10 clearly also shows, the 'optimum compromise' between the contrary requirements of 'low folding frequency' and a 'smooth, high-frequency $\alpha \approx 1$' is given by $\Xi d/\varrho c = 2$.

Good absorbent structures should roughly follow this optimum. The acoustic efficiency is $\alpha > 0.6$ in the frequency range above about $d/\lambda = 0.1$. If an absorption coefficient of $\alpha = 0.6$ at 340 Hz is required, it can just be achieved by a sheet of 10 cm thickness. Tuning absorbers towards even lower frequencies would require a substantially larger sheet thickness. As a matter of principle, porous sheets are therefore absorbers which are useful in the high frequency range.

6.5.3 The porous curtain

For absorption at lower frequencies using porous sheets, less material is required if a thin layer is mounted as a curtain in a certain distance to a wall (Fig. 6.11a).

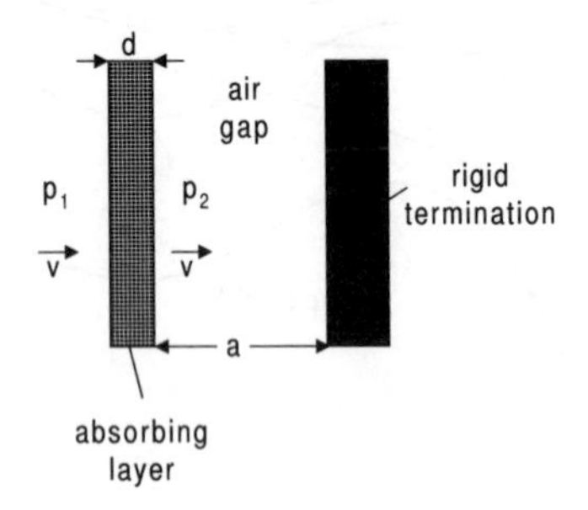

Fig. 6.11. (**a**) Porous curtain in front of a rigid wall

The absorption coefficient for this arrangement can also be estimated graphically. It will be large only if the porous layer is roughly located at a maximum of the velocity in the sound field in front of the wall. The maxima are therefore located at

$$\alpha_{\max}: \quad \lambda \left(\frac{1}{4} + \frac{n}{2} \right) = a \, .$$

The corresponding frequencies are given by

$$f = \frac{c}{a} \left(\frac{1}{4} + \frac{n}{2} \right) \, . \tag{6.60}$$

The minima, where $\alpha \approx 0$, are located at

$$\alpha_{\min}: \quad \lambda\frac{n}{2} = a \; .$$

The corresponding frequencies are given by

$$f = \frac{c}{a}\frac{n}{2} \; . \tag{6.61}$$

The peaks become broader, the larger the curtain thickness d is.

Thin absorbent layers will be assumed in the following calculation. The pressure difference $p_1 - p_2$ of the sound pressure p_1 in front of the curtain and p_2 behind it, can be estimated similar to (6.36) by the flow resistance

$$p_1 - p_2 = \Xi dv \; . \tag{6.62}$$

It was assumed that the layer is rigid itself: v denotes the velocity in the front and in the back of the curtain. The impedance of the air gap $z_2 = p_2/v$ is known from the previous section (using $\sigma = 1$ and $k_{\mathrm{a}}d = kd$ in (6.54)):

$$z_2 = \frac{p_2}{v} = -j\varrho c \cot(ka) \tag{6.63}$$

and by using the impedance of the complete with the aid of (6.62): structure is obtained

$$z = \frac{p_1}{v} = \Xi d + \frac{p_2}{v} = \Xi d - j\varrho c \cot(ka) \; . \tag{6.64}$$

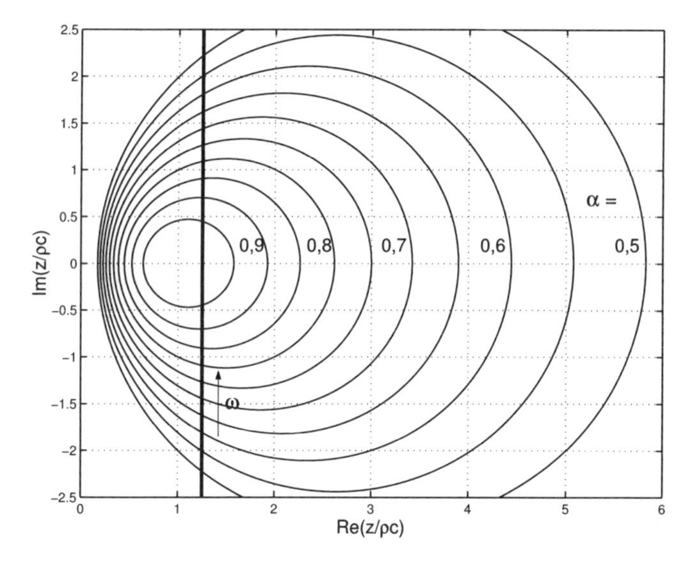

Fig. 6.11. (**b**) Phasor representation of the impedance of a porous curtain

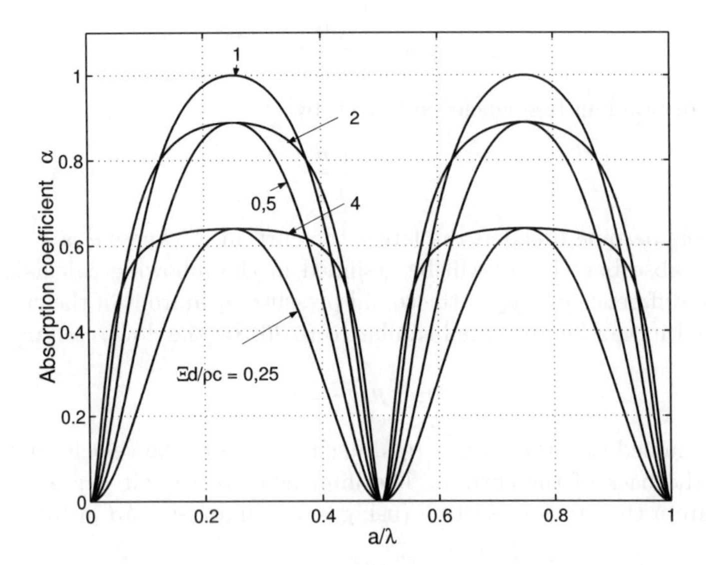

Fig. 6.11. (**c**) Absorption coefficient of a porous curtain

The phasor representation (Fig. 6.11b) is parallel to the imaginary axis which is overlaid several times, due to the periodicity of the cotangent function. As usual, α has a maximum if the phasor representation crosses the real axis. This is the case for

$$\cot(ka) = 0 \; ; \quad \text{therefore} \quad ka = \frac{\pi}{2} + n\pi \; ,$$

which is equal to (6.60). The maxima are given by

$$\alpha_{\text{max}} = \frac{4\frac{\Xi d}{\varrho c}}{\left(\frac{\Xi d}{\varrho c} + 1\right)^2} = \frac{4}{\frac{\Xi d}{\varrho c} + \frac{\varrho c}{\Xi d} + 2} \; . \tag{6.65}$$

The maxima remain equal if the ratio $\Xi d/\varrho c$ is substituted by its inverse.

By relocating the phasor representation in (Fig. 6.11b), it becomes obvious that

- a small resistance $\Xi d/\varrho c$ has a narrow peak in the absorption. Changing the frequency away from $\cot(k_\text{a} d) = 0$ crosses along several different curves $\alpha = \text{const.}$;
- a large resistance $\Xi d/\varrho c$ has a broader peak in the absorption. Changing the frequency away from $\cot(k_\text{a} d) = 0$ nearly does not cross any other curves $\alpha = \text{const.}$

Summarizing, it can be stated that resistances $\Xi d/\varrho c$, which are the inverse of each other, have the same maximum in the absorption coefficient, but have peaks with very different widths (see also Fig. 6.11c). To eliminate any doubt, a resistance will probably be chosen which is too large. Due to the space they take up (which equals that of the porous sheet of finite thickness) porous curtains are also usable exclusively at higher frequencies.

6.5.4 Resonance absorbers

An effective low-frequency absorber is obtained when a mass is added to the porous curtain. As will be shown in the following, the additional mass characteristics compensate the spring characteristics given by the cotangent in (6.64). Given the same demand for space, a wall impedance 'without an imaginary component' is created at a lower frequency than without the additional mass. This is the advantage of the resonance absorber.

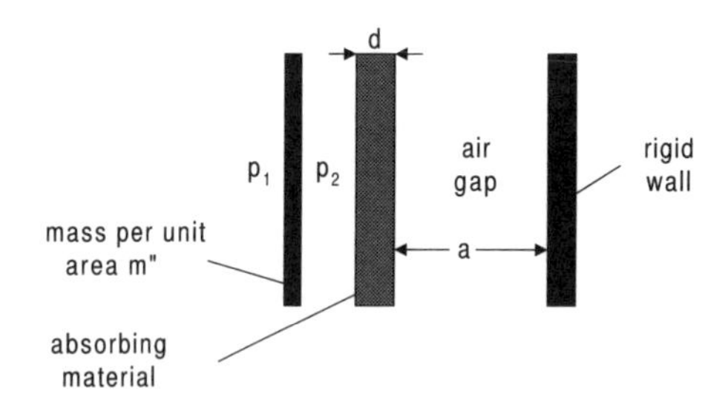

Fig. 6.12. Resonance absorber (schematic)

The wall impedance of the structure depicted in Fig. 6.12 can be determined using the inertia law. The pressure difference in the front and behind the mass is given by

$$p_1 - p_2 = j\omega m'' v \tag{6.66}$$

(m'' = mass per unit area), and this results in

$$z = \frac{p_1}{v} = j\omega m'' + \frac{p_2}{v} = j\omega m'' + z_2 ,$$

where z_2 represents the impedance of the porous curtain (6.64). For the porous curtain including the additional mass, it is

$$z = j\omega m'' - j\varrho c \cot(ka) + \Xi d . \tag{6.67}$$

The characteristics of the phasor representation are the same as for the porous curtain (Fig. 6.11b), but this time the frequencies are assigned differently. Instead of the maxima $\cot(kd) = 0$ of the absorption coefficient for the porous curtain, the maxima for the resonance absorber are given by:

$$\cot(ka) = \cot\frac{\omega a}{c} = \frac{\omega m''}{\varrho c} . \tag{6.68}$$

The transcendental equation (6.68) used to determine the frequency points with $\alpha = \alpha_{\max}$ can easily be solved graphically. This equation involves the intersections of the cotangent and a straight line with a gradient proportional to m''. As shown in Fig. 6.13, the frequencies with a maximum α are lower, the larger the mass per unit area m'' is. As already mentioned, this is the main advantage of the resonance absorber compared to simple porous curtains. By using an additional mass, the absorption coefficient can be tuned to lower frequencies, especially the first maximum, without changing the depth of the structure. A typical frequency response function of the absorption coefficient of a resonance absorber is shown in Fig. 6.14. A comparison with Fig. 6.11c shows a noticeable shift in the maxima, whereas their height $\alpha_{\max}$ remains unchanged. Just as in the case of the porous curtain,

$$\alpha_{\max} = \frac{4}{\frac{\Xi d}{\varrho c} + \frac{\varrho c}{\Xi d} + 2}$$

is given by (6.67).

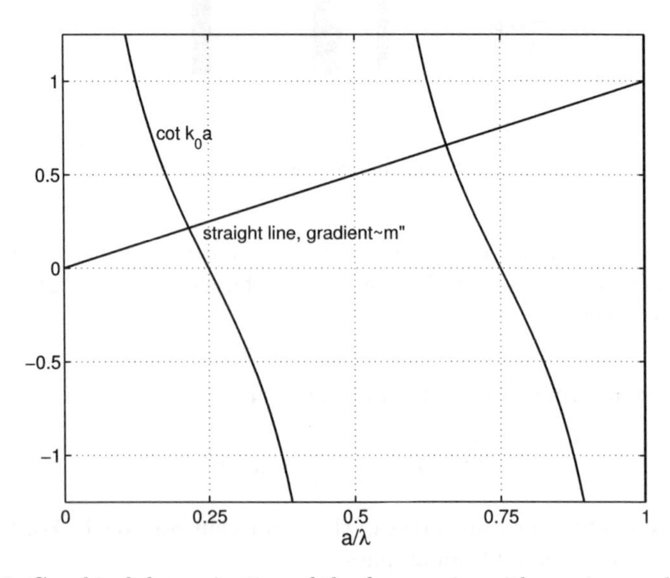

Fig. 6.13. Graphical determination of the frequencies with maximum absorption

The lowest maximum is mainly relevant for practical purposes. Under the assumption that the wavelength is large compared to the depth a of the air gap, the cotangent function can be replaced by its reciprocal argument. In the frequency range of the lowest maximum, the impedance is approximately given by

$$z = \Xi d + j \left(\omega m'' - \frac{\varrho c^2}{\omega a} \right) . \tag{6.69}$$

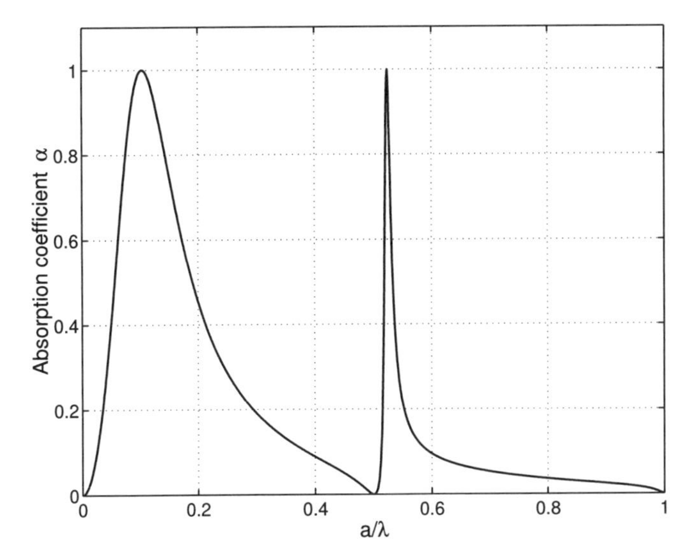

Fig. 6.14. Absorption coefficient of a resonance absorber (calculated using $\Xi d/\varrho c = 1$ and $m''/\varrho a = 2$)

The tuning frequency of the maximum absorption is equal to the resonance frequency

$$\omega_{\text{res}} = \sqrt{\frac{\varrho c^2}{am''}} \tag{6.70}$$

of the simple mass-spring-oscillator, comprised of mass per unit area m'' and the air spring with the stiffness $\varrho c^2/a$.

The required tuning frequency is the product of the depth of the gap a and the mass per unit area m''. It is therefore equivalent to either producing lightweight mass linings with a larger air gap, or to heavy linings that demand less space. It should be kept in mind that the choice has a consequence for absorption bandwidth. What is the effect of an exciting frequency which differs from the resonance frequency on the absorption coefficient? This question can easily be answered by simply allowing small changes around the resonance frequency. The imaginary part of the impedance in (6.69) can be expressed by the first term of the Taylor series

$$\omega m'' - \frac{\varrho c^2}{\omega a} = (\omega - \omega_{\text{res}}) \left. \frac{\mathrm{d}(\omega m'' - \varrho c^2/\omega a)}{\mathrm{d}\omega} \right|_{\omega=\omega_{\text{res}}}$$

$$= (\omega - \omega_{\text{res}}) \left(m'' + \frac{\varrho c^2}{\omega_{\text{res}}^2 a} \right) = 2m''(\omega - \omega_{\text{res}}) \,, \tag{6.71}$$

consequently resulting in

$$\alpha = \frac{4\frac{\Xi d}{\varrho c}}{\left(\frac{\Xi d}{\varrho c}+1\right)^2 + \left[\frac{2m''}{\varrho c}(\omega-\omega_{\mathrm{res}})\right]^2} . \tag{6.72}$$

Usually, the bandwidth of the maximum is expressed by the frequency distance $\Delta\omega$ of two points, one to the left and one to the right of the maximum, where the absorption coefficient has fallen to half of the maximum value

$$\alpha(\omega = \omega_{\mathrm{res}} \pm \Delta\omega/2) = \frac{1}{2}\alpha_{\mathrm{max}} .$$

The absorption coefficient is equal to half the maximum value if the two terms in the denominator of (6.72) are equal, resulting in

$$\left[\frac{2m''}{\varrho c}(\omega_{\mathrm{res}} \pm \Delta\omega/2 - \omega_{\mathrm{res}})\right]^2 = \left[\frac{\Xi d}{\varrho c}+1\right]^2 ,$$

or

$$\Delta\omega = \frac{\Xi d + \varrho c}{m''} . \tag{6.73}$$

Therefore, the half-bandwidth is inversely proportional to the mass per unit area m''. For that reason, if a broad bandwidth is preferred, small masses are typically used. A slightly larger air gap must be allowed in order to be able to tune to lower frequencies.

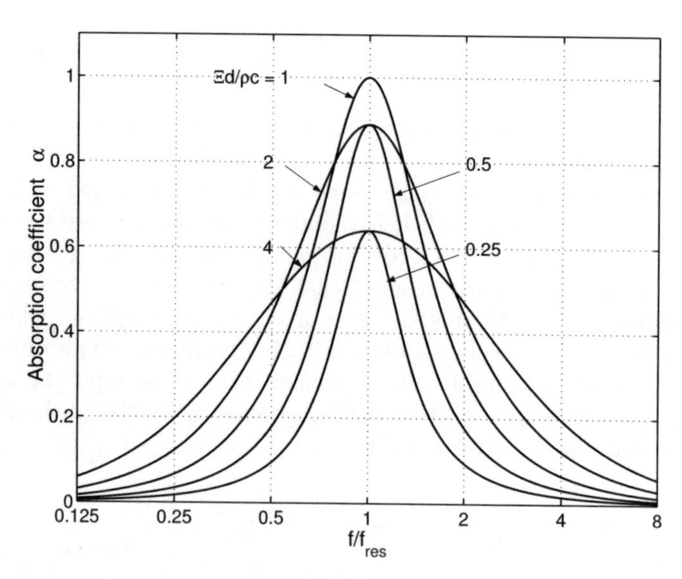

Fig. 6.15. (a) Absorption coefficients of resonance absorbers (calculated using $m''\omega_0/\varrho c = 2$)

The dependence of the (lowest) maximum absorption coefficient on its parameters is again summarized in Fig. 6.15.

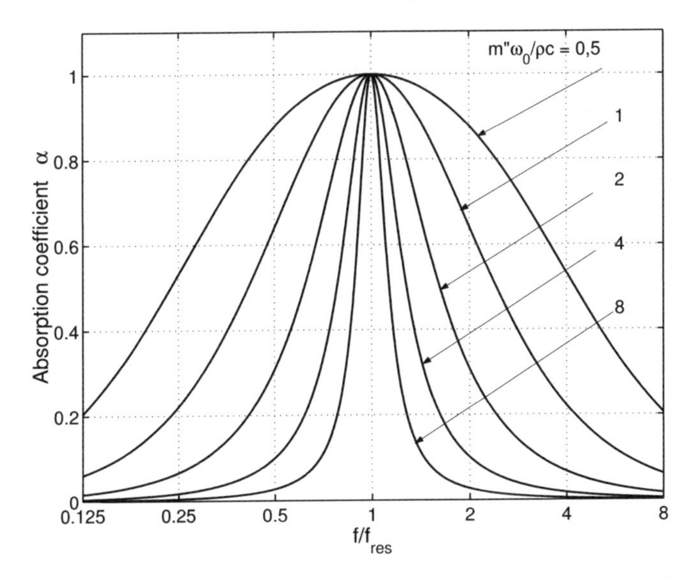

Fig. 6.15. (**b**) Absorption coefficients of resonance absorbers (calculated using $\Xi d/\varrho c = 1$)

As can be clearly seen (and observed in (6.73)):

- the peaks with constant m'' become broader for larger flow resistances $\Xi d/\varrho c$. The maxima are equal for reciprocal flow resistances.
- the peaks with constant $\Xi d/\varrho c$ are narrower, the larger m'' is.

The order of magnitude of the applicable mass linings is relatively small (compared to windows, sheet metals or even walls). This will be illustrated by means of an example, where a resonance absorber shall be tuned to 200 Hz. The acoustician would like to use a large air gap a in order to be able to use a small mass and to broaden the bandwidth. For other reasons, he is certainly not allowed to 'steal' a lot of the room's volume (an exception to this might be the ceiling of high rooms). It is therefore assumed that the air gap is restricted to a depth of 10 cm (for a porous curtain or an absorbent sheet, 40 cm would ultimately be necessary!). This results in

$$m'' = \varrho c^2/\omega_{\mathrm{res}}^2 a,$$

which is a value of approximately $m'' = 0.850\,\mathrm{kg/m^2}$. The half-bandwidth approximately results in $\Delta f = 150\,\mathrm{Hz}$. The absorber is therefore 'efficient' in a range of about 125 Hz to 275 Hz. Halving the mass (and therefore doubling the bandwidth and the air gap a as well) would be more beneficial from the acoustician's point of view. Thus, resonance absorbers also have a substantial demand for space if they are supposed to be effective at low frequencies in a wide bandwidth.

Small amounts of air oscillating in the small holes of perforated plates are intended to compensate for the small mass linings. As depicted in Fig. 6.16, the absorbent structure is constructed of a rigid plate with holes that lead to the absorber's backing; the plate is mounted at a certain distance from a rigid wall.

The force balance equation for uniform mass distribution

$$p_1 - p_2 = j\omega m'' v$$

must be substituted by an equation which applies to the case of holes.

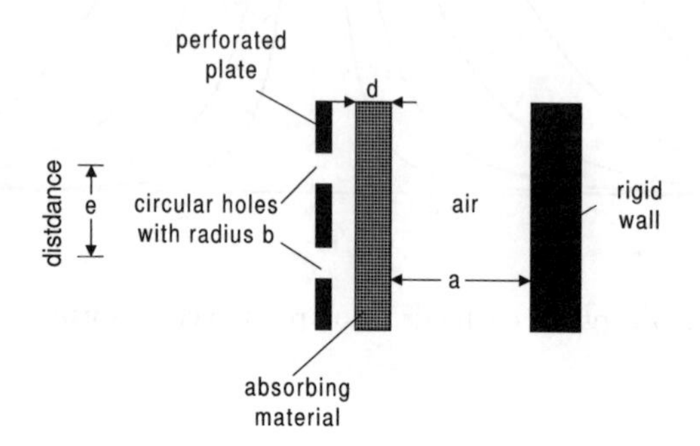

Fig. 6.16. Resonance absorber with perforated plates

The mass, corresponding to an individual hole, is focused here. The forces $p_1 S_L$ and $p_2 S_L$ act on either side of the mass (S_L is the cross-sectional area of the hole). The inertia law therefore requires

$$p_1 - p_2 = \frac{M}{S_L} j\omega v_M \tag{6.74}$$

where M is the mass moved in the hole and v_M its velocity. Apart from the movement of the air mass, which is located in the volume of the hole, volumes of adjacent air in front of and behind it will move along with the mass. For circular holes with radius b, it can be assumed that the correction, which is established by this fact, is given by a semi-sphere with a radius equal to the radius of the hole on both sides of the hole. Thus

$$M = \varrho \left(\pi b^2 W + \frac{4\pi}{3} b^3 \right) ,$$

or

$$\frac{M}{S_L} = \varrho \left(W + \frac{4}{3} b \right) \tag{6.75}$$

where W is the 'thickness of the hole' or the plate thickness. Now it has to be taken into account that the 'external' velocity v in the sound field and the velocity in the hole v_M may differ significantly. The mass impinging onto the perforated plate during the time interval of one second is 'forced' through the total perforated surface area S_{Ltot} included in S. Based on mass conservation, it follows that

$$S_{\mathrm{Ltot}} v_{\mathrm{M}} = S v \,,$$

or

$$v_{\mathrm{M}} = \frac{S}{S_{\mathrm{Ltot}}} v = \frac{v}{\sigma} \,, \tag{6.76}$$

where σ represents the ratio of the hole area to the plate surface area:

$$\sigma = \text{Surface area of the holes/Total surface area} \,.$$

After inserting (6.75) and (6.76) in (6.74) it becomes

$$p_1 - p_2 = j\omega \frac{\varrho \left(W + \frac{4}{3} b\right)}{\sigma} v \,. \tag{6.77}$$

The previous conditions concerning resonance absorbers still apply when calculating the 'effective mass' per unit area of the perforated plate only, based on the data, which yields

$$m'' = \frac{\varrho (W + 4b/3)}{\sigma} \tag{6.78}$$

As can be seen, the moved mass m'' is *not* equal to the mass stored in the holes, as is sometimes erroneously assumed. Since σ is always considerably smaller than 1 (typically σ between 0.1 and 0.3), the effective mass m'' is a lot larger than ϱW.

It should be mentioned that, in contrast to the previous considerations, the end-correction is actually defined by 'the 1.25-fold of a sphere'. In better accordance with most practical setups, (6.78) is replaced by

$$m'' = \frac{\varrho (W + 5b/3)}{\sigma} \tag{6.79}$$

The example of $\sigma = 0.1$, $W = b = 1\,\mathrm{cm}$ with $m'' = 350\,\mathrm{g/m^2}$ shows that the required mass linings are easy to produce.

There are a variety of setups for resonance absorbers. The aforementioned example of a perforated plate with a backing absorber material is often used in acoustic ceilings, because these allow for enough buffer for low frequency tuning. Other constructions, for instance, consist of a cellular foam with a coated surface, which acts as the additional mass. Foil absorbers have been used for many years. The sound field in air induces them to perform membrane and bending vibrations, thereby extracting energy from the sound field. They can also be tuned to the lowest frequencies.

6.6 Oblique sound incidence

In practice, sound waves rarely ever impinge normally to the absorber's surface. In fact, diffuse sound incidents coming from all directions are the more realistic assumption for actual rooms. For this reason, we will proceed to examine how oblique sound incidents affect sound absorption.

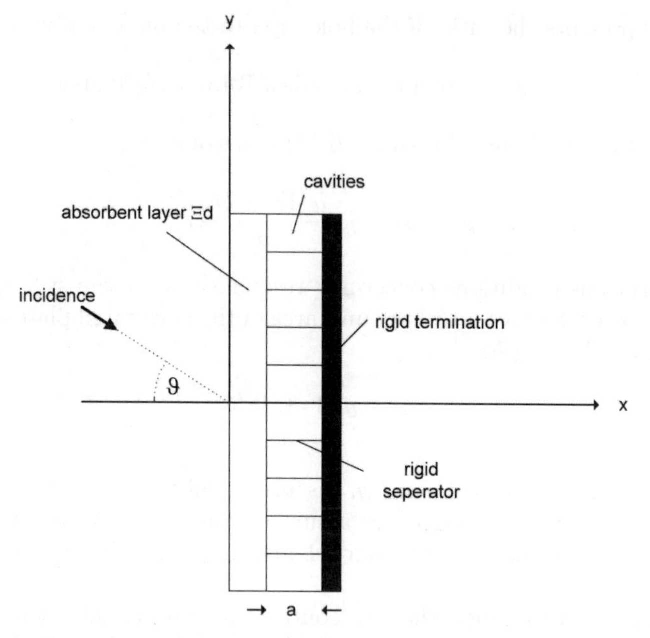

Fig. 6.17. A system consisting of an absorbent layer in front of a partition separated by a segmented cavity can be assessed as locally effective, since cross coupling is suppressed in the y-direction.

The answer is simple, presuming locally effective absorbent structures. By definition, in structures such as these, transients in the y-direction never occur. In the case of porous sheets in front a rigid partition at a sufficient distance, this requires the segmentation of the cavity region in likewise rigid partitions (see Figure 6.17). Only then can coupling between the separated parallel cavities be eliminated. If there is no segmentation, an oblique field likewise propagates in the cavity that are in the presence of skew wave incidence, causing the effect to no longer be strictly local. The systems shown in the scheme in Figure 6.17 can therefore be understood as locally effective and therefore, once again, can be described by a space-dependent impedance. Even absorbent sheets mounted directly to rigid partitions can be understood as locally reactive, as long as their internal attenuation is large in respect to sufficiently high flow resistance.

In the case of a wave impinging upon the absorbent system below the angle ϑ

$$p_{ein} = p_0 e^{-jkx cos\vartheta + jky sin\vartheta}, \tag{6.80}$$

the net field comprised of sound incident and reflected components is composed of

$$\begin{aligned} p &= p_0 (e^{-jkx cos\vartheta + jky sin\vartheta} + r e^{jkx cos\vartheta + jky sin\vartheta}). \\ &= p_0 e^{jky sin\vartheta} (e^{-jkx cos\vartheta} + r e^{jkx cos\vartheta}) \end{aligned} \tag{6.81}$$

The space-dependent wall impedance z is derived from

$$z = \frac{p}{\frac{j}{\omega\varrho}\frac{\partial p}{\partial x}} = \frac{\rho c}{cos\vartheta}\frac{1+r}{1-r}, \tag{6.82}$$

leading to the definition

$$r = \frac{\frac{z}{\varrho c}cos\vartheta - 1}{\frac{z}{\varrho c}cos\vartheta + 1}. \tag{6.83}$$

All preceding considerations for normal sound incidents apply here as well. The only difference from skew incidences is that the wall impedance is multiplied by $cos\vartheta$. According to eq.(6.83), impedance effects which are comparatively large in respect to the structural adjustment implemented are mitigated, as can be seen in the absorption coefficient of the porous sheet in Figure 6.18. Too small impedances, on the other hand, are further reduced, and the absorption coefficient diminishes with decreasing incidence angle (Figure 6.19).

A tendency similar to porous sheets is exhibited by porous layers mounted directly to a rigid substrate. The particular properties of such structures are somewhat more complex, because in this case, real and imaginary components of the impedance are influenced by the flow resistance. As shown in Figures 6.20 and 6.21, at high levels of flow resistance, absorption only increases with incidence angle at very low frequencies, while sound absorption is somewhat hindered at high frequencies. At low levels of flow resistance, the absorption properties decrease along with the incidence angle.

In conclusion, in order to reconstruct diffuse sound incidence, it is recommended to take the mean of the absorption coefficient over several incidence angles.

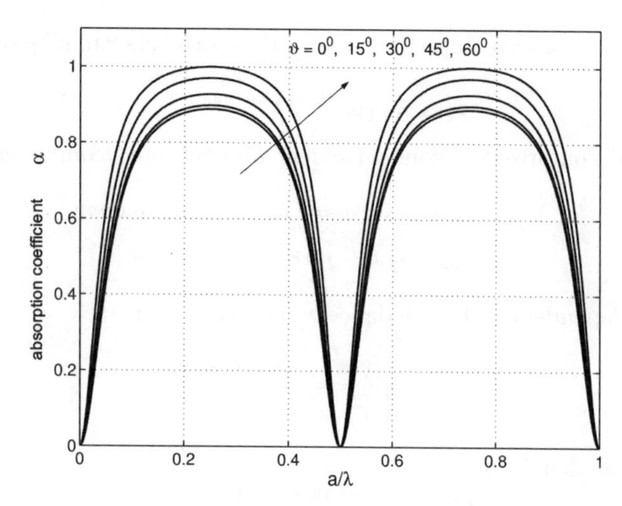

Fig. 6.18. Absorption coefficient of porous curtain with $\Xi d/\rho c = 2$ for oblique sound incidence.

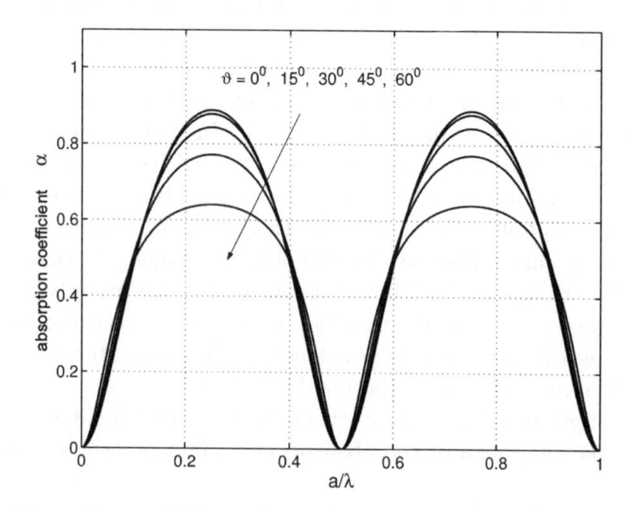

Fig. 6.19. Absorption coefficient of porous curtain with $\Xi d/\rho c = 0.5$ for oblique sound incidence.

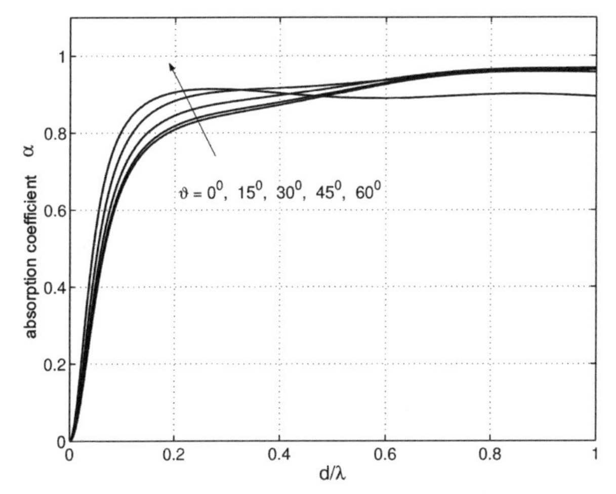

Fig. 6.20. Absorption coefficient of porous layer with $\Xi d/\rho c = 5$ for oblique sound incidence.

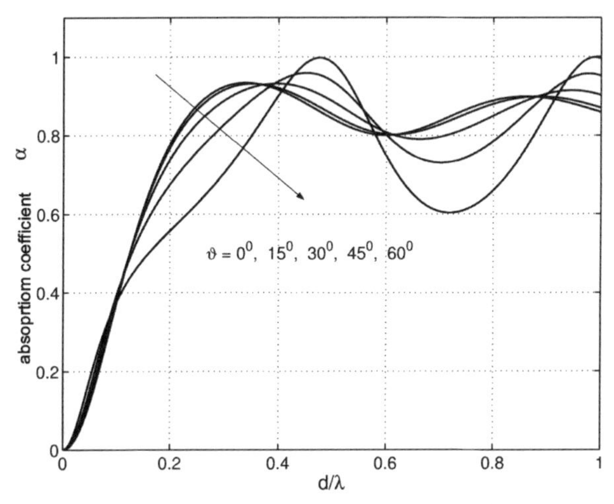

Fig. 6.21. Absorption coefficient of porous layer with $\Xi d/\rho c = 1$ for oblique sound incidence.

6.7 Summary

Measurements of the absorption coefficient of normal sound incidences in wall structures are taken in an impedance tube. The lowest cut-on frequency of the higher transverse modes in the tube determines the upper limit of the frequency range for that technique. The measurement is based on the principle that the sound field is comprised of both progressive and standing waves, which are dependent on the reflection factor of the test sample at the end of the tube. A completely sound absorbent termination produces only progressive waves. In the case of partial reflection, the test sample produces a local sound pressure space-dependence with minima and maxima. The ratio of pressure minimum to pressure maximum represents the scale of the wall absorption coefficient. A ratio that is close to zero indicates low sound absorption while a ratio close to one correlates to a high absorption coefficient.

The wall impedance z was introduced in order to describe the properties of wall structures. z is equal to the ratio of pressure and particle velocity along the wall's surface. The 'matching law' describes the correlation between absorption coefficient α and wall impedance z. It says that the imaginary components of the wall impedance are detrimental to absorption and that $\alpha = 1$ only becomes valid where $z = \rho c$ holds. A high internal damping property within the absorber material is only effective if some portion of the external sound field is able to permeate the material without being reflected at the wall's surface.

Low frequency sound absorption is only partially achieved by using porous layers. Resonance absorbers can provide a certain degree of improvement in this situation.

6.8 Further reading

A whole series of questions and problems arise which cannot all be handled in one textbook. Just to mention a few examples:

- What happens at oblique sound incidence? Is it necessary to take coupling effects parallel to the surface into account?
- Is it always correct to assume a rigid absorber skeleton, or is it required to take their elastic properties into account?
- How are micro-perforated structures, which are sometimes implemented, treated?
- What are the design rules for membrane and foil absorbers?

The answers to these (and other) questions are discussed in other books. Many, mainly theoretical problems are treated in the work 'Schallabsorber' of F.P. Mechel. Practical hints, and even descriptions of individual products are included in the book by H.V. Fuchs 'Schallabsorber und Schalldämpfer' (Springer, Berlin and Heidelberg, 2007). Finally, the corresponding chapter in the 'Taschenbuch der Technischen Akustik' (i.e. Handbook of Engineering Acoustics) is a rich source of knowledge on sound absorption. Numerous other references, which can be used to get more detail on this topic, can be found there as well.

6.9 Practice exercises

Problem 1

Refer to the diagrams for an overview of the space and time dependencies of the sound pressure in the impedance tube. Its terminations have a reflection coefficient of $r = 0,25$, $r = 0,5$ and $r = 0,75$, respectively. Graph each space-dependency under the condition of harmonic periodic oscillations for the times $t = nT/20$ ($n = 0, 1, 2, 3, ...; T$=duration of a period).

Problem 2

Calculate the frequency response of the absorption coefficient and impedance of a sound system encased in an $8\,cm$–thick material made out of a mixture of wood fiber and cement, and located just before the tube's rigid termination (total sound reflection). The table indicates measurements of maximum level L_{max}, minimum level L_{min} and the distance $|x_{min}|$ of the first pressure minimum from the object's surface.

Frequency/Hz	L_{max}/dB	L_{min}/dB	$\lvert x_{min}\rvert/cm$
200	76.9	62.1	34.3
300	70.3	62.5	20
400	75.8	72.2	18
500	71.2	62.7	19.5
600	64.5	52.5	15
700	71	56.1	12.3
800	72.3	56.3	10
900	66.9	52.3	8.8
1000	70.2	56.6	7.3
1100	73.4	61.4	6.5
1200	76	67.3	5.5
1300	76.5	69.4	5.7
1400	71.6	64.2	5.7
1500	56.9	50.4	5.3
1600	61.1	52.2	4.8
1700	60.5	51.1	4.4
1800	65.6	54.7	3.9

Problem 3

Calculate the absorption coefficient, the phase φ of the reflection factor, and the location of the first minimum when measuring the following wall impedances in the impedance tube:

- $z/\varrho c = 1 + j$,

- $z/\varrho c = 2 + j$,
- $z/\varrho c = 1 + 2j$,
- $z/\varrho c = 3 + j$, and
- $z/\varrho c = 1 + 3j$.

Problem 4

How great are the values of the wall impedance and the absorption coefficient of one layer of porous material with the following specifications:

- $\Xi = 10^4 Ns/m^4$, $\sigma = 0.97, \kappa = 2$,
- $\Xi = 10^4 Ns/m^4$, $\sigma = 0.97, \kappa = 1$,
- $\Xi = 2\,10^4 Ns/m^4$, $\sigma = 0.97, \kappa = 2$, and
- $\Xi = 2\,10^4 Ns/m^4$, $\sigma = 0.97, \kappa = 1$,

each layer with a thickness of $10\,cm$ and installed right in front of a rigid wall at the frequencies of $200 Hz$, $400 Hz$, $800 Hz$, and $1600 Hz$?

Problem 5

Suppose the absorption coefficients from Problem 4 are to be specified for aquatic silencers ($c_{water} = 1200\,m/s$, $\varrho_{water} = 1000\,kg/m^3$). State the drag and layer thicknesses required to achieve these absorption coefficient values, without changing the porosity σ or structure coefficient κ.

Problem 6

A resonance absorber is to be tuned to resonance frequencies of $250\,Hz$, $350\,Hz$, and $500\,Hz$, respectively, with a relative bandwidth of $0.5 = \Delta f / f_{res}$. How deep do the cavity and the lining of the mass have to be in order to obtain $\alpha = 1$ in the resonance frequency itself?

Problem 7

The thinnest lining from Problem 6 ($m'' = 0,51 kg/m^2$) is to be achieved by implementing a very thin perforated plate (thickness W is negligibly small). The holes make up 0.05 (0.1) of the surface. How large do the radii of the hole have to be?

Problem 8

How far do the circular holes have to be apart if they are to comprise 0.05 (0.1)? Assume an equidistant arrangement of the holes in a grid pattern.

Problem 9

State the lowest cut-on frequencies of tubes with a square cross-section and diameters of $5\,cm$ and $7\,cm$ ($6\,cm$ and $9\,cm$), respectively.

Problem 10

Show the dependency of the absorption coefficient, in the case of oblique sound with an angle of incidence for a thick porous absorber (hemisphere) with $\Xi = 10^4 Ns/m^4$, $\sigma = 0,9$ and frequencies of $1000\,Hz$ and $500\,Hz$ each, using an array of curves with the parameter $\kappa = 1, 2, 4, 8$ and 16.

Problem 11

A perforated plate is covered with the most circular holes possible so that the average distance between the holes is equal to the diameter of each hole. How great is the maximum hole coverage on the plate? Again, assume a grid-like arrangement of equidistant holes.

Problem 12

How great is the damping of the highest modes ($n > 1$) in an impedance tube terminated by two parallel, rigid plates, if the modes are excited by frequencies below their cut-on frequencies?

Problem 13

State the lowest cut-on frequencies of tubes with a diameters of $5\,cm$, $10\,cm$ and $15\,cm$ each.

Problem 14

In a thick layer made out of absorber material, a damping effect of $1\,dB/cm$ ($=$ level decrease over a distance of $1\,cm$) is measured using a sensor. The frequency is higher than the absorber's folding frequency ω_f, the porosity is measured at $\sigma = 0.95$, and the structural coefficient at $\kappa = 2$. How great is the material's drag for the given length? Express the value in $Rayl/cm$.

7

Fundamentals of room acoustics

A reverberation can be heard in an enclosed room if a sound source, which was in operation for a longer period, is suddenly turned off. Its duration depends on the room volume and the room's interior design; the reverberation time is short in small rooms and in rooms which have large absorbing surfaces. Large volumes with less absorption have reverberation times which can easily extend to a couple of seconds. During a period of 2 s, for instance, the sound travels a distance of nearly 700 m and hits the room surfaces several times; the sound waves are reflected several times at the walls under different angles. Each reflection at a (rigid) plane surface can be regarded as coming from a sound source which is mirrored at the wall. To represent multiple reflections, higher order mirror sources have to be assigned to the mirror sources as well.

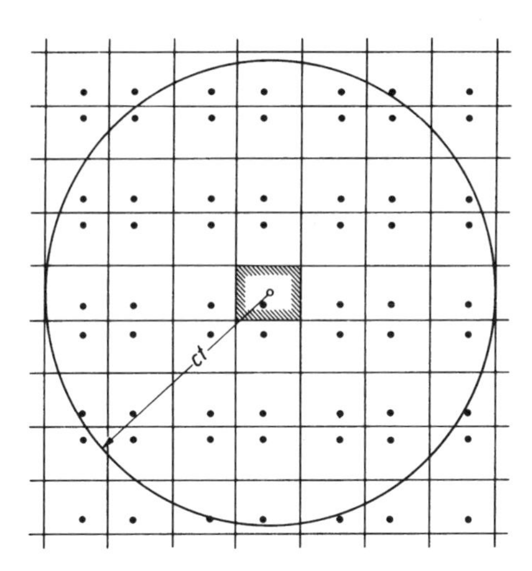

Fig. 7.1. Mirror sources in a rectangular room

M. Möser, *Engineering Acoustics*, DOI 10.1007/978-3-540-92723-5_7,

For a rectangular room, a 'starry sky' of substitute sources is obtained, which is depicted in Fig. 7.1 for one plane. The three-dimensional extension is derived similarly. The sound field in the room can be replaced by the sum of the sound waves starting at the same time at the source and all the mirror sources. The delays between the individual sound waves are expressed by the distances of the mirror sources to the observation point.

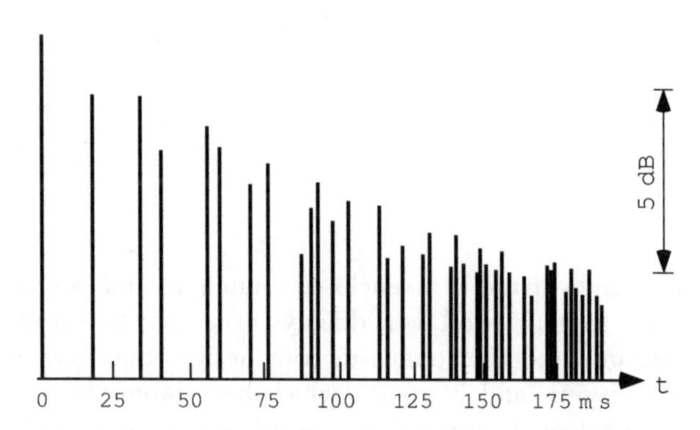

Fig. 7.2. Time series of reflections in a room enclosed by plane walls

If the original sound source emits a short impulse, an echogram is obtained as shown in Fig. 7.2. Only for the first few reflections is the arrival time of the impulses mainly determined by the actual distance of sender and receiver and their position in the room. For higher order reflections (analogue to higher order mirror sources) the differences become more and more blurred, because the dimensions of the room become irrelevant at great distances to the mirror sources. The number N of the impulses which have arrived at the time t ($t = 0$ corresponds to the sending time of the source) can be estimated by the number of mirror sources located in a sphere with the radius $r = ct$. The number of reflections is roughly equal to the ratio of the sphere volume V_k and the room volume V

$$N = \frac{V_s}{V} = \frac{4\pi}{3}\frac{(ct)^3}{V} \ . \tag{7.1}$$

For a (not even very large) room volume of $V = 200\,\mathrm{m}^3$, about 800 000 reflections result in the first second.

As can be seen from (7.1) and Fig. 7.2, the density of the incident impulses

$$\frac{\Delta N}{\Delta t} \approx \frac{\mathrm{d}N}{\mathrm{d}t} = 4\pi c \frac{(ct)^2}{V} \tag{7.2}$$

increases with time, whereas the magnitude of the incident energy impulses decreases, due to the larger distance of the mirror sources

$$E_{\text{in}} \approx \frac{1}{(ct)^2} \ .$$

The time average of the detected energy E at one point in the room is given by the product of the number of impulses, arriving in a time interval and their magnitude

$$E = E_{\text{in}} \frac{\Delta N}{\Delta t} = \text{const} .$$

The energy density will soon reach a constant value in time. The same applies to spatial energy distribution: the more the room dimensions decrease compared to the distances of the mirror sources, the smaller the effect of the position of the observation point. The spatial and temporal energy distribution are therefore expected to have constant characteristics.

Experience teaches that even in very reverberant rooms the same sound intensity can be observed at each room position not too close to the source; however, the reverberation always decays with time. The reason for that is given by the attenuation the sound waves undergo due to damping effects along the propagation route and absorption at the walls (and the room furnishing). The following discussions on sound propagation in rooms must therefore essentially take losses into account. This does not, however, change the fact that the sound field is uniformly spatially distributed. Such a sound field – graphically denoted as 'diffuse' – will likewise exist in the initial 'statistical' approximation for the case of low damping. In accordance with the model derived by the aid of mirror sources, the diffuse sound seems to arrive from all directions at each room position. Therefore, a diffuse sound field is understood as a field which is uniformly distributed with respect to the incidence angles as well as the distributed sound level.

Of course, such an 'ideal-diffuse' sound field in rooms are, once again, hypothetical; real rooms certainly differ from this idealized case. The larger the absorption in a room and at the walls is, the more the actual relationships will contradict the assumptions made. For non-uniformly distributed absorbers – e.g. large absorption at the walls, but not at the floor and the ceiling – the assumption of uniform directivity is not fulfilled. A so-called 'fluttering echo' will occur in this example. On the other hand, assuming uniform statistical distribution will at least provide a general assessment of sound fields in rooms, an approximation which is otherwise difficult and tedious to determine using more precise methods. At least in the case of a rectangular room with its simple parallel boundaries, some basic statements can be made for the case of total reflection at the walls, using wave theory. If partial sound absorption occurs and even if it is spatially distributed and dominated by larger objects in the room, objects which can scatter or diffract the sound, a 'rigorous' calculation by means of wave theory is impossible due to its complexity. Some simplified investigations will thus be made under the aforementioned assumptions in the following. It should be clear by now that the simplifications in detail – e.g. for spatial distribution – are only reasonable when regarded statistically. A spatial average is actually implied when discussing a constant sound level distribution of a diffuse sound field. This spatial average is what need to be determined by taking multiple measurements at multiple positions.

Before beginning the discussion, which will be based on the assumption of uniform distributions in space and time, we will illustrate the complexity of the wave theory for the simple, lossless rectangular room. For a room with the dimensions l_x, l_y and l_z, the boundary conditions require a sound pressure of the form

$$p(x,y,z) = \sum_{n_x=0}^{\infty} \sum_{n_y=0}^{\infty} \sum_{n_z=0}^{\infty} p_{n_x n_y n_z} \cos\left(\frac{n_x \pi x}{l_x}\right) \cos\left(\frac{n_y \pi y}{l_y}\right) \cos\left(\frac{n_z \pi z}{l_z}\right),$$

because pressure maxima must be present at all boundary surfaces. Each of the spatial, three-dimensional modes is related to a resonance frequency

$$k = \frac{\omega}{c} = \sqrt{\left(\frac{n_x \pi x}{l_x}\right)^2 + \left(\frac{n_y \pi y}{l_y}\right)^2 + \left(\frac{n_z \pi z}{l_z}\right)^2}$$

which is obtained based on the wave equation. The resonances can be represented graphically in a 'frequency room' by means of a three-dimensional grid (Fig. 7.3), where each cube element of the grid has the dimensions $c/2\,l_x$, $c/2\,l_y$ and $c/2\,l_z$. The number M of the resonance frequencies, which occur up to a certain frequency f, is approximately given by the volume of an eighth of a sphere with the radius f, divided by the volume of a cube

$$M = \frac{\frac{\pi}{6} f^3}{\frac{c^3}{8 l_x l_y l_z}} = \frac{4\pi}{3} \left(\frac{f}{c}\right)^3 V\,. \tag{7.3}$$

In a (small) room volume of $V = 200\,\mathrm{m}^3$, we already find about 800 resonances at just a frequency of 340 Hz! The density of the so-called eigenfrequencies is given by

$$\frac{\Delta M}{\Delta f} \approx \frac{\mathrm{d}M}{\mathrm{d}f} = \frac{4\pi}{c} \left(\frac{f}{c}\right)^3 V\,. \tag{7.4}$$

In this example $M/f = 60/\mathrm{Hz}$ for $f = 1000\,\mathrm{Hz}$. About 60 resonances can be found in an interval of 1 Hz bandwidth for $f = 1000\,\mathrm{Hz}$. These numbers illustrate very well that only the assumption of statistical distributions can provide a reasonable overview on sound fields in enclosed rooms.

A diffuse sound field, which by definition constitutes a (roughly) constant sound level distribution and a sound incidence uniformly distributed over all incidence angles, can only be produced with the aid of broadband signals. Harmonic excitation in a lightly damped room inevitably results in standing waves with distinct nodes and antinodes. Only if multiple standing waves are excited at the same time, can they form a spatially independent, diffuse sound field.

Usually, a relationship between the signal bandwidth Δf and the room volume V must be established in such a way that

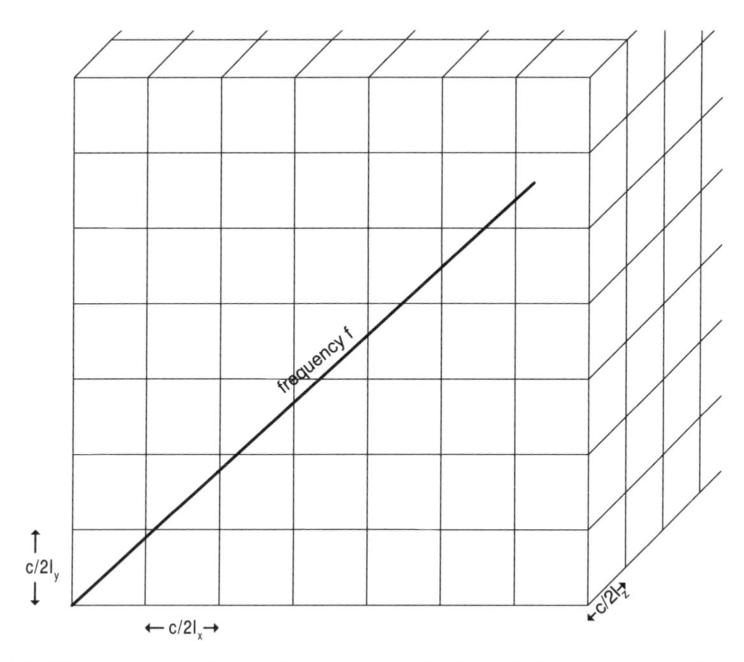

Fig. 7.3. Resonance grid: graphical representation of the resonance frequencies. Each node on the grid represents a resonance frequency

$$\Delta M/\Delta f \approx 1/\mathrm{Hz}$$

is fulfilled when using octave or third-octave bands for signal excitation. Equation (7.4) then specifies the permitted frequency range for the measurement, which is given by

$$f \geq \sqrt{\frac{1}{\mathrm{Hz}} \frac{c^3}{4\pi V}} \approx \frac{1800\,\mathrm{Hz}}{\sqrt{V/\mathrm{m}^3}} .$$

For $V = 200\,\mathrm{m}^3$, for example, it would only be possible to measure above about 125 Hz. For instance, the practical implications of using the condition $\Delta M/\Delta f > 1/\mathrm{Hz}$ are especially evident when taking measurements on the bandwidth of third-octave band noise, which is typically the bandwidth used in room-acoustic measurements. As is generally known, it is $\Delta f = 0.23 f_\mathrm{m}$ (f_m = center frequency). The condition therefore requires that in the third-octave band with $f_\mathrm{m} = 125\,\mathrm{Hz}$ at least $M = 30$ ($f_\mathrm{m} = 200\,\mathrm{Hz}$: at least $M = 50$) resonances must be present.

7.1 Diffuse sound field

The sound processes in a room can be imagined as a leaking container filled with water (Fig. 7.4). The water supply pipe, after being switched on, fills the container with water in the same way as the sound source fills the room with acoustic energy until a certain balance is achieved. The balanced level (water or sound level) can be explained by a compensation of the inflow by the outflow exiting through the 'leaks,' which are energy loss through absorption in the room. If the source is switched off after attaining the balanced, steady state, the level falls off again. The water, or the acoustic energy respectively, flows out.

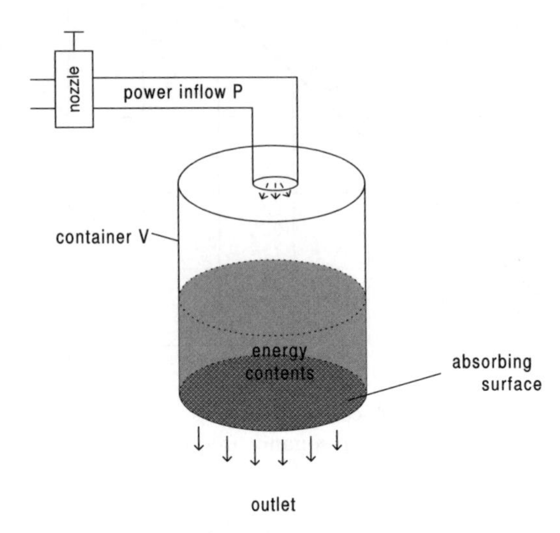

Fig. 7.4. Analogy between water level in a leaking container and the acoustic energy in a room

A time characteristic of the diffuse sound field, as schematically shown in Fig. 7.5, is expected, which for obvious reasons is described by dividing it into the intervals 'onset', 'steady state' and 'reverberation'. The three intervals can be described by an energy balance which is equal to a mass balance in the container analogy. As the mass inflow during the time Δt is divided between the change of the water level and the outflow, which occurs at the same time, the sound power P flowing in from the sender during Δt is composed of the change of energy ΔE stored in the room and the sound power loss P_{L} flowing out during Δt:

$$P\Delta t = V\Delta E + P_{\mathrm{L}}\Delta t \,, \tag{7.5}$$

where E denotes the spatial energy density and V the volume of the room.

It seems to be reasonable to assume that the power loss P_{L} is proportional to the momentarily stored energy EV. As in the water analogy, the more sound (water) flows out, the higher the water level is at the moment. This is easily verified by seeing what happens with a container filled with water and which has a leak at the bottom. Thus, it is also

$$P_{\mathrm{L}} = \gamma EV \tag{7.6}$$

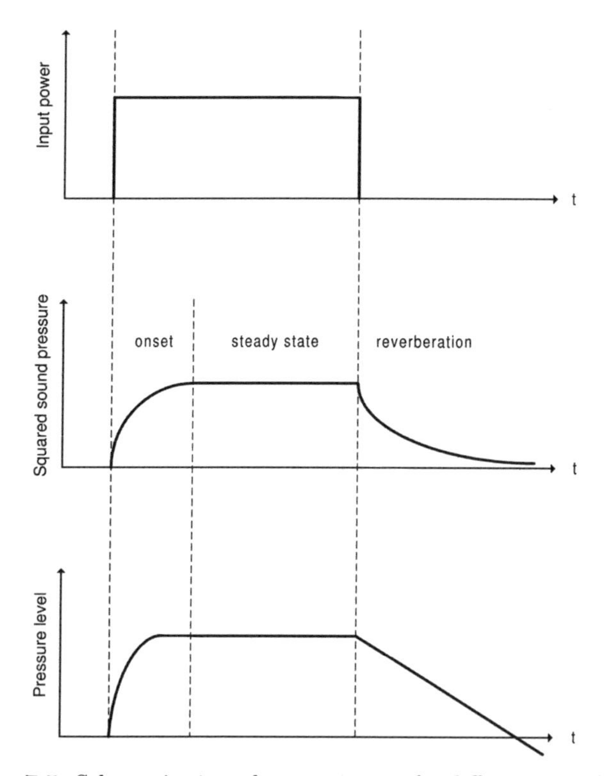

Fig. 7.5. Schematic time characteristics of a diffuse sound field

where γ is a 'room loss factor' related to the absorbing surface. In the water analogy, γ would describe the properties and the position of the outlets, i.e. the outflow surface.

Using (7.5) and (7.6) in the limiting case $\Delta t \to 0$, we obtain the energy balance

$$\frac{\mathrm{d}E}{\mathrm{d}t} = \frac{P}{V} - \gamma E. \tag{7.7}$$

The analogy of the 'container' room is expressed in (7.7) by means of an equation. However, the energy density cannot be determined by direct measurements of the sound pressure. Therefore the relation between sound pressure and energy density needs to be clarified. Under the assumption of uniformly distributed incidence angles, it can be stated that the sound velocity in a short-term time- and space-average equals zero. Therefore, mainly potential energy is stored in the room. Thus

$$E = \frac{\tilde{p}^2}{\varrho c^2}\,, \tag{7.8}$$

where $\tilde{p}$ denotes the rms-value of the sound pressure in the diffuse field.

7.1.1 Reverberation

If the power inflow of the container 'room', or, in other words, the power inflow of the sound energy is switched off, the room is gradually emptied. The duration of this process depends on the outlet surface area, expressed in the loss factor γ. Large outlet surface areas result in a quick outflow, whereas small areas result in a slow outflow with a long duration. It is therefore obvious that the way to quantify the loss properties of the room is to calculate the reverberation time.

For the sender switched off at $t = 0$, (7.7) yields the astonishing fact that the sound energy exponentially decays with

$$E = E_0 \mathrm{e}^{-\gamma t} \tag{7.9}$$

after switch-off. According to (7.8) the rms-value of the sound pressure becomes

$$\tilde{p}^2(t) = \tilde{p}^2(0)\mathrm{e}^{-\gamma t} \, ,$$

and the sound pressure level

$$L(t) = 10 \lg \frac{\tilde{p}^2}{p_0^2} = L(t=0) - \gamma t 10 \lg \mathrm{e} \, . \tag{7.10}$$

The level falls linearly with time. As can be seen from the example in Fig. 7.6, these time characteristics of the level can also be found in experiments in sufficiently diffuse sound fields.

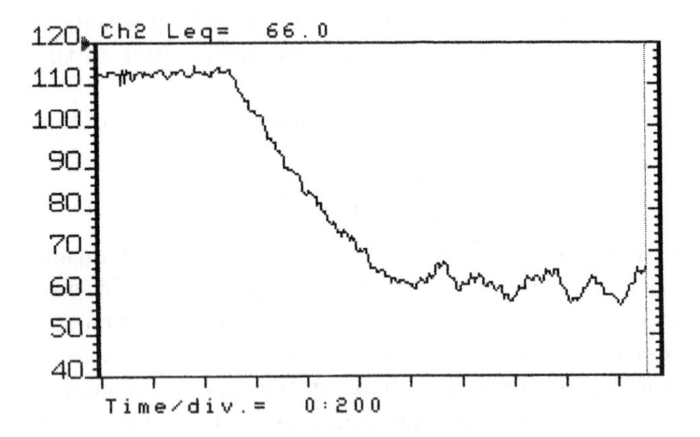

Fig. 7.6. Level-time characteristics of a reverberation process

The so far undetermined loss factor γ can now be calculated based on the slope of the decay line in the level-time characteristics. However, we do not use the 'mathematical slope' of the curve, since experimental data is very seldom smooth enough that a differentiation would lead to a reasonable result. Rather, we utilize what is known as the reverberation time T to determine the time needed for the energy fall to a millionth of its initial value after switching off the sender. This is equivalent to the time it takes for the level to drop by 60 dB. According to (7.10) it is

$$60 = \gamma T 10 \lg \mathrm{e} \, ,$$

or

$$\gamma = \frac{13.8}{T} \, . \tag{7.11}$$

Calculating the time is a actually quite a simple experiment. One only needs to graph the level-time characteristics using a plotter once the source is switched off. Typically, only half the reverberation time (30 dB difference) is used for the measurements rather than using the whole level difference of 60 dB. Otherwise, it would be necessary to maintain too great of a distance to the actual or electrical noise interferences (which can be discerned in Figure 7.6). As already mentioned, the sound absorption can significantly depend on the frequency. Therefore, the measurements have to be performed for several frequency bands, usually octave or third-octave bands. An example of the frequency characteristics of the reverberation time in a reverberation room is given in Fig. 7.7. The purpose of this will be discussed later.

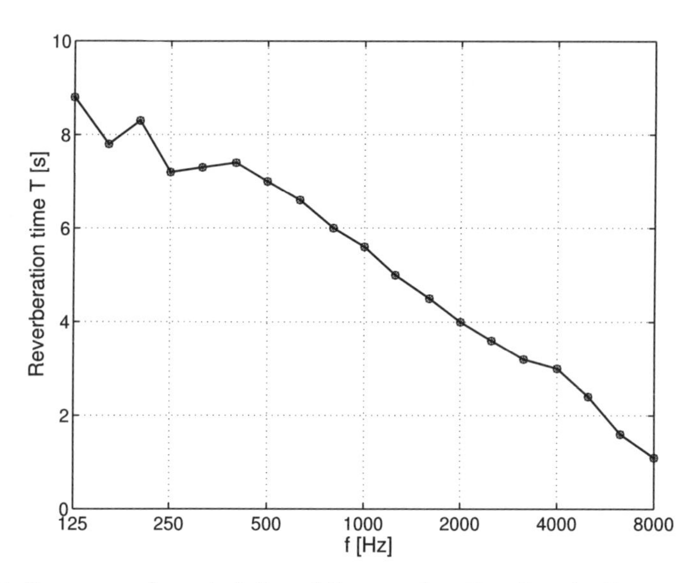

Fig. 7.7. Frequency characteristics of the reverberation time in the reverberation room of the Institute of Engineering Acoustics, Polytechnical University of Berlin

7.1.2 Steady-state conditions

The reverberation decay is used to characterize the losses of the room. However, for the acoustical engineer, the central question is how losses affect the sound intensity in the room and how the latter can be influenced by changing the room attributes. An intuitive approach to the problem is conceptualizing 'continuously operated sources' representing the so-called steady state.

After an initial onset of the reverberation, which is not of interest here, the energy content of the room does not change anymore, the steady state condition is reached at

$$\frac{\mathrm{d}E}{\mathrm{d}t} = 0 \,.$$

The power inflow from the sender is needed just to compensate the losses. According to (7.7), (7.8) and (7.11), we need

$$\frac{P}{V} = \gamma E = \frac{13.8}{T} E = \frac{13.8}{T} \frac{\tilde{p}^2}{\varrho c^2} \tag{7.12}$$

With the aid of (7.12) the sound pressure level can be calculated by the room attributes volume V, reverberation time T and source power P.

It is of interest for several reasons to adjust the reverberation time of the room to a certain value. The design goal may be to use high absorption in order to make a room quiet. This is especially true for rooms which are intended for a certain purpose (like offices and factory halls). In many other cases, a reverberation time is targeted which fulfills certain requirements for 'good audibility' – which may vary depending on the particular intended purpose of the room. Concert halls, for instance, should have a reverberation time of about 2 s, whereas lecture rooms should have a T of about 0.5 s.

The relationship between the reverberation time and the absorbers present in the room must be examined when adjusting the reverberation time, as it is the absorbers which primarily influence the reverberation time. First, all absorbent surfaces in the room are split into separate areas with constant uniform attributes. Then the sound power P_{in} impinging on an area S from one partial room or hemisphere is considered (if an object has a back and a front – like a person, for example – it is assumed to have two or more separate areas). If the absorbing attributes of the surface are known and expressed in the absorption coefficient, the absorbed power P_{ab} can be calculated using the incident sound power

$$P_{ab} = \alpha P_{in} \,. \tag{7.13}$$

If the absorption coefficient of a structure depends on the incident angle, the average value of all angles is used for α.

Where the angle of the surface under consideration is positioned in 'the starry sky' of mirror sources in Fig. 7.1 is irrelevant. Only the contribution of half the sources to the single-sided power incidence matters. For the sake of simplicity, the sound pressure of this half space of sources is denoted by $p_{1/2}$.

If the sound would impinge only at a certain angle ϑ, measured normal to the surface, the power flow onto the surface would be related to the total sound pressure of the relevant half space of sources by

$$P_\vartheta = SJ_\vartheta = S\frac{\tilde{p}^2_{1/2}}{\varrho c}\cos\vartheta\ . \tag{7.14}$$

Since a diffuse sound field is assumed, sound impinges uniformly from all directions. The average over all incidence angles has to be taken into account to obtain the relationship between the sound power impinging on the surface and the sound pressure. Since $\cos\vartheta$ ranges uniformly between zero and unity, the average can be set to $\cos\vartheta = 1/2$ and for the diffuse field, we obtain

$$P_{\mathrm{in}} = \frac{\tilde{p}^2_{1/2}}{2\varrho c}S. \tag{7.15}$$

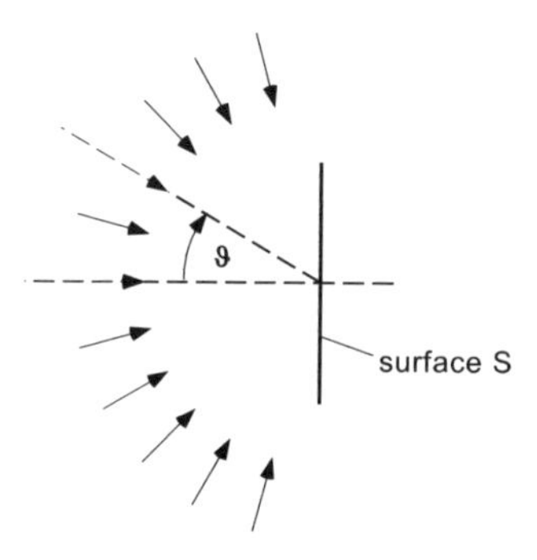

Fig. 7.8. Sound intensity impinging uniformly on the surface S from a half space

It is also fairly easy to express the sound pressure only produced by one half of the sources by the total sound pressure of the diffuse field. As shown earlier in Sect. 3.4, coherent sound sources can be regarded as incoherent as well, if the distance is large compared to the wavelength and the squared sound pressure is understood as a spatial average, given that the squared sound pressure of half the sources is therefore half that of all sources

$$\tilde{p}^2_{1/2} = \frac{1}{2}\tilde{p}^2\ . \tag{7.16}$$

The one-sided sound power incidence P_{in} on the surface S is thus given by the sound pressure in a diffuse sound field

$$P_{\mathrm{in}} = \frac{\tilde{p}^2 S}{4\varrho c}\ . \tag{7.17}$$

The sound power extracted from the sound field by that surface is therefore

$$P_{ab} = \alpha P_{in} = \frac{\tilde{p}^2}{4\varrho c}\alpha S \; .$$

Finally, we evaluate all individual surfaces S_i of the room which come into question. Altogether, it follows that

$$P_{ab} = \frac{\tilde{p}^2}{4\varrho c}A \; , \tag{7.18}$$

where A is already taken as the sum of all individual absorbing surfaces

$$A = \sum_i \alpha_i S_i \; . \tag{7.19}$$

The quantity A, resulting from the product of all individual surfaces and their corresponding absorption coefficients, is called 'equivalent absorption area'. Since the absorbent effect can also be imagined as being replaced with a smaller area with the absorption coefficient $\alpha = 1$, the term 'open window area' is also quite common for A. The relationship to the reverberation time T results from the discussion of the steady state condition in the room, where the sound power inflow P from the source is equal to the total absorbed sound power P_{ab}. In comparing (7.12) to (7.18), we arrive at the Sabine equation for reverberation time, named after its discoverer Sabine:

$$\frac{13.8V}{cT} = \frac{A}{4}$$

Typically, we use the 'dimensionless' form which applies to air

$$T/\mathrm{s} = 0.163 \frac{V/\mathrm{m}^3}{A/\mathrm{m}^2} \; . \tag{7.20}$$

The proportionality between reverberation time and room volume contained in the Sabine equation corresponds to the graphical representation. It is quite obvious that a large volume V reacts with a larger reverberation time than a smaller volume, if both are equipped with the same equivalent absorption area A.

The example of the 'cubic room' $V = 200\,\mathrm{m}^3$ (length $a = 5.85\,\mathrm{m}$) also illustrates that realistic reverberation times can be calculated using the Sabine equation. If all surfaces of the cube $6a^2$ are assumed to have a small absorption $\alpha = 0.05$, the reverberation time is $T = 3.3\,\mathrm{s}$. Important is that the total energy transformation of sound into heat which occurs in the room and at the boundary surfaces is entirely accounted for in the equivalent absorption area A. Actually, a finite reverberation time would even be detected in rooms with total reflection on all the walls due to the (negligible but existing) losses along the propagation paths. Therefore A may be divided into two parts

$$A = A_\alpha + A_\mathrm{L} \tag{7.21}$$

where A_α represents the part which can be intentionally adjusted utilizing absorbent surfaces, and A_L, the inevitable loss in the medium air. For most practical applications the propagation losses A_L can be neglected. It should only be mentioned that the inevitable loss is larger, the larger the addressed volume is:

$$A_\mathrm{L}/\mathrm{m}^2 = \nu\, V/\mathrm{m}^3 \tag{7.23}$$

Here, ν is a 'material parameter' which mainly depends on the frequency and the air humidity. An empirical equation provides an approximation

$$\nu = \frac{80}{\varphi/\%}\left(\frac{f}{\mathrm{kHz}}\right)^2 10^{-3}\,, \tag{7.24}$$

where φ specifies the relative humidity in air in percent. Without any other absorption in the room, the maximum possible reverberation time is given by

$$T_\mathrm{max} = 0.163\, V/A_\mathrm{L} = 0.163/\nu = 80/(f/\mathrm{kHz})^2,$$

assuming a humidity of $\varphi = 40\%$ (which is typically present in interior rooms). It is easy to see that the given natural limiting reverberation time can only play a role at higher frequencies.

The initially posed question of the qualitative relationship between the sound pressure in the steady state and the absorption in the room remains to be answered. The answer can be found in the fact that the power inflow P under steady state conditions is equal to the power P_{ab} which is lost at the same time. It is to say that, based on (7.18)

$$\frac{P_{ab}}{P_0} = \frac{\tilde{p}^2}{4p_0^2}\frac{A}{\mathrm{m}^2}$$

(where $P_0{=}p_0^2/\varrho c\, 1\,\mathrm{m}^2$ is the characteristic value of the sound power, and p_0 the characteristic value of the sound pressure) and thus the sound pressure level L and the source power level L_p are described by

$$L = L_p - 10\lg A/\mathrm{m}^2 + 6\,\mathrm{dB}\,. \tag{7.25}$$

Therefore, the spatial average of the sound pressure can be calculated for a given source power, based on a known absorption area. Obviously, the diffuse field level in rooms can be reduced by 3 dB by doubling the absorption area. Equipping the room with additional absorption only guarantees positive results in level reduction if the original reverberation times were relatively long. Shorter reverberation times (in the range of about 1 s) only rarely provide any elbow room for additional level reductions.

Equation (7.25) can also be used to determine the radiated sound power of a source (e.g. a machine) by measuring the averaged sound pressure level of the room, if the reverberation time of the experimental room is known. In order to meet the requirements of a diffuse field, a room with a substantially

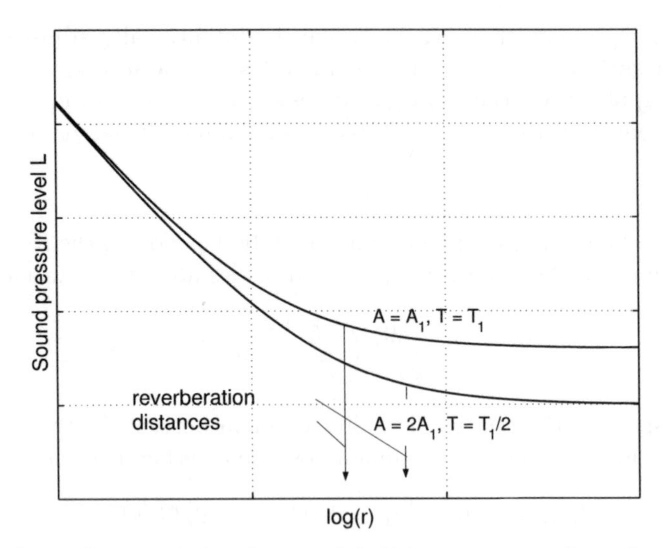

Fig. 7.9. Space characteristics of a sound field in two rooms of equal volume with the absorption areas A_1 and $2A_1$

large and easily reproduced reverberation time must be used for sound power measurements in a reverberation room. Therefore, a certain distance from the source has to be maintained in order to measure the individual sound pressure levels required for the spatial average. In close vicinity to the source the direct field dominates the sound pressure level compared to the reverberant field which is (almost entirely) composed of reflections; a local sound pressure gradient depending on the distance r as shown in Fig. 7.9 should result. The transition point depends on the level of the diffuse field. For small absorbent areas, it can reach within close vicinity to the source. An estimate of the transition point can be derived by using the free-field equation which is valid for the direct-field $P = 4\pi r^2 p_{dir}^2/\varrho c$, and (7.18) $P = Ap_{diff}^2/4\varrho c$. If the 'reverberation radius' r_r is defined in such a way that it denotes the distance from the source where the sound pressure of direct and diffuse field are equal, we obtain $4\pi r_r^2 = A/4$, or

$$r_r = \frac{1}{7}\sqrt{A}. \tag{7.26}$$

For distances $r > r_r$ the total sound field is dominated by the diffuse component, for $r < r_r$ it is dominated by the direct-field. Thus, measurements explicitly requiring a diffuse field must always be taken outside of the reverberation radius.

In 'rooms with distributed communication', such as cafés and restaurants, it is very important to keep in mind that the 'typical conversation partners' are not bombarded with sound primarily coming from the diffuse field, which means other conversations or interfering noise. In such a bad acoustical situations, each individual often solves the problem intuitively by raising the own sound intensity gradually in order to be better understood. Such rooms are only bearable, because of the good times one has in them. Nonetheless, with respect to acoustics, they are small catastrophes. Perhaps a sort of 'individuality radius' r_I would be an appropriate measure in that case, defined as follows. Suppose there are N persons in the room, speaking with the same sound power at the same time. The power NP of these N incoherent sources produces the sound pressure

$$p_{diff}^2 = 4\varrho c N P/A$$

in the diffuse field. The individuality radius is now defined as the distance at which the direct field produced by each individual speaker

$$p_{dir}^2 = \varrho c P/4\pi r^2$$

is equal to the diffuse field of N persons

$$r_I = \frac{1}{7}\sqrt{\frac{A}{N}} \, .$$

If 'undisturbed conversation' is desired at a radius of about 0.4 m, an absorption area of about 8 m^2 per person is required ($A = 8N\,\mathrm{m}^2$ in total), a fairly high requirement. For restaurants in which tables are not too close to each other this can be realized by covering the whole ceiling with absorption. Popular pubs full of guests leave little room for the acoustician. This is changed very little by the fact that N guests provide about $N\,\mathrm{m}^2$ of absorption area. In the limiting case of a huge number of persons, their absorption area of $A = N\,\mathrm{m}^2$ takes precedence over the room furnishing. It thus follows that $r_I = 14\,\mathrm{cm}$, with the corresponding consequences.

7.1.3 Measurement of the absorption coefficient in the reverberation room

It is often required for the application of absorbent linings for room acoustic purposes to determine their absorption coefficient under the condition of sound impinging from many distributed directions in the laboratory. These measurements can be taken in an empty reverberation room which has the reverberation time

$$T_{empty} = 0.163\, V/A_{empty} \tag{7.27}$$

(T in s, V in m^3, A in m^2). All possible losses of the empty reverberation room, including those occurring during propagation, are accounted for in A_{empty}. If, subsequently, an absorbing area A is introduced to the room (the area should be about 10 m^2 for a typical reverberation room of $V = 200\,\mathrm{m}^3$), the absorbing area is increased to

$$A = A_{empty} + \Delta A, \tag{7.28}$$

if it is correctly assumed that lining only a part of the surface of the reverberation room with absorbent surfaces is insufficient. In the example of a room with $V = 200\,\mathrm{m}^3$, a boundary surface of $S = 200\,\mathrm{m}^2$ and a sample area of $10\,\mathrm{m}^2$ for the absorber, strictly speaking, A_{empty} would have to be corrected by 5%. However, the reverberation times cannot be determined that precisely, the measurement uncertainty is usually considerably larger. The 'coverage error' can therefore be safely neglected. The measured reverberation time, which is related to the absorption area enlarged by the sample, is given by

$$T = 0.163\, V / \left(A_{empty} + \Delta A\right) \ . \tag{7.29}$$

Hence, the absorption area of the sample

$$\Delta A = 0.163 \frac{V}{T} - A_{empty} = 0.163\, V \left(\frac{1}{T} - \frac{1}{T_{empty}}\right) \tag{7.30}$$

is obtained by measuring the reverberation times T and T_{empty} with and without the sample. The absorption coefficient can thus be calculated by

$$\alpha = \frac{\Delta A}{S}$$

(where S is the sample surface).

It is possible that absorption coefficients $\alpha > 1$ are found and these typically cannot physically occur, the reason being that the condition of the spatially uniform distribution is not stringently met. Along the edges of the material samples, which always have a finite thickness, diffraction effects surface, leading to pressure accumulation near the edges, even when the edges are covered by rigid, reflecting plates. In this way, we slightly overestimate the actual absorption coefficient in our calculations.

The above measurements should be taken under consideration of Norm EN ISO 354: 'Acoustics - Measuring Sound Absorption in Reverberation Chambers' (2003).

7.2 Summary

Under conditions of sufficient signal bandwidth and low absorption, a diffuse sound field will develop in closed rooms. At distances from the source that lie beyond the reverberation radius, the diffuse field level present there will be about the same everywhere. The sound incidence is equal emanating from all directions. Here, the sound energy is characterized by equal spatial distribution and reacts similarly to a liquid in a leaking container. An increase in sound power input leads to an increase in contained energy and therewith an increase in overall sound level in the room, resulting in a stationary equilibrium, wherein the sound pressure level is still dependent on the 'acoustic leaks' inherent in the properties of the room. These leaks can be quantified in the 'equivalent absorption area' or the 'open window area', values which contain all significant loss mechanisms in the room. As the equivalent absorption area doubles, the sound level in the room decreases in a balanced steady state by $3\,dB$.

After a source is switched off, the sound level decreases linearly with time. The absorption area is proportional to the gradient in the level-time characteristics. This relationship is used to measure the losses in the room, expressed by the reverberation time T. This indicates the time that it takes for the level to decrease by $60\,dB$. The relationship between reverberation time and absorption area is expressed through the Sabine equation $A = 0,163\,V/T$ (absorption area A in m^2, volume V in m^3, T in s). The equation indicates how the reverberation time of a room can be specifically adjusted. It can also be used to determine the absorption area based on the reverberation times measured in a given room.

7.3 Further reading

The work "Room Acoustics" by Heinrich Kuttruff (Elsevier Science Publishers, London 1991) contains a highly instructive and interesting store of knowledge on room acoustics. Furthermore, it is very easy to read and comprehensibly written.

7.4 Practice exercises

Problem 1

Determine the sound power in a source where the level has been measured in the diffuse sound field of a reverberation chamber with $V = 200\ m^3$. The locally averaged levels (a moving microphone was used to take the measurements) are listed with their corresponding reverberation times in the following table.

f/Hz	$L_{third-octave}/dB$	T/s
400	78.4	5
500	80.6	4.8
630	79.2	4.1
800	80	3.6
1000	84.4	3.6
1250	84.2	3.5

How high is the A-weighted sound pressure level? How high is the sound power level registered in third-octaves? How high is the A-weighted sound power level of the source (the A-weight is given in Problem 1 of Chapter 1)?

The same sound source is introduced in a living room with $V = 100\,m^3$ and an (average) reverberation time of $0.8\,s$. How high is the A-weighted sound pressure level in this room? How large is the corresponding reverberation radius?

Problem 2

The following reverberation times have been measured in a café with a floor space of $110\ m^2$ and $3\ m$–high walls:

f/Hz	T/s
500	3.8
1000	3.2
2000	2.8

How large are the equivalent absorption areas and the reverberation radii? How would the reverberation times and the diffuse sound pressure levels change, given the same source, if the entire ceiling of the room were to be outfitted with an absorber with absorption coefficients $\alpha = 0.6$ at $500\,Hz$, $\alpha = 0.8$ at $1000\,Hz$ and $\alpha = 1$ at $2000\,Hz$?

Problem 3

In a reverberation chamber with $V = 200\,m^3$, a sample area of $10\,m^2$ in an absorbent structure is measured. The reverberation times in the empty chamber (control) and in the chamber containing the sample are:

f/Hz	T_{empty}/s	T_{sample}/s
500	5.8	3.2
630	5.2	2.8
800	4.8	2.3
1000	4.6	2.0

How large are the absorption areas and absorption coefficients of the absorbent sample system?

Problem 4

Calculate the first ten resonance frequencies of a square room with the dimensions $6\,m$, $5\,m$ and $4\,m$.

How many resonance frequencies are in one octave band for the center frequencies $200\,Hz$, $400\,Hz$ and $800\,Hz$?

Problem 5

In a room with V $= 1000\,m^3$, a reverberation time of $1.8\,s$ is registered at mid-frequency. A machine produces a given sound pressure level L_1 in the room's diffuse field. By how much must the room's equivalent absorption area be adjusted in order to preserve the existing sound level when operating N similar sources?

Problem 6

A sound pressure level of $100\,dB$ has been registered in the diffuse field of a room with $V = 500m^2$. How great is the energy density and the overall energy stored in the room? How long would a light bulb with the power of $1\,Watt$ burn if one could supply it electrically with this same amount of energy?

Problem 7

Two rooms with the volumes V_1 and V_2 are joined by a door opening with the area S_T when the door is open. When the door – which has high sound insulation – is closed, the equivalent absorption areas A_1 and A_2 are able to be determined by the reverberation times already given. A sound source with the power level L_P is present in Room 1 when the joining door S_T is open.

a) How much is the difference in levels $\Delta L = L_1 - L_2$ between both rooms
b) How great is the sound pressure level L_1 in Room 1?

Find the values for $V_1 = 200\,m^3$, $V_2 = 100\,m^3$, $A_1 = 20\,m^2$, $A_2 = 16\,m^2$ and $L_P = 95\,dB$.

Problem 8

In a room with a volume of $80\,m^3$, the following reverberation times listed in the table below are measured in third-octave intervals. A sound source is present in the room. Its third-octave levels L_P are also listed in the table. How high are the A-weighted sound pressure levels in the room? The A-weighted values are given in the last column of the table to make the calculations for this problem easier.

f/Hz	T/s	L_P/dB	Δ_i/dB
400	1.8	78	-4.8
500	1.6	76	-3.2
630	1.4	74	-1.9
800	1.2	75	-0.8
1000	1.0	74	0
1250	1.0	73	0.6

8 Building acoustics

This chapter deals with sound transmission between rooms in a building (or from the outside into the building). This is a topic of significant practical importance which concerns noise control of indoor rooms with respect to traffic noise and residential noise. Noise, penetrating a room from the exterior, can have two possible reasons:

1. There are forces which exert a direct effect on walls and ceilings such a neighbor walking on a floor in the apartment above or operating a machine in a building. Such forces induce vibrations in building structures and structure-borne sound develops which is transported to other floors. The vibrating structures excite the surrounding air and radiate sound. This sound development mechanism can be summarized by the terms 'force – structure-borne sound – airborne sound' (Fig. 8.1).
2. The airborne sound in a room, such as speech or the operation of consumer electronic devices or machinery, in respect to the surrounding walls and ceilings also represents an exciting force which is now spatially distributed. It no longer represents a point force as described above. Vibrations are also generated in the structures. The transfer path can be described by 'airborne sound – structure-borne sound – airborne sound' (Fig. 8.1).

What both forms of excitation in the rooms of buildings have in common is that the sound is not necessarily transmitted via the 'direct' path (Fig 8.2). The propagation of vibrations can take multiple paths because adjacent partitions are able to mutually exchange vibration energy. In addition to the direct transfer path, which leads from the partition wall (or the ceiling) to the adjacent room, there are many other transfer paths, so-called flanking paths. Generally speaking, it is basically impossible to even be able to determine the dominant path without taking measurements first. The direct partition wall, for instance, can have such high sound insulation that the path with flanking transmission represents the dominating path. Therefore additional acoustic improvements made to the partition wall will not necessarily achieve better results in the total sound insulation.

M. Möser, *Engineering Acoustics*, DOI 10.1007/978-3-540-92723-5_8,

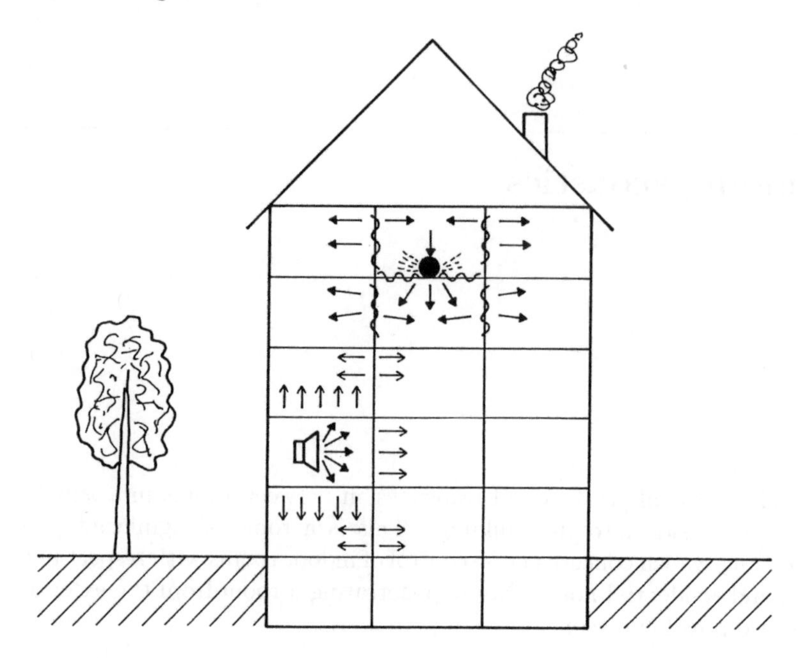

Fig. 8.1. Transmission and generation of airborne sound in buildings

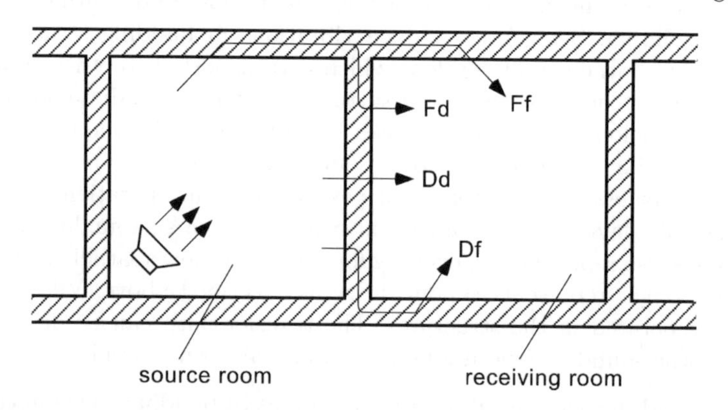

Fig. 8.2. Sound transfer paths: Ff: flanking–flanking, Fd: flanking–direct, Dd: direct–direct, Df: direct–flanking

These remarks show that the problem of sound transmission in buildings is in fact quite complex. Only the basic principles, of course, can be discussed here. Thus, the following considerations focus on sound transmission through the direct partition wall. In many, but not all cases can the main transfer path be characterized in this way. Windows, for example, are the weak point in the sound insulation towards the exterior, and usually the transmission through other partitions can be neglected. Apart from some extreme requirements, it can be assumed that for heavy flanking structures (e.g. walls with a mass per unit area of more than 300 kg/m^2) that the direct path is also the most important one. According to tradition, we will first begin this chapter with an introduction to the methods which are commonly used in measuring sound insulation of building partitions.

8.1 Measurement of airborne transmission loss

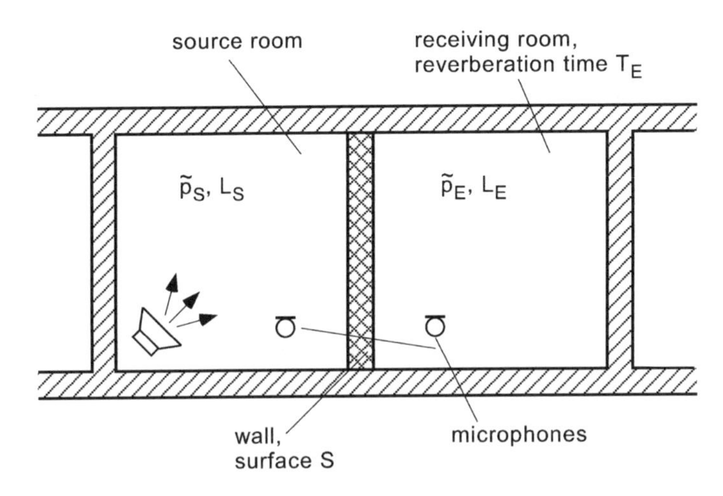

Fig. 8.3. Experimental setup for measuring the transmission loss of a partition wall between two rooms

In the determination of transmission loss, the sample is constituted by a partition wall between two rooms (Fig. 8.3), which will be henceforth referred to as the source and receiving rooms. As explained in more detail in Chap. 7 on room acoustics, the sound pressure level present in a room, not only depends on the sound power incidence, but also on the acoustic design of the room. If only the level difference between source and receiving room were used as a gauge for sound insulation of a wall with respect to the airborne sound, the resulting value would not only characterize the wall attributes, but also the room properties as well. For this reason, as a basic principle, the sound power transmission coefficient τ is principally used to describe the wall properties

$$\tau = P_{\mathrm{E}}/P_{\mathrm{S}} \,. \tag{8.1}$$

The transmission coefficient represents the ratio of the power P_{E}, passing through the receiving side, and the power incidence P_{S} on the source side. If diffuse sound fields are assumed on both sides of the partition element, the incident sound power, according to (7.17), is given by

$$P_\mathrm{S} = \frac{\tilde{p_\mathrm{S}}^2 S}{4\varrho c},$$

where p_S represents the rms-value of the sound pressure in the source room and S, the surface area of the partition (for a description refer to Fig. 8.3). Under steady-state conditions the power inflow in the receiving room is equal to the absorbed sound power (7.18)

$$P_\mathrm{E} = \frac{\tilde{p_\mathrm{E}}^2 A_\mathrm{E}}{4\varrho c},$$

where A_E represents the equivalent absorption area of the receiving room. The sound power transmission coefficient

$$\tau = \frac{\tilde{p_\mathrm{E}}^2}{\tilde{p_\mathrm{S}}^2} \frac{A_\mathrm{E}}{S} \tag{8.2}$$

is a component of the transmission loss R (or otherwise known as the sound reduction index)

$$R = 10 \lg 1/\tau = L_\mathrm{S} - L_\mathrm{E} - 10 \lg \frac{A_\mathrm{E}}{S} \tag{8.3}$$

which by definition results in large values of R in the case of small transfers. The sound pressure levels in the source and receiving room L_S and L_E are, of course, spatially averaged mean values (which means the level of the squared sound pressure in a spatial average). For the reasons already discussed, the reverberation time has to be measured in the receiving room. The reverberation time is required to calculate the equivalent absorption area with the aid of the Sabine equation.

As the examples in Figs. 8.8, 8.9 and 8.10 show, the transmission losses of partitions are frequency dependent. They show a more or less increasing tendency toward higher frequencies. Thus, measurements are performed with varying frequency, normally in octave or third-octave increments. Typically, noise in the appropriate bandwidths is used as the test signal. The frequency response of R measured in the so-called frequency range of 'building acoustics', which is between 100 Hz and 3.15 kHz, is obtained. Higher frequencies are of minor importance because, as a matter of fact, the sound reduction is large. At lower frequencies, there is a rapid decline in the sensitivity of the ear, making measurements difficult and less accurate.

Basically, transmission loss is clearly described by its frequency response which is represented by a certain number of values. For several reasons, it is prudent to transform this 'multiplicity of values' into one 'single value' which subsumes 'the' transmission loss by means of at least a single number. Comparing different partition elements with respect to their sound transmission is certainly much easier then.

The measured curve **R** is compared to the reference curve **B** as follows. The characteristic curve is shifted in 1 dB-steps towards the measured curve until the 'sum of negative differences' of the measured curve compared to the reference curve is smaller than 32 dB (Fig. 8.4):

$$\text{average negative difference} = \frac{1}{N}\sum \text{negative differences}$$

Only negative differences are counted (where the shifted reference curve is located above the measured curve). Positive differences are not considered. The 500 Hz - point of the shifted characteristic curve specifies the so-called 'measured' transmission loss R_w. . If the characteristic curve, as in a special case experimental results, were not able to be shifted for example, then $R_w = 52\,dB$ (Bild 8.4). Positive differences are not taken into account. The mentioned sum of negative differences is roughly equal to the 'average negative difference' of 2 dB for $N = 16$, which represents the number of measured frequency bands.

Shifting the reference curve in 1 dB-steps is performed until the right one is found by trial and error. This is simplified by computers nowadays. If the reference curve is shifted downwards, the result is negative (AIM < 0 dB); upward shifts count positive. Some examples are presented in Fig. 8.4.

Thus, R_{W} represents the 'average' transmission loss in the 'mid'-frequency range. If $A_{\mathrm{E}} \sim S$ is assumed in (8.3) (which is approximately valid for 'typical' conditions in dwellings and for walls, but not for windows), the level difference can roughly be approximated by

$$L_{\mathrm{S}} - L_{\mathrm{E}} = R_{\mathrm{W}}\ . \tag{8.4}$$

This – not even very accurate – estimation is often required in practice. The actual noise impact in a room when transmission loss and the external impact sound level are both known is a question which often arises. The approximation in (8.4) can be quite off if the frequency response R of the transmission loss significantly deviates from the reference curve and the 'dominant' frequency is substantially below 500 Hz. (8.3) is used for making more accurate assessments of the receiving room level L_{E}, where the frequency response of L_{S}, R and, as a matter of principle, A_{E} must be known. The response function of L_{E} obtained in this way can be transformed into single values (such as dB(A), etc). In everyday engineering practices, the aforementioned is seldom known (or too expensive to measure). (8.4) gives at least an initial approximation.

8.2 Airborne transmission loss of single-leaf partitions

As already mentioned at the beginning, sound transmission through a wall or a ceiling of a room (henceforth called the source room) into another room (henceforth called the receiving room) is based on simple chain of events: the incoming air wave elastically 'bends' the wall. The wall vibrations act as a sound source in the receiving room.

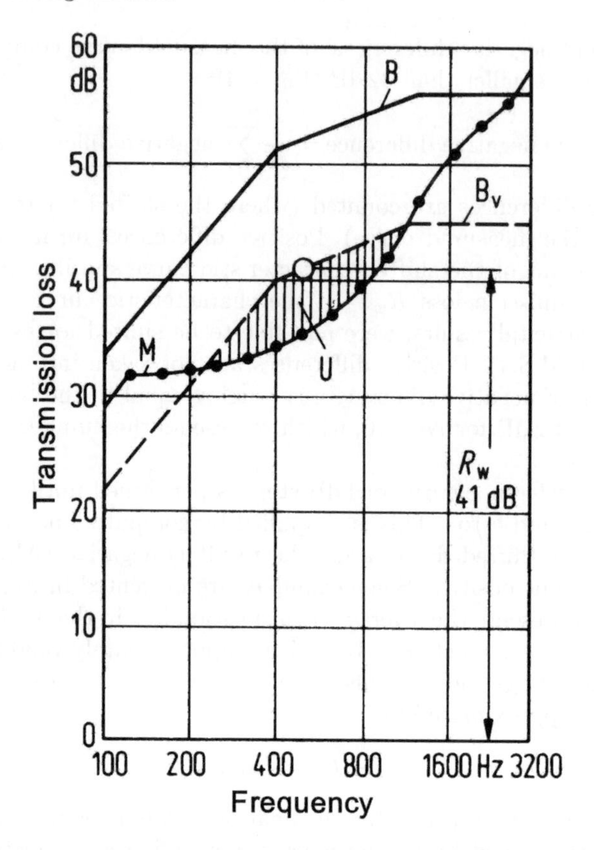

Fig. 8.4. Definition of the weighted sound reduction index R_W. The diagram shows the transmission loss R plotted versus frequency f. **B**: reference curve, **Bv**: shifted reference curve, **M**: measured values, **U**: negative differences of **M** – **Bv** (from : K. Gösele and E. Schröder: Schalldämmung in Gebäuden, Chap. 8 in "Taschenbuch der Technischen Akustik", Springer, Berlin and Heidelberg 2004, edited by G. Müller and M. Möser)

A model, as simple as possible, shall reveal the influence of the wall parameters (mass, thickness, bending stiffness. . .) on airborne transmission loss.

As depicted in Fig. 8.5, the model consists of three elements:

1. The 'source room' 1, regarded in this case as a half space filled with air – The sound field is given by a wave incident from the angle ϑ

$$p_\mathrm{a} = p_0 \mathrm{e}^{-jkx\cos\vartheta} \mathrm{e}^{jkz\sin\vartheta} \tag{8.5}$$

   and a reflected sound field

$$p_\mathrm{r} = r p_0 \mathrm{e}^{jkx\cos\vartheta} \mathrm{e}^{jkz\sin\vartheta} \; . \tag{8.6}$$

   The resulting sound field in room 1 consists of the two elements

$$p_1 = p_a + p_r = p_0 e^{jkz \sin\vartheta} \left(e^{-jkx\cos\vartheta} + r e^{jkx\cos\vartheta} \right) . \tag{8.7}$$

2. The 'receiving room' 2 also regarded as a half space filled with air – For simplicity's sake, it is assumed that the space dependence of the incident wave p_a with respect to the z-direction exists in (8.5). The sound power radiated into the receiving room is therefore described by

$$p_2 = \tau p_0 e^{-jkx\cos\vartheta} e^{jkz\sin\vartheta} , \tag{8.8}$$

where τ represents the sound power transmission coefficient.
3. Finally, the wall is induced by the pressure difference $p_1(0,z) - p_2(0,z)$ to vibrations which represent solutions to the bending wave equation (see (4.47) in Chap. 4.5, p. 133)

$$\frac{1}{k_B^4} \frac{d^4 v_W}{dz^4} - v_W = \frac{j}{m''\omega} \left(p_1(x=0,z) - p_2(x=0,z) \right) . \tag{8.9}$$

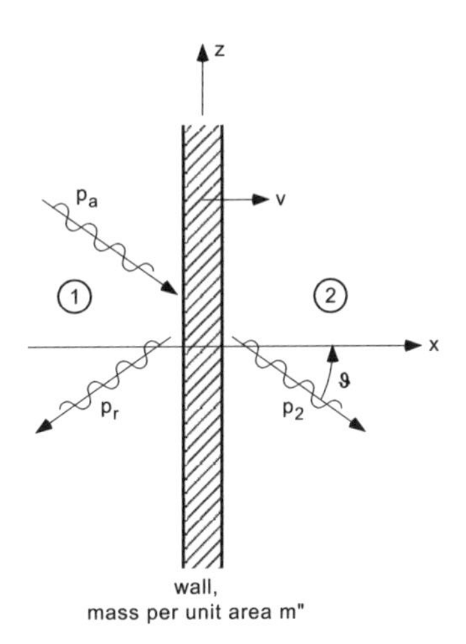

Fig. 8.5. Model used to calculate the transmission loss of a single-leaf partition. p_a: incident sound field, p_r: reflected sound field, $p_1 = p_a + p_r$: total sound field in front of the wall, p_2: transmitted sound field

If it is also assumed that the wall vibrates with respect to the z-axis 'in the same direction as the sound field incidence'

$$v_W = v_0 e^{jkz\sin\vartheta} . \tag{8.10}$$

It follows from (8.9) using (8.7) and (8.8) that

$$v_0 = \frac{1 + r - t}{\frac{k^4}{k_B^4}\sin^4\vartheta - 1} \frac{jp_0}{m''\omega} \tag{8.11}$$

is true for the amplitude v_0 of the wave in the wall v_W. The sought-after but still unknown quantities are the reflection coefficient r and the sound pressure transmission coefficient t, since they describe the sound fields in front of the wall and behind it. These quantities are determined by taking advantage of the simple fact that the velocities on both sides of the wall are equal to the wall velocity v_W

$$v_1(x=0) = \frac{j}{\omega\varrho}\frac{\partial p_1}{\partial x}\bigg|_{x=0} = v_W \tag{8.12}$$

and

$$v_2(x=0) = \frac{j}{\omega\varrho}\frac{\partial p_2}{\partial x}\bigg|_{x=0} = v_W \; . \tag{8.13}$$

Equations (8.12) and (8.13) are equivalent to

$$\frac{p_0}{\varrho c}\cos\vartheta(1-r) = v_0 \tag{8.14}$$

and

$$t\frac{p_0}{\varrho c}\cos\vartheta = v_0 \; . \tag{8.15}$$

Thus $t = 1 - r$, or

$$r = 1 - t \; . \tag{8.16}$$

Using Equations (8.11) and (8.15), we finally arrive at

$$t\cos\vartheta = \frac{j\varrho c}{m''\omega}\frac{1+r-t}{\frac{k^4}{k_B^4}\sin^4\vartheta - 1} \; . \tag{8.17}$$

Next, r is eliminated, according to (8.16), and the sound pressure transmission coefficient

$$t = \frac{\frac{2j\varrho c}{m''\omega}}{\left(\frac{k^4}{k_B^4}\sin^4\vartheta - 1\right)\cos\vartheta + \frac{2j\varrho c}{m''\omega}} \tag{8.18}$$

obtained, which is used to determine the sound power transmission coefficient

$$\tau = |t|^2$$

and the airborne transmission loss

$$R = 10\lg 1/\tau \; .$$

When interpreting the result (8.18), the ratio of bending wavelength λ_B and air wavelength λ is of particular significance. The bracketed term in the denominator of (8.18) is

$$\left(\frac{k^4}{k_B^4}\sin^4\vartheta - 1\right) = \left(\frac{\lambda_B^4}{\lambda^4}\sin^4\vartheta - 1\right) = \left(\frac{f^2}{f_{cr}^2}\sin^4\vartheta - 1\right) \; .$$

Below the coincidence frequency $\lambda_B \ll \lambda$ (corresponding to $f \ll f_{cr}$) this term is nearly independent of the incidence angle ϑ and roughly equal to -1. Within the frequency range $f > f_{cr}$ ($\lambda_B > \lambda$) on the other hand, the term in brackets strongly is linked to ϑ. It particular, it can also become zero. This is the cases $f \ll f_{cr}$ and $f > f_{cr}$ must be defined.

a) Frequency range below the critical frequency $f \ll f_{\mathrm{cr}}$

is for this case,

$$t \approx \frac{\frac{2j\varrho c}{m''\omega}}{\frac{2j\varrho c}{m''\omega} - \cos\vartheta} \,. \tag{8.19}$$

The ratio c/m'' is almost always a very small number. The specific impedance is $\varrho c = 400\,\mathrm{kg/m^2s}$, so even for only 100 Hz and $m'' = 10\,\mathrm{kg/m^2}$ it is $m'' = 6300\,\mathrm{kg/m^2s}$. If we assume that the grazing incidence, $\vartheta = 90°$, with $t = 1$, rarely occurs,

$$\tau = |t|^2 \approx \left(\frac{2\varrho c}{m''\omega}\right)^2 \frac{1}{\cos^2\vartheta} \tag{8.20}$$

and

$$R = 10\lg\left(\frac{m''\omega}{2\varrho c}\right)^2 + 10\lg\cos^2\vartheta \tag{8.21}$$

can be written. If we additionally assume diffuse sound incidence from all directions, an average incidence angle of $\vartheta = 45°$ can be inserted into (8.21), resulting in

$$R = 10\lg\left(\frac{m''\omega}{2\varrho c}\right)^2 - 3\,\mathrm{dB}. \tag{8.22}$$

Equation (8.22) is called the 'mass law' or 'Berger's mass law' . It states that R increases at 6 dB per octave and also with 6 dB per doubling of mass.

Obviously, the bending stiffness of the wall is unimportant in the frequency range $f \ll f_{\mathrm{cr}}$. The walls are therefore called 'flexible walls'. Usually, a partition or a wall is 'flexible' if the critical frequency is above the frequency range of interest.

b) Frequency range above the critical frequency $f > f_{\mathrm{cr}}$

Above the critical frequency, a specific 'critical' incidence angle ϑ_{cr} exists, resulting in a total transmission $t = 1$ of the sound field (that is, for the simplest case). For $\vartheta = \vartheta_{\mathrm{cr}}$ using

$$\sin\vartheta_{\mathrm{cr}} = \frac{k_{\mathrm{B}}}{k} = \frac{\lambda}{\lambda_{\mathrm{B}}} = \sqrt{\frac{f_{\mathrm{cr}}}{f}}, \tag{8.23}$$

the wall seems to be 'acoustically transparent'. The reason for this result is given by the fact that the incident sound field in air and the wall vibration are perfectly matched for $\vartheta = \vartheta_{\mathrm{cr}}$. The trace wavelength λ_{s} of the sound field in air, directly located upon the wall (see Fig. 8.6),

$$\lambda_{\mathrm{s}} = \lambda/\sin\vartheta \tag{8.24}$$

matches the bending wavelength for $\vartheta = \vartheta_{\text{cr}}$

$$\text{for} \quad \vartheta = \vartheta_{\text{cr}}: \quad \lambda_{\text{s}} = \lambda / \sin \vartheta_{\text{cr}} = \lambda_{\text{B}} \,. \tag{8.25}$$

This effect is called 'coincidence effect'. Obviously, it also occurs above the lowest coincidence frequency. The latter is often simply called the 'limiting frequency' or 'critical frequency'.

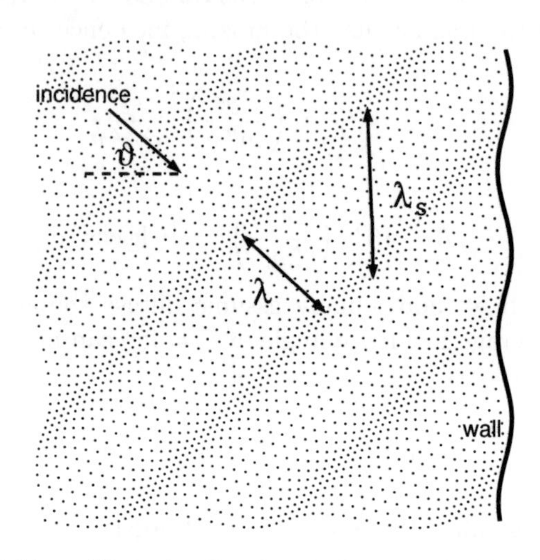

Fig. 8.6. A sound field at oblique incidence with the wavelength λ in its propagation direction produces a sound pressure with the wavelength $\lambda_{\text{s}} = \lambda / \sin \vartheta$ at the wall plane $x = 0$

In the previously described (and simplified) model, a sound transmission of $t = 1$ is obtained for $\vartheta = \vartheta_{\text{cr}}$, in accordance with (8.18). Apart from the fact that this points to an important physical effect in the transmission of sound, the result $t = 1$ might not be satisfactory in practice. It is impossible to observe total transmission through a wall even under ideal measurement conditions. The simplest explanation is internal damping, which is always present in the partition. Similar to the derivation of the complex spring stiffness in Chap. 5, the losses of the wall can be expressed by a complex bending stiffness

$$B' \to B' \,(1 + j\eta) \tag{8.26}$$

where η is the loss factor of the wall. Therefore, the bending wave number likewise becomes complex

$$k_{\text{B}}^4 = \frac{m''}{B'} \omega^2 \to \frac{m''}{B'(1 + j\eta)} \omega^2 = \frac{k_{\text{B}}^4}{1 + j\eta} \,. \tag{8.27}$$

Equation (8.18) then becomes

$$t = \frac{\frac{2j\varrho c}{m''\omega}}{\left(\frac{\lambda_B^4}{\lambda^4}\sin^4\vartheta\,(1+j\eta)-1\right)\cos\vartheta + \frac{2j\varrho c}{m''\omega}} \; . \tag{8.28}$$

At coincidence $\vartheta = \vartheta_{\mathrm{cr}}$ (using $\lambda_{\mathrm{B}}\sin\vartheta/\lambda = 1$) it is therefore

$$t(\vartheta = \vartheta_{cr}) = \frac{\frac{2\varrho c}{m''\omega}}{\frac{2\varrho c}{m''\omega} + \eta\cos\vartheta_{cr}} \; . \tag{8.29}$$

At sufficiently large frequencies, the sound transmission now depends on the loss factor.

To simulate more realistic conditions, omnidirectional and uniform 'diffuse' sound incidence is assumed. This situation is described by the averaged sound power transmission coefficient

$$\bar{\tau} = \frac{1}{\pi/2}\int_0^{\pi/2} \tau(\vartheta)\,\mathrm{d}\vartheta = \frac{1}{\pi/2}\int_0^{\pi/2} |t(\vartheta)|^2\,\mathrm{d}\vartheta \tag{8.30}$$

and the corresponding transmission loss

$$R = -10\lg\bar{\tau} \; . \tag{8.31}$$

The integration in (8.30) can only be approximated analytically. This somewhat lengthy procedure for approximating the solution of the integral will be described in the next subsection. Readers who are not interested in the details may skip over this section and the following equations numbered with (I) without missing substantial information and may safely proceed to equation (8.44) on p. 249.

Approximating transmission loss above the Critical Frequency

The integral (8.30) can only be approximately calculated given with the following two prerequisites:

1. The frequency range under consideration is 'far above' the critical frequency, $f \gg f_{\mathrm{cr}}$. Hence, it is $\lambda_{\mathrm{B}} \gg \lambda$.
2. It is assumed that angles in the range $\vartheta \approx \vartheta_{\mathrm{cr}}$ dominate the value of the integral.

Using

$$\sin\vartheta_{\mathrm{cr}} = \frac{\lambda}{\lambda_{\mathrm{B}}} \; , \tag{8.32}$$

it follows from $\lambda_{\mathrm{B}} \gg \lambda$ that ϑ_{cr} is a small angle. Thus, $\cos\vartheta \approx \cos\vartheta_{\mathrm{cr}} \approx 1$ can be used. Furthermore, the imaginary part of the denominator in (8.28) is

$$j\left[\frac{2\varrho c}{m''\omega} + \eta\frac{\lambda_{\mathrm{B}}^4}{\lambda^4}\sin^4\vartheta\cos\vartheta\right] \approx j\left[\frac{2\varrho c}{m''\omega} + \eta\right] \tag{8.33}$$

because $\sin\vartheta \approx \sin\vartheta_{\mathrm{cr}} \approx \lambda/\lambda_{\mathrm{B}}$ and $\cos\vartheta \approx 1$ can be inserted in this case. It is therefore approximately

$$\bar{\tau} = \frac{\tau_0}{\pi/2} \int_0^{\pi/2} \frac{\mathrm{d}\vartheta}{\left(\frac{\lambda_{\mathrm{B}}^4}{\lambda^4}\sin^4\vartheta - 1\right)^2 + \left[\frac{2\varrho c}{m''\omega} + \eta\right]^2}\,, \tag{8.34}$$

abbreviating using

$$\tau_0 = \left(\frac{2\varrho c}{m''\omega}\right)^2 . \tag{8.35}$$

For sufficiently large frequencies it is

$$\eta \gg \frac{2\varrho c}{m''\omega}$$

and thus the averaged transmission coefficient becomes

$$\bar{\tau} = \frac{\tau_0}{\pi/2} \int_0^{\pi/2} \frac{\mathrm{d}\vartheta}{\left(\frac{\lambda_{\mathrm{B}}^4}{\lambda^4}\sin^4\vartheta - 1\right)^2 + \eta^2}\,. \tag{8.36}$$

Using the variable substitution

$$\begin{aligned} u &= \frac{\lambda_{\mathrm{B}}}{\lambda}\sin\vartheta \\ \mathrm{d}u &= \frac{\lambda_{\mathrm{B}}}{\lambda}\cos\vartheta\mathrm{d}\vartheta \approx \frac{\lambda_{\mathrm{B}}}{\lambda}\mathrm{d}\vartheta \\ \mathrm{d}\vartheta &\approx \frac{\lambda}{\lambda_{\mathrm{B}}}\mathrm{d}u \end{aligned}$$

it results in

$$\bar{\tau} = \frac{\tau_0}{\pi/2}\frac{\lambda}{\lambda_{\mathrm{B}}} \int_0^{\lambda_{\mathrm{B}}/\lambda} \frac{\mathrm{d}u}{\left(u^4 - 1\right)^2 + \eta^2}\,. \tag{8.37}$$

As already mentioned, it is assumed that only the angles in the range $\vartheta \approx \vartheta_{\mathrm{cr}}$ are critical. This is equivalent to using the range $u \approx 1$ and approximately results in

$$u^4 - 1 = \left(u^2 - 1\right)\underbrace{\left(u^2 + 1\right)}_{\approx 2} \approx 2\left(u^2 - 1\right) = 2\left(u - 1\right)\underbrace{\left(u + 1\right)}_{\approx 2} \approx 4\left(u - 1\right) \tag{8.38}$$

where one obtains

$$\bar{\tau} = \frac{\tau_0}{\pi/2}\frac{\lambda}{\lambda_{\mathrm{B}}} \int_0^{\lambda_{\mathrm{B}}/\lambda} \frac{\mathrm{d}u}{16\left(u - 1\right)^2 + \eta^2} = \frac{\tau_0}{\pi/2}\frac{\lambda}{\lambda_{\mathrm{B}}}\frac{1}{16} \int_0^{\lambda_{\mathrm{B}}/\lambda} \frac{\mathrm{d}u}{\left(u - 1\right)^2 + \left(\eta/4\right)^2}\,. \tag{8.39}$$

Finally, using the substitution $y = u-1$ with $\mathrm{d}u = \mathrm{d}y$, we obtain the tabulated integral

$$\bar{\tau} = \frac{\tau_0}{8\pi}\frac{\lambda}{\lambda_\mathrm{B}} \int\limits_{-1}^{\lambda_\mathrm{B}/\lambda-1} \frac{\mathrm{d}y}{y^2 + (\eta/4)^2}, \tag{8.40}$$

which can be looked up in an integral table. This yields

$$\bar{\tau} = \frac{\tau_0}{8\pi}\frac{\lambda}{\lambda_\mathrm{B}}\frac{1}{\eta/4}\left[\arctan\left(\frac{4}{\eta}\left(\frac{\lambda_\mathrm{B}}{\lambda} - 1\right)\right) + \arctan\left(\frac{4}{\eta}\right)\right] \tag{8.41}$$

(where arctan is the inverse tangent). For small η and due to $\lambda_\mathrm{B} \gg \lambda$, both arctan-terms have the value $\pi/2$. This results in

$$\bar{\tau} = \tau_0 \frac{\lambda}{\lambda_\mathrm{B}}\frac{1}{2\eta} . \tag{8.42}$$

If τ_0 and

$$\frac{\lambda}{\lambda_\mathrm{B}} = \sqrt{\frac{f_\mathrm{cr}}{f}} \tag{8.43}$$

are inserted, we arrive at the simple and clear result of

$$\bar{\tau} = \left(\frac{2\varrho c}{m''\omega}\right)^2 \sqrt{\frac{f_{cr}}{f}}\frac{1}{2\eta} \tag{8.44}$$

for the sound power transmission coefficient. The transmission loss for frequencies above the critical frequency $f > f_\mathrm{cr}$ is thus given by

$$R = 10\lg\left(\frac{m''\omega}{2\varrho c}\right)^2 + 5\lg\frac{f}{f_\mathrm{cr}} + 10\lg 2\eta . \tag{8.45}$$

Here, the gradient of the transmission loss R rises at a higher slope than it does in the range, with $f < f_\mathrm{cr}$ at 7.5 dB per octave. R is influenced by the loss factor of the wall. No statement is made here regarding the actual reason for the losses. The loss factor consists of all loss effects which in fact occur, from internal damping to vibration energy transport to adjacent partitions.

A summary of the frequency characteristics of the transmission loss of single-leaf partitions below and above the critical frequency is given in Fig. 8.7. Approaching the critical frequency, the two methods described above fail to predict the behavior of the transmission loss. The asymptotes in Fig. 8.7 were arbitrarily connected. Hence, it is obvious that the transmission loss 'drops' in that frequency range, and that this effect depends on the loss factor (and – as can be shown – on the wall dimensions not to be taken into account here).

Some practical examples are shown in Figs. 8.8, 8.9 and 8.10. The characteristics have a similar structure as theoretically predicted. In particular, the drop at the coincidence can be discerned quite well. It is conspicuous that the

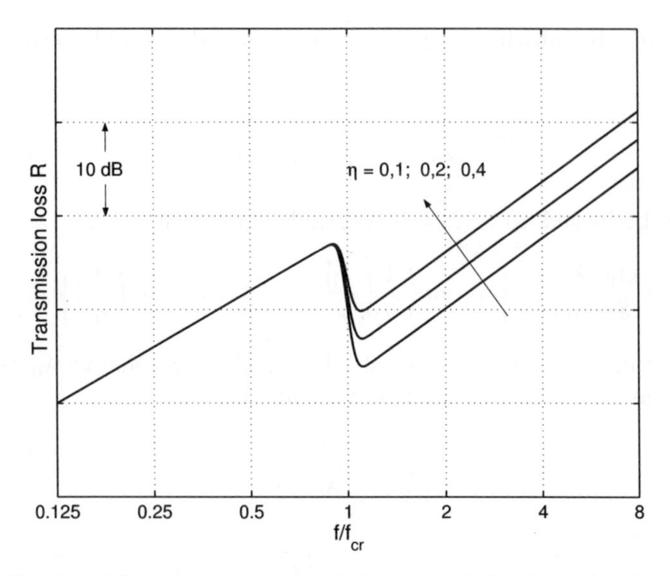

Fig. 8.7. Predicted frequency response of the transmission loss of a single-leaf partition

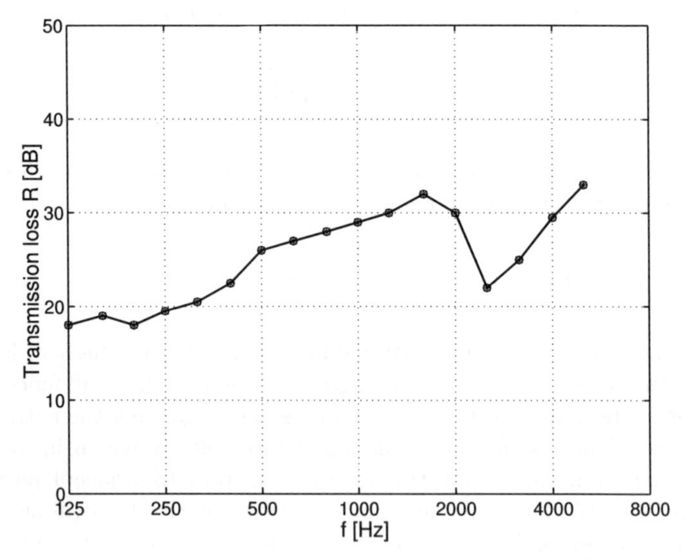

Fig. 8.8. Transmission loss of a window pane, $m'' = 15\,\mathrm{kg/m^2}$, $f_{\mathrm{cr}} = 2500\,\mathrm{Hz}$

coincidence drops more drastically the higher the critical frequency is. Theoretical investigations on walls with finite dimensions confirm this assumption.

The correlation between theory and practice is not always as strong as shown in Figs. 8.8 to 8.10. There are a lot of other factors which reduce the measured transmission loss in respect to the theory, for instance,

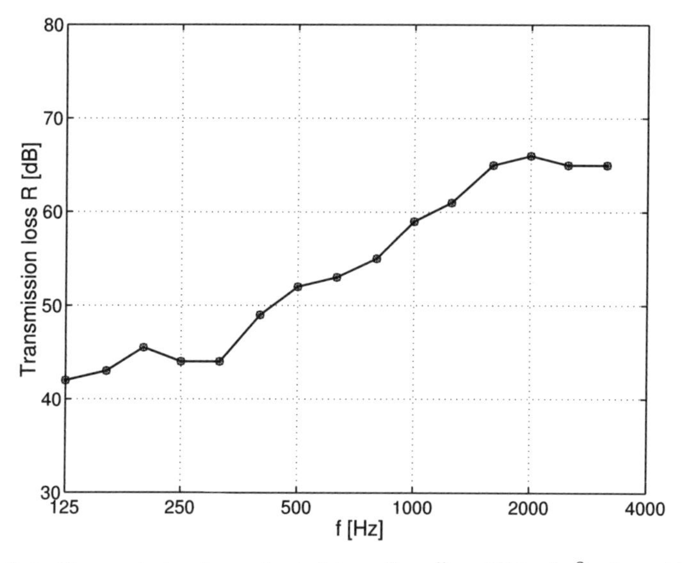

Fig. 8.9. Transmission loss of a brick wall, $m'' = 400\,\mathrm{kg/m^2}$, $f_{\mathrm{cr}} = 130\,\mathrm{Hz}$

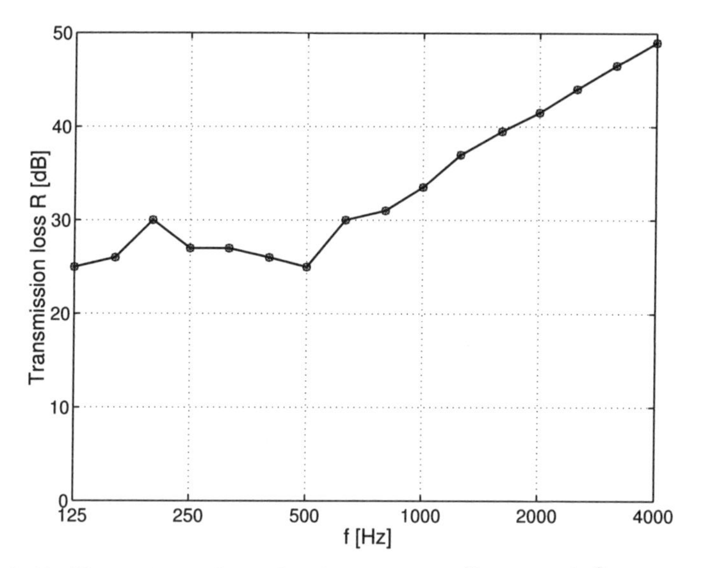

Fig. 8.10. Transmission loss of a plaster wall, $m'' = 60\,\mathrm{kg/m^2}$, $f_{\mathrm{cr}} = 350\,\mathrm{Hz}$

- leakage (important for doors and windows),
- internal inhomogeneities (e.g. cracks in the masonry, but also a brick structure with local resonances as in lightweight hollow building blocks),
- resonances in the thickness of the wall (where the front and back of the wall are not uniformly moving, thus invalidating the bending wave theory),

- porous construction materials, where sound transmission is more likely to occur through the pores than through the skeleton (for example, a concrete wall – especially aerated concrete – has a substantially higher sound reduction if the surface is plastered).

For these and other reasons – it is often required in practice to rely on experience as opposed to the above equations. Some indications for that are given in Fig. 8.11, which illustrates the transmission loss of walls and single-leaf structures for the most commonly used materials.

The 'plateau' in the graph describes the transitions between partitions, from those which are flexible over the entire frequency range (i.e. a very high critical frequency with a very small mass) to stiff partitions where the critical frequency decreases with the thickness (and thus with the mass). Consult the literature for field reports to find answers to questions on specific problems.

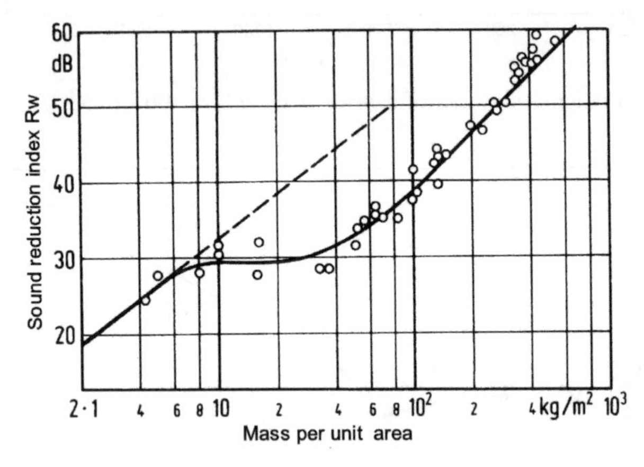

Fig. 8.11. Weighted sound reduction index R_W, depending on the mass per unit area, for commonly used building materials (from: K. Gösele and E. Schröder: Schalldämmung in Gebäuden, Chap. 8 in "Taschenbuch der Technischen Akustik", Springer, Berlin and Heidelberg 2004, edited by G. Müller and M. Möser)

8.3 Double-leaf partitions (flexible additional linings)

A simple and cheap method to increase the transmission loss of a wall is to cover it with a second, additional lining (an additional lining made of plaster boards costs about as much as an expensive carpet). As the discussions in the previous section show, the additional lining cannot be installed in the first wall: a small additional mass per unit area would only result in a negligible increase of the transmission loss. The second partition has to be mounted separately. Generally, space usage should remain minimal. Therefore, we assume a cavity between the two partitions, where the thickness d is small compared to the wavelength. The air in the cavity acts as a spring with the stiffness per unit area

$$s'' = \frac{\varrho c^2}{d} \tag{8.46}$$

(where d is the width of the cavity). The pressure p_i in the cavity between the two partitions is thus given by

$$p_\mathrm{i} = \frac{s''}{j\omega}\left(v_1 - v_2\right) \tag{8.47}$$

(see also Fig 8.12). Indeed, this implies that the transverse coupling of the volume elements in the cavity parallel to the walls can be neglected. Otherwise standing waves would occur parallel to the lining. The aforementioned assumption of a 'pure stiffness' already implies a cavity is already implied which is damped with mineral wool. The flow resistance of the damping material acts as a transverse decoupling of the 'flat room' between the walls.

Generally, the additional lining will be thin. It can also be assumed that the critical frequency is above the frequency range of interest. As explained in the previous section, the dynamic bending of the wall is unimportant in this case. The flexible additional lining only reacts to an exciting pressure in the cavity with an inertia force. This can be expressed as

$$p_\mathrm{i} = j\omega m_2'' v_2, \tag{8.48}$$

where m_2'' is the mass per unit area of the additional lining. It was also assumed in (8.48) that the sound pressure in the 'receiving room', located to the right of the additional lining (Fig. 8.12), is a lot smaller in order of magnitude than p_i

$$|p_2(x=0)| \ll |p_\mathrm{i}| .$$

Furthermore, if only heavy, solid 'original walls' (with the index 1) are implied, it can be assumed that their vibrations are not substantially influenced by the additional lining. Therefore the results obtained earlier in this chapter pertaining to single-leaf partitions can be adopted for the vibrations of wall 1 in a first-order approximation.

Using (8.15), (8.10) results in

$$v_1 = \frac{p_0}{\varrho c} t_1 \cos\vartheta e^{jkz\sin\vartheta} \tag{8.49}$$

where, according to (8.18),

$$t_1 = \frac{\frac{2j\varrho c}{m_1''\omega}}{\left(\frac{k^4}{k_1^4}\sin^4\vartheta - 1\right)\cos\vartheta + \frac{2j\varrho c}{m_1''\omega}} . \tag{8.50}$$

This represents the sound pressure transmission coefficient of the 'single-leaf partition' 1 without the additional lining.

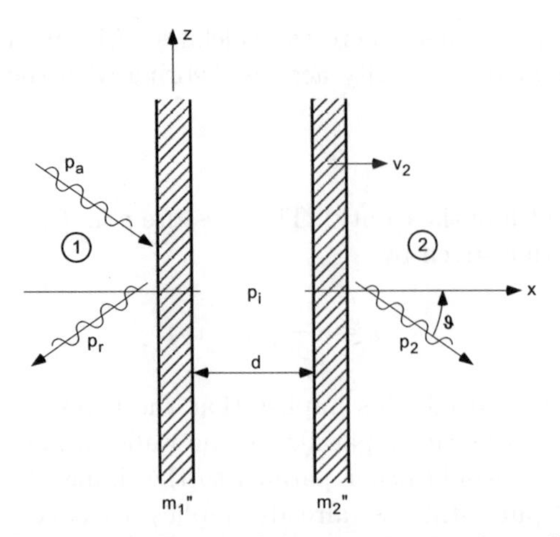

Fig. 8.12. Model for the calculation of the transmission loss of a double-leaf partition. p_{a}: incident sound field, p_{r}: reflected sound field, $p_1 = p_{\mathrm{a}} + p_{\mathrm{r}}$: total sound field in front of the wall, p_2: transmitted sound field, p_{i}: sound pressure in the filled cavity

The vibrations v_2 of the additional lining can now easily be calculated. Inserting (8.47) into (8.48) results in

$$j\omega m_2'' v_2 = \frac{s''}{j\omega}(v_1 - v_2)$$

or

$$v_2 = \frac{v_1}{1 - \frac{\omega^2 m_2''}{s''}} = \frac{v_1}{1 - \frac{\omega^2}{\omega_0^2}} . \tag{8.51}$$

The air spring in the cavity s'' and the mass of the additional lining m_2'' constitute a simple resonator with the resonance frequency ω_0, which is given by

$$\omega_0^2 = \frac{s''}{m_2''} .$$

At $\omega = \omega_0$, the additional lining could theoretically perform infinitely large vibrations because of the lack of damping. Adding a loss factor to the spring stiffness would prevent this.

If (8.49) is inserted into (8.51), we obtain

$$v_2 = \frac{p_0}{\varrho c} \frac{t_1}{1 - \frac{\omega^2}{\omega_0^2}} \cos\vartheta \mathrm{e}^{jkz\sin\vartheta} \tag{8.52}$$

For sound pressure radiated by the additional lining into the receiving room, we take the same basic approach as in (8.8)

$$p_2 = t p_0 \mathrm{e}^{-jkx\cos\vartheta} \mathrm{e}^{jkz\sin\vartheta} , \tag{8.53}$$

where this time, t represents the sound pressure transmission coefficient of the double-leaf partition (transmission of the incident sound field p_0). As before, t adheres to the condition

$$\frac{j}{\omega\varrho}\frac{\partial p_2}{\partial x}\bigg|_{x=0} = v_2 \,.$$

It is simply

$$t = \frac{t_1}{1-\frac{\omega^2}{\omega_0^2}} \tag{8.54}$$

and thus

$$R = 10\lg\frac{1}{\tau} = 10\lg\frac{1}{t^2} = 10\lg\left(1-\frac{\omega^2}{\omega_0^2}\right)^2 + 10\lg\frac{1}{\tau_1} \tag{8.55}$$

or

$$R = R_1 + 10\lg\left(1-\frac{\omega^2}{\omega_0^2}\right)^2 \tag{8.56}$$

where R_1 denotes the transmission loss of the single-leaf partition (with the index 1).

In a first-order approximation, the transmission loss of the double-leaf partition is composed of the sum of the transmission loss of the heavier wall and a frequency-dependent term, which will be referred to as the improvement of the additional lining R_E

$$R = R_1 + R_E \tag{8.57}$$

with

$$R_E = 10\lg\left(1-\frac{\omega^2}{\omega_0^2}\right)^2 \,. \tag{8.58}$$

It is clear that the additional lining

- is ineffective below the resonance $\omega \ll \omega_0$ where $R_E = 0$,
- decreases the transmission loss at the resonance $\omega = \omega_0$ (of course, the magnitude depends on the loss factor of the spring) and
- gains a real advantage of $R_E \approx 40\lg(\omega/\omega_0)$ at resonances above $\omega \gg \omega_0$, rising at a gradient of 12 dB per octave.

For the purposes of scaling it can be said that from a noise control perspective, the resonance frequency should be 'as low as possible'. It follows from

$$f_0 = \frac{1}{2\pi}\sqrt{\frac{\varrho c^2}{m_2'' d}} \approx \frac{60\,\text{Hz}}{\sqrt{\frac{m_2''}{\text{kg/m}^2}\frac{d}{\text{m}}}}$$

that it is better to use heavy additional linings at a large distance d from the original single-leaf partition. Using $m_2'' = 10\,\text{kg/m}^2$ and $d = 10\,\text{cm} = 0.1\,\text{m}$ as an example results in $f_0 = 60\,\text{Hz}$, proving that tuning the resonance below

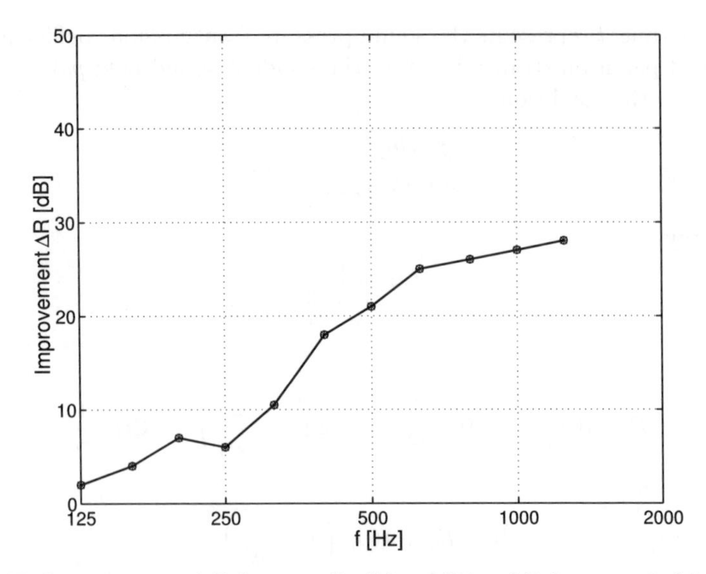

Fig. 8.13. Improvement ΔR due to a flexible additional lining mounted in front of a plaster wall (80 mm). $m_2'' = 4\,\text{kg/m}^2$, cavity width 65 mm damped with mineral wool

the 'frequency range of building acoustics' (which usually starts at 100 Hz) can be achieved with a reasonable effort.

Experience with flexible additional linings shows that the estimation described above for the improvement of transmission loss tends not to be so bad, provided that the cavity is damped with enough mineral wool and that no fixed connections (so-called structure-borne sound bridges) between the linings exist. Fig. 8.13 shows a measured improvement. Fairly high values are obtained for R_{E} which at least in the mid-frequency range adhere to (8.58).

Flexible additional linings (usually plaster boards of more than 10 mm thickness) are the most important means of the acoustician during the redevelopment of buildings. For light-weight double-leaf partitions (which also includes double-glazed windows) the rough estimates made above more or less agree with experience gained in practice. The reason for this is partially given by effects not discussed here. For instance, cavity damping does not take place in double-glazed windows. In the case of nearly equal partitions, it can no longer be assumed that one is vibrating independently of the other. The importance of cavity damping is shown in experiments with double-leaf plaster structures (Fig. 8.14). The lack of absorption allows the sound reduction to decrease dramatically. Therefore, in the case of windows, attempts were made to focus the damping at the frame by equipping the edges of the cavity with absorbent material (Fig. 8.15), which ultimately produced appreciable improvements. Other experiments testing the influence of the cavity on sound

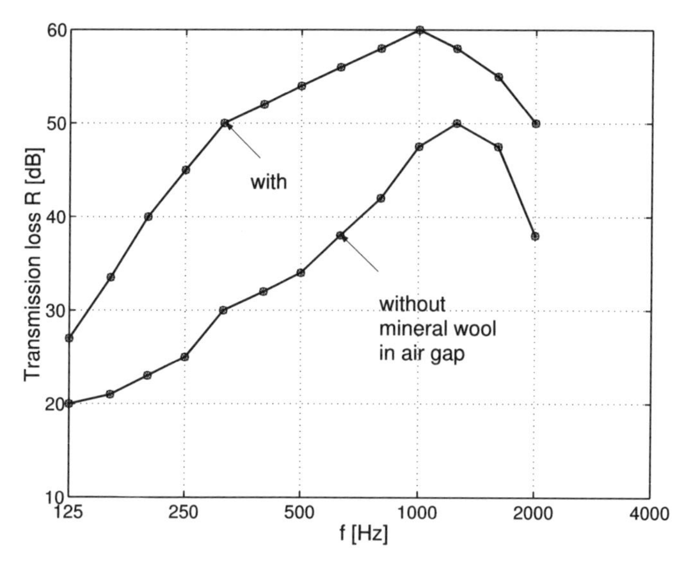

Fig. 8.14. Transmission loss of a double-leaf plaster partition ($f_{cr} = 3000\,\text{Hz}$) with and without mineral wool in the cavity

transmission were carried out with helium (Fig. 8.16) which shows similar effects.

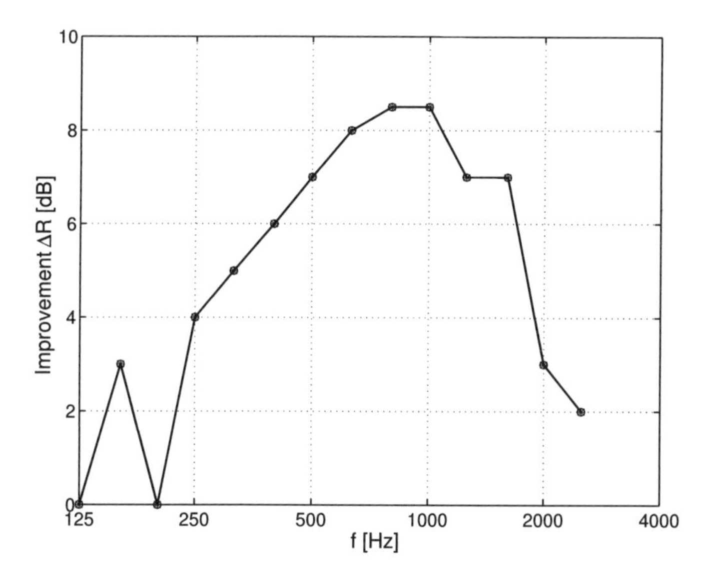

Fig. 8.15. Improvement made by lining the cavity edges of a double-glazed window frame (5 mm and 8 mm glass, cavity width 24 mm) with 50 mm mineral wool

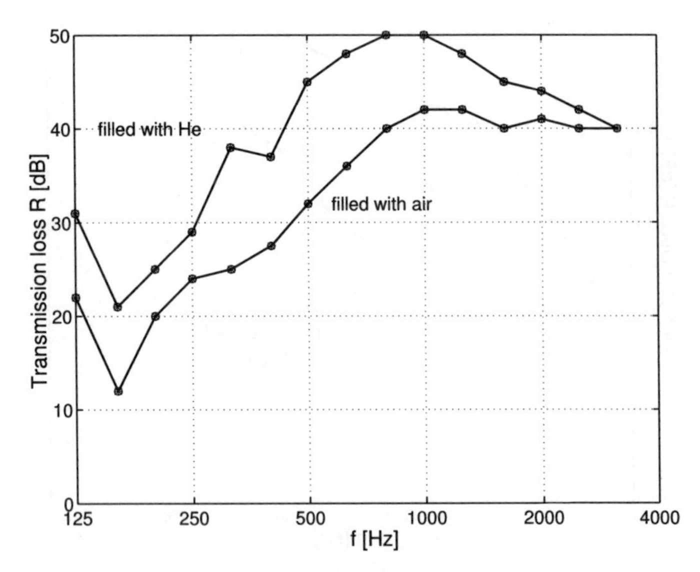

Fig. 8.16. Transmission loss of a double-glazed window with 8 mm and 4 mm glass, 16 mm cavity filled with an air-helium mixture

Such effects and other details important in practice unfortunately cannot be extensively treated here. Yet, it is possible to derive some construction criteria and dependencies based on the above discussions and practical experience. In order to obtain a high transmission loss, it is prudent to

- use a large total mass and distribute it non-uniformly over the lining if possible (i.e. avoid partitions with the same mass),
- consider a large cavity width for the design,
- damp the whole cavity with absorbent material, if possible,
- avoid leakage (sometimes the window frame may be more important than the pane), and
- by all means avoid structure-borne sound bridges (i.e. connections between the partitions).

The last item particularly should be stressed in the actual construction process. Additional linings must be mounted on separate studs, only soft-resilient mountings are allowed on the 'original wall' in order to improve the static stability of the structure. Sound bridges relocate the critical frequency of the whole structure in such a way that the transmission loss can be lower than without the additional lining! Often, existing connections between original wall and additional lining are underestimated, resulting in an additional, worsening resonance effect directly in the middle of the critical frequency range. This ill effect is sometimes caused by mounting facings too tightly to a house wall.

8.4 Impact sound reduction

Sound impact in building acoustics is understood as the air-borne sound radiated by a wall or a ceiling, when the ceiling of another room is induced to structure-borne vibrations by walking, moving of chairs, operating kitchen equipment or similar. It is thus not only the sound radiation into the room under the ceiling. It is, for example, quite common to determine the 'impact sound level' in the flat *above* a restaurant and recommend solutions to reduce it, whenever it is required to do so.

8.4.1 Measuring impact sound levels

A standardized source of force used to excite the corresponding floor enables the unified classification of impact noise radiated by a structure. This source is generated by the so-called tapping machine (Fig. 8.17). It consists of five hammers (each weighing 500 g), mechanically lifted by a motor, which fall onto the floor from a defined height one after another (tapping frequency 5 Hz). This process is repeated periodically as long as the motor operates (and makes an enormous racket). The impact sound levels L_{E} are measured in the receiving room (in spatially averaged octave or third-octave bands).

Fig. 8.17. Tapping machine used to measure the normalized impact sound level

Since the absorption of the room has to be taken into account (the impact noise sounds quieter in a highly damped room than in a lightly damped room), the normalized impact sound level

L_n is given by

$$L_n = L_E + 10 \lg A_e / A_0 \ , \tag{8.59}$$

where A_e is the equivalent absorption area of the receiving room. The reference value is $A_0 = 10\,\mathrm{m}^2$, thus, L_E is scaled to a room with an absorption area of $A_e = 10\,\mathrm{m}^2$.

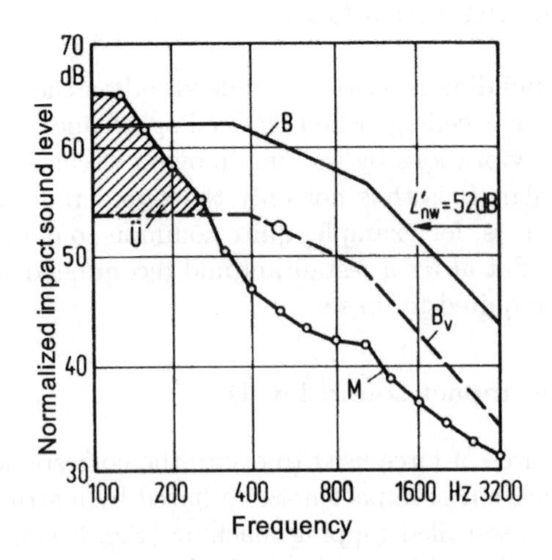

Fig. 8.18. Definition of the normalized impact sound level L_{nw} drawn versus frequency. **B**: reference curve, **Bv**: shifted reference curve, **M**: measurement, **Ü**: positive difference of **M** with respect to **Bv** (from :K. Gösele and E. Schröder: Schalldämmung in Gebäuden, Chap. 8 in "Taschenbuch der Technischen Akustik", Springer, Berlin and Heidelberg 2004, edited by G. Müller and M. Möser)

In order to specify sound impact protection levels,, the third-octave band impact sound levels are weighted with a reference curve. This follows the same principle as the determination of the AIM from the transmission loss R by shifting the impact sound reference curve, here (Fig. 8.18). The shift in dB denotes the impact protection margin. The details are standardized in ISO 140 and ISO 717.

The TSM is still a well-cited resource nowadays. Hence, the weighted standard impact sound level L_{nw} is usually used which is given by

$$L_{nw} = 63 - \text{TSM} . \tag{8.60}$$

8.4.2 Improvements

Some normalized impact sound levels of bare floors are shown in Fig. 8.19. They range approximately from 70 dB to 90 dB. It is clear that the level falls off at 10 dB per doubling of mass (a physical explanation for this effect is given in the book "Structure Borne Sound" by Lothar Cremer and Manfred Heckl (translated by B.A.T. Petersson, Springer, Berlin and New York 2004)).

Compared to the minimum requirement found in DIN 4109 (impact sound level < 63 dB) the values of the bare floors are far too high. Yet, there are some simple solutions which improve impact noise protection with some justifiable effort without significantly increasing the mass. The order of magnitude of the improvements is shown in Table 8.1.

Table 8.1. Normalized impact sound level reduction of single-leaf bare floors and corresponding mass per unit area (from: K. Gösele and E. Schröder: Schalldämmung in Gebäuden, Chap. 8 in "Taschenbuch der Technischen Akustik", Springer, Berlin and Heidelberg 2004, edited by G. Müller and M. Möser)

Coatings	
Linoleum, PVC-coating without base	3 to 7 dB
Linoleum on 2 mm Cork	15 dB
PVC-floor with 3 mm felt	15 to 19 dB
Needled felt	18 to 22 dB
Carpet, thick	25 to 35 dB
Floating floors	
Cement layer	
on corrugated board	18 dB
on hardened foam board, stiff	about 18 dB
on hardened foam board, soft	about 25 dB
on mineral fibre boards	27 to 33 dB

The coatings can be regarded as an 'elastic layer' mounted onto the floor. Together with the thudding mass M of the tapping machine hammers or the foot of a walking person, they form a mass-spring-system with the resonance frequency

$$\omega_{\text{res}}^2 = \frac{s}{M}$$

As in double-leaf partitions, decoupling goes into effect above the resonance frequency, explaining the improved results of soft elastic layers. Since a short pulse of a force uniformly contains all frequencies, a substantial amount of spectral energy is affected by decoupling if the resonance frequency is low enough. Thus, the improvement is high.

It is noteworthy in these remarks that the impinging force itself determines the result. Thus, the tapping machine does not simulate the walking on a floor very precisely, since the masses involved in walking are certainly larger. The improvements achieved with the tapping machine only vaguely correspond to those achieved by level reduction, which is actually perceived using different force excitation mechanisms are used. On the other hand, the tapping machine represents a practical compromise, because the noise from falling objects, the excitation of the ceiling by loudspeaker boxes, etc. also constitute the impact sound level.

In the class of floating floors in Table 8.1, the decoupling effect of a resonator has become a construction rule for floors. As sketched in Fig. 5.3, an elastic layer carrying a mass per unit area (usually a cement floor) is built on top of the raw floor. The level reduction similar to that in airborne sound reduction is given by

$$\Delta L = 20 \lg \left(\frac{\omega}{\omega_{\text{res}}} \right)^2$$

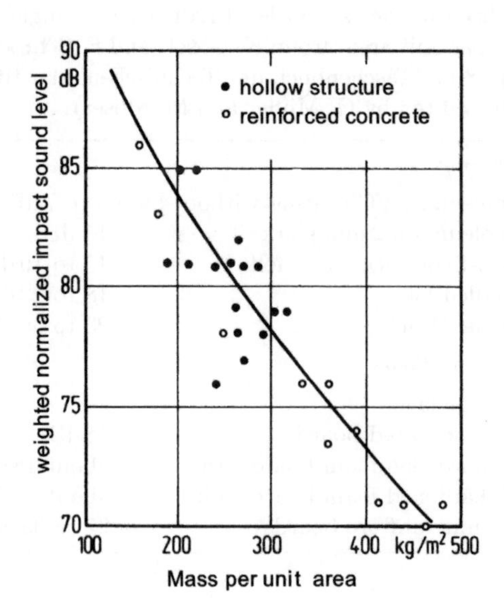

Fig. 8.19. Weighted normalized impact sound level of heavy, single-leaf bare floors shown versus their mass per unit area (from : K. Gösele and E. Schröder: Schalldämmung in Gebäuden, Chap. 8 in "Taschenbuch der Technischen Akustik", Springer, Berlin and Heidelberg 2004, edited by G. Müller and M. Möser). The filled circles represent ceilings with a hollow structure, the empty circles represent ceilings made of reinforced concrete

where ω_{res} is the resonance frequency

$$\omega_{\mathrm{res}}^2 = \frac{s''}{m_2''}$$

(s'': stiffness per unit area of the elastic layer, m_2'': mass per unit area of the floating floor).

The experimental data of an example is shown in Fig. 8.20. The construction rules are similar to those for the improvement of the transmission loss with flexible additional linings. Structure-borne sound bridges have to be avoided. One should bear in mind that the floating floor (the cement layer) has no connections to the flanking walls. This can be ensured by hoisting the elastic layers at the flanking walls or by using insulating stripes at the edges.

The thorough and careful placement of the isolating layer is even more important. Holes develop very easily and are filled during cementation. These form sound bridges, rendering improvement ineffective. The intrusion of cement at the joints of the webs can be avoided by a flushed covering of the plastic foil before filling it with cement or by overlapping the webs during the placement. Sometimes problems arise by heating pipes which are installed in

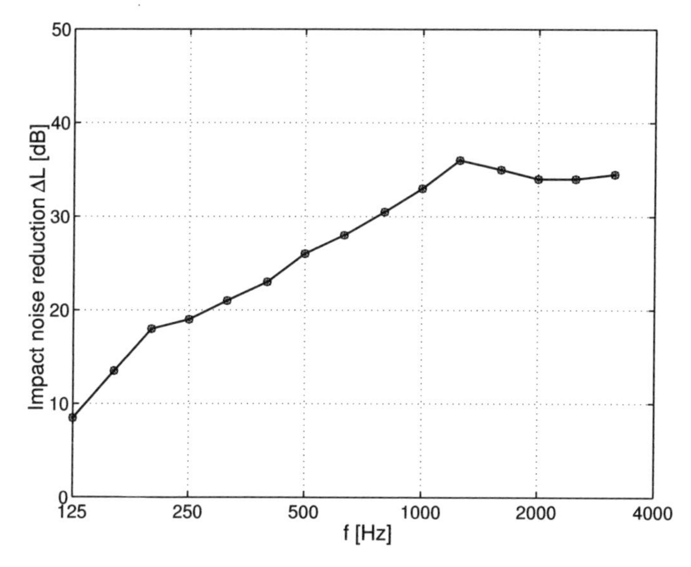

Fig. 8.20. Impact sound level reduction of a floating floor

the isolating layer. During the construction process, mandatory use of soft-coated pipes is a given.

It should finally be mentioned that ceilings suspended underneath the excited component are a lot less effective than floating floors. With respect to the suspended ceiling, which acts as a flexible additional partition, the transmission is not mitigated by the likewise exciting flanking walls; for a correctly constructed floating floor the flanking transmission is also reduced. Therefore, suspended ceilings only promise to be successful for light-weighted bare floors with heavy flanking walls.

8.5 Summary

Air-borne transmission loss of single-leaf structures such as walls, ceilings, and windows, is determined by specifying the first trace match or coincidence effect between the air wave hitting the structure at an oblique angle of incidence and the plate-bending wave above a critical frequency f_{cr}. The trace matching or coincidence effect does not occur at frequencies below the critical frequency $f < f_{cr}$, referred to as simply the critical frequency or the coincidental critical frequency, because the bending waves are shorter than the air waves. Elastic tensions in the wall or partition are therefore irrelevant, as the partition only reacts to excitation by air waves with inert forces. The transmission loss R is only dependent on the ratio of the mass impedance of the wall $j\omega m"$ to the specific resistance of the surrounding medium ϱc. Therefore, R increases by $6\,dB$ per doubling of frequency or mass.

In frequencies above the critical frequency $f > f_{cr}$, the bending wavelength is greater than the air wavelength. This condition leads to the occurrence of a trace match between the wave types present at the incidence angle $sin\,\vartheta = \lambda(air)/\lambda(bending)$. The simplest theoretical model, not taking into account losses in the wall, reflects the unhindered passage of the sound wave striking at this given incidence angle. If, on the other hand, these losses are taken into account and the intensities are averaged over all sound incidence trajectories for the purposes of modelling diffuse sound field conditions, the resulting transmission loss increases by $7.5\,dB$ per octave and is dependent on the loss factor η. For values close to the critical frequency f_{cr} itself, the transmission loss contains a deviation, which is flatter for low critical frequencies and more pronounced for higher ones. The critical frequency itself is inversely proportional to the thickness h of the partition structure, $f_{cr} \sim 1/h$ (see Chapter 4).

Wall structures can be improved for the purposes of better sound insulation using flexible additional linings, each elevated on separate bearings. The improvement first becomes noticeable above the spring-mass resonance frequency of the resonator given by the cavity between the walls and the lining. Theoretically, the mitigating effect increases thereafter by $12\,dB$ per octave. Impact sound can be reduced, for instance, by installing floating floors.

It is necessary to determine the average source and receiving sound levels as well as the equivalent absorption area (derived from the reverberation time in the receiving room) in order to measure air-borne transmission loss. This is because the levels occurring in the receiving room are not only dependent on the transmission loss through the wall, but also on the room loss. For this same reason, in order to specify impact sound level standards, initial level measurements have to be corrected in order to take into account the equivalent absorption area of the receiving room.

8.6 Further reading

A helpful and valuable support to this chapter is provided by A.C.C. Warnock and W. Fasold: "Sound Insulation: Airborne and Impact", Chap. 93 in "Encyclopedia of Acoustics" (editor Malcolm Crocker, John Wiley, New York 1997)

Please note the measurement standards DIN EN ISO 140 "Measurement of sound insulation in buildings and of building elements".

In addition, two German books are recommended for further reading: Gösele/Schüle/Künzel: "Schall. Wärme. Feuchte." (Bauverlag, Wiesbaden 1997) and W. Fasold, E. Sonntag and H. Winkler: "Bauphysikalische Entwurfslehre" (Verlag für Bauwesen, Berlin 1987).

8.7 Practice exercises

Problem 1

Find the frequency-dependent transmission loss of a wall with $10\,m^2$ surface if the following sound pressure levels are measured in the source and receiving room. The reverberation time of the receiving room ($V = 140\,m^3$) is also given below .

f/Hz	L_S/dB	L_E/dB	T_E/s
400	78.4	48.2	2
500	76.6	43.8	1.,8
630	79.2	42.2	1.7
800	80.0	41	1.6
1000	84.4	40	1.6
1250	83.2	40.6	1.5

Problem 2

How high is the resonance frequency of a double-leaf partition with an additional lining, which has a surface mass of $12,5\,kg/m^2$ ($25\,kg/m^2$) at a distance of $5\,cm$ (in $10\,cm$) from the original wall, which is much heavier?

Problem 3

How much is the transmission loss of the steel plate of a car ($0.5\,mm$ thick) at $100\,Hz$ ($200\,Hz, 400\,Hz, 800\,Hz$)? First, find the coincidence critical frequency.

Problem 4

How much is the transmission loss of a heavy cement wall which is $35\,cm$ thick with a loss factor of $\eta = 0.1$ at $200\,Hz, 400\,Hz, 800\,Hz$ and $1600\,Hz$?

Problem 5

A window with an area of $3\,m^2$ and a transmission loss of $30\,dB$ is set into a wall which has an area of $15\,m^2$ (without any other windows) and a transmission loss $60\,dB$. How much is the combined transmission loss?

If the window were to take up a larger amount of the wall's surface at 50 percent, how large would the combined transmission loss be?

Problem 6

The following transmission losses have been measured for a double-pane window (9/16/13 mm) at third-octave intervals:

Frequency/Hz	$R_{thirdoctave}/dB$	reference curve/dB
100	29.4	33
125	36.9	36
160	41.8	39
200	38.2	42
250	42.4	45
315	40.9	48
400	39.4	51
500	41.9	52
630	46.6	53
800	44.5	54
1000	42.4	55
1250	43.8	56
1600	47.1	56
2000	51.0	56
2500	49.6	56
3150	39.9	56

How great is the weighted transmission loss R_w ?

9

Silencers

Generally speaking, silencers are technical devices which attenuate a sound field travelling through them along its propagation path.

The most commonly known silencer is used in vehicles. It consists of pipes filled with gas which connect the engine with a 'pot' or several pots with each other. It finally ends in an outlet, the so-called exhaust pipe. Basically, such a silencer consists of a pipeline with sudden changes in the cross-sectional area.

Silencers are also frequently used in heating, ventilation and air-conditioning systems. Nearly every duct used for ventilating concert halls, operas and other auditoria (like congress halls) is equipped with silencers. In a lot of industrial plants, where, for instance, air-gas mixtures are transported in pipes by way of blowers, severe noise problems arise which can at least partially be solved by appropriate 'acoustic treatment' of the sound transmission inside the pipes.

Mainly, two basic principles can be considered to reduce the sound propagation along tubes and ducts. Either rigid tubes are used, like in vehicle exhaust pipes, where the cross-section along the tube-axis is suddenly expanded or reduced. As a matter of principle, such constructions simply take effect by reflecting the sound at their inlet, since the dissipation of sound energy into heat is neither desired nor of major importance.

The second, alternative basic principle is exemplified by tubes, where the walls of the tube represent an arbitrary acoustical surface which, for example, is characterized by absorbent linings. The required effect of such 'lined silencers' can be described by the sound energy which is lost into the wall lining of the tube. One may be tempted to describe this principle by the term 'absorbent silencers'. A more detailed investigation of the subject in the following will shortly reveal that the main effect of lined silencers is not always attained by absorption. Even the case of an acoustically soft lining with $z = 0$ results in a high silencing effect which is obviously not produced by absorption: no sound power penetrates a surface $z = 0$, as is generally known. Therefore, it can be deduced that even lined silencers can work by means of reflection,

M. Möser, *Engineering Acoustics*, DOI 10.1007/978-3-540-92723-5_9,

verifying that absorbent duct linings are not always the main contributing physical factor behind insertion loss.

For these reasons the common classifications made of reflecting and absorbing silencers is not very precise. The actual effective principle behind the attenuation is only inadequately described by this distinction, because the dominating effect of these two 'types of silencers' is not necessarily the dissipation of sound energy into heat. Even the term 'silencer' might be misleading if it is used equivalently in the sense of 'damping' or 'attenuation'. The terms 'damping' and 'attenuation' shall be used in a more colloquial meaning in this chapter, simply as a level reduction along a distance. As already mentioned, the reason for this is not always dissipation.

9.1 Changes in the cross-section of rigid ducts

The discussions in this section focus on ducts with a small cross-sectional area or on correspondingly low frequencies: the cross-section of the duct is assumed to be always small compared to the wavelength. Some typical applications are therefore

- exhaust pipes of vehicles, the radius of the pipe is approximately 5 cm and
- the low frequency buzzing tone (100 Hz) of a fan in a ventilation system with a width of 30 cm to 50 cm.

9.1.1 Abrupt change in cross-section

The simplest way to realize a reflector for a one-dimensional waveguide is to build abrupt changes into the cross-section of a duct (Fig. 9.1).

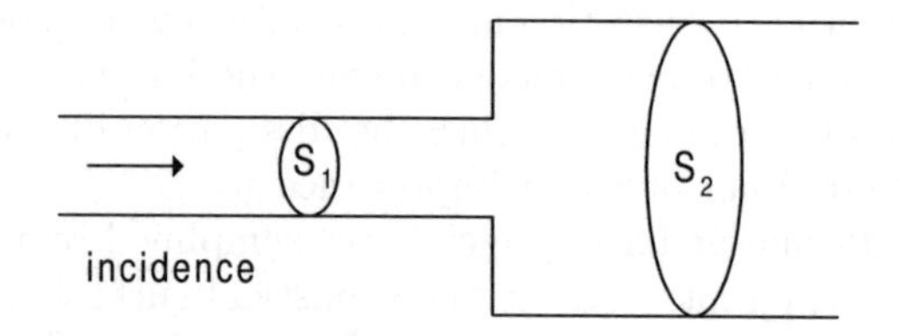

Fig. 9.1. Abrupt change in cross-sectional area

Reflection and transmission can be determined by some simple considerations. To do this, consider in the left (semi-infinite) branch of the duct an incoming wave, originating from the source, and a wave reflected at the change in the cross-section

$$p_1 = p_0 \left(e^{-jkx} + r e^{jkx} \right) \tag{9.1}$$

with the particle velocity

$$v_1 = \frac{j}{\omega\varrho}\frac{\partial p_1}{\partial x} = \frac{p_0}{\varrho c}\left(e^{-jkx} + re^{jkx}\right) . \tag{9.2}$$

The right branch of the duct is assumed to be non-reflecting, therefore only one transmitted wave

$$p_2 = tp_0 e^{-jkx} \tag{9.3}$$

and

$$v_2 = t\frac{p_0}{\varrho c}e^{-jkx} \tag{9.4}$$

occurs. The still unknown reflection coefficient r and the likewise unknown pressure transmission coefficient t can be determined by the two boundary conditions at the surface $x = 0$ of the change of cross-section. On both sides of the boundary $x = 0$, the pressure is the same

$$1 + r = t, \tag{9.5}$$

and the mass, flowing through the cross-sectional area in a time interval on the left side $\varrho S_1 v_1(0)$, and the mass, flowing through the cross-sectional area on the right side $\varrho S_2 v_2(0)$ are equal to

$$S_1 v_1(0) = S_2 v_2(0), \tag{9.6}$$

or using (9.2) and (9.4),

$$(1 - r) S_1 = tS_2 . \tag{9.7}$$

Equations (9.5) and (9.7) formulate the consequences of the boundary conditions with respect to transmission and reflection. If inserted into one another, they result in the pressure transmission coefficient

$$t = \frac{2}{1 + \frac{S_2}{S_1}} \tag{9.8}$$

and the reflection coefficient

$$r = t - 1 = \frac{1 - S_2/S_1}{1 + S_2/S_1} . \tag{9.9}$$

The acoustic effect of silencers is characterized by the transmitted sound power. Therefore, the power transmission coefficient is mainly of interest

$$\tau = \frac{\text{transmitted power}}{\text{incident power}} = \frac{S_2 p_0^2 |t|^2}{S_1 p_0^2} = \frac{S_2}{S_1}|t|^2 = \frac{4\frac{S_2}{S_1}}{1 + 2\frac{S_2}{S_1} + \left(\frac{S_2}{S_1}\right)^2} . \tag{9.10}$$

The expression for the power transmission coefficient becomes a little clearer if it is multiplied by S_1/S_2:

$$\tau = \frac{4}{\frac{S_2}{S_1} + \frac{S_1}{S_2} + 2} \tag{9.11}$$

The insertion loss R related to an unchanged duct with $S_2 = S_1$ results in

$$R = 10 \lg 1/\tau = 10 \lg \frac{1}{4} \left(\frac{S_2}{S_1} + \frac{S_1}{S_2} + 2 \right) . \tag{9.12}$$

Thus, an expansion of the cross-section has the same effect as a contraction if the ratio of the larger to the smaller cross-section stays the same. This also implies that the insertion loss during excitation 'from the left' or 'from the right' is the same.

Fig. 9.2 shows the characteristics of R versus the ratio of the cross-sections. The achievements are not very large in spite of the considerable effort. For circular ducts with a ratio of 1:2 in the radii and thus a ratio of 1:4 in the cross-sectional areas, only $R = 1.9\,\text{dB}$ is obtained. A ratio of 1:4 in the radii and of 1:16 in the cross-sectional areas is needed to achieve an insertion loss of $R = 6.5\,\text{dB}$, representing a considerable effort. Beginning our discussion

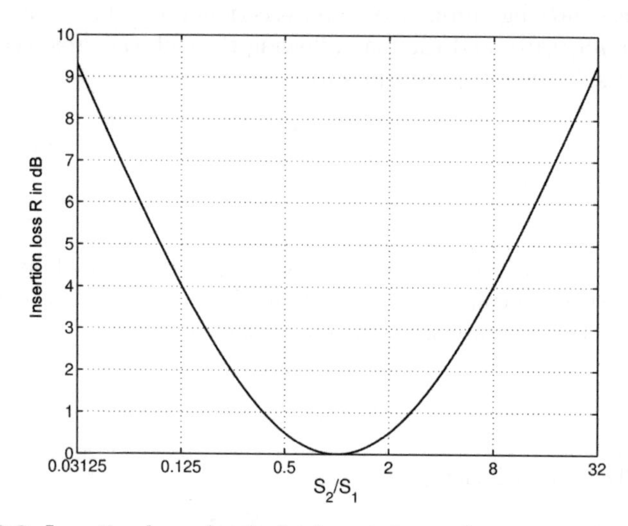

Fig. 9.2. Insertion loss of a single abrupt change in cross-sectional area

in subsequent sections, it should be noted that the reflection coefficient r, according to (9.9), becomes negative for expanding ducts with $S_2 > S_1$ (which corresponds to a soft reflection in principle), whereas it takes positive values for contracting ducts $S_2 < S_1$ (which corresponds to a rigid reflection in principle).

9.1.2 Duct junctions

A single duct branching off from a system of ducts can be regarded as a simple change in cross-sectional area. (Figure (9.3).

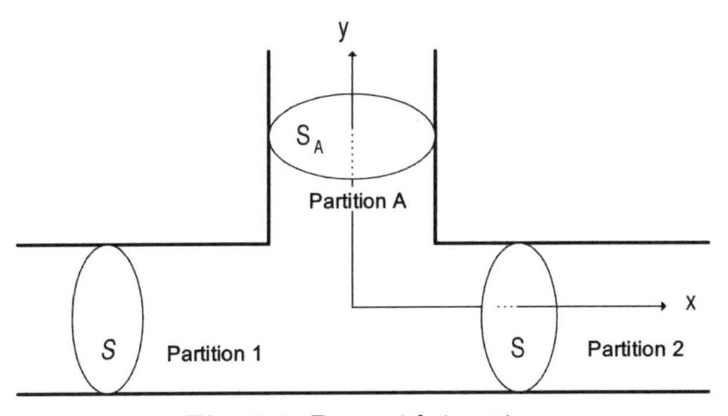

Fig. 9.3. Duct with junction

Tube branches often occur in practice and provide an excellent introduction to the following treatment of lined ducts. There are certainly enough reasons to discuss duct branches in the following sections. To simplify things, assume that there are no abrupt changes in the cross-sectional area of a straight duct. That is, the pipe branches to the left and to the right of the bifurcation all possess the same cross-sectional area S. In the case of lower frequencies where only the basic modes are present in the duct branches, the inlets to the partitions 1 and 2 to the left and to the right of the bifurcation have to be the same as in the previous section

$$p_1 = p_0 \left(e^{-jkx} + re^{jkx}\right) . \tag{9.13}$$

For the particle velocity, using $v = \frac{j}{\omega\varrho}\frac{\partial p}{\partial x}$ and $k = \omega/c$, it follows that

$$v_1 = \frac{p_0}{\varrho c}\left(e^{-jkx} - re^{jkx}\right) . \tag{9.14}$$

Once again, assume that the right duct branch is anechoic, producing solely the transmitted wave

$$p_2 = tp_0 e^{-jkx} \tag{9.15}$$

with

$$v_2 = t\frac{p_0}{\varrho c}e^{-jkx}. \tag{9.16}$$

Consider both anechoically terminated duct segments in the branch along the y-axis (in this case, with input impedance $z_A = p_A(y=0)/v_A(y=0) = \varrho c$) as well as finitely long segments with rigid terminations at $y = l_A$. For the rigid termination at $y = l_A$, the input impedance of the branch (see Chapter 6) is $z_A = p_A(y=0)/v_A(y=0) = -j\varrho c\, ctg(kl_A)$. In other words, in this case, we assume that the input impedance z_A of the diverging branch is known, whereby both extreme cases are particularly relevant. The two unknowns, the reflection factor r at the inlet and the transmission factor t at the extension, result from transitions occurring at the joints $x = 0$ and $y = 0$. All three pressures must be equal

$$p_1(0) = p_A(0) = p_2(0) . \tag{9.17}$$

Again, the result is

$$1 + r = t . \tag{9.18}$$

Based on the conservation of mass, the volume flowing into the joint $Sv_1(0)$ is equal to the sum of all volume flows out $S_A v_A(0) + Sv_2(0)$:

$$Sv_1(0) = S_A v_A(0) + Sv_2(0) . \tag{9.19}$$

Divided by $Sp_1(0)$, one obtains

$$\frac{v_1(0)}{p_1(0)} = \frac{S_A}{S}\frac{v_A(0)}{p_1(0)} + \frac{v_2(0)}{p_1(0)} = \frac{S_A}{S}\frac{v_A(0)}{p_A(0)} + \frac{v_2(0)}{p_2(0)} , \tag{9.20}$$

using eq.(9.17). The expression $p_A(0)/v_A(0)$ is equal to the input impedance z_A of the duct branch. The ratio $p_2(0)/v_2(0)$ describes the impedance ϱc in the extension 2, meaning

$$\frac{v_1(0)}{p_1(0)} = \frac{1}{\varrho c}\frac{1-r}{1+r} = \frac{S_A}{S}\frac{1}{z_A} + \frac{1}{\varrho c} , \tag{9.21}$$

or when applying eq.(9.18),

$$\frac{2-t}{t} = \frac{S_A}{S}\frac{\varrho c}{z_A} + 1 . \tag{9.22}$$

Finally, we obtain the transmission factor t

$$t = \frac{1}{1 + \frac{1}{2}\frac{S_A}{S}\frac{1}{z_A/\varrho c}} . \tag{9.23}$$

The following section mainly deals with the cases of primary interest that were described above: a duct junction where one branch is in itself anechoically terminated, and the other one which has a dead-end termination.

Junction of anechoically-terminated ducts

As mentioned previously, if the branch with a cross-sectional area of S_A is anechoically terminated at its end, the input impedance of the branch z_A is, , $z_A = \varrho c$. So, according to (9.23), the transmission factor t only depends on the surface area ratio S_A/S

$$t = \frac{1}{1 + \frac{1}{2}\frac{S_A}{S}} . \tag{9.24}$$

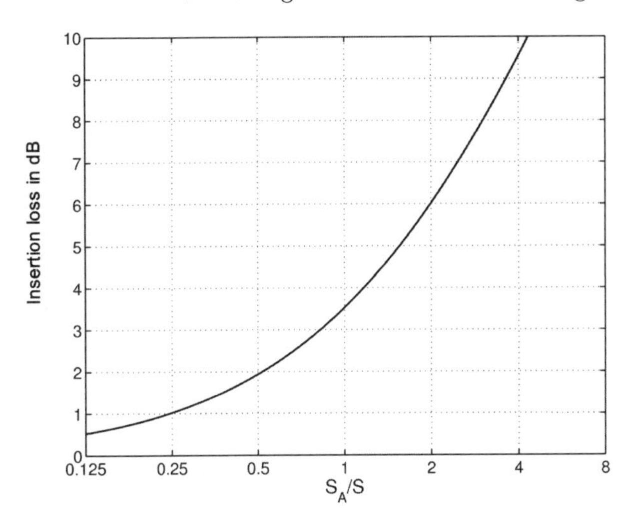

Fig. 9.4. Insertion loss of the junction

If all three duct branches have an equal cross-sectional area of $S_A = S$, the sound fields in the extension of the straight duct are the same as in the branch. The fact that the sound field branches off is simply the result of conservation principles , completely uninfluenced by the angles between the branch, its extension and the inlet duct. For this reason, the waves dispersed after the junction are identical. $t_A = t = 2/3$ for $S_A = S$, the insertion loss is therefore $R = 10lg(1/t^2) = 3,5\,dB$ (t_A = transmission loss of the branch).

The insertion loss resulting in the other surface ratios is shown in 9.4. Take into consideration, however, that for larger surface area ratios S_A/S, the condition that no higher modes do not exist, even at lower frequencies, no longer needs to be met. In that case, the general theory derived here would no longer apply.

Junction of duct segments with reflective terminations

In the case of duct junctions to a segment with rigid termination at a length of l_A, the input impedance of this dead-end termination (see Chapter 6.5) is

$$\frac{z_A}{\varrho c} = -j\, ctg(kl_A). \tag{9.25}$$

This describes the impedance frequency response of what is known as the '$\lambda/4$ resonator,' which shows roots for

$$kl_A = \pi/2 + n\pi$$

(n=0,1,2,...), or the equivalent,

$$l_A = \lambda/4 + n\lambda/2. \tag{9.26}$$

These roots indicate resonances. As a rule, impedance values of $z = 0$ signal that the structure thus described is 'easy to oscillate' to any extent. The transmission factor of the duct extension

$$t = \frac{1}{1 - j\frac{1}{2}\frac{S_A}{S}\frac{1}{ctg(kl_A)}} \tag{9.27}$$

is also zero in the resonance frequencies indicated by (9.26). This implicates a complete blocking of the resonances in the diverging duct extension! In contrast, the impedance z_A in the 'anti-resonances' between two resonances grows infinitely large. This is why, in theory, the duct branch can in theory be omitted and replaced by a rigid surface whose transmission factor is $t = 1$. The bandwidth of the valve ratio around the resonance frequencies depends on the relationship between the cross-sectional areas S_A/S, as can be seen in Figure (9.5). That the duct branch blocks all of its resonances lends itself to

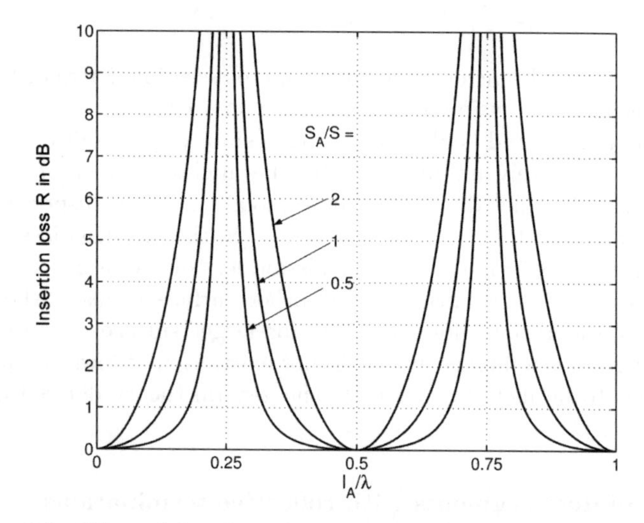

Fig. 9.5. Effect of duct branches as exemplified in a $\lambda/4$ resonator

the fact that cross-sectional dependencies in the sound pressure along the x axis of the deviating branch were cut off to begin with. The following section on lined ducts, which can also be used to produce a wall impedance of zero, will proceed to examine the spatial dependency of the sound field, which has been left out of the discussion so far. These space-dependencies result in high, but finite insertion losses.

This section concludes with a final short mention of two intersecting ducts (Figure 9.6). We will assume, for the sake of simplicity that all duct branches are rigidly terminated and all cross-sectional areas are equal. In such cases, identical wave fields with identical transmission factors t develop in all three branches. Because all pressures are the same at the intersecting point,

$$1 + r = t \tag{9.28}$$

applies, where r indicates the reflection factor in the inlet duct. The divergence of the volume flow requires that

$$1 - r = 3t\,m \tag{9.29}$$

resulting in $t = 1/2$. The insertion loss at the duct intersection is obviously only $6\,dB$.

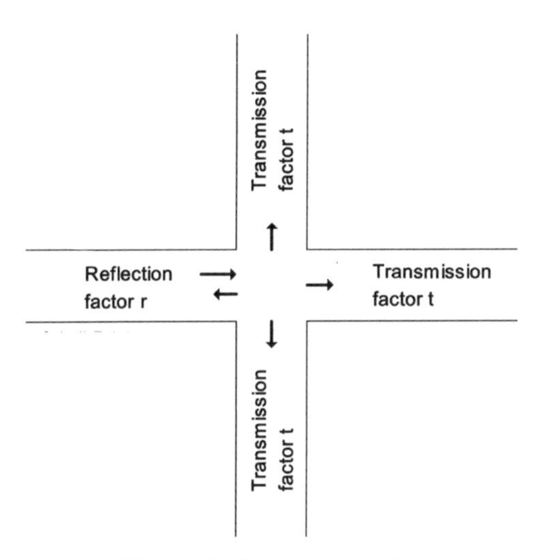

Fig. 9.6. Duct intersection

9.1.3 Expansion chambers

Simple abrupt changes in the cross-sectional area are not very effective. With the simple combination of two reflectors, forming a chamber of length l (Fig. 9.7), the effect can be increased substantially, as will be shown in the following discussion.

The sound field at the inlet branch is likewise composed of travelling and reflecting waves:

$$p_1 = p_0 \left(e^{-jkx} + r e^{jkx}\right) . \tag{9.30}$$

Thus follows that for the sound propagation speed with $v = \frac{j}{\omega\varrho}\frac{\partial p}{\partial x}$ and $k = \omega/c$,

$$v_1 = \frac{p_0}{\varrho\, c}\left(e^{-jkx} - r e^{jkx}\right) . \tag{9.31}$$

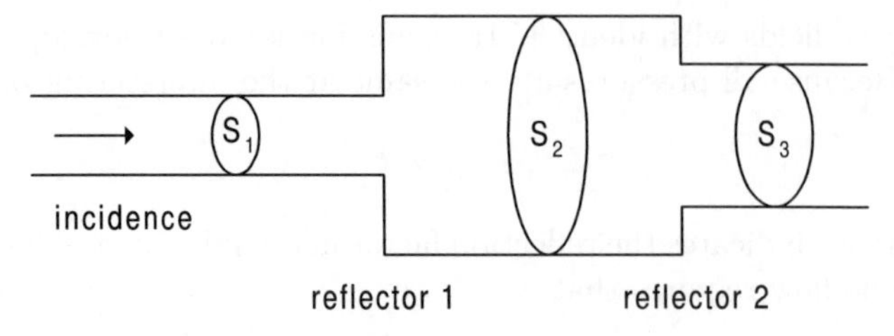

Fig. 9.7. Expansion chamber

Of course a similar approach using waves travelling in opposite directions could be made for the expansion chamber. Yet, the following calculation is somewhat easier if a linear combination

$$p_2 = p_0 \left(\alpha \sin kx + \beta \cos kx\right) \tag{9.32}$$

is used, with

$$v_2 = \frac{jp_0}{\varrho c} \left(\alpha \cos kx - \beta \sin kx\right) \tag{9.33}$$

as a solution to the wave equation.

The outlet 3 represents, once again, an anechoic termination. Therefore, only the transmitted wave

$$p_3 = p_0 t \mathrm{e}^{-jk(x-l)} \tag{9.34}$$

exists, with

$$v_3 = \frac{p_0}{\varrho c} t \mathrm{e}^{-jk(x-l)} \ . \tag{9.35}$$

The boundary conditions $p_1 = p_2$ and $S_1 v_1 = S_2 v_2$ at the reflector $x = 0$ yield

$$1 + r = \beta \tag{9.36}$$

and

$$S_1 \left(1 - r\right) = j\alpha S_2 \ . \tag{9.37}$$

Using $p_2 = p_3$ and $S_2 v_2 = S_3 v_3$ at $x = l$ results in

$$\alpha \sin kl + \beta \cos kl = t \tag{9.38}$$

and

$$S_2 \left(\alpha \cos kl - \beta \sin kl\right) = -jtS_3 \ . \tag{9.39}$$

Equations (9.36) to (9.39) represent a set of linear equations with the four unknown variables α, β, t and r. Since mainly the transmission loss is of interest, (9.38) and (9.39) are solved for α and β

$$\alpha = t \left[\sin kl - j\frac{S_3}{S_2} \cos kl\right] \tag{9.40}$$

$$\beta = t \left[\cos kl + j\frac{S_3}{S_2} \sin kl\right] \ . \tag{9.41}$$

The reflection coefficient is eliminated, using (9.36) and (9.37):

$$\beta + j\alpha \frac{S_2}{S_1} = 2 \,. \tag{9.42}$$

Inserting (9.40) and (9.41) into (9.42) finally yields the pressure transmission coefficient t

$$t\left[\cos kl + j\frac{S_3}{S_2}\sin kl + j\frac{S_2}{S_1}\left(\sin kl - j\frac{S_3}{S_2}\cos kl\right)\right] = 2$$

or

$$t = \frac{2}{\cos kl\left(1+\frac{S_3}{S_1}\right) + j\sin kl\left(\frac{S_3}{S_2}+\frac{S_2}{S_1}\right)} \,, \tag{9.43}$$

respectively. The power transmission coefficient τ is the ratio of transmitted and incident sound power, and therefore given by

$$\tau = \frac{S_3}{S_1}|t|^2 = \frac{4\frac{S_3}{S_1}}{\cos^2 kl\left(1+\frac{S_3}{S_1}\right)^2 + \sin^2 kl\left(\frac{S_3}{S_2}+\frac{S_2}{S_1}\right)^2} \,. \tag{9.44}$$

For a simple interpretation it is suggested to assume that the cross-sectional area of inlet and outlet are equal using $S_3 = S_1$. This results in

$$\tau = \frac{4}{4\cos^2 kl + \sin^2 kl\left(\frac{S_1}{S_2}+\frac{S_2}{S_1}\right)^2} \,.$$

This expression can still be simplified even more:

$$\tau = \frac{4}{4+\sin^2 kl\left\{\left(\frac{S_1}{S_2}+\frac{S_2}{S_1}\right)^2 - 4\right\}} = \frac{1}{1+\frac{1}{4}\left(\frac{S_1}{S_2}-\frac{S_2}{S_1}\right)^2 \sin^2\, kl} \,. \tag{9.45}$$

Thus, the insertion loss amounts to

$$R = 10\lg 1/\tau = 10\lg\left\{1+\frac{1}{4}\left(\frac{S_1}{S_2}-\frac{S_2}{S_1}\right)^2 \sin^2 kl\right\} \,. \tag{9.46}$$

The insertion loss of the expansion chamber is frequency-dependent – in contrast to the abrupt change in cross-sectional area (see also Figs. 9.8 and 9.9a,b,c). It has a periodic shape, where alternating minima and maxima

$$R = R_{\min} = 0\,\text{dB} \quad \text{for} \quad kl = 0, \pi, 2\pi, \ldots$$

and

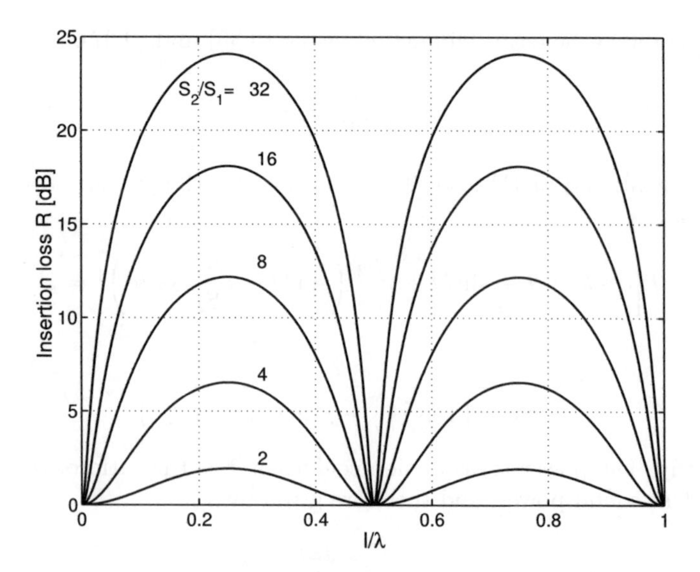

Fig. 9.8. Insertion loss of an expansion chamber for different ratios of cross-sectional areas

$$R = R_{\max} = 10\lg\left\{1+\frac{1}{4}\left(\frac{S_1}{S_2}-\frac{S_2}{S_1}\right)^2\right\} \quad \text{for} \quad kl = \frac{\pi}{2}, \frac{3\,\pi}{2}, \frac{5\,\pi}{2}, \ldots \tag{9.47}$$

occur. The maxima take on higher values this time. For example, $R_{\max} = 6.5\,\text{dB}$ for $S_2/S_1 = 4$ (simple change in cross-section: $R = 1.9\,\text{dB}$) and $R_{\max} = 18.1\,\text{dB}$ for $S_2/S_1 = 16$ (simple change in cross-section: $R = 6.5\,\text{dB}$).

The frequencies f_n, where the maxima $R = R_{\max}$ are located, are given by

$$f_n = \frac{1}{4}\frac{c}{l}, \frac{3}{4}\frac{c}{l}, \frac{5}{4}\frac{c}{l}, \ldots \tag{9.48}$$

the corresponding wavelengths $\lambda = c/f_n$ are given by

$$l = \frac{1}{4}\lambda, \frac{3}{4}\lambda, \frac{5}{4}\lambda, \ldots . \tag{9.49}$$

At the maxima the length of the expansion chamber is an uneven multiple of a quarter wavelength. It can be graphically shown that the frequencies with 'maximum effect' result from the condition 'length = $\lambda/4 + n\lambda/2$'. The maximum effect can be expected if the wave which is reflected at the outlet of the chamber and reflected again at the chamber inlet (thus already reflected twice) is *out of phase* with the actually incident wave at the chamber inlet. Since the travelled distance of the double-reflected wave is $2l$ and an opposite phase constitutes a shift of $\lambda/2 + n\lambda$, it likewise follows from this discussion that $l = \lambda/4 + n\lambda/2$. Overall, the expansion chamber represents a silencer which can be tuned in frequency by the chamber length and where the maximum insertion loss can be influenced by the ratio of the cross-sectional areas. But the 'fairly good' effect of the silencer is limited to certain frequency intervals. Usually, the first low-frequency maximum $n = 0$ is mainly of interest in practice. The effective bandwidth can be described by the distance $\Delta f = f_{1+} - f_{1-}$ of the two frequencies f_{1-} and f_{1+} which are located to the left and to the right of the maximum, where $R = R_{\max} - 3\,\text{dB}$. Apart from the minima $R = 0\,\text{dB}$ and for ratios S_2/S_1 of the cross-sections which are not too close to the value $S_2/S_1 = 1$, the insertion loss can be approximated by

$$R \approx 10 \lg \left\{ \sin^2 kl \left(\frac{S_1}{S_2} - \frac{S_2}{S_1} \right)^2 \right\} - 6\,\text{dB} .$$

f_{1+} and f_{1-} are thus given by

$$\sin^2 \left(\frac{2\pi\, f_{1\pm} l}{c} \right) = \frac{1}{2}$$

or, of course,

$$\frac{2\pi\, f_{1+} l}{c} = \frac{3\pi}{4} \quad \text{and} \quad \frac{2\pi\, f_{1-} l}{c} = \frac{\pi}{4} .$$

It is therefore

$$\Delta f = f_{1+} - f_{1-} = \left(\frac{3}{8} - \frac{1}{8} \right) \frac{c}{l} = \frac{1}{4} \frac{c}{l} = f_1 . \tag{9.50}$$

The 3-dB bandwidth is then equal to the center frequency f_1 (frequency of the first maximum).

Figs. 9.9a,b,c show that theory and experiment correlate fairly well.

The small shift in frequency between the calculated and measured curves can be explained by the fact that the 'acoustic length' of expansion chambers is slightly smaller than the geometric length (the end correction would extend the in- and outlets and thus reduce the chamber length).

The following example shows that a fairly good effect can be achieved by a 'silencer without absorption'. Implementing a duct diameter of 5 cm and a realistic chamber diameter of 20 cm with a chamber length of 25 cm can achieve an insertion loss of 18 dB at 340 Hz, which decreases to 15 dB at 170 Hz and 510 Hz. The same characteristics are also found in the range of 850 Hz to 1190 Hz (center frequency: 1020 Hz).

It is obvious that the gap in the frequency band around 680 Hz can be closed by a second chamber, as depicted in Fig. 9.10. The next section is therefore dedicated to the combining chambers.

9.1.4 Chamber combinations

A silencer is consists of N pipe elements with different cross-sectional areas S_i, $i = 1, 2, \ldots N$ and one inlet and one outlet pipe. Naturally, the theoretical discussion consists of two steps:

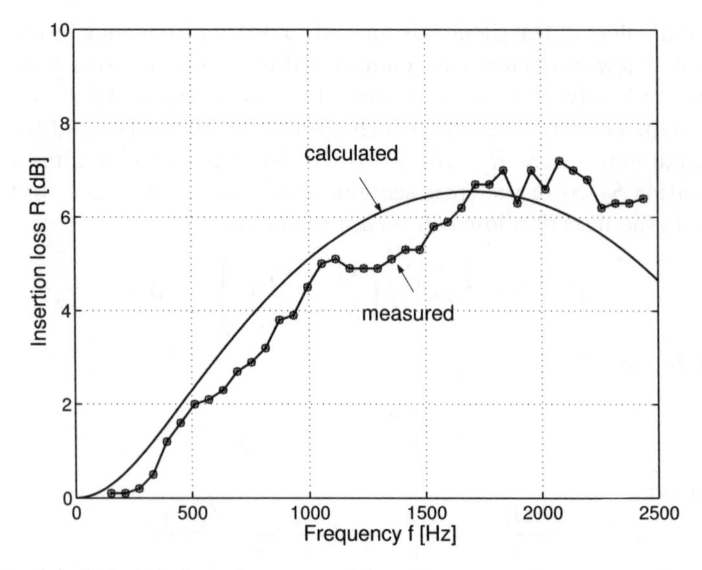

Fig. 9.9. (**a**) Calculated and measured insertion loss of an expansion chamber $S_2/S_1 = 4$, $l = 5\,cm$

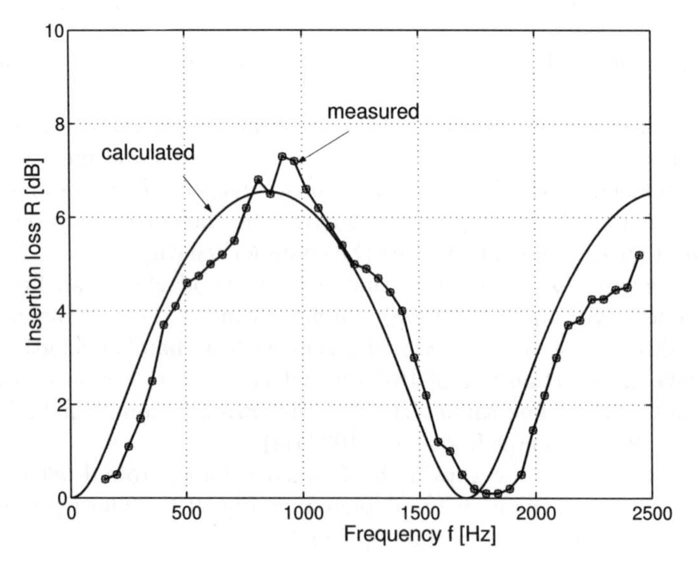

Fig. 9.9. (**b**) Calculated and measured insertion loss of an expansion chamber $S_2/S_1 = 4$, $l = 10\,cm$

1. The transmission at an abrupt change in cross-sectional area and
2. The transmission along a duct element with a constant cross-section

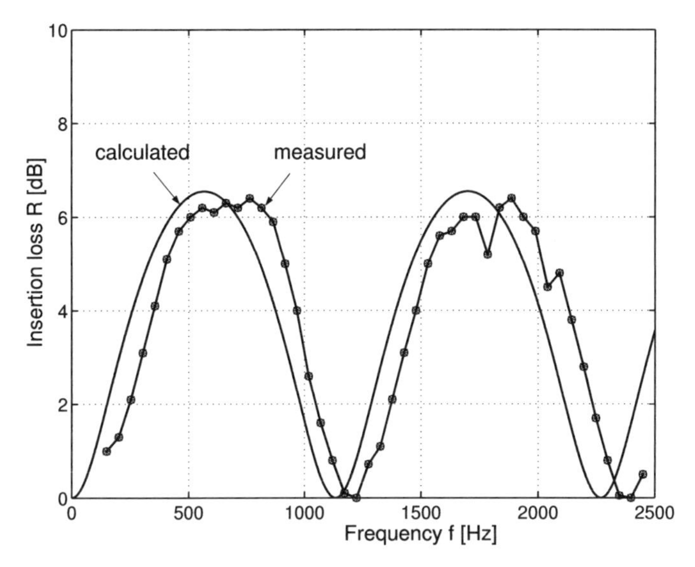

Fig. 9.9. (**c**) Calculated and measured insertion loss of an expansion chamber $S_2/S_1 = 4$, $l = 15\,cm$

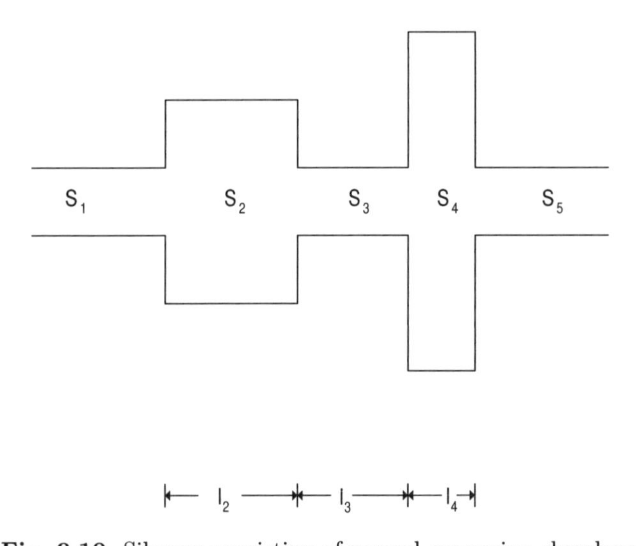

Fig. 9.10. Silencer consisting of several expansion chambers

A coordinate x_i is associated with each duct element i, where $x_i = 0$ defines the inlet and $x_i = l_i$ defines the outlet (see Fig. 9.11). Sound pressure and velocity at the inlet are thus denoted by $p_i(0)$ and $v_i(0)$ and by $p_i(l_i)$ and $v_i(l_i)$ at the outlet.

The outlet pipe of the silencer (the element $N + 1$ which is semi-infinite) is assumed to be non-reflecting. Its impedance is therefore given by

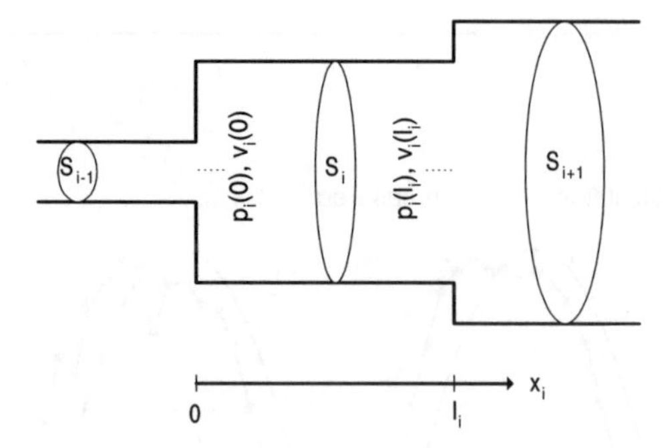

Fig. 9.11. Definitions for the description of transmission along a tube segment

$$z_{N+1}(0) = \frac{p_{N+1}(0)}{v_{N+1}(0)} = \varrho c \,. \tag{9.51}$$

With the aid of the aforementioned boundary conditions at a change in cross-sectional area,

$$p_i(l_i) = p_{i+1}(0)$$

and

$$S_i v_i(l_i) = S_{i+1} v_{i+1}(0) \;,$$

the impedance at the end of a duct element with a given order can be determined by the impedance of the consecutive element as

$$z_i(l_i) = \frac{p_i(l_i)}{v_i(l_i)} = \frac{S_i}{S_{i+1}} \frac{p_{i+1}(0)}{v_{i+1}(0)} = \frac{S_i}{S_{i+1}} z_{i+1}(0) \;. \tag{9.52}$$

Starting with the last outlet, as exhibited by (9.51), we can calculate the final impedance $z_N(l_N)$ in the last duct element with the aid of (9.52). The next step consists of calculating the ratio of the field quantities $z = p/v$ at the inlet of the duct element $z_N(0)$ based on the ratio at the end $z_N(l_N)$. The procedure is repeated thereafter, using (9.52) to calculate $z_{N-1}(l_{N-1})$ from $z_N(0)$, and $z_{N-1}(0)$ from $z_{N-1}(l_{N-1})$, until the inlet of the duct is reached.

The last step is ot calculate the correlation between $z_i(0)$ and $z_i(l_i)$. This is done using the approach

$$p_i(x_i) = \alpha_i \cos k(x_i - l_i) + \beta_i \sin k(x_i - l_i) \;. \tag{9.53}$$

By Inserting $x_i = l_i$ shows that α_i is simply the sound pressure at l_i, $\alpha_i = p_i(l_i)$, and, due to

$$v_i(x_i) = \frac{j}{\omega\varrho} \frac{\partial p_i}{\partial x_i} = -\frac{j}{\varrho c} \left\{ \alpha_i \sin k(x_i - l_i) - \beta_i \cos k(x_i - l_i) \right\} \;, \tag{9.54}$$

β_i is proportional to the velocity in $x_i = l_i$

$$\frac{j\beta_i}{\varrho c} = v_i\left(l_i\right) \ .$$

At each arbitrary position x_i in the ith duct element, sound pressure and velocity, $p_i(x_i)$ and $v_i(x_i)$ can completely be expressed by the field quantities at the end of the duct element $x_i = l_i$

$$p_i\left(x_i\right) = p_i\left(l_i\right)\cos k\left(x_i - l_i\right) - j\varrho c v_i\left(l_i\right)\sin k\left(x_i - l_i\right)$$

and

$$v_i\left(x_i\right) = -\frac{j}{\varrho c} p_i\left(l_i\right)\sin k\left(x_i - l_i\right) + v_i\left(l_i\right)\cos k\left(x_i - l_i\right) \ .$$

The impedance at the inlet of the duct element $z_i(0) = p_i(0)/v_i(0)$ can now easily be calculated by the impedance at the end of the duct element $z_i(l_i) = p_i(l_i)/v_i(l_i)$:

$$\frac{z_i\left(0\right)}{\varrho c} = \frac{\frac{z_i(l_i)}{\varrho c}\cos kl_i + j\sin kl_i}{\frac{jz_i(l_i)}{\varrho c}\sin kl_i + \cos kl_i} \tag{9.55}$$

As already mentioned, one 'calculates one's way' along the silencer 'from the back to the front'. By alternately applying (9.52) and (9.55), one finally ends up at the inlet of the first duct element $z_1(0)$. The final step is given by the basic approach for the (semi-infinite) inlet duct

$$p_{\text{E}} = p_0\left(e^{-jkx_1} + r_E e^{jkx_1}\right) \tag{9.56}$$

$$v_{\text{E}} = \frac{p_0}{\varrho c}\left(e^{-jkx_1} - r_E e^{jkx_1}\right), \tag{9.57}$$

where the same coordinate system as in the first duct element 1, x_1 is used. As usual, the x_1-axis meets the end of the inlet duct at $x_1 = 0$. It is

$$\frac{z_{\text{E}}\left(0\right)}{\varrho c} = \frac{1 + r_E}{1 - r_E} = \frac{S_o}{S_1}\frac{z_1\left(0\right)}{\varrho c} \tag{9.58}$$

or

$$r_{\text{E}} = \frac{\frac{S_o}{S_1}\frac{z_1(0)}{\varrho c} - 1}{\frac{S_o}{S_1}\frac{z_1(0)}{\varrho c} + 1} \ . \tag{9.59}$$

The energy balance equation requires

$$\tau = 1 - \left|r_{\text{E}}\right|^2 \ . \tag{9.60}$$

The insertion loss is, once again, $R = -10\lg\tau$. Calculated examples are shown in Fig. 9.12 which targets a broadband effect. Utilizing the above procedure, ducts that exhibit an arbitrary axial change in the cross-sectional area can be calculated as shown in Fig. 9.13. $S = S(x)$ The continuous characteristics of $S(x)$ are decomposed into multiple small 'steps' (see Fig. 9.13). The theoretical prediction as well as the practical measurement of such 'structured tubes' show

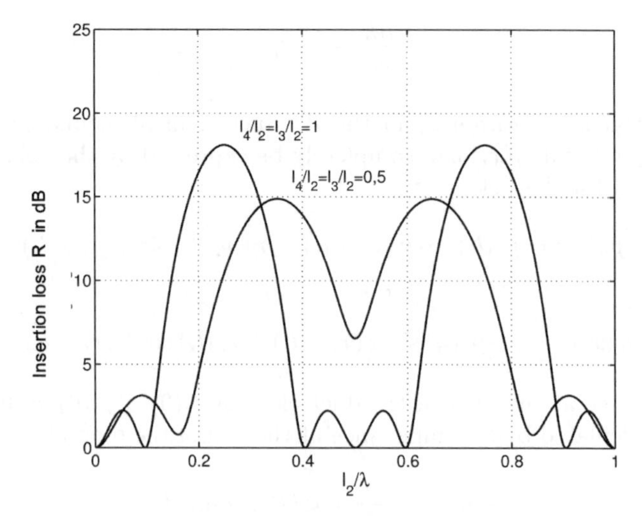

Fig. 9.12. Effect of a silencer consisting of three duct elements as depicted in Fig. 9.10 using $S_1 = S_3 = S_5$ and $S_2 = S_4 = 4S_1$

that a pretty good effect can be achieved at ratios of the cross-sectional areas pretty close to unity. Fig. 9.14b shows the result based on a sample. Fig. 9.14a shows the theoretical prediction of the step function $S(x)$ shown in Fig. 9.13.

Only in rare cases can the fairly large but very narrow-banded insertion loss actually be used in practise. It pays off to only use pure tones, whose frequency is not subject to changes (such as those caused by the speed of an engine, for instance). The frequency with the maximum effect results from $\lambda_\mathrm{w} = \lambda/2$ (wall period=air wavelength/2), so that only high, sometimes mid-frequencies can be considered for the sake of economy.

9.2 Lined ducts

This section deals with silencers which are realized by lining the duct wall with an appropriately chosen impedance. In practice, usually a wider duct has to be split into several smaller ducts, as shown in Fig. 9.15. In ventilation systems, for instance, large amounts of fresh air need to be transported. Since, on the other hand, only ducts with a small distance between the limiting boundaries provide large attenuation, constructions like the splitter attenuator (Fig. 9.15) are often required.

The basic principle of such attenuator constructions shall be explained here. The two-dimensional duct is treated as the simplest possible model ($\partial/\partial z = 0$ in Fig. 9.18), where the wall at $y = 0$ is a rigid plate and the parallel wall at $y = h$ is given by the impedance in air.

Before discussing the model configuration it is reasonable to recapitulate the information outlined in the previous Chap. 6 on the simplest special case –

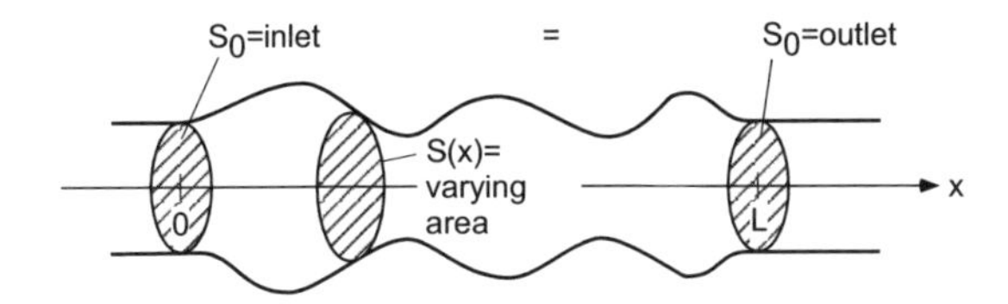

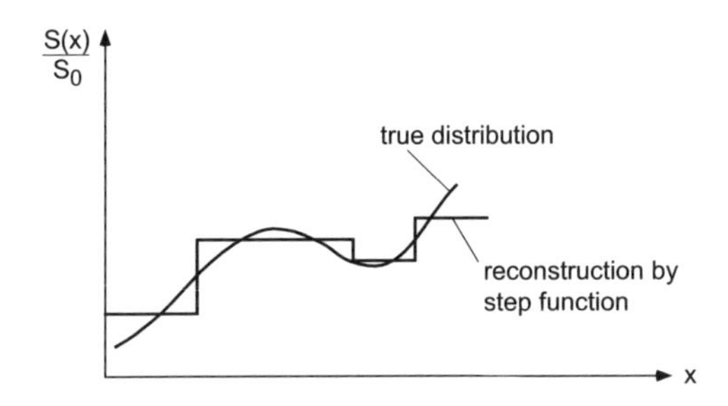

Fig. 9.13. Duct with variable cross-section (top) and reconstruction of the area function by a step function (bottom)

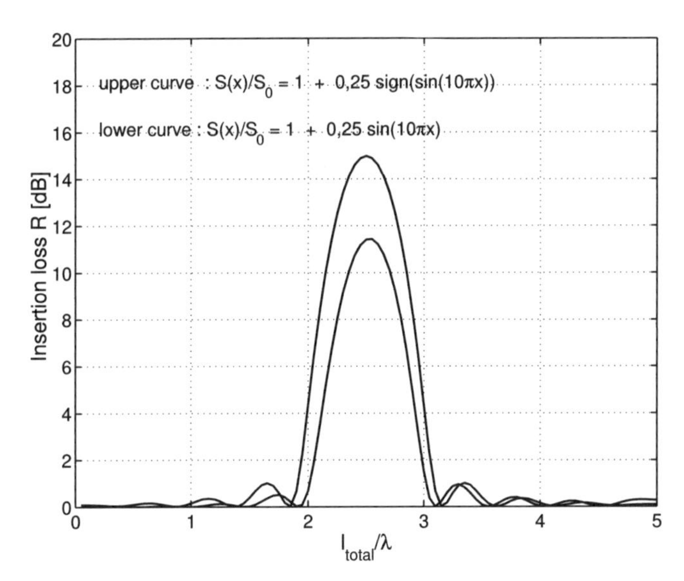

Fig. 9.14. (**a**) Calculated insertion loss of a structured tube for $\varepsilon = 0.25$. Total length: $5\lambda_{\mathrm{w}}$

the rigid-walled duct – and to augment the knowledge gained in the previous chapter with respect to the attenuation that occurs. After that, the 'new' simplest case is treated: the duct lining at $y = h$ with a soft impedance $z = 0$. As will be seen, this would represent a highly effective silencer. Yet,

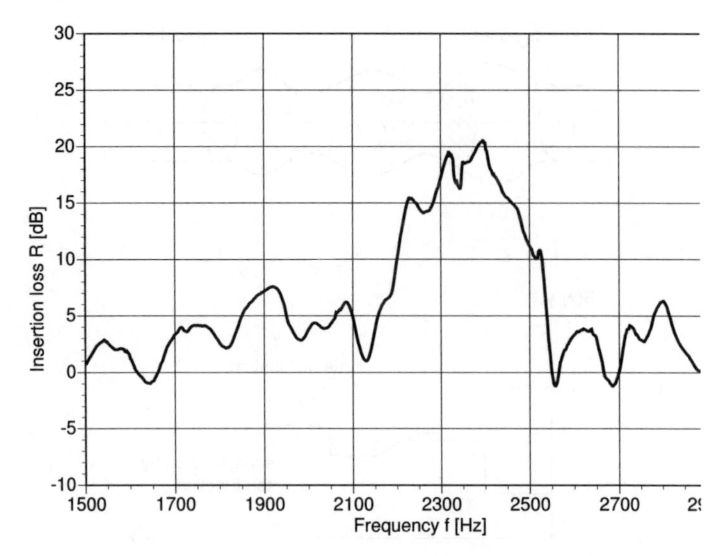

Fig. 9.14. (b) Measured insertion loss of a structured tube for $\varepsilon = 0.25$. Total length: $5\lambda_{\mathrm{w}}$

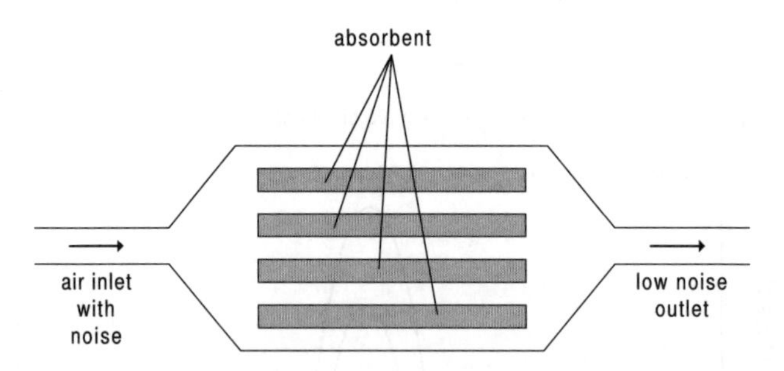

Fig. 9.15. Sketch of a splitter attenuator

the realization of $z = 0$ is difficult indeed. Nevertheless, it is worth discussing it here, because it clarifies some basic principles. Afterward, we will conduct an approximation for general impedances, which will finally be compared to the exact solution of the problem.

9.2.1 Ducts with rigid walls

The duct with rigid walls was already discussed in Chap. 6. It is therefore sufficient to recapitulate the results which are summarized in (6.3) and (6.4) (see p. 173). The sound pressure is composed of cosine-shaped transverse distributions – the modes. Each mode has a certain wave number k_x in the propagation direction along the duct axis, which specifies its propagation. The sound pressure is given by

$$p = \sum_{n=0}^{\infty} p_n \cos \frac{n\pi y}{h} \mathrm{e}^{-jk_x x} \, . \tag{9.61}$$

The modal amplitudes p_n must be determined by the sound source (if they are of interest). This requires a detailed knowledge of the source (which is often not available). Each mode has a different wave number k_x along the propagation direction, which is calculated by inserting the members of (9.61) into the wave equation (see also (6.4), p. 174). Combined with the fact that a non-reflecting sound field always decays exponentially with increasing distance from the source, the resulting equations for the modal wave numbers come out to be:

$$k_x = \begin{cases} +\sqrt{k^2 - \left(\frac{n\pi}{h}\right)^2} & ; |k| \geq \frac{n\pi}{h} \\ -j\sqrt{\left(\frac{n\pi}{h}\right)^2 - k^2} & ; |k| \leq \frac{n\pi}{h} \end{cases} \tag{9.62}$$

The mode $n = 0$ is characterized by a purely real wave number $k_x = k_0$. The lowest mode $n = 0$ – the plane wave – is basically transmitted without attenuation. Generally, the imaginary part of a wave number (regardless of its nature) describes the damping of the characterized waveform. By decomposing it into a real and an imaginary part

$$k_x = k_\mathrm{r} - jk_\mathrm{i}, \tag{9.63}$$

the field of this wave form behaves in the manner of

$$p \sim \mathrm{e}^{-jk_x x} = \mathrm{e}^{-jk_\mathrm{r} x} \mathrm{e}^{-k_\mathrm{i} x} \, . \tag{9.64}$$

The corresponding level distribution

$$L = 10 \lg |p|^2 \sim 10 \lg \mathrm{e}^{-2k_\mathrm{i} x} = -k_\mathrm{i} x \, 20 \lg \mathrm{e} = -8.7 k_\mathrm{i} x \tag{9.65}$$

falls linearly with x. Usually, the damping is characterized by the level difference D_h along an element of the x-axis whose length is equal to the duct's cross-section. The modal damping, which is also called 'duct attenuation', is generally given by

$$D_\mathrm{h} = 8.7 k_i h \, . \tag{9.67}$$

In the rigid-walled duct, the modes $n > 0$ are damped by

$$D_\mathrm{h} = 8.7 \sqrt{(n\pi)^2 - (kh)^2} \tag{9.68}$$

(for frequencies below the cut-off frequency, see Chap. 6). Therefore, at low frequencies $kh \ll n$, the duct attenuation is approximately given by

$$D_\mathrm{h} \approx 8.7 n\pi = 27.3 n \tag{9.69}$$

which indicates very high modal damping. The mode $n = 1$ is attenuated by 27.3 dB ($n = 2$: 54.6 dB) per duct width along the duct. Therefore, after

4 duct widths away from the source, this mode is reduced by more than 100 dB ($n = 2$: 200 dB)!

A practical application is realized in the section on the measurement technique the impedance tube, where higher modes interfere, while the lowest mode $n = 0$ is desirable. When implementing them in silencers, the rigid wall would only be useful if the source does not excite the lowest mode $n = 0$. This would require specialized source configurations which would probably play little role in noise control applications. General statements, quoting the 'feasibility' for *each* source, can only be made using the 'worst case' which is given by the mode of lowest damping. Thus, here and in the following sections, we consider the lowest modal damping D_{h} of the modes. For ducts with rigid walls, the lowest damping is simply given by $D_{\mathrm{h}} = 0\,\mathrm{dB}$. Such ducts are certainly not very useful as silencers.

9.2.2 Ducts with soft boundaries

As will be shown in the following, a simple soft lining with the boundary condition

$$p(y = h) = 0 \tag{9.70}$$

achieves a highly effective silencer.

The transverse modes are again approached by cosine distributions due to the boundary condition $\partial p/\partial y = 0$ at $y = 0$:

$$p \sim \cos qy \tag{9.71}$$

Based on the boundary condition (9.70) at the impedance plane $y = h$, the eigenvalues q are now

$$qh = \pi/2 + n\pi \quad \text{for} \quad n = 0, 1, 2, 3, \ldots \,. \tag{9.72}$$

The sound field is therefore

$$p = \sum_{n=0}^{\infty} p_n \cos((\pi/2 + n\pi)\frac{y}{h})\mathrm{e}^{-jk_x x} \,. \tag{9.73}$$

The transverse modes of the form $\cos((\pi/2 + n\pi)\frac{y}{h})$ are shown in Figure 9.16.

The modal wave numbers k_x result from the wave equation

$$k_x = \begin{cases} +\sqrt{k^2 - \left(\frac{(n+1/2)\pi}{h}\right)^2} & ; |k| \geq \frac{(n+1/2)\pi}{h} \\ -j\sqrt{\left(\frac{(n+1/2)\pi}{h}\right)^2 - k^2} & ; |k| \leq \frac{(n+1/2)\pi}{h} \end{cases} \,. \tag{9.74}$$

In ducts with soft boundaries, all modes have thus a non-zero cut-on frequency given by

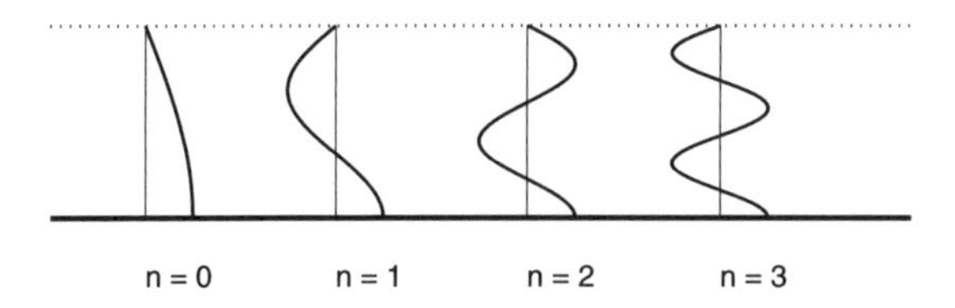

Fig. 9.16. Sound pressure transverse modes in a two-dimensional insulation-lined duct, above $z = 0$ and $z \to \infty$ below.

$$f_n = \frac{n + 1/2}{2h} \, . \tag{9.75}$$

After all, the lowest cut-on frequency is 1700 Hz for a duct width of 5 cm. At frequencies below that, the modes do not propagate. All transverse distributions represent exponentially decaying near-fields. This is also now valid for the mode $n = 0$ with the lowest cut-on frequency. At low frequencies $kh \ll \lambda/2$, it is approximately

$$D_{\mathrm{h}} = 8.7(n + 1/2)\pi = 13.7 + 27.3n, \tag{9.76}$$

according to (9.67) and (9.74).

In the worst case, at n=0, the total sound field at low frequencies falls at 13.7 dB per duct width along the duct. For practical applications, this is an exceedingly high value which is only slightly below the maximum possible value of $D_{\mathrm{h,max}} = 19.1$ dB (see the section after the next for more information).

It would therefore be highly desirable for practical applications to realize a soft surface in a frequency range as broad as possible. However, it is not easy to create a surface with the impedance $z = 0$ after all. As it was explained in Sect. 6.5.4 on resonance absorbers, this can only be realized by an undamped structure in the shape of a resonator, which has the impedance

$$z = j\omega m'' - s''/j\omega$$

with a zero-crossing $z = 0$ at the resonance frequency $\omega_{\mathrm{res}} = \sqrt{s''/m''}$. In contrast, finite impedance values are obtained at frequencies below or above the resonance frequency with stiffness characteristics ($\mathrm{Im}\{z\} < 0$) or with mass characteristics ($\mathrm{Im}\{z\} > 0$). Ultimately, the question of practical interest is, which corresponding damping D_{h} can be obtained by a duct whose walls are lined by resonators and how broad in frequency this effect is.

For frequencies f above the lowest cut-on frequency $f > f_0$ the mode $n = 0$ changes to an non-attenuated wave. If the frequency is still below the cut-on frequency of the next consecutive mode $n = 1$, at a certain distance to the source, a diffuse transverse distribution of sound pressure levels occurs, as shown in Figure 9.17 (of course, given that this mode is, in fact, excited by the source.) It is clear that the sound field is radiated in the form of an acoustic ray or sound beam. The level decreases at the middle of the beam at $y = 0$ - at the rigid duct wall - in the direction of the soft side. For a duct lined on both sides with soft walls, the maximum of the beam would be located in the middle, between the surfaces of the boundaries. Above the lowest cut-on frequency, a sound beam develops, whereby the beam virtually passes the boundaries' surface without touching them. The transmitted sound power is therefore concentrated at the middle of the channel.

From a principle standpoint, it is worth noting that the soft boundaries do not incur any loss of energy below the lowest cut-on frequency, due to the very high damping in the duct D_h. Naturally, no power penetrates the soft surface $p(y = h) = 0$. In other words, the duct lining does not extract any internal energy from the sound field. Therefore, because of the reflections occurring at the inlet, even ducts with soft linings will work.

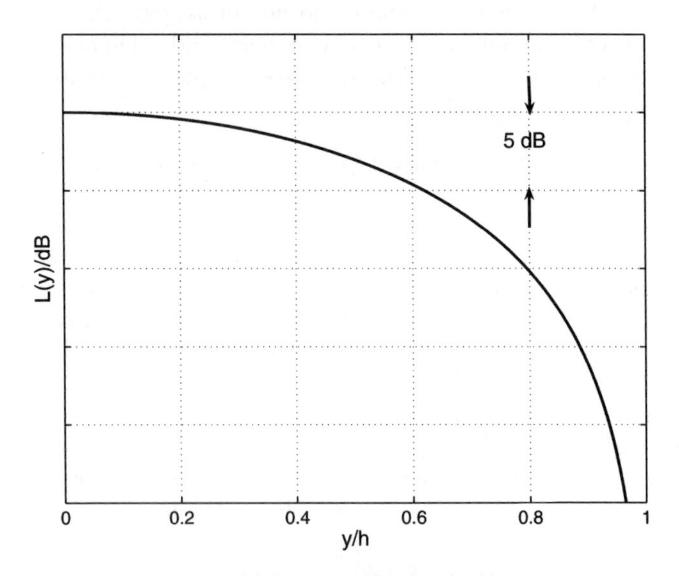

Fig. 9.17. Sound level $L(y) = 10\,lg\,|cos(\pi y/2h)|$ of the lowest mode

9.2.3 Silencers with arbitrary impedance boundaries

This section deals with a two-dimensional attenuator model. As it is shown in Fig. 9.18 it contains a rigid plate at $y = 0$ and a parallel plane at $y = h$ with the impedance z. The field quantities at $y = h$ are thus given by

$$p\,(y = h) = z v_y\,(y = h)\ . \tag{9.77}$$

The beginning of the considerations is an approximation, which shows not only the tendencies of the working principle, but often yields viable results in the order of magnitude. As every approximation it has its limitations; the following section tries to indicate, what lies beyond.

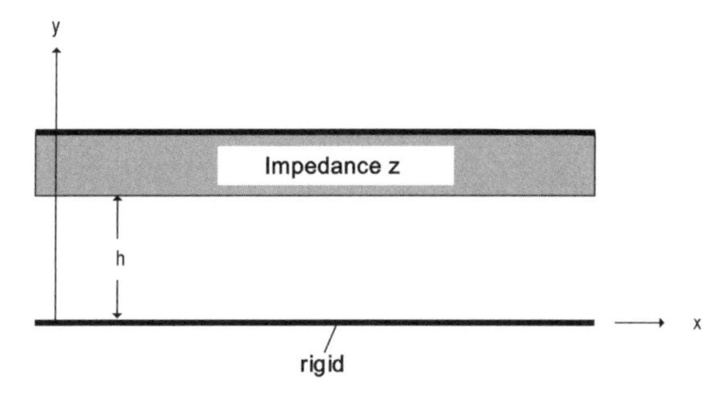

Fig. 9.18. Model arrangement for calculating duct attenuation by impedance bounding with z

Approximation for the lowest mode

The main difference between a rigid boundary and a boundary with a finite impedance is simply given by the fact that a mass flow penetrates the impedance plane in the latter case; for $z \to \infty$ this is impossible. To account for this effect, it makes sense to use the 'acoustic' mass conservation law (2.53) (see p. 39)

$$\frac{\partial v_x}{\partial x} + \frac{\partial v_y}{\partial y} = -\frac{j\omega}{\varrho c^2} p \, . \tag{9.78}$$

For small duct widths $h \ll \lambda$ and for the lowest mode, the differential quotient $\partial v_y/\partial y$ in (9.78) can be replaced by the difference quotient

$$\frac{\partial v_x}{\partial x} + \frac{v_y\,(h) - v_y\,(0)}{h} = -\frac{j\omega}{\varrho c^2} p \, . \tag{9.79}$$

For large impedances z, the transverse pressure distribution will approximately remain constant. Under such conditions, the velocity can be expressed by the pressure and the impedance as

$$v_y\,(h) = p/z \, . \tag{9.80}$$

Using $v_y(0) = 0$, (9.79) results in

$$\frac{\partial v_x}{\partial x} = -\left(\frac{j\omega}{\varrho c^2} + \frac{1}{zh}\right) p \, . \tag{9.81}$$

As already mentioned, p/zh accounts for the mass penetrating the impedance plane (per unit time and per unit area).

The force balance equation

$$v_x = \frac{j}{\omega \varrho} \frac{\partial p}{\partial x} \tag{9.82}$$

is indeed independent of the impedance at the boundary. It can be used to express the velocity in (9.81) in terms of the sound pressure. Differentiating (9.82) with respect to x, one arrives at

$$\frac{j}{\omega\varrho}\frac{\partial^2 p}{\partial x^2} = -\left(\frac{j\omega}{\varrho c^2} + \frac{1}{zh}\right)p,$$

or simplified,

$$\frac{\partial^2 p}{\partial x^2} + \frac{\omega^2}{c^2}\left(1 - j\frac{1}{\frac{z}{\varrho c}kh}\right)p = 0\,. \tag{9.83}$$

Equation (9.83) represents a one-dimensional wave equation for the sound pressure. It provides a simplified representation of the propagation of the lowest mode for 'large' impedances z and small duct dimensions $h \ll \lambda$. What exactly is meant by 'large z' can principally be explained by the more exact calculation presented in the next section.

The statements regarding duct attenuation contained in (9.83) are confirmed by the corresponding wave number k_x, as it amounts to

$$k_x = k\sqrt{1 - \frac{j}{\frac{z}{\varrho c}kh}}\,. \tag{9.84}$$

As already explained the damping is contained in the imaginary part of the wave number $k_x = k_\mathrm{r} - jk_\mathrm{i}$ (see (9.64) and (9.65)) and given by $D_\mathrm{h} = 8.7k_\mathrm{i}h$.

For the practical construction of wall structures, absorbent linings of porous material in front of a rigid wall or a boundary lined with resonators are commonly used. Several preliminary remarks on the effect of these three possible impedance types are beneficial for the proceeding discussions of the attenuation produced by these two constructions. In the case of the (nearly) lossless resonator the impedance is always imaginary, with a negative imaginary part in the stiffness-controlled region below the resonance frequency and with a positive imaginary part in the mass- controlled region above the resonance frequency. For porous sheets with not too small of a flow resistance, the impedance tends towards (positive) real values with increasing frequency.

First and foremost, either imaginary impedances with a positive or negative imaginary part or real impedances are essentially relevant in practice.

a) stiffness-controlled impedance $z = -j|z|$

For stiffness-controlled impedances, the wave number

$$k_x = k\sqrt{1 + \frac{1}{\frac{|z|}{\varrho c}kh}} \tag{9.85}$$

is always real; the duct is undamped.

In the stiffness-controlled region, the wall lining is completely useless. From a physical point of view it should be noted that the speed of sound c_x in the duct is reduced compared to the unbounded propagation. From $k_x = \omega/c_x$ it follows that

$$c_x = \frac{c}{\sqrt{1+\frac{1}{\frac{|z|}{\varrho c}kh}}} . \tag{9.86}$$

A realization would be a (thin) layer of porous material which always acts like a spring at low frequencies (see Sect. 6.5.2, p. 194)

$$\frac{z}{\varrho c} = -\frac{j}{kd}$$

(d: layer thickness). In that particular case, the propagation speed becomes

$$c_x = \frac{c}{\sqrt{1+\frac{d}{h}}} . \tag{9.87}$$

b) mass-controlled impedance $z = j|z|$

Impedances with mass characteristics result in a duct wave number

$$k_x = k\sqrt{1-\frac{1}{\frac{|z|}{\varrho c}kh}} . \tag{9.88}$$

The wall lining only causes duct attenuation if the argument of the square-root is negative, i.e.

$$\frac{|z|}{\varrho c}kh < 1 . \tag{9.89}$$

Basically, impedances with mass characteristics increase with increasing frequency. Thus, a band limit is denoted by (9.89), where the duct attenuation is non-zero only below it.

Averaged over time, no sound power penetrates the surface when the wall impedance is imaginary. Attenuation is not achieved by extracting energy to the wall. The working principle behind attenuation, similar to the case of ducts with soft boundaries, is to generate a lowest mode incapable of propagation.

c) Real Impedances $z = |z|$

Real impedances at the duct wall result in a duct wave number

$$k_x = k\sqrt{1-j\frac{1}{\frac{|z|}{\varrho c}kh}} , \tag{9.90}$$

which always leads to damping. Furthermore, if, instead of a 'large impedance'

$$\frac{|z|}{\varrho c}kh > 1$$

is postulated, (9.90) can be approximated by

$$k_x \simeq k\left(1 - j\frac{1}{2}\frac{1}{\frac{|z|}{\varrho c}kh}\right) . \tag{9.91}$$

Equation (9.91) correctly represents the transition to the rigid boundary with increasing $|z|$; when $|z|/\varrho c$ increases the damping decreases. In contrast, it cannot be deduced from (9.91) that for small $|z|$, one can obtain any arbitrary D_h. The calculations leading to (9.91) explicitly required large impedances.

Based on the basic principles explained pertaining to the three types of impedances the frequency characteristics of the attenuation can be estimated for the purposes of practical constructions. As already mentioned, either absorbent linings with a rigid termination or linings using resonators can be implemented in practice. Both will be discussed in the next section. Here, the boundaries are fluid. The initially discussed porous layer moves into the realm of a resonator at small flow resistance.

a) Boundaries with absorbent linings

The frequency characteristics of the impedance of a porous layer of thickness d with a rigid termination

$$\frac{z}{\varrho c} = -j\frac{k_a}{k}\cot(k_a d) \tag{9.92}$$

with the wave number

$$k_a = k\sqrt{1 - j\frac{\Xi}{\omega\varrho}}$$

in the porous material were already discussed extensively in Sect. 6.5.2. To recapitulate the facts presented there with respect to the duct attenuation (see also Fig. 6.10, p. 194), we will provide some calculated samples given in Figs. 9.19a,b,c, as seen in (9.84), where the flow resistance $\Xi d/\varrho c$ and the layer thickness d are varied. These serve as an illustration of the basic principles introduced in the following.

As already mentioned, the wall impedance acts as a spring at low frequencies $d \ll \lambda$, due to $\cot(k_a d) \approx 1/k_a d$ it is $z/\varrho c - j/kd$. Therefore, at low frequencies, no duct attenuation is expected for a duct with an absorbent wall lining, thus $D_\mathrm{h} \approx 0\,\mathrm{dB}$ (see also Figs. 9.19a,b,c). The onset of the damping effect begins when the impedance with $d \approx \lambda/4$ crosses the real axis. As it can be seen from Fig. 6.10, the impedance at this frequency point is small for a small flow resistance $\Xi d/\varrho c$. For this reason, the duct attenuation has an abrupt onset with high values for low $\Xi d/\varrho c$. In the limiting case $\Xi = 0$, it is $z = 0$ and thus $D_\mathrm{h} = 13.5\,\mathrm{dB}$. With increasing flow resistance the attenuation decreases, as described in (9.91).

The additional frequency characteristics of D_h are easily discussed. One octave above the first $\lambda/4$-resonance, the impedance takes on higher values

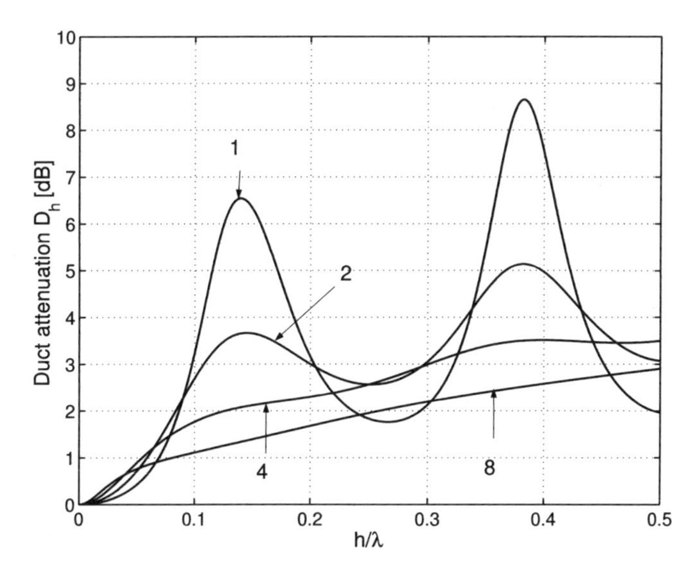

Fig. 9.19. (**a**) Duct attenuation D_{h} of a duct lined with an absorbent layer $d/h = 2$. The values at the curves specify $\Xi d/\varrho c$

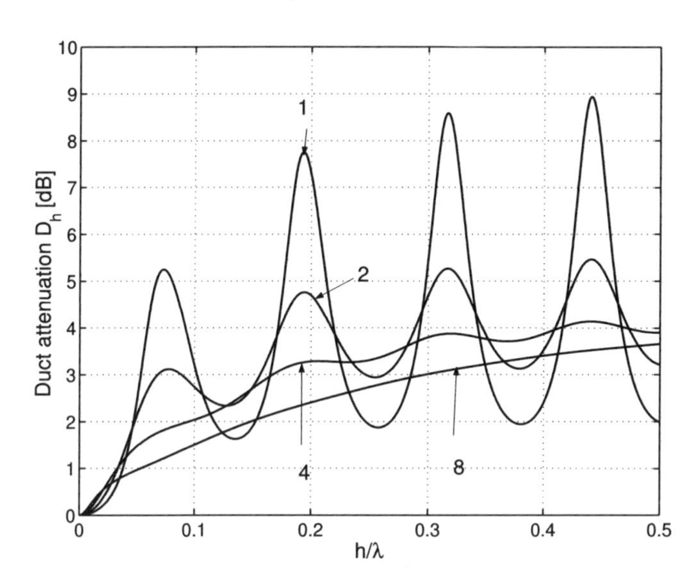

Fig. 9.19. (**b**) Duct attenuation D_{h} of a duct lined with an absorbent layer $d/h = 4$. The values at the curves specify $\Xi d/\varrho c$

for $d \simeq \lambda/2$, the smaller the flow resistance is (compare Fig. 6.10). The attenuation is smaller here, the smaller $\Xi d/\varrho c$ is. As a rule of thumb, the frequency characteristics of D_{h} are repeated after that. In it's continued progression, the impedance completes a circle, so that a 'quasi-periodic' structure is obtained, as is also shown in Figs. 9.19a,b,c. Although the large and narrow-banded maximum values of D_h at the smallest flow resistance are not quite accurate (they occur at low impedances, where (9.84) actually no longer holds), the basic principle can be observed. Either a fairly high attenuation in a narrow band (up to $D_\mathrm{h} \approx 13.5\,\mathrm{dB}$ at points $z = 0$) can be achieved and one has to be satisfied with smaller values of D_h outside of these bands; or one uses a comparatively small, but broadband attenuation D_h for medium flow resistances $\Xi d/\varrho c$.

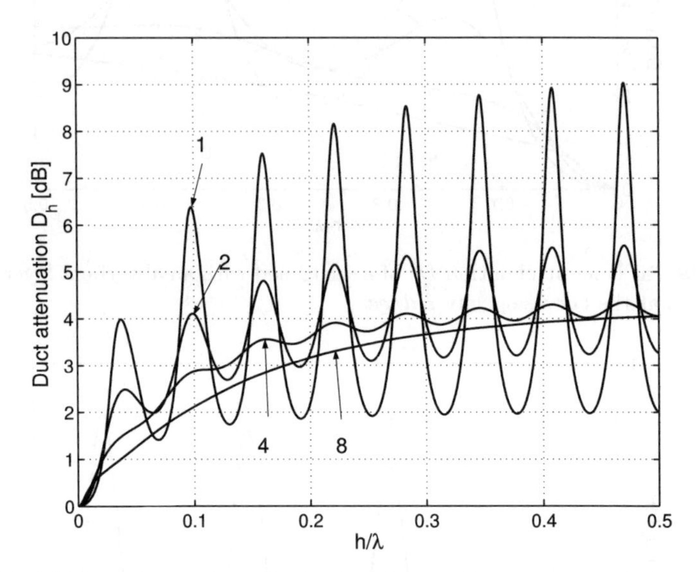

Fig. 9.19. (**c**) Duct attenuation D_h of a duct lined with an absorbent layer $d/h = 8$. The values at the curves specify $\Xi d/\varrho c$

'Comb filters', realized by lightly damped resonators, are useless for broadband noises: it is not very useful to filter a small bandwidth of the signal and let the 'remaining bulk' of it pass without any reduction. Applications using lightly damped resonators only occur in some special cases where tonal noise disturbances are present, which contain a single frequency component only. The next section will discuss this possible case in more detail, where, for instance, only the first tonal component of fan noise (e.g. in the ventilation of subterranean garages) has to be reduced.

On the other hand, a lot of silencers have to meet the requirements of a broad frequency band. Ventilation ducts of a concert hall, for example, have to be free of external noise; also, silencers of vehicles (car mufflers, for example) must be designed for broadband attenuation, due to the continuously altering engine speed. In such cases a medium flow resistance with values 'not too large and not too small' has to be provided in the wall lining, which ranges approximately in the interval $2 \leq \Xi d/\varrho c \leq 4$.

This design criterion can slightly change if the wall linings are not formed by one rigid wall and one impedance plane parallel to it. Some remarks on arbitrary duct cross sections can be found at the end of this section.

b) Resonator linings

We will consider two methods of outfitting duct linings with resonators. One way is to use an array of pipes, as shown in Fig. 9.20. For this, the pipe elements must have the same length and be open at the side adjoining the duct wall and be rigidly terminated at the other end. The resonance phenomenon associated with the lowest resonance frequency occurs if the length of the pipe segment is one-quarter the wavelength. This is why we will subsequently refer to this structure as a $\lambda/4$ resonator.

Another method to do this is to isolate the aforementioned pipe elements using mass linings m'' on their open sides form the duct. This can mean installing perforated plates (See the section on 6.5.4). This construction allows for the tuning of lower resonance frequencies for the same duct length a. In other words, mass linings can save installation space a for longer duct systems which would otherwise have to be more widely constructed to provide the same insulation for the same resonance frequency without a lining.

$\lambda/4$ Resonators

The wall impedance produced is given in the same way as in (9.92) by

$$\frac{z}{\varrho c} = -j\cot(kd) \tag{9.93}$$

where k now represents the wave number in air.

The characteristics of the attenuation D_h in the limiting case of a small flow resistance result from the previous section and have already been mentioned there. These are narrow-banded characteristics with an effect at the resonance $ka = k_\mathrm{R}a = \pi/2$ of the duct with the absorbent layer, where $D_\mathrm{h} = 13.7\,\mathrm{dB}$. The upper limit of the bandwidth k_E can be deduced by (9.84) and (9.93)

$$k_x = k\sqrt{1 + \frac{1}{kh\cot(kd)}} = k\sqrt{1 + \frac{\tan(kd)}{kh}}\,.$$

The upper band limit k_E is reached, when the duct wavelength becomes real, i.e.

$$\tan k_\mathrm{E}d = -k_\mathrm{E}h = -\frac{h}{d}k_\mathrm{E}d\,. \tag{9.94}$$

The last conversion in (9.94) was done so that it would be more compatible to a graphical solution to this so-called transcendental equation. As Fig. 9.21

Fig. 9.20. Photograph of a duct wall lined with resonators which are realized by rigidly terminated pipes of depth d

shows, $k_{\mathrm{E}}d$ is given by the point of intersection between the tangent function $\tan k_{\mathrm{E}}d$ and the straight line $-h/d\, k_{\mathrm{E}}d$. The smaller the gradient of the line is (i.e., the ratio h/d), the farther the intersection located to the right and the larger the bandwidth are. Naturally, the maximum bandwidth is at most an octave wide due to $k_{\mathrm{E}} = 2k_{\mathrm{R}}$.

The straight line depicted as an example in Fig. 9.21 is given by $h/d = 1$. The intersection can easily be seen in the graph at $k_{\mathrm{E}}a/\pi \approx 0.65$ Thus, we can determine the effective bandwidth by

$$\frac{f_E}{f_R} = \frac{k_{\mathrm{E}}a}{k_{\mathrm{R}}a} = \frac{0.65\pi}{0.5\pi} \approx 1.3.$$

As a matter of fact, a suitable silencer for the case of $a = h$ does not yet exist, because the cut-on frequency of a soft-lined duct at the mode $n = 0$ corresponds to the resonance frequency of the pipe elements. That there has yet to be any suitable solutions for $a = h$ can also be verified by the maximum damping D_{max} in the resonance frequency, which can estimated based on eq.(9.74) (with $n = 0$) as follows:

$$D_{max} = 8.7k_i h = 8.7h\sqrt{\left(\frac{\pi}{2h}\right)^2 - k^2} = 8.7\sqrt{\frac{\pi^2}{4} - (kh)^2} =$$

$$8.7\sqrt{\frac{\pi^2}{4} - (ka)^2\frac{h^2}{a^2}} = 8.7\frac{\pi}{2}\sqrt{1 - \frac{h^2}{a^2}}$$

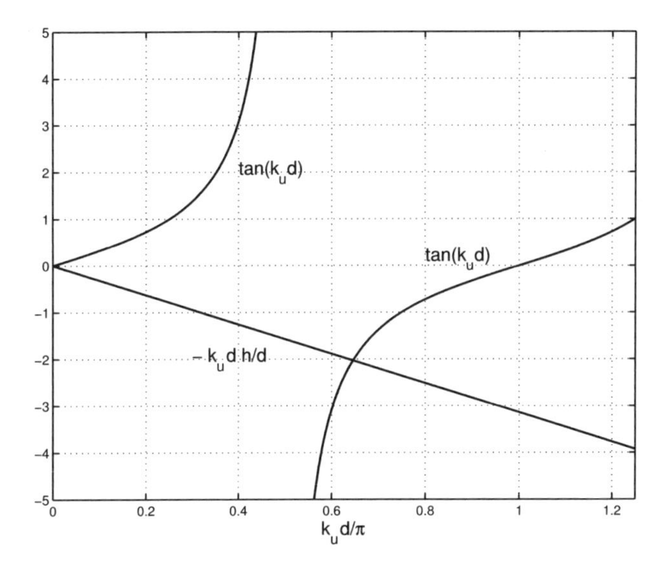

Fig. 9.21. Graphical solution of the transcendental equation (9.94)

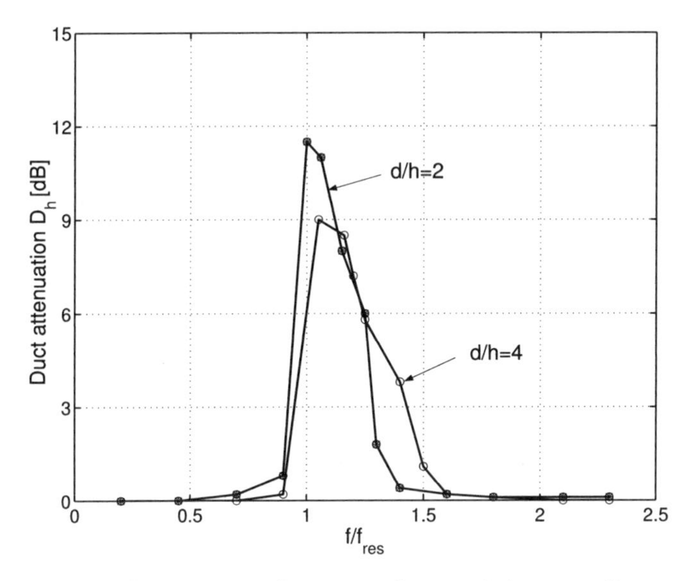

Fig. 9.22. Measured attenuation frequency characteristics according to the setup in Fig. 9.20. Experiments by T. Kohrs

(due to $ka{=}\pi/2$ in the resonance frequency.) For $h = a$, the maximum damping is clearly $D_{max}{=}0$. Once $a = 2h$, the highest possible damping ($13.7\,dB$) with $D_{max} = 11.9\,dB$ has been attained.

The attenuation D_{h}, measured in the experimental arrangement pictured in Fig. 9.20, is shown in Fig. 9.22. It can be seen that the connecting line representing a straight line between the 'acoustically soft' point at the resonance frequency and the 0 dB-point at the upper band limit already results in a useful approximation.

Outfitting duct linings with resonators is evidently useful only in the case of small-banded noise interference. It does not work well for broad-banded noise control. Noise control measures are accommodated by first determining which frequency is to be reduced as well as the duct length a. The duct width h is adjusted depending on the bandwidth needed (which can be deduced by investigating the speed variation or temperature differences in summer vs. winter). Since large installation spaces are required for low frequencies, they are, in effect, undesirable. An example would be an installation space of $a = 85\,cm$ which would be required for noise reduction at a resonance frequency of $100\,Hz$. The structures described in the next section are also designed to save construction space.

Resonators with mass linings

Tuning the resonance frequency by means of pipe lengths smaller than $a = \lambda/4$ can be achieved by covering the pipe elements with a mass lining m'' (see also Sect. 6.5.4, p. 201). In that case

$$\frac{z}{\varrho c}kh = j\left(\frac{\omega m''}{\varrho c} - \frac{c}{\omega d}\right)kh = j\left(\frac{\omega^2 m'' d}{\varrho c^2} - 1\right)\frac{h}{d} = j\left(\frac{\omega^2}{\omega_{res}^2} - 1\right)\frac{h}{d}$$

is given. In estimating the bandwidth, it can be assumed that the impedance at the upper band limit is sufficiently large and that

$$k_x = k\sqrt{1 - j\frac{1}{\frac{z}{\varrho c}kh}} = k\sqrt{1 - \frac{1}{\left(\frac{\omega^2}{\omega_{res}^2} - 1\right)\frac{h}{d}}}$$

is given. The band limit ω_{E} is then denoted by

$$\left(\frac{\omega_{\mathrm{E}}^2}{\omega_{res}^2} - 1\right)\frac{h}{d} = 1$$

or by

$$\frac{\omega_{\mathrm{E}}}{\omega_{\mathrm{res}}} = \sqrt{\frac{d}{h} + 1}\,. \tag{9.95}$$

To complete our discussion of ducts attenuated by wall linings, other duct cross-sections will be treated (Fig. 9.23). When applying the mass conservation principle to a small duct element of length Δx, rather than (9.79), we arrive at the general equation

$$\frac{\partial v_x}{\partial x} + \frac{U}{S} v_n = -\frac{j\omega}{\varrho c^2} p. \tag{9.96}$$

Here, S is the cross-sectional area of the duct and U is the circumference lined with the impedance. If $v_n = p/z$ is inserted, one obtains

$$\frac{\partial v_x}{\partial x} = -\left(\frac{j\omega}{\varrho c^2} + \frac{1}{zS/U}\right) p\,. \tag{9.97}$$

In eq.(9.81), zh is substituted for zS/U. This is the only change that has been made, transforming eq.(9.84) to

$$k_x = k\sqrt{1 - \frac{j}{\frac{z}{\varrho c} k \frac{S}{U}}}\,. \tag{9.98}$$

All previous deliberations can be reworked the same way.

Thus, zh is replaced by zS/U, but apart from that, everything else stays the same and all considerations made earlier are transferrable. Two cases mentioned earlier may serve as a double-check: the rigid surface and the surface equipped with z, where $S/U = h$ is obtained, using $S = hl$ and $U = l$ (l is the lined transverse length). If linings are present on both surfaces with $U = 2l$, we thus obtain $S/U = h/2$. It has the same effect as halving the impedance. The attenuation is thereby approximately doubled. For a duct with a double-sided soft lining, for instance, the maximum duct attenuation becomes $D_{\mathrm{h,max}} \approx 27\,\mathrm{dB}$.

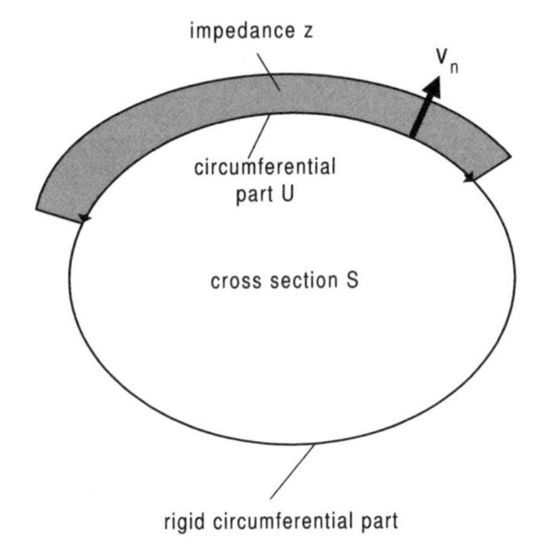

Fig. 9.23. Definition of the quantities in (9.97)

For circular cross-sections of radius b with a complete lining along the circumference, it is $S = \pi b^2$ and $U = 2\pi b$, thus $S/U = b/2$.

Finally, it is worth mentioning that duct damping can be estimated by the absorption coefficient of the duct lining, albeit in a largely simplified form (defined only for normal sound incidence). In everyday practice, porous sheets are generally used for duct linings. For higher frequencies and greater drag resistance Ξd (d=sheet thickness), the corresponding wall impedance is basically real (see the chapter on sound absorbers, particularly Figure 6.10 therein). In practice, one can apply eq.(9.91) and obtain

$$k_x \simeq k\left(1 - j\frac{1}{2}\frac{1}{\frac{|z|}{\varrho\, c}k\frac{S}{U}}\right). \tag{9.99}$$

For the level decrease along a given segment l lengthwise as in eq.(9.67),

$$D_l = 8,7 k_i l = 4,35\frac{l}{\frac{|z|}{\varrho c}k\frac{S}{U}} . \tag{9.100}$$

On the other hand, based on the matching law eq.(6.29), one arrives at

$$\alpha = \frac{4\mathrm{Re}\left\{z/\varrho\, c\right\}}{\left[\mathrm{Re}\left\{z/\varrho\, c\right\}+1\right]^2 + \left[\mathrm{Im}\left\{z/\varrho\, c\right\}\right]^2} \approx \frac{4\mathrm{Re}\left\{z/\varrho\, c\right\}}{\left[\mathrm{Re}\left\{z/\varrho\, c\right\}+1\right]^2} \approx$$

$$\frac{4|z|/\varrho c}{(|z|/\varrho c + 1)^2} \approx \frac{4}{|z|/\varrho c} \tag{9.101}$$

in the case of real impedances, if larger impedance ratios $|z|/\varrho c$ exist. The 'Piening formula', named after its author, is derived from eq.(9.100) and eq.(9.101):

$$D_l \approx 1,1\frac{Ul}{S}\,\alpha \tag{9.102}$$

This formula serves as an initial orientation for roughly estimating the expected effects of attenuators.

Exact calculation with arbitrary impedances

In nearly all practical applications, conducting approximations for the duct attenuation discussed in the previous section is sufficiently precise. One either strives for broadband attenuation and thus a wall impedance with a high impedance, or one implements a narrow-banded but highly effective resonator lining which starts with high and 'soft' attenuation at the resonance and then rapidly decreases in the direction of the upper band limit. Since it might not be enough to simply rely on approximations and since the following exact calculation of the sound field between two parallel planes $y = 0$ (with $v = 0$) and $y = h$ (with the impedance z) is not very difficult, we will proceed to discuss calculations with arbitrary impedances in the following.

The boundary condition $\partial p/\partial y = 0$ for $y = 0$ requires the basic approach

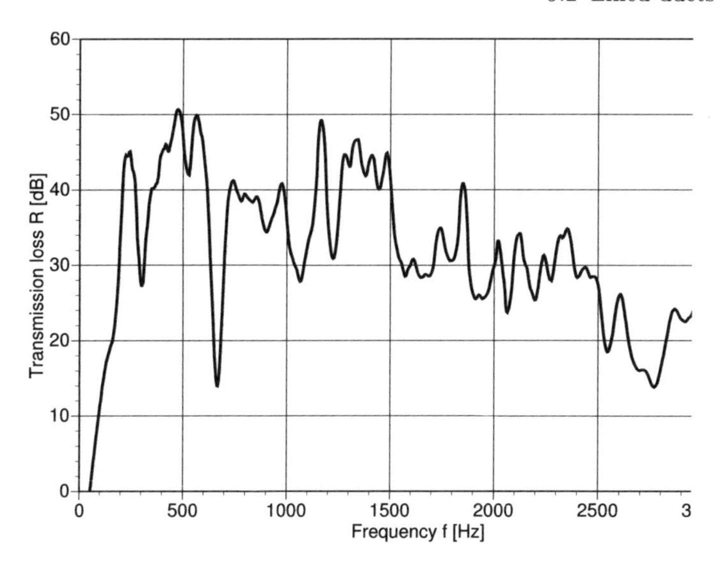

Fig. 9.24. Measured transmission loss (level difference between inlet and outlet using an anechoic termination) of a circular silencer (inner diameter 50 mm) with an absorbent lining of mineral wool (thickness 100 mm). Experiments by J. Feldmann

$$p = A\cos(k_y y)\mathrm{e}^{-jk_x x} \tag{9.103}$$

for the modes of the transverse distribution. The wave numbers k_y follow from the boundary condition in the $z = h$ plane $p(h) = zv(h)$

$$\cos(k_y h) = -\frac{jk_y z}{\omega\varrho}\sin k_y h = -jk_y h\frac{z/\varrho c}{kh}\sin k_y h$$

or

$$-j(k_y h)\tan(k_y h) = \frac{kh}{z/\varrho c}\,. \tag{9.104}$$

Equation (9.104) forms the so-called eigenvalue equation for sound propagation in ducts. The solutions to (9.104) indicate all appearing transverse wave numbers k_y. The resulting axial wave numbers k_x are based on the wave equation, which, for (9.103), requires

$$k_x h = \sqrt{(kh)^2 - (k_y h)^2}\,. \tag{9.105}$$

The modal wave properties are included in k_x, in particular,

$$D_{\mathrm{h}} = -8.7\,\mathrm{Im}\{k_x h\}\,.$$

The special cases already dealt with need to be found in the eigenvalue equation. Indeed, for $z \to \infty$ (9.104) tends to $\sin k_y h = 0$ and thus to $k_y h = n\pi$ ($n = 0, 1, 2, \ldots$). This cross-check for $z = 0$, using $\cos k_y h = 0$, also results in the eigenvalues $k_y h = \pi/2 + n\pi$ $(n = 0, 1, 2, \ldots)$ equal to the result derived earlier.

These examples reveal some aspects about the basic principles included in (9.104): this transcendental equation has not only one but many (infinite) solutions. The propagation is generally described by a multitude of modes, where the wave numbers are general solutions for (9.104). If the modal amplitudes are unknown, the duct attenuation is always calculated by the mode with the smallest attenuation D_h which is called the 'principal mode'.

The approximation already discussed in the previous section for the lowest mode results from (9.104) also applies for a large impedance. Under the assumption of $|z/\varrho c| \gg kh$ in (9.104) for the lowest mode, $\tan k_y h \approx k_y h$ can be approximated and

$$(k_y h)^2 = j\frac{kh}{z/\varrho c}$$

is obtained and thus

$$k_x h = \sqrt{(kh)^2 - j\frac{kh}{z/\varrho c}}$$

results in the same way as in (9.84). The only case, sometimes relevant in practice, but not yet dealt with here, are small imaginary impedances which occur close to resonance, whenever the walls are lined with resonators. For the lowest mode and $|z/\varrho c| \ll kh$

$$k_y h = \frac{\pi}{2} + \Delta$$

$(|\Delta| \ll \pi/2)$ can be assumed. Equation (9.104) is then approximately given by

$$-j\left(\frac{\pi}{2} + \Delta\right)\frac{\sin\left(\frac{\pi}{2} + \Delta\right)}{\cos\left(\frac{\pi}{2} + \Delta\right)} \approx j\frac{\frac{\pi}{2} + \Delta}{\sin \Delta} \approx j\frac{\frac{\pi}{2} + \Delta}{\Delta} = j\left(1 + \frac{\pi}{2}\frac{1}{\Delta}\right) = \frac{kh}{z/\varrho c}\,.$$

The last equation is solved for Δ,

$$\Delta = -\frac{\frac{\pi}{2}}{1 + j\frac{kh}{z/\varrho c}}$$

and the transverse wave number k_y finally results in

$$k_y h = \frac{\pi}{2} + \Delta \approx \frac{\pi}{2}\frac{j\frac{kh}{z/\varrho c}}{1 + j\frac{kh}{z/\varrho c}} = \frac{\pi}{2}\frac{1}{1 - j\frac{z/\varrho c}{kh}} \approx \frac{\pi}{2}\left(1 + j\frac{z/\varrho c}{kh}\right)\,.$$

The last step incorporated $1/(1-x) \approx 1 + x$ for $|x| \ll 1$, resulting in the axial wave number

$$k_x h = \sqrt{(kh)^2 - \frac{\pi^2}{4}\left(1 + j\frac{z/\varrho c}{kh}\right)^2} \tag{9.106}$$

or for sufficiently low frequencies $kh \ll \lambda/2$,

$$k_x h \approx -j\frac{\pi}{2}\left(1 + j\frac{z/\varrho c}{kh}\right) = \frac{\pi}{2}\left(\frac{z/\varrho c}{kh} - j\right) .$$

The duct attenuation is thus given by

$$D_{\mathrm{h}} = -8.7\,\mathrm{Im}\left\{k_x h\right\} = 13.5\left[1 - \mathrm{Im}\left\{\frac{z/\varrho c}{kh}\right\}\right] . \tag{9.107}$$

To derive (9.107) an approximation of the tangent around its pole is used, which implies – as already mentioned – small impedance values. 'Approximations close to poles' are always very sensitive to small changes and thus (9.107) quickly lose their validity with increasing impedance. In any case, small absolute values assist in estimating the effect of different impedances.

Obviously, the duct attenuation decreases with increasing (small) mass-like impedance $z = j|z|$. In contrast, D_{h} increases with increasing (small) spring-like impedance and can obviously be larger than $D_{\mathrm{h}} = 13.5\,\mathrm{dB}$ (as in the sound field with $z = 0$).

On the other hand, no duct attenuation at all ($D_{\mathrm{h}} = 0$) can be expected for a larger stiffness impedance. D_{h} must therefore increase up to a maximum value for an increasing magnitude of the stiffness impedance and subsequently decrease rapidly and tend toward zero. Obviously, an optimum impedance $D_{\mathrm{h,opt}}$ exists in the stiffness region of the impedance.

The question regarding the optimum impedance can simply be answered using a graphical representation of the eigenvalue equation

$$-\,jw\tan w = \beta \tag{9.108}$$

which is first briefly discussed in general.

For simplicity, $w = k_y h$ and $\beta = kh/(z/\varrho c)$ was defined in (9.104). Equation (9.108) describes a transcendental equation, where β is given and solutions of w are needed. The easiest way to find these solutions is often a table of the complex function $F(w) = -jw\tan(w)$, which can be computer-generated fairly easily. A matrix of complex function values F can be calculated, for instance, where the imaginary part of $w = w_{\mathrm{r}} + jw_{\mathrm{i}}$ is kept constant in one row, while the real part w_{r} is systematically varied with the column number. In this way, we obtain a table description of $F(w)$, where the rows yield the function values for $w_{\mathrm{i}} = $ const., and the columns for $w_{\mathrm{r}} = $ const. For a given value of β finding $F(w) = \beta$ in the table would result in the solution to the eigenvalue equation. It is, by the way, recommended to actually use this technique in a numerical program.

The properties of the matrix can also be represented graphically. The complex values of $F(w)$, for example, which are obtained by $w_{\mathrm{i}} = $ const. and a variable w_{r} in one row of the matrix, can be connected by a line in the graph. Thus, lines $w_{\mathrm{i}} = $ const. are obtained in the complex plane which are

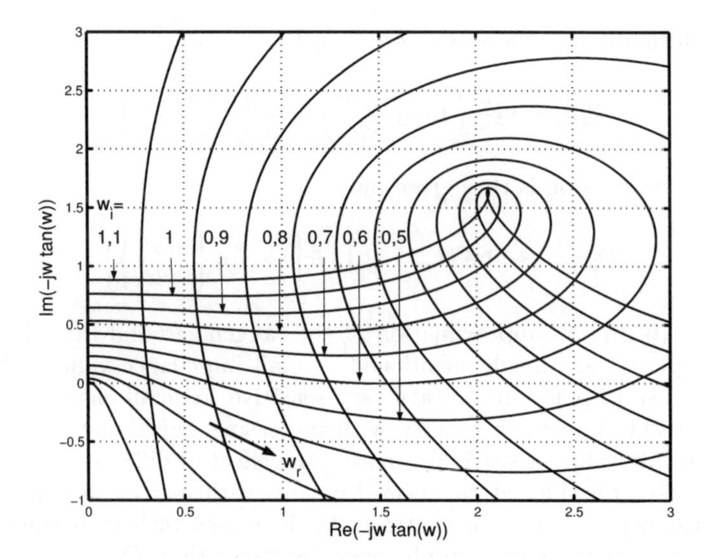

Fig. 9.25. Lines $w_{\mathrm{i}} = \mathrm{const.}$

plotted in Fig. 9.25. Along one line, $w_{\mathrm{i}} = \mathrm{const.}$, w_{r} is varied, increasing in the arrow direction in the range $0 \leq w_{\mathrm{r}} \leq \pi$ (the end of the curve $w_{\mathrm{r}} = \pi$ may be outside of the depicted plot range). The curves $w_{\mathrm{i}} = \mathrm{const.}$ can intersect with each other: this only means that the eigenvalue equation (9.108) has several solutions. If a larger interval of w_{r} than the one used here were to be specified, the complex plane would be crossed multiple times.

The graphical representation of $F(w)$ must also enable the determination of the solution to the problem of the optimum impedance. It is firstly stated that due to

$$k_x h = \sqrt{(kh)^2 - w^2} \approx -jw = w_{\mathrm{i}} - jw_{\mathrm{r}}$$

(for $kh \ll |w|$) the real part w_{r} of the solution w dominates the duct attenuation which is approximately given by

$$D_{\mathrm{h}} \approx 8.7 w_{\mathrm{r}} \ .$$

As already mentioned, the curves $w_{\mathrm{i}} = \mathrm{const.}$ intersect each other. The intersection denotes two modes $w = w_1 = w_{\mathrm{r}1} + jw_{\mathrm{i}}$ and $w = w_2 = w_{\mathrm{r}2} + jw_{\mathrm{i}}$ with different attenuations for the corresponding value of β. Yet, w_1 is the value of w, where the curve passes the point of intersection in the arrow direction for the first time, w_2 is the value, where the curve passes the point of intersection for the second time. Thus, $w_{\mathrm{r}1} < w_{\mathrm{r}2}$ is given. The duct attenuation D_{h} is always calculated using the mode with the lowest attenuation which is thus given by $D_{\mathrm{h}} \approx 8.7 w_{\mathrm{r}1}$.

If the development of the array of curves $w_{\mathrm{i}} = \mathrm{const.}$ is pursued for increasing w_{i}, it can be observed that the 'loop', which is carried by the curves as they return to the point of intersection, becomes increasingly narrower. With increasing w_{i}, $w_{\mathrm{r}1}$ also increases, while $w_{\mathrm{r}2}$ decreases simultaneously. Finally the loop collapses to a point. The latter is called the winding point, where $w_1 = w_2$ is given and both w-values fall together. The winding point yields the maximum possible duct attenuation $D_{\mathrm{h,opt}}$: surely, $w_{\mathrm{r}1}$ can increase further, but simultaneously $w_{\mathrm{r}2}$ must drop below the optimum value and the smallest possible attenuation would thus decrease.

The winding point $w_1 = w_2$ obviously represents two zeros of

$$G(w) = \beta + jw \tan w \; . \tag{9.109}$$

In Fig. 9.25,

$$\beta_{\mathrm{opt}} \approx \frac{kh}{z_{\mathrm{opt}}/\varrho c} = 2 + j\,1.6 \tag{9.110}$$

can be approximated as the winding point. Thus, the optimum impedance becomes

$$\frac{z_{\mathrm{opt}}}{\varrho c} \simeq \frac{h}{\lambda}\,(1.9 - j\,1.5) \; , \tag{9.111}$$

which is given by a small stiffness impedance with an additional real part of about the same value.

The resulting maximum attenuation $D_{\mathrm{h,opt}}$ for $z = z_{\mathrm{opt}}$ can not be simply looked up in Fig. 9.25 without further action. The maximum attenuation is $w_{r,opt} = 2.1$, leading to the obvious conclusion $D_{h,opt} = 8.7 w_{r,opt} = 18.3\,dB$.

As mentioned, the condition for the two zeros of (9.109) can be found in the winding point. As in a real equation, this is the case if

$$G(w) = 0 \tag{9.112}$$

and

$$\frac{\mathrm{d}G\,(w)}{\mathrm{d}w} = 0 \tag{9.113}$$

are met. After applying some elementary algebra, (9.113) leads to

$$\sin 2w = 2w \; . \tag{9.114}$$

The solution of (9.114) can now be easily solved numerically. It is given to the third digit of accuracy as

$$w_{\mathrm{opt}} = 2.19 + j1.12, \tag{9.115}$$

which results in the optimum attenuation $D_{\mathrm{h,opt}} = 19.1\,\mathrm{dB}$.

The question of the maximum possible value is always interesting from a scientific point of view. In practice, the discussed optimum impedance is entirely meaningless. It could only be realized in a narrow band close to the resonance of a resonator.

In conclusion, an additional remark should be made as to the numerical solution to the eigenvalue function (9.108), which will not prove to be too difficult. Solving functions that contain poles using numerical methods is cumbersome. To avoid this problem, we transform the eigenvalue function to

$$H(w) = jw\, sin\, w + \beta\, cos\, w = 0. \tag{9.116}$$

This function can be examined for its roots and be easily programmed. It is sufficient to examine a narrow band of about $0 < w_i < 5$ above the real axis. This band must contain all the roots and the higher modes, because beyond the band for larger values of w_i, $j \sin w = -\cos w$ holds. It is enough to base the numerical root findings simply on the rms-value $|H(w)|$.

9.3 Summary

Duct silencers work based on the principles of reflection and absorption. Pure reflection silencers are abrupt changes in cross-section, bifurcations and chamber insertion (as in mufflers). If there are imaginary wall impedances, pure reflectors can also be constructed by lined silencers.

Generally, that high insertion losses in a narrow band, or, alternatively, smaller ones in a broader frequency band, are apparently obtained with the same amount of effort. Some examples of narrow-band silencers are ducts with periodically changing cross-sections along the length at a small hub, which it becomes wider or narrower, the wall lining with slightly attenuated resonators, and the optimal lining impedance occurring only at a single frequency point. Wide-band but smaller damping effects D_h are obtained at boundary impedances from absorbent linings with about $\Xi d/\varrho c = 4$. Therefore, significant level reductions can only be achieved with proper silencer length.

9.4 Further reading

F.P. Mechel dedicated an important part of his work 'Schallabsorber' to the subject of absorbent and reactive duct linings.

9.5 Practice exercises

Problem 1

A cylindrical chamber silencer with circular terminations (without absorbent linings) is to be used to produce a minimum insertion loss of $7\,dB$ between $200\,Hz$ and $600\,Hz$. What dimensions does the pot have to have (diameter and length) if it is to be inserted in a pipe that is $5\,cm$ in diameter? What is the lowest cut-on frequency of the pot?

Problem 2

The low-frequency attenuation D_a of a duct with a square cross-section and length a along the edges, which has an impedance ϱc $(2\varrho c)$ along the entire girth, has been determined. What is the attenuation D_b for a circular cross-section with a radius b?

Problem 3

The wall of a silencer has been outfitted on one side with non-damped resonators to reduce network noise (at a frequency of $50 Hz$). The depth of the resonator is $50\,cm$ $(1\,m)$. What mass lining is required for it?

Problem 4

Suppose a network frequency shifts around $50\,Hz$ by $5\,Hz$, either to $55\,Hz$ or $45\,Hz$. How would the wall impedance defined in Problem 2 change? How would these changes affect the duct attenuation? How great is the maximum possible duct attenuation? For this problem, assume that the duct cross-section measures $h = 0,25\,m$.

10

Noise barriers

Everyone recalls their own experience at attempting to do away with interfering noise by trying to put a barrier between the source and one's ear. One tries to escape a pneumatic hammer or a lawn mower behind the next house. Wenn hiking in a quite forest, for example, one quickly takes another path over the next hill, whenever a lumberjack is heard cutting into the stillness with his chain saw.

Likewise, everyone knows how fruitless such attempts to escape noise can be. Although the source is screened by large objects, the noise still gets around to the ear more or less attenuated. Obviously, the sound diffracts around the barrier and thus deviates from the straight propagation. The physical effect is therefore called 'diffraction'.

In our noisy environment. is a a major question as to how acoustic screens (buildings, walls, dams...), either already existing or to be newly constructed, can be used to prevent noise impact (and thereby, protect one's long-term health) and what level of reduction can be achieved. In Germany alone, the length of noise barriers along roads and railway tracks adds up to thousands of kilometers. How effective they are and what level reductions they provide are certainly typical problems which belong to the basic subjects of engineering acoustics. For this reason dealt with in this chapter. However, not all phenomena of diffraction can be discussed here.

One aspect of diffraction not discussed here, for instance, is dependent on the shape of the barrier, an effect not discussed here. The following will concentrate on the basic principles by discussing the most simple arrangements like the diffraction at the rigid, semi-infinite screen, onto which a plane wave impinges obliquely (Fig. 10.1). The problem of diffraction was discussed in the context of light about 100 years ago by Sommerfeld in his book "'Lectures on Theoretical Physics"' (Sommerfeld, A.: "Lectures on Theoretical Physics", Akademische Verlagsgesellschaft, Leipzig 1964). The quantitative examination of diffraction phenomena is relatively new, compared to the great physical discoveries in this century, such as exemplified by the work of Albert Einstein. Until recently, the topic has not necessarily been fully integrated into the

M. Möser, *Engineering Acoustics*, DOI 10.1007/978-3-540-92723-5_10,

foundations course for acoustical engineering, perhaps partly because much of this topic can be comprehensibly and easily expressed using equations, as the following sections show.

10.1 Diffraction by a rigid screen

The model arrangement which underscores the following discussions is introduced in Figure 10.1. The model consists of a volume line source at a distance of a in front of the end of a semi-infinite screen which simultaneously constitutes the origin of the coordinate system. The coordinates of the circular cylinder are used here, the angle φ mathematically counts as positive in relation to the upper side of the arbitrarily thin acoustic screen. The angle φ only spans the interval $0 \leq \varphi \leq 360°$; values outside of the defined interval are excluded.

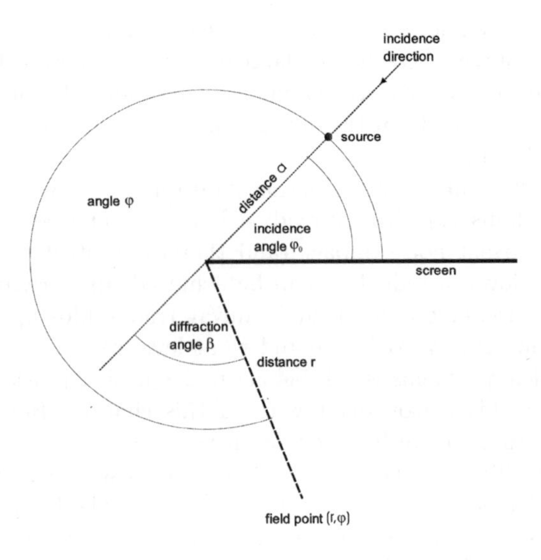

Fig. 10.1. Geometrical quantities at the edge of the rigid screen. Diffraction angle $\beta = \varphi - \pi - \varphi_0$

The location of the source is defined by the angle φ_0. Let us proceed to discuss the two-dimensional case. There are no changes along the axis normal to the plotting plane ($\partial/\partial z = 0$). The area within the cylinder with radius a is defined as 'subspace 1'. The area beyond the radius of the cylinder $r > a$ is defined as 'subspace 2'.

The wave equation for the sound pressure takes on the following structure in the cylinder coordinate system:

$$r^2 \frac{\partial^2 p}{\partial r^2} + r \frac{\partial p}{\partial r} + \frac{\partial^2 p}{\partial \varphi^2} + k^2 r^2 p = 0 \tag{10.1}$$

($k = \omega/c = 2\pi/\lambda$ = wave number of free waves). The wave equation for the cylindrical coordinate system can be found in numerous mathematical reference resources.

The solution functions in the direction of φ are now applied to create standing waves propagating in the circumferential direction, so that the constraints on the screen's upper and lower edges ($\partial p/\partial\varphi = 0$ for $\varphi = 0$ and for $\varphi = 360°$) are fulfilled. Obviously, all possible φ dependencies must be accounted for by

$$p \sim R(r) \cos(n\varphi/2) \tag{10.2}$$

(n=0,1,2,3,..), because it is precisely these cosine gradients which fulfill the required parameters for the upper and lower edge of the rigid screen. This simplifies the wave equation to the general differential equation

$$r^2 \frac{\partial^2 R}{\partial r^2} + r \frac{\partial R}{\partial r} + (k^2 r^2 - (n/2)^2) R = 0\,. \tag{10.3}$$

The solutions to these so-called Bessel differentials are made up of the Bessel functions $J_{n/2}(kr)$ and the Neumann functions $N_{n/2}(kr)$, with partial-number orders of $n/2$. For the first ten orders n, the functions are shown in the Figures 10.2 and 10.3. The functions are listed in a table and also appear as a standard routine in MATLAB.

The figures depict the overall sine functions with gradually decreasing amplitudes. This characteristic can also be seen in the following approximations for large arguments x:

$$J_{n/2}(x) \simeq \sqrt{\frac{2}{\pi x}} \cos(x - n\pi/4 - \pi/4) \tag{10.4}$$

and

$$N_{n/2}(x) \simeq \sqrt{\frac{2}{\pi x}} \sin(x - n\pi/4 - \pi/4) \tag{10.5}$$

It is plain to see that progressive waves are made up of sums of Bessel and Neumann functions. The so-called Hankel functions of the first and second type are defined as both

$$H^{(1)}_{n/2}(x) = J_{n/2}(x) + jN_{n/2}(x) \tag{10.6}$$

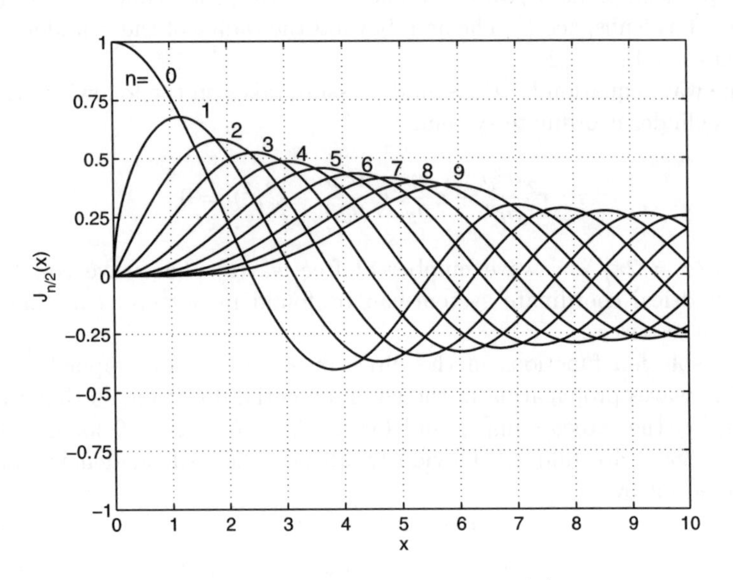

Fig. 10.2. Bessel functions $J_{n/2}(x)$

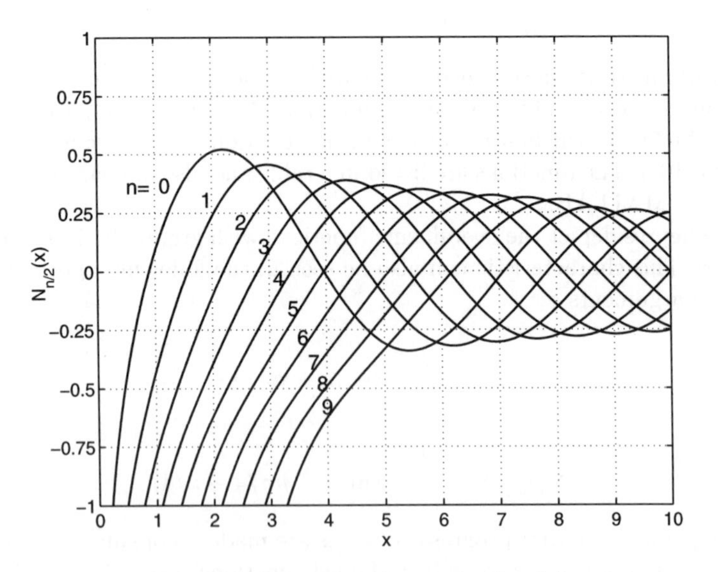

Fig. 10.3. Neumann functions $N_{n/2}(x)$

and

$$H_{n/2}^{(2)}(x) = J_{n/2}(x) - jN_{n/2}(x)\,. \tag{10.7}$$

Obviously, the first type of Hankel function basically describes waves travelling inward in a coordinate system in the direction of $-r$ and the second type of Hankel function describes a wave travelling outward in the direction of $+r$.

An important difference between Neumann and Bessel functions is the fact that the former has poles in $x = 0$, which behave like $ln(x)$ for $n = 0$, and otherwise like $x^{-n/2}$. On the other hand, the Bessel functions contain finite values ($J_0(0) = 1, J_{n/2}(0) = 0$).

Now we must outline several theoretical projections regarding the sound pressure for the subspaces shown in Figure 10.1.

Subspace 1

In subspace 1 between the wall's edge and the circle with the radius a where the source is located, the pressure cannot grow without bound at the barrier's edge. On the contrary, the sound field must have finite values everywhere (except for at the source itself). For this reason, we cannot use the Neumann functions for our ansatz here. Instead, for subspace 1 we use the ansatz

$$p_1 = \sum_{n=0}^{\infty} a_{n/2} J_{n/2}(kr) \cos(n\varphi/2) \tag{10.8}$$

Here, the dimensions $a_{n/2}$ are unknown coefficients which are yet to be defined based on the constraints of the separation plane between the subspaces and the source.

Subspace 2

The wave field in subspace 2 can only consist of waves travelling in the direction of $+r$. It is impossible for infinity to deflect a sound field. Since no other reflectors are present, isolated standing waves cannot occur either. Therefore, the ansatz below must be formulated for subspace 2:

$$p_2 = \sum_{n=0}^{\infty} b_{n/2} H_{n/2}^{(2)}(kr) \cos(n\varphi/2) \tag{10.9}$$

As mentioned before, the coefficients $a_{n/2}$ and $b_{n/2}$ must now be derived from the constraints on the circle with the radius. To this end, the circle is circumnavigated one time inside $r = a$ in subspace 1 in the $r = a - \varepsilon$ direction and one time outside of $r = a$ in subspace 2 in the $r = a + \varepsilon$ directions (each time with a very small $\varepsilon \to 0$). In so doing, the same pressures result at the same location φ, precisely at $\varphi = \varphi_0$. Indeed, the sound field can grow very large at this juncture, but because the assumed omnidirectional sound source produces equal sound pressure at both sides, the pressures $p_1(a-\varepsilon,\varphi_0)$ and $p_2(a+\varepsilon,\varphi_0)$ are equal, regardless of how large they actually are. In sum, without any constraints,

$$p_1(a,\phi) = p_2(a,\phi) \tag{10.10}$$

holds. Even for the r-unidirectional sound velocity v_r, the same values must be present at every location φ on both circles, therefore $v_{r1} = v_{r2}$. The exception to this, of course, is the location of the source, which produces equally large sound velocities at both sides in opposite directions. This fact leads us to the conclusion, that for an arbitrarily small point source at any distance to the source on both circles, both sound velocities at the location of the source differ by an infinitely large amount. The difference function $v_{r2}-v_{r1}$ is therefore equal to zero everywhere, except for the location $\varphi=\varphi_0$. In $\varphi=\varphi_0$ the difference function is infinitely large. Such a function, which is zero everywhere but infinite at a single point is known as the Dirac delta function $\delta(\varphi-\varphi_0)$. Therefore, we can say that

$$v_{r1}(a,\varphi) - v_{r2}(a,\varphi) = Q_0\delta(\varphi-\varphi_0)\,. \tag{10.11}$$

Here, Q_0 signifies a source dimension whose meaning will be elaborated in the following sections.

We only need some basic calculations to find the coefficients $a_{n/2}$ and $b_{n/2}$ based on the constraints eq.(10.10) and eq.(10.11). The simplest approach is perhaps to use the constraint of equal pressures in $r=a$ to reformulate the equations. This constraint is already fulfilled by the newly formulated equations

$$p_1 = \sum_{n=0}^{\infty} d_{n/2} J_{n/2}(kr) H_{n/2}^{(2)}(ka)\cos(n\varphi/2) \tag{10.12}$$

and

$$p_2 = \sum_{n=0}^{\infty} d_{n/2} H_{n/2}^{(2)}(kr) J_{n/2}(ka)\cos(n\varphi/2). \tag{10.13}$$

The difference function $v_{r1}(a,\varphi)-v_{r2}(a,\varphi)$, resulting from the restructuring based on $v_r = j/(\omega\varrho)\,\partial p/\partial r$, becomes

$$\frac{j}{\varrho c}\sum_{n=0}^{\infty} d_{n/2}[H_{n/2}^{(2)}(ka)J'_{n/2}(ka) - H_{n/2}^{\prime(2)}(ka)J_{n/2}(ka)]\cos(n\varphi/2) = Q_0\delta(\varphi-\varphi_0)\,. \tag{10.14}$$

$'$ denotes the derivative according to the argument (that is, $\partial/\partial kr$). The brackets can be simplified

$$[H_{n/2}^{(2)}J'_{n/2} - H_{n/2}^{\prime(2)}J_{n/2}] = j[J_{n/2}N'_{n/2} - J'_{n/2}N_{n/2}] = \frac{2j}{\pi ka} \tag{10.15}$$

(in the last step, an addition theorem for Bessel and Neumann functions was used. For more information, refer to Taschenbuch der Mathematik, z.B.

Abramowitz, M. (Hrsg.); Stegun, I. A.(Hrsg.): 'Handbook of Mathematical Functions', 9th Dover Printing, New York 1972). Finally, this gives us

$$\sum_{n=0}^{\infty} d_{n/2} \cos(n\varphi/2) = -\frac{\pi k a Q_0}{2} \delta(\varphi - \varphi_0) \tag{10.16}$$

as the resulting conditional equation for the coefficients $d_{n/2}$. This can be solved easily, as shown in the following using a specific unknown $d_{m/2}$. To do this, both sides are multiplied by $\cos(m\varphi/2)$ and then integrated over the interval $0 < \varphi < 2\pi$. This leaves us with the integrals on the left-hand side

$$\int_0^{2\pi} \cos(n\varphi/2)\cos(m\varphi/2)d\varphi \, .$$

such integrals are always equal to zero, except for the case $n = m$, where

$$\int_0^{2\pi} \cos^2(m\varphi/2)d\varphi = \begin{cases} \pi, & m \neq 0 \\ 2\pi, & m = 0 \end{cases}$$

applies. This leaves us only with the summand from the sum in eq.(10.16) with $n = m$, which is

$$d_{m/2} = -\frac{kaQ_0}{2\varepsilon_m} \cos(m\varphi_0/2) \, , \tag{10.17}$$

and abbreviating,

$$\varepsilon_m = \begin{cases} 1, & m \neq 0 \\ 2, & m = 0 \end{cases} .$$

Eq.(10.17) applies to any coefficient $d_{m/2}$ and thereby for every index m. Eq(10.17) therefore contains all coefficients $d_{m/2}$.

Thereby, the sound field in the entire room can be described by both

$$p_1 = -\frac{kaQ_0}{2} \sum_{n=0}^{\infty} \frac{1}{\varepsilon_n} J_{n/2}(kr) H_{n/2}^{(2)}(ka) \cos(n\varphi/2)\cos(n\varphi_0/2) \tag{10.18}$$

and

$$p_2 = -\frac{kaQ_0}{2} \sum_{n=0}^{\infty} \frac{1}{\varepsilon_n} H_{n/2}^{(2)}(kr) J_{n/2}(ka) \cos(n\varphi/2)\cos(n\varphi_0/2). \tag{10.19}$$

First, the source size Q_0 has to be considered. This is most easily done by examining a special case scenario by rotating the source at the angle of $\varphi_0 = 180°$. The rigid screen now turns its side of zero thickness toward the source, making the screen acoustically invisible from the standpoint of the source.

The sound field is therefore distributed in such as a way as if no rigid screen even existed, creating a special-case scenario where the source's undisturbed sound field exists "in the open air". The best way to describe the source is, of course, by the sound pressure the sound field creates at a distance a from the source. This pressure is located at the origin of the coordinate system and will henceforth be referred to as $p_Q(0)$. This should make clear that $p_Q(0)$ only stands for the source pressure that exists in the origin. Since only the zero-order Bessel function affects the sum for p_1 based on $J_{n/2}(0) = 0$ for $n > 0$, with $J_0(0) = 1$,

$$-\frac{kaQ_0}{4} H_0^{(2)}(ka) = p_Q(0)$$

applies. The results are then applied to eq.(10.18) and eq.(10.19). The local sound pressure distribution is therefore

$$p_1 = p_Q(0) \sum_{n=0}^{\infty} \frac{2}{\varepsilon_n} J_{n/2}(kr) \frac{H_{n/2}^{(2)}(ka)}{H_0^{(2)}(ka)} \cos(n\varphi/2) \cos(n\varphi_0/2) \tag{10.20}$$

and

$$p_2 = p_Q(0) \sum_{n=0}^{\infty} \frac{2}{\varepsilon_n} J_{n/2}(ka) \frac{H_{n/2}^{(2)}(kr)}{H_0^{(2)}(ka)} \cos(n\varphi/2) \cos(n\varphi_0/2) \,. \tag{10.21}$$

Although these equations are certainly sufficient to describe the sound field, they are quite complex and can, in any case, be proved using computer programs. The following step-by-step simplifications are intended to aid in the understanding of the statements these equations make.

An initial examination should be based on the extreme case of a far-away sound source. The sound incident should simply be regarded as a plane progressive wave impinging at an oblique incidence angle φ_0. The approximations made by equations(10.4) and (10.5) result in the following for the Hankel function of the second order:

$$H_{n/2}^{(2)}(x) \simeq \sqrt{\frac{2}{\pi x}} e^{-j(x - n\pi/4 - \pi/4)} \tag{10.22}$$

transforming (10.20) into

$$p(r,\varphi) = p_1 = p_Q(0) \sum_{n=0}^{\infty} \frac{2e^{jn\pi/4}}{\varepsilon_n} J_{n/2}(kr) \cos(n\varphi/2) \cos(n\varphi_0/2) \tag{10.23}$$

The application of eq.(10.23) becomes more exact with ever increasing distance to the source. Given the outlying case, where the source is at an infinite distance and only a plane wave is present, this approximation then becomes the correct and exact solution. The subspace 2 has now expanded without bound into infinity and has therefore become irrelevant. p_1 is now replaced by p in the eq.(10.23), which indicates the sound field in the entire room. We will use this abbreviation hereafter.

We can now express the sound pressure in simpler terms, due to the fact that a seemingly unimportant parameter (the distance to the source) has been eliminated. However, this mathematical model still does not produce self-evident and comprehensible results. Moreover, the convergence of the series to the right is primarily dependent on the distance defined r. Precisely at the large distances useful for calculations in practice (such as residential areas near streets with noise barriers), the numerical calculations become extremely complicated because the Bessel functions of increasing orders only become negligible (small) when the order becomes larger than the argument kr. This is reason enough to consult the reference tables in the relevant reference resources for other ways to describe the problem. In fact, other options for expressing the right-hand expansion term by way of so-called Fresnel integrals are explored in the highly-recommended work of authors Gradshteyn, Ryzhik: Table of Integrals, Series and Products (Academic Press, New York 1965), on page 973, Nr. 8.511.5. Ultimately, using the equations given in this reference is up to the reader. After performing simple calculations, we obtain

$$p\left(r,\varphi\right) = p_{\mathrm{Q}}\left(0\right)\frac{1+j}{2}\left\{\mathrm{e}^{jk_0 r\cos(\varphi-\varphi_0)}\phi_+ + \mathrm{e}^{jk_0 r\cos(\varphi+\varphi_0)}\phi_-\right\}, \qquad (10.24)$$

whereby

$$\phi_+ = \frac{1-j}{2} + C\left(\sqrt{2k_0 r}\cos\frac{\varphi-\varphi_0}{2}\right) - jS\left(\sqrt{2k_0 r}\cos\frac{\varphi-\varphi_0}{2}\right) \qquad (10.25)$$

and

$$\phi_- = \frac{1-j}{2} + C\left(\sqrt{2k_0 r}\cos\frac{\varphi+\varphi_0}{2}\right) - jS\left(\sqrt{2k_0 r}\cos\frac{\varphi+\varphi_0}{2}\right) \qquad (10.26)$$

are used for a better representation of the sound pressure. The functions which are defined herein are defined by

$$C\left(x\right) = \sqrt{\frac{2}{\pi}}\int\limits_0^x \cos\left(t^2\right)dt \qquad (10.27)$$

and

$$S\left(x\right) = \sqrt{\frac{2}{\pi}}\int\limits_0^x \sin\left(t^2\right)dt \qquad (10.28)$$

These are known as Fresnel integrals. Their characteristics are shown in Fig. 10.4 (the code of a MATLAB program for the calculation of C and S is printed in the appendix of this chapter for the use in the public domain).

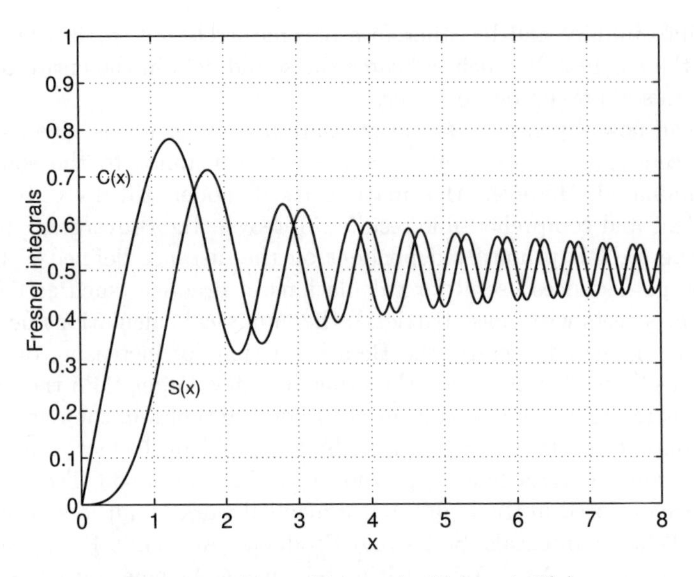

Fig. 10.4. Fresnel integrals $S(x)$ and $C(x)$

Representing sound pressure using Fresnel integrals has some advantages over the Bessel series expansion in eq.(10.23). The Fresnel integrals are not only simply to program (see the appendix to this chapter), but they can also be quite accurately approximated for the more practically relevant large distances r. Namely, they enable a direct and immediate assessment of sound fields.

Obviously, C and S are functions which, for increasing argument, alternate around $1/2$ with decreasing amplitude and thus approximations for the Fresnel integrals for $x \gg 1$ exist as

$$C(x) \simeq \frac{1}{2} + \frac{1}{\sqrt{2\pi x}} \sin\left(x^2\right) \tag{10.29}$$

$$S(x) \simeq \frac{1}{2} - \frac{1}{\sqrt{2\pi x}} \cos\left(x^2\right) . \tag{10.30}$$

For negative arguments, the symmetry following from the definitions (10.27) and (10.28)

$$C(-x) = -C(x) \tag{10.31}$$

$$S(-x) = -S(x) \tag{10.32}$$

can be observed.

It should be explicitly stated that the permissible range for the circumferential angle in (10.24) to (10.28) is restricted to $0 < \varphi < 2\pi$. Angle values outside of this interval, especially negative values, are excluded. They produce errors in the results during the analysis. For the incidence angle φ_0, positive values are also assumed. In (10.24), $p_Q(0)$ is understood as the sound pressure, which the incident plane wave would produce without the screen (in the free-field) at the coordinate origin $r = 0$ (see also (10.33)).

The incident wave is described by

$$p_{\mathrm{ein}} = p_Q\,(0)\,\mathrm{e}^{jk(x\cos\varphi_0 + y\sin\varphi_0)}\ ,$$

or, with $x = rcos\varphi$ and $y = rsin\varphi$ for the coordinate systems (x, y) and (r, φ), and due to $cos\varphi\, bycos\varphi_0\ +\ sin\varphi\, sin\varphi_0\ = cos(\varphi - \varphi_0)$,

$$p_{\mathrm{ein}} = p_Q\,(0)\,\mathrm{e}^{jkr\cos(\varphi-\varphi_0)}\ . \tag{10.33}$$

Since the amplitude of a plane wave is spatially independent and is equal to $p_Q(0)$ everywhere, the insertion loss of the semi-infinite screen is given by

$$R_{\mathrm{E}} = -10\,\mathrm{lg}\left|\frac{p\,(r,\varphi)}{p_Q\,(0)}\right|\ . \tag{10.34}$$

The insertion loss can, of course, be different at different positions.

Instead of the mathematical proof by an exact calculation – which is extensive and honestly, a bit lengthy and dry – the results will be represented graphically in order to at least deliver some plausibility. For that reason, we calculate the displacement of the field points in the elastic continuum, which consists of gas, instead, as in

$$\xi_x = \frac{1}{\varrho\omega^2}\frac{\partial p}{\partial x}; \qquad \xi_y = \frac{1}{\varrho\omega^2}\frac{\partial p}{\partial y} \tag{10.35}$$

and drawn in a field pattern of points (Fig. 10.5 to 10.7). The derivatives can be replaced approximately by difference quotients, for instance, $\mathrm{d}p/\mathrm{d}x \approx (p(x + \Delta x) - p(x))/\Delta x$ ($\Delta x = \lambda/100$ was used in the figures, a choice which is also useful otherwise), where each p is calculated according to (10.24). The resulting pattern of motion can be easily interpreted: 'too high density' of points (compared to the equidistant pattern 'without sound') indicates sound density and sound pressure above (or for 'low density', below) the atmospheric quantities, the distance between two areas of high (low) compression indicates the wavelength. The sound field in Fig. 10.5 to 10.7 is shown for a constant (frozen) time; multiple snapshots (e.g. for $t/T = 0; 1/50; 2/50; \ldots, 49/50$, where T is the time period) in a series would result in an animation which would document the time history of the wave propagation.

The snapshot of the sound field produced in this way (Figures10.5 to 10.7) show reasonable tendencies. Aside from the fact that this obviously deals with waves occurring everywhere,

- the boundary conditions on both sides of the rigid screen are fulfilled,
- the reflection on top of the screen can be observed with resulting standing waves in the area $\varphi < \pi - \varphi_0$,

- the total sound field in the 'light zone' consists of the undisturbed by-passing incident plane wave and, finally,
- the diffracted wave expected in the shadow region can be observed.

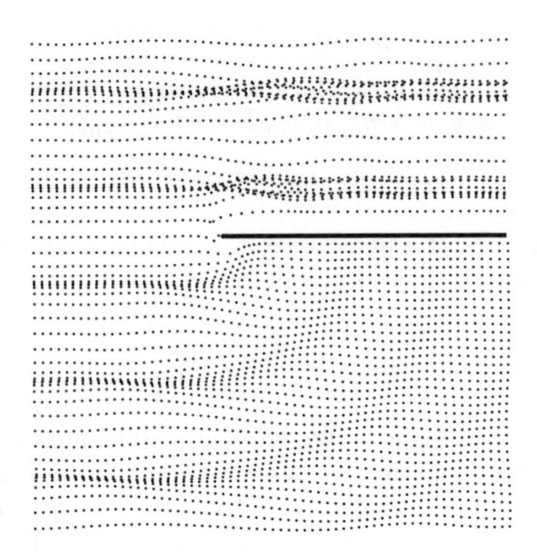

Fig. 10.5. Particle motion in the sound field in front of semi-infinite screen, incidence angle $\varphi_0 = 90°$

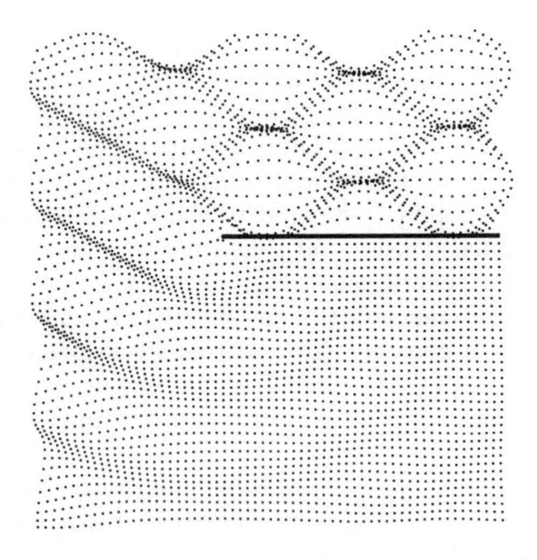

Fig. 10.6. Particle motion in the sound field in front of semi-infinite screen, incidence angle $\varphi_0 = 60°$

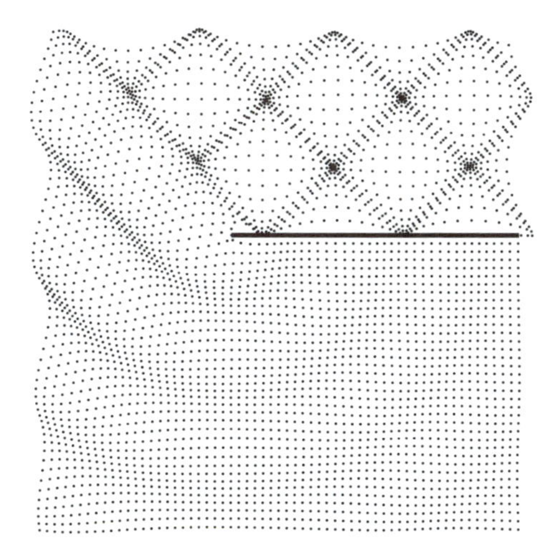

Fig. 10.7. Particle motion in the sound field in front of semi-infinite screen, incidence angle $\varphi_0 = 45°$

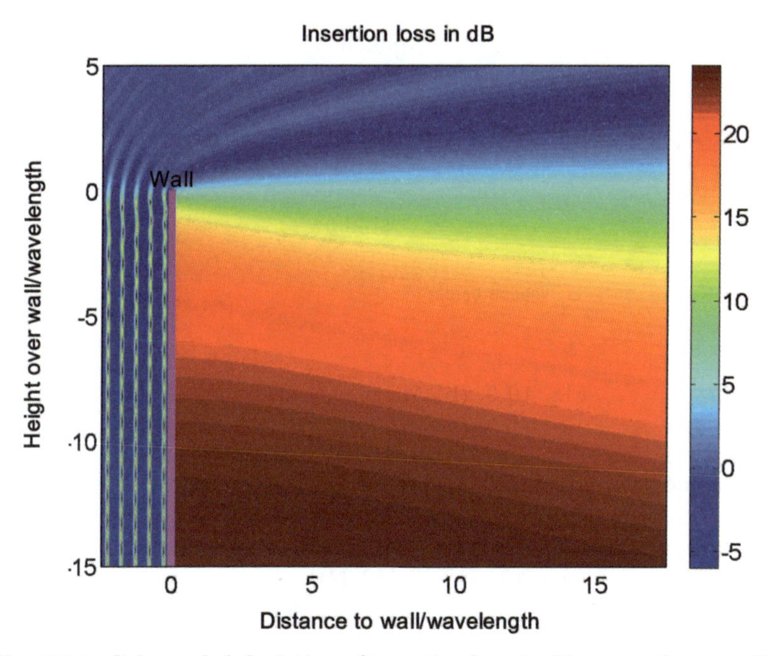

Fig. 10.8. Color-coded depiction of insertion loss, incidence angle $\varphi_0 = 90°$

It can be said in terms of the shadow region that the visible dynamic of this type of representation in particle motion snapshots as shown in Figures

10.5 to 10.7 encompasses approximately $10\,dB$, which is why the insertion losses of $R_{\mathrm{E}} > 10\,dB$ are not able to be adequately depicted this way. A more 'readable' diagram more accurately reflecting the insertion loss can be seen in Figure 10.8, which provides a color-coded schema of the insertion loss (whereby the coordinate system is simultaneously rotated so that the barrier – shown here in pink – is 'upright' and the sound incidence is impinging from the left.)

As mentioned previously, one of the advantages of representing the sound field using eq.(10.24) is the rather simple treatment of its approximation. This discussion is not only to explore the primary subject of interest – shadow region – but we will also examine the reflection domain, or 'light zone,' and finally, the shadow border, so that we can verify the result. The spatial zones of interest here are shown in Figure 10.9. Since the area surrounding the edge of the screen $r \approx 0$ is of little interest, the parameter $kr >> 1$ applies to the discussion which follows.

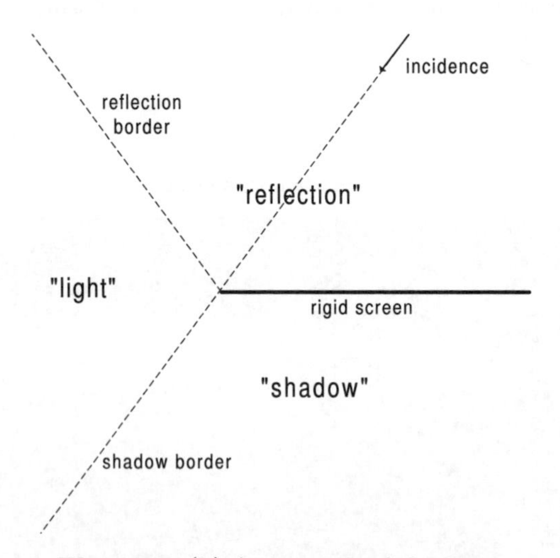

Fig. 10.9. (b) Assignment of the zones

The principal characteristics of the quantities ϕ_+ and ϕ_- are determined by the sign of the argument in the corresponding Fresnel integral; they alternate around the value 1/2 for positive arguments and around $-1/2$ for negative arguments (see (10.31) and (10.32)).

If the argument of the Fresnel integral is denoted by u, which means $u = \sqrt{2k_0 r}\cos(\varphi - \varphi_0)/2$ for ϕ_+ and $u = \sqrt{2k_0 r}\cos(\varphi + \varphi_0)/2$ for ϕ_-, it follows from (10.29) to (10.32) that

$$\phi \approx 1 - j \quad \text{for} \quad u > 0 \quad \text{and} \quad |u| \gg 1 \tag{10.36a}$$

and

$$\phi \approx \frac{j\mathrm{e}^{-ju^2}}{\sqrt{2\pi}\,|u|} \quad \text{for} \quad u < 0 \quad \text{and} \quad |u| \gg 1\,. \tag{10.36a}$$

Using the simplifications mentioned in (10.36) and (10.36), the principal characteristics of the sound field in the zones of interest are easily discussed.

a) Reflection zone

The reflection zone is characterized by

$$\varphi < \pi - \varphi_0\,.$$

Inside of it

$$\frac{\varphi - \varphi_0}{2} < \frac{\pi}{2} - \varphi_0$$

and

$$\frac{\varphi + \varphi_0}{2} < \frac{\pi}{2}$$

are given. It follows that

$$\cos\frac{\varphi - \varphi_0}{2} > 0$$

and

$$\cos\frac{\varphi + \varphi_0}{2} > 0\,.$$

Hence, *both* arguments of the occurring Fresnel integrals are positive. The approximation for ϕ_+ as well as ϕ_- is thus given by (10.36). Therefore, in the reflection zone, according to (10.24) and using $(1-j)(1+j) = 2$,

$$p\,(r,\varphi) \approx p_{\mathrm{Q}}\,(0)\left\{\mathrm{e}^{jk_0 r\cos(\varphi-\varphi_0)} + \mathrm{e}^{jk_0 r\cos(\varphi+\varphi_0)}\right\} \tag{10.37}$$

applies. The first term describes the incident field (see (10.33)), the second term describes the field reflected at $\varphi = 0$.

b) Light zone

The 'light zone' describes the region in space, where the undisturbed incident wave is an expected result occurrence. Here, it is

$$\pi - \varphi_0 < \varphi < \pi + \varphi_0,$$

and therefore

$$\frac{\pi}{2} - \varphi_0 < \frac{\varphi - \varphi_0}{2} < \frac{\pi}{2}$$

and

$$\frac{\pi}{2} < \frac{\varphi + \varphi_0}{2} < \frac{\pi}{2} + \varphi_0\,.$$

For this reason

$$\cos\frac{\varphi-\varphi_0}{2}>0$$

and

$$\cos\frac{\varphi+\varphi_0}{2}<0$$

are given. It is therefore $\phi_+ \approx 1-j$, under the assumption of large distances $k_0 r \gg 1$. In contrast, ϕ_- is small according to (10.36) and can be neglected in comparison to ϕ_+. Consequently, the total sound field according to (10.24) correctly consists of the incident wave alone

$$p(r,\varphi)=p_Q(0)\,\mathrm{e}^{jk_0 r\cos(\varphi-\varphi_0)}\ .$$

The discussions in the reflection and light zone served more as a cross-check of the equations derived earlier. The following discussion specifies the benefit which can be expected by a screen.

c) Shadow border

Along the shadow border given by

$$\varphi=\pi+\varphi_0\ ,$$

which can also be written as

$$\frac{\varphi-\varphi_0}{2}=\frac{\pi}{2}\ ,$$

it is

$$\frac{\varphi+\varphi_0}{2}=\frac{\pi}{2}+\varphi_0\ .$$

The argument of the Fresnel integrals for ϕ_+ is likewise zero, due to $\cos(\varphi-\varphi_0)/2=0$ and, using $S=C=0$,

$$\phi_+=\frac{1-j}{2}\ .$$

The argument of the Fresnel integrals for ϕ_- is negative because of

$$\cos\frac{\varphi+\varphi_0}{2}<0\ .$$

According to (10.36), ϕ_- can again be neglected in comparison to ϕ_+ and

$$p(r,\varphi)=p_Q(0)\,\frac{1}{2}\mathrm{e}^{-jk_0 r} \qquad (10.38)$$

is obtained. Hence, at larger distances from the screen's edge, half the sound field incidence is obtained along the shadow border. This interesting fact could be interpreted as the distant source being 'half covered by the screen', similar to what happens during a sunset, when only half the sun is visible. With increasing distance, the insertion loss along the shadow border tends to

$$R_E=6\,\mathrm{dB}\ . \qquad (10.39)$$

d) Shadow region

In the shadow region, given by

$$\varphi > \varphi_0 + \pi,$$

it is likewise

$$\frac{\varphi - \varphi_0}{2} > \frac{\pi}{2}$$

and

$$\frac{\varphi + \varphi_0}{2} > \frac{\pi}{2} + \varphi_0 .$$

In this case, the arguments of all Fresnel integrals are negative and, according to (10.36),

$$\phi_+ \approx \frac{j e^{-j2k_0 r \cos^2 \frac{\varphi-\varphi_0}{2}}}{\sqrt{2\pi}\left|\sqrt{2k_0 r}\cos\frac{(\varphi-\varphi_0)}{2}\right|}$$

and

$$\phi_- \approx \frac{j e^{-j2k_0 r \cos^2 \frac{\varphi+\varphi_0}{2}}}{\sqrt{2\pi}\left|\sqrt{2k_0 r}\cos\frac{(\varphi+\varphi_0)}{2}\right|}$$

are given. Consequently, the sound pressure is given by

$$p = p_Q(0)\,\frac{j-1}{2\sqrt{2\pi}}\,\frac{e^{-jk_0 r}}{\sqrt{2k_0 r}}\left\{\frac{1}{\left|\cos\frac{(\varphi-\varphi_0)}{2}\right|} + \frac{1}{\left|\cos\frac{(\varphi+\varphi_0)}{2}\right|}\right\}, \tag{10.40}$$

(where $\cos(\alpha) - 2\cos^2(\alpha/2) = \cos(\alpha) - (1+\cos(\alpha)) = -1$ was used in the arguments of the exponential functions).

In discussing the shadow region, it becomes clear that the distance of a point to the shadow border $\varphi = \varphi_0 + \pi$ is crucial. For that reason, we introduce what is termed the diffraction angle β. It simply counts relative to the shadow border and is given by

$$\varphi = \pi + \varphi_0 + \beta .$$

The two angle-dependent expressions in (10.40) become

$$\left|\cos\frac{\varphi - \varphi_0}{2}\right| = \sin\frac{\beta}{2}$$

and

$$\left|\cos\frac{\varphi + \varphi_0}{2}\right| = \sin\left(\frac{\beta}{2} + \varphi_0\right) .$$

Note, that for small diffraction angles, the approximations derived for the shadow region are not valid (see the remarks on the shadow border above), therefore, we assume 'medium to large' diffraction angles. For the angular ranges of $30° < \beta < 120°$ and $0° < \varphi_0 < 90°$, $\sin(\beta/2 + \varphi_0)$ differs only slightly from $\sin(\beta/2)$. The second term in (10.40) can thus be approximated by the first:

$$p \approx p_{\mathrm{Q}}(0) \frac{(j-1)}{\sqrt{2\pi}} \frac{\mathrm{e}^{-jk_0 r}}{\sqrt{2k_0 r} \sin\frac{\beta}{2}} \tag{10.41}$$

The insertion loss is then given by

$$R_{\mathrm{E}} = 10 \lg \left| \frac{p_{\mathrm{Q}}(0)}{v} \right|^2 \approx 10 \lg \left(4\pi^2 r \sin^2 \frac{\beta}{2} \right) . \tag{10.42}$$

The term $2r \sin^2(\beta/2)$ can be interpreted geometrically. It is equal to the difference U between the path which a sound ray takes from the distant source to the field point, 'bending' over the screen edge, versus the 'direct' path of the sound ray to the field point in the absence of the barrier (Fig. 10.10). This path difference is called detour U and is, according to Fig. 10.10, given by

$$U = r - D = r - r\cos\beta = r(1 - \cos\beta) = 2r \sin^2 \frac{\beta}{2}$$

and thus the insertion loss is given by

$$R_{\mathrm{E}} \approx 10 \lg \left(2\pi^2 \frac{U}{\lambda} \right) . \tag{10.43}$$

Equation (10.43) is called 'detour law', because it states that the insertion loss which is produced by sound protection screens only depends on the ratio of detour and wavelength.

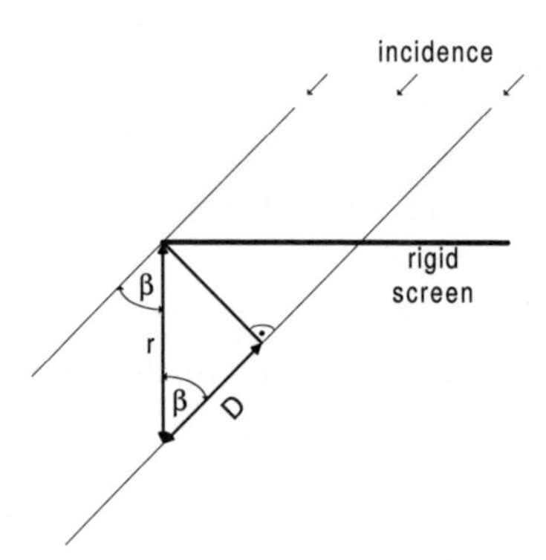

Fig. 10.10. Detour U = path over the edge r – direct path D

These days, practically all measurements of the effects of noise barriers are still conducted with the aid of eq.(10.43) or other such methods of approximation. This guideline generally treats the detour in a somewhat more complex manner, referring to the detour as the 'z value'. The detour law is also applied to sources with a finite distance to the wall. Detours will furthermore be evaluated in the following sections in terms of their geometry. Ground reflection will be eliminated from the discussion.

According to eq.(10.43), the following basic tenets apply to noise barriers:

- The insertion loss is frequency-dependent. This effect is minimal at low frequencies, more significant at higher frequencies.
- The highest noise barriers possible are required for greater detour effect.
- Low-lying sound sources, lying directly on the street or train tracks, are more easily mitigated by way of shadow effects as sound sources at higher altitudes.

In sum, the noise of truck tires is better shadowed than an open exhaust pipe, which is farther from the ground. Noise emission from the walls of the locomotive is greater than that of the attached coach trains, because the coach only only experiences noise emission from the wheels on the track, whereas the locomotive receives additional noise impact from the engine air box.

10.2 Approximation of insertion loss

Although the detour law likely represents one of the most fundamental aspects of noise control, it does not provide very precise data regarding insertion loss. The following approximation equation provides a more accurate estimation of the information of insertion loss data

$$R_E = 20\lg\left(\frac{\sqrt{2\pi N}}{th(\sqrt{2\pi N})}\right) + 5\,dB\,. \tag{10.44}$$

N is what is referred to as the Fresnel number

$$N = 2U/\lambda \tag{10.45}$$

and th refers to the hyperbolic tangent. Eq.(10.44) is taken from the 'Handbook of Acoustic Engineering' (Springer-Verlag, Berlin 2004, edited by G. Müller and M. Möser).

The accuracy of the statement made by the approximation (10.44) can be subsequently examined by comparing the approximation to the result of the exact calculation based on equation (10.24). To do this, we choose an independent quantity U/λ for a constant incidence angle φ_0 and draw the array of curves for the insertion loss, using the diffraction angle as parameter. With the aid of (10.24) the necessary quantities for a numerical evaluation are then obtained by both

$$\frac{r}{\lambda} = \frac{U}{\lambda}\frac{1}{2\sin^2\beta/2}$$

and $\varphi = \varphi_0 + \pi + \beta$. The calculated array of curves is depicted in Fig. 10.11. As can be seen, an maximum margin of error of $2\,dB$ results in the interval of the diffraction angles shown. To achieve more accurate results, it is either necessary to execute the eq.(10.24) itself using a computer program or to use the curves in Figure 10.12 as a reference (these curves are calculated with the aid of eq.(10.24).

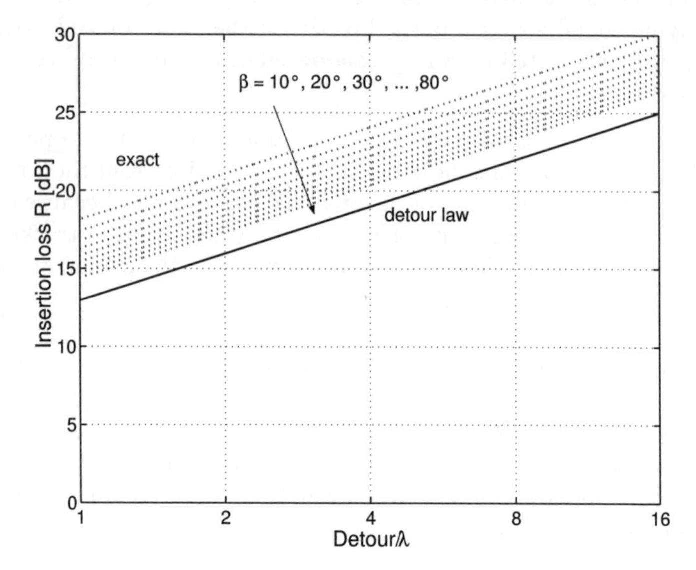

Fig. 10.11. Comparison of the detour law (10.43) with the exact calculation (10.24)

As can be seen in Figure 10.11, eq.(10.44) indicates a rather precise approximation even for small diffraction angles β. At the shadow border $N = 0$, $R_E = 5\,dB$ (because of $th(x) = x$ for small x), based on eq.(10.44), approaching the aforementioned correct value of $6\,dB$.

For Fresnel numbers $N > 0.36$, $0.9 < th(\sqrt{2\pi N}) \leq 1$, with less than $1\,dB$ margin of error for $N > 0.36$ if the hyperbolic tangent in (10.44) is set at 1. At an accuracy level standard for the field of acoustics, the following applies in the case of $N > 0.36$:

$$R_E = 10\lg(2\pi N) + 5dB \tag{10.46}$$

Determining the insertion loss is therefore reduced to simple geometric considerations, which has qualitative and quantitative implications in practice and which will be elaborated in the following. Figure (10.13) shows a typical arrangement of source (a_Q from the wall), an acoustic screen h_S high (above the source) and the position of impact E, which is h_E above the source and a_E from the wall. The main sources are located in the roadway, h_S and h_E indicates the heights relative to the street or train tracks. In respect to the

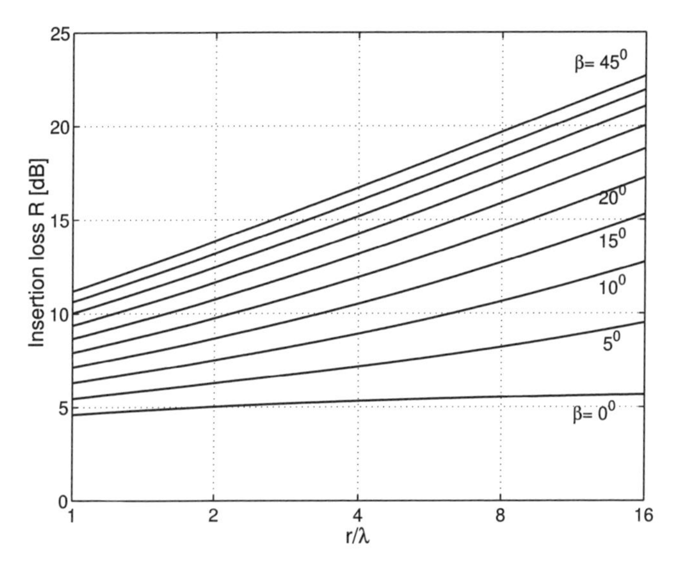

Fig. 10.12. Insertion loss according to (10.24) for small diffraction angles

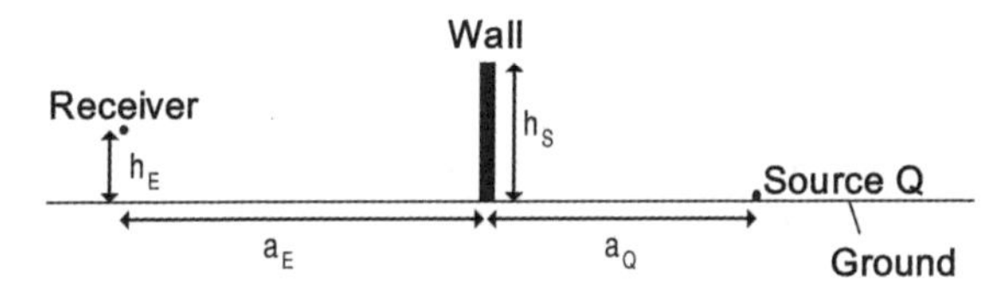

Fig. 10.13. Arrangement of source Q, noise barrier with height of h_S and position of impact ('receiver')

pathway over the edge of the barrier K (=Emission path from Q to E over the acoustic screen), the following results from applying the Pythagorean theorem twice

$$K = \sqrt{h_S^2 + a_Q^2} + \sqrt{(h_S - h_E)^2 + a_E^2}\,,$$

and for the direct path

$$D = \sqrt{h_E^2 + (a_E + a_Q)^2}\,.$$

The detour U is $U = K - D$. In practice, noise impacted areas tend to be far enough away so that $a_E >> h_S$ applies. Typical distances a_E are anywhere from at least 100 or so meters and barrier heights which, in contrast, seldom reach more than 5 meters. Therefore, $a_E >> h_S$ almost always applies. The expressions containing $(h_S - h_E)^2$ and h_E^2 are therefore far smaller than a_E^2, as long as h_E does not greatly surpass the barrier height h_S. If, for example, $a_E = 20(h_S - h_E)$ were true, $(h_S - h_E)^2 = a_E^2/400$ would, of course, be a comparatively tiny quantity, which would have very little impact on the corresponding radical. The quadratically small terms in the radicals can therefore be eliminated, resulting in

$$K = \sqrt{h_S^2 + a_Q^2} + a_E$$

and

$$D = a_E + a_Q \, .$$

For large distances a_E, the detour U is almost independent of the measured distance a_E and height h_E, as evident in

$$U = K - D = \sqrt{h_S^2 + a_Q^2} - a_Q \, . \tag{10.47}$$

The insertion loss is therefore almost exclusively defined by the source distance a_Q and barrier height h_S, while remaining virtually independent of the specified position of impact E.

A realistic measurement of the insertion loss quantity can also be made based on these considerations. Take a somewhat typical example of a $5\,m$–wall along a wide, three-lane highway. The width of each lane is approx. $3.5\,m$. The shoulder accounts for an additional $3\,m$ of distance from the edge of the highway to the noise barrier. One can assume that the source is located in the middle of the highway, at about $8\,m$ distance to the sound wall. This results in a detour of $1.43\,m$. The center frequency for traffic noise is approx. $1000\,Hz$ with $\lambda = 0.34\,m$. The Fresnel number is thereby $N = 8.44$. Since the Fresnel number is $N > 0.36$, we can use eq.(10.46), resulting in $R = 22.2\,dB$.

It is also worth noting that, at a distance between the wall and the source greater than a_Q, the detour U monotonically attenuates. The noise barrier is therefore more effective the closer the source is to the wall.

10.3 The importance of height in noise barriers

Of course, the question has been raised as to the advantages of building higher noise barriers. All other factors notwithstanding, if a higher noise barrier is built, h_2 as opposed to the previous height h_1, then

$$\Delta R = 10 lg(\frac{U_2}{U_1}) \tag{10.48}$$

shows the advantage for the insertion loss. U_1 and U_2 are the detours for h_1 and h_2, respectively. The advantage gained furthermore depends on the source distance. The interval where the gain ΔR has to be, however, is not easy to ascertain. A minimal gain only becomes apparent if the wall has already been shown to be effective. This is always the case when the wall and the source are not far apart, whereby, according to eq.(10.47), the detours are equal to the heights, resulting in

$$\Delta R_{min} = 10lg(\frac{h_2}{h_1}) . \tag{10.49}$$

Otherwise, where a is much greater than the height of the wall h, the following

$$\sqrt{a^2 + h^2} \approx a + \frac{h^2}{2a} , \tag{10.50}$$

is a good approximation for far-away sources, resulting in the detour $U \approx h^2/2a$. For this reason,

$$\Delta R_{max} = 20lg(\frac{h_2}{h_1}) = 2\Delta R_{min} \tag{10.51}$$

applies.

As can be seen above, constructing higher noise barriers results in a greater attenuation gain for sources at a greater distance to the wall as for those close to the wall. However, the improvements are not all that significant for most real scenarios. If, for example, a $6\,m$– instead of a $5\,m$–high wall is built, an increase in attenuation has only been obtained somewhere between $0.8\,dB$ and $1.6\,dB$, depending on the location of the source. Even if the height of the sound wall is doubled, the improvement will only be somewhere between 3 and 6 dB. The high insertion losses required in the construction of noise barriers is quite expensive.

As outlined in the example in the previous section, noise barriers can guarantee roughly $20\,dB$ insertion attenuation. Larger attenuations would require truly gigantic wall structures. Noise barriers are an important aspect in noise control, but cannot create wonders.

10.4 Sound barriers

The simple model of a barrier in the shape of a semi-infinite screen resulted in a lucid description of the diffraction and reflection principles. But some questions remained unanswered. For instance, can the calculations be readily transferred to other geometric shapes like sound protection dams instead of walls? The effect of wedge-shaped sound barriers can also be calculated theoretically. The derivation of the formulas would be beyond the scope of this book, but a numerical evaluation is given in Fig. 10.14. (See Figures 10.14 and 10.15).

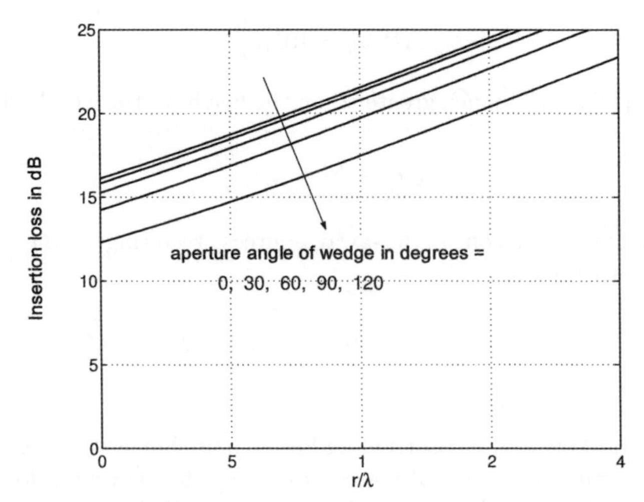

Fig. 10.14. Insertion loss of wedge-shaped sound walls, computed for an incidence angle $\varphi_0 = 60^0$ and diffraction angle $\beta = 60^0$

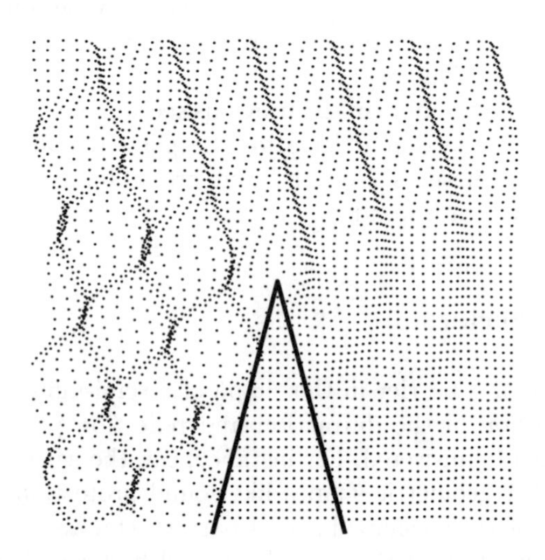

Fig. 10.15. Reflection and diffraction of wedge-shaped sound walls

As shown in Figure 10.14 the aperture angle below 90° is only of minor importance, because the detour law can readily be used for $\gamma < 90°$. In contrast, the deviation from the detour law quickly becomes significant above 90°. The insertion loss R_{E} is worse in this case than it is for semi-infinite screens. Aperture angles of more than 120° can be found on roofs and raised dams. Compared to walls of the same height, a considerable influence of their geometric shape on the insertion loss has to be taken into account.

10.5 Absorbent noise barriers

In practice, noise barriers always have absorbent qualities along the side facing the source. This section attempts to explain the advantages this feature has for sound insulation.

As illustrated by the simple example below, an absorber mounted to the sound wall is only effective when the relevant barriers are close to one another. Such is the case when large vehicles, such as trains or a trucks, pass by. In other words, that the distance between the barrier and the outside of these vehicles is relatively short. Indeed, the actual large-scale sound emitters mentioned above are located relatively low, as they mainly consist of the noise of the tire traction against the surface of the highway. In the case of a reflecting noise barrier, however, the sound field is sent upwards along a zig-zag path (see Figure 10.16) before striking the edge of the barrier at an unfavorable angle in terms of desired sound insulation effect. This effect leads to a virtual raising of the sound source emission to the much higher level, far above the actual location on the highway or train tracks. This causes a considerable reduction in the desired attenuation effect.

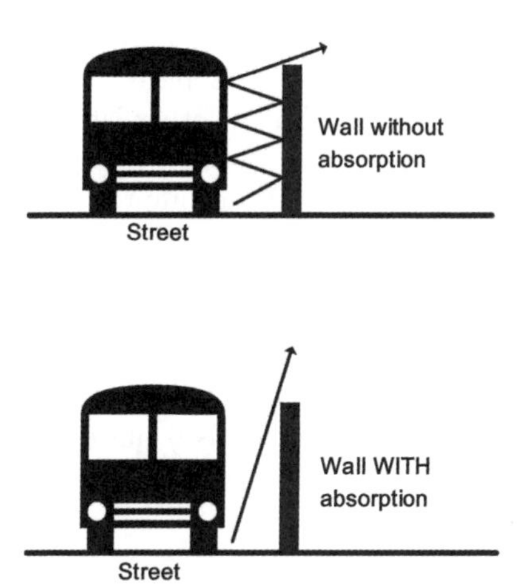

Fig. 10.16. Main pathway of the sound propagation from the source to the barrier's edge for sound walls with or without source-side absorption

In the case of smaller vehicles passing by the barrier, the reflection off the vehicle itself does not play much of a role. Multiple reflections can only occur if noise barriers are built along both sides of the highway or train tracks. For the most part, however, the distances are so great that the sound incidence from the original sources as well as the reflective sources has a more or less horizontal striking effect.

In the case of large noise emitters, the implementation of source-side absorbent noise barriers can prevent the so-called zigzag effect. A simulation calculation can illustrate the difference they make. Two examples are shown in Figure 10.17. The simulation is drawn on the assumption that sound fields resulting from reflective sources are diffracted on each wall. To keep it simple, let us assume that the level of the source outdoors at a distance of 1 m is 100 dB. The center frequency of the traffic noise has thereby been measured at 500 Hz. The strong spatial dependency of the improvements achieved by the absorption are plain to see here.

For more extreme cases, great improvements can be attributed to wall absorption. Sample measurements of the effect of absorbent linings in the shadow region behind the wall are given in Figure 10.18. This figure shows the level reductions obtained by the implementation of the source-side wall absorption as compared to the level reductions of a simple reflective barrier. To measure the level reductions, sheet metal plates were mounted to the absorbent wall lining. The level measurements of the wall covered by the sheet metal plates were then compared with the levels without the plates. The measured improvements were considerable as Figure 10.18 shows. The quantitative relationship between the sound insulation results and sound absorption usage were proven by these measurements. The highest absorption factor correlates with the greatest level difference. Here, however, the distance between the wall and the reflector (a passing truck) was only 1 m, whereby the wall and the truck were both 4 m high. This is a typical situation which actually occurs in the construction industry, but in other situations, the distances are considerably greater.

As already mentioned, these significant improvements diminish when the reflector and the noise barrier are close together (see the vehicle in Figure 10.16). For distances equal to the height of the wall, only the absorption itself is attributable to a slight improvement of approx. $3\,dB$. Such improvements directly correlated with the absorption only work for large vehicles, such as trucks, but for small sources like passenger cars and motorcycles, the effect of sound absorption on overall emission levels is quite minimal.

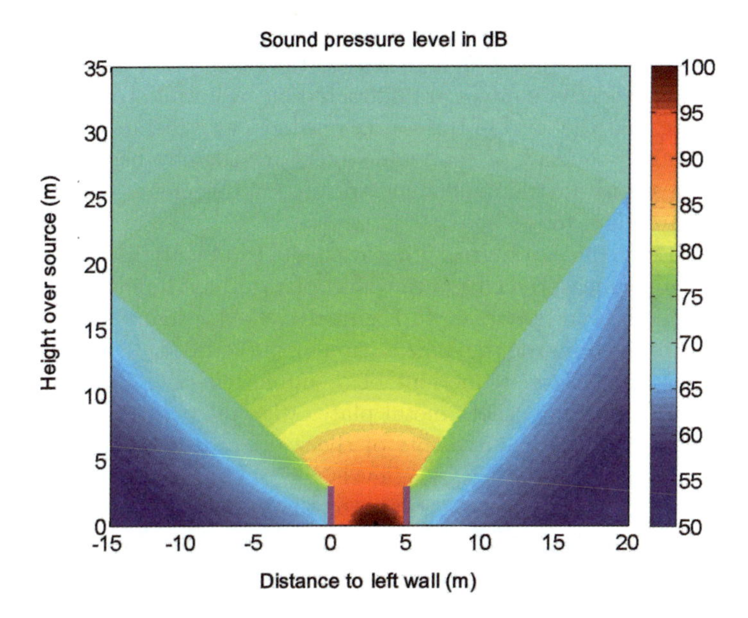

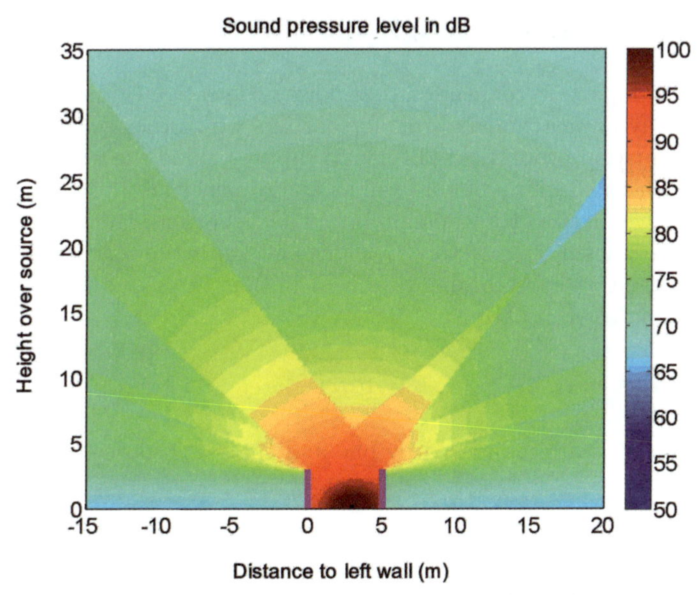

Fig. 10.17. Sound pressure levels for fully absorbent (above) and fully reflective (below) walls

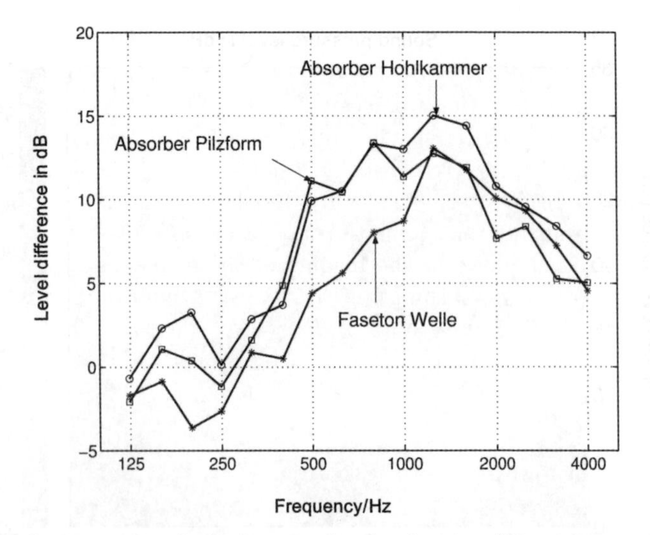

Fig. 10.18. Space-averaged third-octave level reduction (30 positions of measurement) due to absorbent linings, measured behind a 3.5 *m*–high wall as a truck passes by (height of truck likewise 3.5 *m*) 1 *m* from the wall for three different absorbent linings. Walls belonging to the Rieder family in Maishofen, Austria.

10.6 Transmission through the barrier

Finally, it is important to mention that noise penetration through sound walls can itself play an instrumental role for barriers which themselves have insufficient attenuation properties. The sound supply behind the wall results from the aforementioned diffraction as well as due to penetration through the barrier itself. Overall, two transmission paths must be considered. They are defined by the transmission factors τ_B along the diffraction path and τ_D along the penetration path. According to this definition, the sound reduction index corresponding to each path is

$$R = 10\,lg(\frac{1}{\tau})\,. \tag{10.52}$$

Since both paths are influenced by one and the same source, the net transmission factor is

$$\tau_{ges} = \tau_B + \tau_D\,. \tag{10.53}$$

This leads to the definition of the net sound reduction index in

$$R_{ges} = 10\,lg\frac{1}{\tau_{ges}} = 10\,lg\frac{1}{\tau_B + \tau_D} = 10\,lg\frac{1}{10^{-R_B/10} + 10^{-R_D/10}}\,. \tag{10.54}$$

A simpler notation for this is

$$R_{ges} = 10\,lg\frac{10^{R_B/10}}{1+10^{(R_B-R_D)/10}} = R_B - 10\,lg(1+10^{(R_B-R_D)/10}) \qquad (10.55)$$

Accordingly, the penetration reduction index of $R_D = R_B$ results in a marked decline of $3\,dB$ in the net effect compared to a well insulating barrier. For $R_D = R_B + 6\,dB$ the net reduction factor is $R_{ges} = R_B - 1\,dB$ and finally, for $R_D = R_B + 10\,dB$, it is $R_{ges} = R_B - 0.4\,dB$. As shown above, even the slightest improvements in attenuation are often only feasible in conjunction with considerable increase in the heights of the sound barriers. This underscores the importance of combating the loss of insulation through potential noise penetration, aiming for approx. $R_D = R_B + 6\,dB$.

10.7 Conclusion

Other important and frequently asked questions regarding the practical implications and optimization of acoustic screens pertain to:

- Ground reflection
- Wind and weather
- Propagation attenuation over large distances
- Plant cover
- Unconventional geometry, such as overhangs
- and other influential factors not discussed here

10.8 Summary

The diffracted sound field in the geometric shadow behind a very long, rigid and thin barrier behaves according to the 'detour law.' This describes the fact that the insertion loss of the wall is derived from the difference between the 'sound path over the edge of the screen' and the 'direct path.' In everyday situations, losses exist, which depend only on the height of the wall and the distance of the wall from the source. Insertion losses seldom surpass $20\,dB$.

Absorption material is installed on the source-side to prevent multiple reflections between the source and the noise barrier, which could impede optimal sound protection.

The transmission loss of walls used as barriers should be at least $6\,dB$ more than the diffraction loss.

10.9 Further reading

To deepen your knowledge on this subject, it is recommended that you read the work of E. Skudrzyk "The Foundations of Acoustics" (Springer, Wien 1971). As a reference book for mathematical functions, the book by M. Abramowitz and I.A. Stegun "Handbook of Mathematical Functions" (9th Dover Printing, New York 1972) is a great resource.

10.10 Practice exercises

Problem 1

Calculate the sound levels reduced by increasing the height of noise barriers from $4\,m$ to $5.5\,m$, from $4\,m$ to $7.5\,m$, from $5.5\,m$ to $7.5\,m$ and from $7.5\,m$ to $10\,m$. Note that the emission positions are far away (distances more than one hundred meters) and the sources are out of sight range. The middle of a three-lane freeway should be used to gauge the distance between the sources (source distance from wall $6.7\,m$, $10.5\,m$ and $15\,m$, sources are on the freeway). Assume that the main frequency of the traffic noise is $1000\,Hz$.

Problem 2

How great are the insertion losses of noise barriers constructed at heights of $4\,m$, $5.5\,m$, $7.5\,m$ and $10\,m$ for distant emission stations out of sight range to the sources? The sources are at distance of $6.7\,m$, $10.5\,m$ and $15\,m$ from the wall. Again, calculate using $f = 1000\,Hz$.

Problem 3

If one considers sources both close to and far away from the barrier, what is the minimum as well as the maximum sound level reduction when increasing noise barriers from $4\,m$ to $5.5\,m$, from $4\,m$ to $7.5\,m$, from $5.5\,m$ to $7.5\,m$, and from $7.5\,m$ to $10\,m$, if the emission stations are far away (at distances of several hundred meters) and not in sight range of the sources?

Problem 4

By how much does the overall transmission loss R_{total} differ from the diffraction loss R_D, if the transit loss of the noise barrier is $6\,dB$ less than R_D?

Problem 5

As shown in the figure below (not to scale), a source and receiver are separated by two different barriers. This situation occurs in everyday life where, for instance, a railway runs alongside a street protected by noise barriers on both sides. How great is the insertion loss if the main frequency emitted is $500\,Hz$?

Problem 6

If we take the same arrangement from Problem 5 and omit one of the noise barriers ($4\,m$ high), by what amount does the insertion loss change at the receiver?

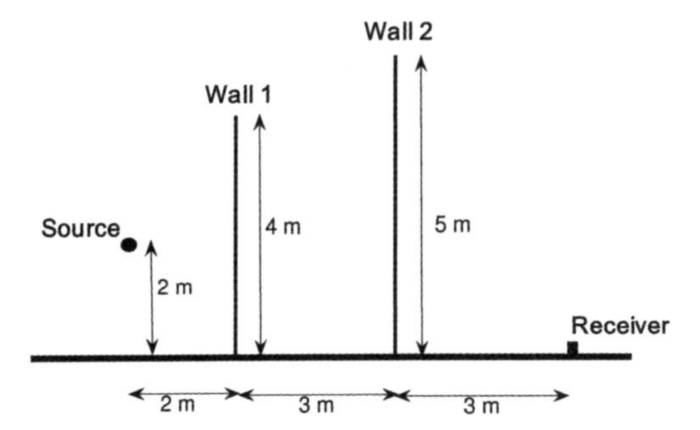

Fig. 10.19. Arrangement of source, receiver, and two barriers in Problem 5

10.11 Appendix: MATLAB program for Fresnel integrals

```
function [cfrenl,sfrenl] = fresnel(xarg)
   x=abs(xarg)/sqrt(pi/2);
   arg=pi*(x^2)/2;
   s=sin(arg);
   c=cos(arg);

   if x>4.4

        x4=x^4;
        x3=x^3;
            x1=0.3183099 - 0.0968/x4;
            x2=0.10132 - 0.154/x4;
            cfrenl=0.5 + x1*s/x - x2*c/x3;
            sfrenl=0.5 - x1*c/x - x2*s/x3;

        if  xarg<0
                        cfrenl=-cfrenl;
                        sfrenl=-sfrenl;
        end

   else

            a0=x;
            sum=x;
            xmul=-((pi/2)^2)*(x^4);
            an=a0;
            nend=(x+1)*20;

            for n=0:1:nend
                        xnenn=(2*n+1)*(2*n+2)*(4*n+5);
                        an1=an*(4*n+1)*xmul/xnenn;
                        sum=sum + an1;
                        an=an1;
            end

            cfrenl=sum;
            a0=(pi/6)*(x^3);
            sum=a0;
            an=a0;
            nend=(x+1)*20;


            for n=0:1:nend
```

```
                        xnenn=(2*n+2)*(2*n+3)*(4*n+7);
                        an1=an*(4*n+3)*xmul/xnenn;
                        sum=sum + an1;
                        an=an1;
            end

            sfrenl=sum;

        if xarg<0
                        cfrenl=-cfrenl;
                        sfrenl=-sfrenl;
        end

end
```

11

Electro-acoustic converters for airborne sound

The basis of each scientific or engineering discipline is the measurement of the quantities of interest. All physical effects mentioned in this book have to be verifiable by experiment. Acoustic measurement techniques which can prove the evidence of emission and impact is the daily bread of engineering acoustics.

This is reason enough to have a closer look at the two most important microphone types for air-borne sound, the condenser microphone and the electrodynamic microphone which form the quintessence of each experimental setup. There are, however, many more types, as for instance the bending membrane piezoelectric oscillator which is found in telephone capsules, the standard type electret condenser microphone in use today, the (old-fashioned) carbon microphone and the electro-magnetic microphone. Yet, these are just a few examples of a long list of microphone types. Several reasons can be given for the choice of the two examples in this chapter:

1. The only type of microphone which meets the standard classification criteria is the condenser microphone. It is indispensable for use conducting absolute (sound) level measurements. Keep in mind that overestimated sound levels may have severe legal and financial consequences. That is why precision is the first priority. However, condenser microphones have a very expendable production and are therefore not cheap.
2. In contrast, more economical microphones can be implemented for taking relative measurements. If only level intervals (e.g. between two rooms) when measuring the transmission loss or decay curves, when determining the reverberation time, are needed, low priced electrodynamic microphones, less precise in absolute value, can also be used. Even for sound studios, electrodynamic microphones can be sufficient.
3. And, finally, discussion can be spared with respect to many other microphone types, since all converters are similar in their mechanical construction (or can be described in a similar way) and for (nearly) all of

M. Möser, *Engineering Acoustics*, DOI 10.1007/978-3-540-92723-5_11,

them, their frequency response, to a large extent, is determined by the mechanical construction, as will be shown in the following.

Sound producing devices, such as the loudspeaker, also play a major role in many measurement procedures as well. As pertains to loudspeakers, we will focus on the most common type, the electrodynamic loudspeaker. The genus for loudspeaker and microphone is the term 'electro-acoustic converter' which indicates acoustic energy being transformed into electric energy and vice versa.

Basically, the principles of these and other converters discussed here are reversible. These principles apply to microphones as well as to loudspeakers.

The coil of the electrodynamic loudspeaker, for instance, which is mounted on the membrane, is forced to vibrate by alternating currents in a magnetic field. In this way, sound is radiated by the membrane. The same device can also serve as a microphone. The alternating forces of the sound pressure set the membrane into motion, inducing a voltage in the voice coil. The specific construction of the device is certainly designed for its purpose.

Naturally, the mechanical construction as well as the electrical circuit have to be taken into account when discussing transducers. The mechanical principle is simply given by the fact that a relative motion between the membrane and the housing is produced by both pressure and electro-magnetic forces. As the mounting of the housing is either heavy or rigid it can be in any case regarded as being at rest. Thus, only the absolute motion of the membrane of mass m is of interest, which also includes the mass of the coil or any other parts moving along with the membrane. The external force

$$F = F_{\mathrm{p}} + F_{\mathrm{e}} , \tag{11.1}$$

acts on the membrane, where F_{p} represents the pressure force and F_{e} the electro-magnetic force.

- For microphones $F_{\mathrm{e}} \ll F_{\mathrm{p}}$
- and for loudspeakers $F_{\mathrm{p}} \ll F_{\mathrm{e}}$

can be assumed. If the restoring electrical force would have the same order of magnitude as the exciting pressure force, the sound receiver would be completely ineffective. Likewise, a loudspeaker is essentially influenced by the sound field it produces, not its radiation. Although restoring forces are always present, these will not be handled in this discussion.

As already mentioned, all converters presented here mainly consist of a mass (including possible additional moving masses). This mass is resiliently mounted into a heavy, nearly immobile housing. The mechanical construction therefore always represents a simple resonator which was already discussed in detail in Chap. 5. According to the results of Sect. 5.1 'Elastic Bearings on a Rigid Foundation', the displacement x of a mass m which is excited by the force F (see (5.6), p. 146) is given by

$$x = \frac{F}{s - m\omega^2 + j\omega r} = \frac{F/s}{1 - \frac{\omega^2}{\omega_0^2} + j\eta\frac{\omega}{\omega_0}} \tag{11.2}$$

where, similar to Chap. 5.1,

$$\omega_0 = \sqrt{\frac{s}{m}}$$

represents the resonance frequency and

$$\eta = \frac{r\omega_0}{s}$$

represents the loss factor of the mechanical converter construction (see also (5.13) and (5.15), p. 147).

As the next two sections will show, the converter principle of the condenser microphone is the displacement x itself; the output voltage U is proportional to the displacement $U \sim x$. In contrast, electrodynamic microphones work on the basis of the induction law, thus $U \sim v = j\omega x$ is given. It is this principal difference in the two types which is ultimately responsible for their different frequency response functions. As can be seen in Fig. 11.1, the frequency response function of $v/F = j\omega x/F$ is simply a 'rotated' version of x/F. The frequency response function of $H_x = x/F$ is constant below the resonance frequency and falls at 12 dB/ per octave above that. In contrast, the frequency response of $H_v = j\omega x/F = j\omega H_x$ rises at 6 dB per octave up to the resonance frequency and falls at 6 dB per octave above it. It is this difference which characterizes the typical frequency response functions of the corresponding transducers, as will be shown in the following.

11.1 Condenser microphones

The essential element of this microphone type is a capacitor, whose one electrode is a lightweight membrane whereas the other electrode is rigid and heavy. An example for the construction of a condenser microphone is shown in Fig. 11.4 (p. 354).

Its converter principle is that the capacity of the capacitor changes with the membrane displacement x. As the capacity of a plate capacitor is inversely proportional to the distance d of the plates

$$C_0 \sim \frac{1}{d}, \tag{11.3}$$

a reduction of the plate distance by x due to an incident sound wave results in a capacity of

$$C \sim \frac{1}{d - x}. \tag{11.4}$$

Since the proportionality constant not written in (11.3) and (11.4) is the same, it follows that

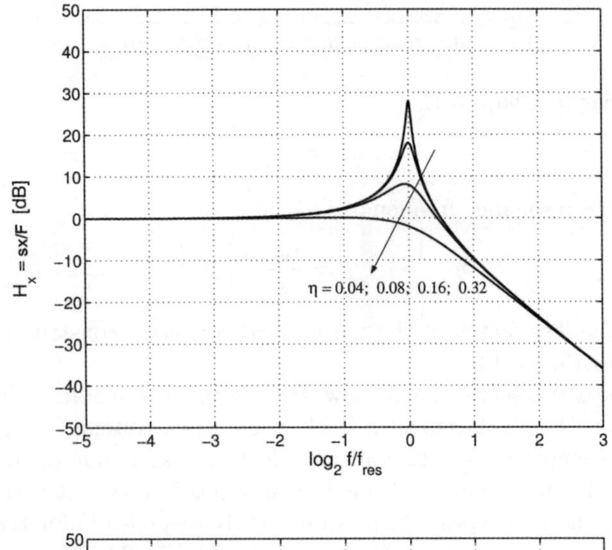

(a)

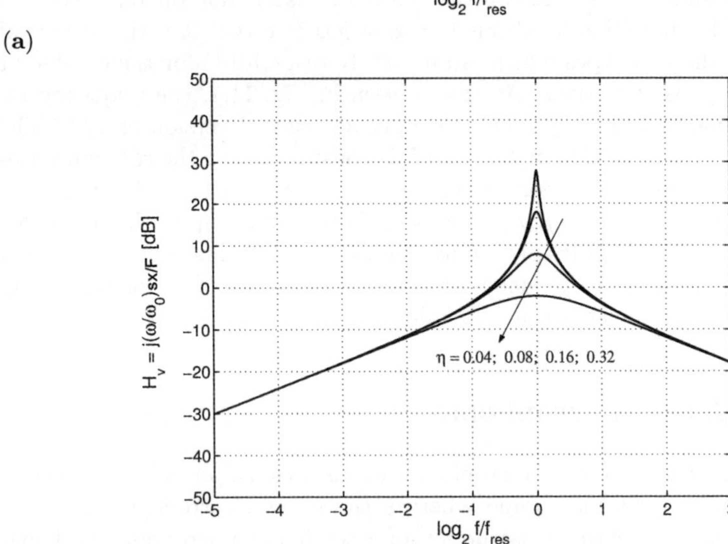

(b)

Fig. 11.1. Frequency response function of a simple resonator. (**a**) Displacement and (**b**) velocity

$$C = C_0 \frac{d}{d-x} = C_0 \frac{1}{1 - x/d} \tag{11.5}$$

where C_0 is the capacity of the capacitor while the membrane is at rest.

First of all, in order to detect any electrical output signal, the microphone capacitor has to be supplied with a DC-voltage. The principal electrical circuit diagram is shown in Fig. 11.2. It consists of a closed circuit formed by a capacitor, a resistor R and a supply voltage U_0. The resistor R represents the high input resistance of the amplifier which processes the extremely small

AC-voltage. The capacity at rest is only about 10 pF (10^{-11} Farad), more is technically not achievable. Common values for U_0 lie in the range of 20 V to 200 V which is limited by the possibility of a short-circuit between the electrodes.

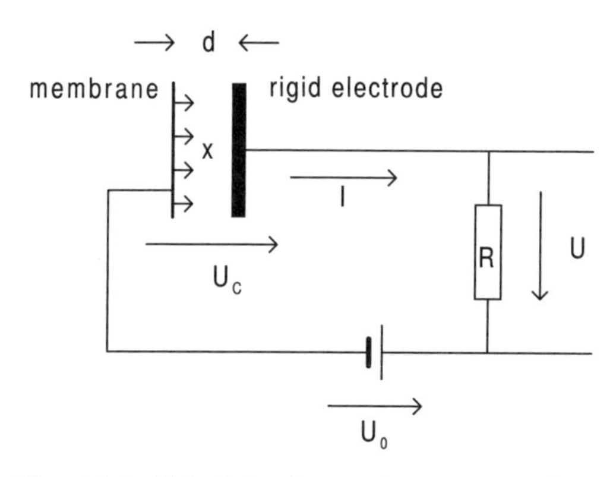

Fig. 11.2. Principle of a condenser microphone

The voltage across the closed circuit capacitor is given by the definition of the term 'capacity' by

$$U_c = \frac{Q}{C} = \frac{Q}{C_0}\left(1 - \frac{x}{d}\right) . \tag{11.6}$$

The output voltage U, using the voltage mesh

$$U_c + U - U_0 = 0$$

results in

$$U = U_0 - U_c . \tag{11.7}$$

A very simple estimation of the microphone frequency response function, which fails only for the lowest frequencies, can be obtained by the following consideration: if the output resistance R is considerably larger than the reactance $1/j\omega C_0$ of the capacitor, the circuit operates in an 'open loop'. For $R \gg 1/\omega C_0$, i.e. for frequencies $\omega \gg 1/RC_0$, nearly no current flows through the circuit. For this reason, the electric charge on the electrodes of the capacitor can be regarded as 'frozen' and invariant. In the frequency range $\omega \gg 1/RC_0$, using $Q \approx Q_0$ in (11.6),

$$U_c = \frac{Q_0}{C_0}\left(1 - \frac{x}{d}\right) = U_0\left(1 - \frac{x}{d}\right) \tag{11.8}$$

is given (due to $U_0 = Q_0/C_0$). According to (11.7), the output voltage U

$$U = U_0\frac{x}{d}$$

is directly proportional to the membrane displacement x. With the aid of (11.2), it can therefore be summarized that in the aforementioned frequency range

$$\frac{U}{p} = \frac{U_0 \frac{S}{sd}}{1 - \frac{\omega^2}{\omega_0^2} + j\eta\frac{\omega}{\omega_0}} \tag{11.9}$$

applies. It was assumed that the membrane force equals the product of pressure and membrane area

$$F = pS \;, \tag{11.10}$$

an assumption which is valid as long as the membrane cross-section is small compared to the wavelength in air.

The frequency response function (11.9) is exclusively determined by that of the mechanical structure $H_x(\omega) = x/F$. Thus, $H_x(\omega)$ is most decisive for the frequency response function of the microphone.

The assumption of a frozen charge on the capacitor is an approximation not only for sufficiently high frequencies, but also for polarized types, where piezo-ceramic layers are used instead of a capacitor. That is to say, they can do without a supply voltage.

As already mentioned, (11.9) is only valid for frequencies $\omega \gg \omega_{\mathrm{f}}$, where ω_{f} is defined as

$$\omega_{\mathrm{f}} = 1/RC_0 \;. \tag{11.11}$$

This 'typical frequency' ω_{f} is very low for all condenser microphones, usually around $f_{\mathrm{f}} = \omega_{\mathrm{f}}/2\pi \simeq 10\,\mathrm{Hz}$. The relevant frequency range is sufficient for accurately describing frequency response function of the microphone. On the other hand, it is not complicated to take into account the charge which is flowing to and from the electrodes. More generally the capacitor charge consists of a static part Q_0 and an alternating part q so that

$$Q = Q_0 + q \;.$$

The voltage across the capacitor U_{c}, according to (11.6), is thus given by

$$U_{\mathrm{c}} = \frac{Q}{c} = \frac{Q_0 + q}{C_0}\left(1 - \frac{x}{d}\right) \approx \frac{Q_0}{C_0}\left(1 - \frac{x}{d}\right) + \frac{q}{C_0} = U_0 - U_0\frac{x}{d} + \frac{q}{C_0} \;.$$

The small quadratic term including xq was thereby neglected. The alternating quantities are small compared to the static parts $q \ll Q_0$ and $x \ll d$. Using (11.7), the output voltage is given by

$$U = U_0\frac{x}{d} - \frac{q}{C_0} \;.$$

Here, only alternating quantities occur. The alternating charge q can additionally be expressed by the alternating current in the circuit $q = I/j\omega$ as

$$U = U_0\frac{x}{d} - \frac{I}{j\omega C_0} \;,$$

or, due to Ohm's law $I = U/R$ for the output resistor R,

$$U = U_0 \frac{\frac{x}{d}}{1 + \frac{1}{j\omega RC_0}} = U_0 \frac{\frac{x}{d}}{1 - j\frac{\omega_k}{\omega}} \,. \tag{11.12}$$

The electrical construction of a condenser microphone acts as a high pass filter with the folding (or typical) frequency ω_f. For $\omega > \omega_\mathrm{f}$ the frequency response function is constant, for $\omega < \omega_\mathrm{f}$ it rises with a gradient of 3 dB per octave spanning from the low frequencies toward the folding frequency.

The total resulting frequency response function, when using $F = pS$ in (11.2) and using (11.12), is given by

$$G_\mathrm{up} = \frac{U}{p} = \frac{\frac{S}{sd}U_0}{\left(1 - j\frac{\omega_k}{\omega}\right)\left(1 - \frac{\omega^2}{\omega_0^2} + j\eta\frac{\omega}{\omega_0}\right)} \,. \tag{11.13}$$

The frequency response represents a band pass which is limited by the electrical folding frequency ω_f at low frequencies and by the mechanical resonance ω_0 at high frequencies (Fig. 11.3). In the transfer range $\omega_\mathrm{f} \ll \omega \ll \omega_0$, the sensitivity is given by

$$G_{\mathrm{up},0} = \frac{S}{sd}U_0 \,. \tag{11.14}$$

The highest operating frequency is usually defined as the point where G_up is 1 dB larger than $G_{\mathrm{up},0}$. At this point it is

$$\left|\frac{G_\mathrm{up}}{G_{\mathrm{up},0}}\right|^2 = 10^{0.1} = 1.25 \,.$$

For sufficiently small η and due to $\omega \gg \omega_\mathrm{f}$, it follows that

$$\left(1 - \frac{\omega^2}{\omega_0^2}\right)^2 = \frac{1}{1.25} = 0.8 \,,$$

or

$$\omega \simeq 0.3\,\omega_0 \tag{11.15}$$

for the operation limit of the microphone. Some technical data of common microphone types are given in Table 11.1.

The sensitivity, quoted in Table 11.1 is measured at the signal amplifier, which is built into the microphone. Obviously, microphones with larger membranes are more sensitive than those with smaller ones, where the ratio of sensitivity roughly corresponds to the ratio of the surface areas. The main difficulty of condenser microphones is the internal noise which stems from the high-ohm resistor. The measurable sound levels (assuming 1 dB uncertainty) are very high for small microphone types; they are not useful in detecting very small sound levels. Here, more sensitive microphones with a large membrane area must be used.

Table 11.1. Technical data of condenser microphones

Diameter (mm)	Sensitivity (mV/Pa)	Frequency range (Hz)	Dynamic range (dB(A))
3.2	1	6.5 – 140 k	55 – 168
6.4	4	4 – 100 k	36 – 164
12.7	12.5	4 – 40 k	22 – 160
23.8	50	4 – 18 k	11 – 146

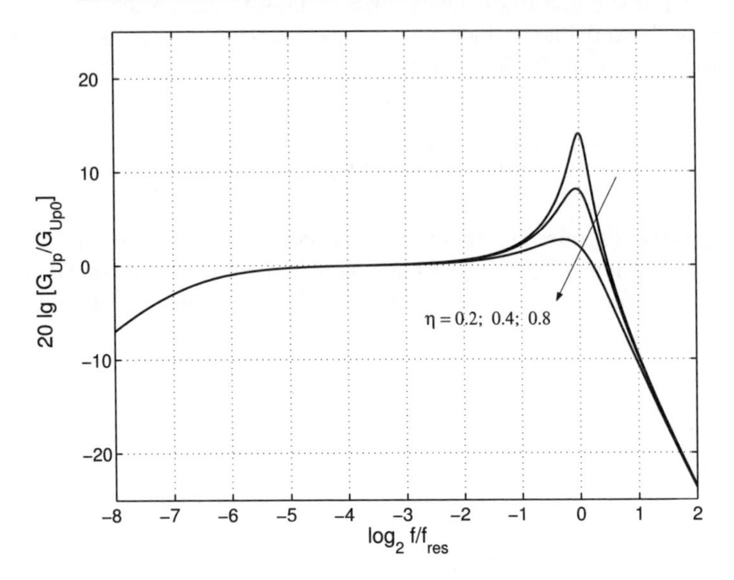

Fig. 11.3. Theoretical frequency response function of a condenser microphone. The electrical folding frequency ω_f is 7 octaves below the resonance frequency, here. Calculated for $\eta = 0.2, 0.4$ and 0.8

As Table 11.1 shows, the resonance frequency decreases with increasing membrane area. This tendency will be discussed in more detail here. The theoretical resonance frequency amounts to

$$\omega_0^2 = \frac{s}{m},$$

where the total stiffness s is composed of two parts

$$s = s_E + s_L\ . \tag{11.16}$$

The stiffness s_E represents the stiffness of the membrane bearing, s_L that of the air cushion between the electrodes and is given by

$$s_L = \frac{\varrho c^2 S}{d}\ . \tag{11.17}$$

During production, the membrane, a thin and prestressed metal foil, is laid upon a supporting ring and affixed there. The stiffness s_E is caused by the internal strain of the membrane. The mass m of the membrane is slightly smaller than the actual mass, because m pertains only to the mass in motion.

It is often cited that the stiffness of the air s_L is larger than that of the bearing stiffness s_E. Since the mass for all sizes (using the same metal foil) increases with the surface area

$$m = m''S ,$$

however, the assumption $s_\mathrm{L} \gg s_\mathrm{E}$ would lead to a resonance frequency

$$\omega_0^2 = \frac{\varrho c^2 S}{dm''S} = \frac{\varrho c^2}{m''d} . \tag{11.18}$$

This would mean that the resonance frequency should be *independent* of the membrane area in contrast to experience. Table 11.1 proves a different line of reasoning. It seems to be much more plausible that the bearing stiffness s_E dominates and is independent of the microphone size; in this case

$$\omega_0^2 = \frac{s_\mathrm{E}}{m''S} \tag{11.19}$$

is given. The resonance frequency would decrease inversely with the diameter of the membrane as confirmed by the actual numbers in the table.

The assumption $s_\mathrm{L} \ll s_\mathrm{E}$ is also underlined by the technical structure of a condenser microphone as sketched in Fig. 11.4, because the perforation of the backing electrode dramatically reduces the stiffness of the air cushion: if the capacitor is compressed, the air escapes through the perforation and is not compressed. The actual purpose of the perforation of the backing electrode is not the reduction of the stiffness, but the attenuation of the resonance peak due to friction losses which the viscous air undergoes when passing the holes. The loss factor η is mainly a result of this effect. In contrast, the resonance frequency is solely affected by the parameters of the membrane.

The frequency response function shown in Fig. 11.4 proves that the attenuation of the resonance can be achieved. A minor elevation of the resonance frequency can be quite useful, especially when oblique sound incidence is considered (see the next section).

The highly sensitive microphone membrane would be destroyed by even the smallest fluctuations of the atmospheric pressure, if there were no a connection between the internal and external air. This coupling, achieved by a capillary, can compensate for pressure difference. Essentially, the air in the capillary acts as a mass m_k in such a way that the pressure difference p between the exterior and the air between the electrodes amounts to

$$\Delta p = j\omega \frac{m_\mathrm{k}}{S_\mathrm{Q}}$$

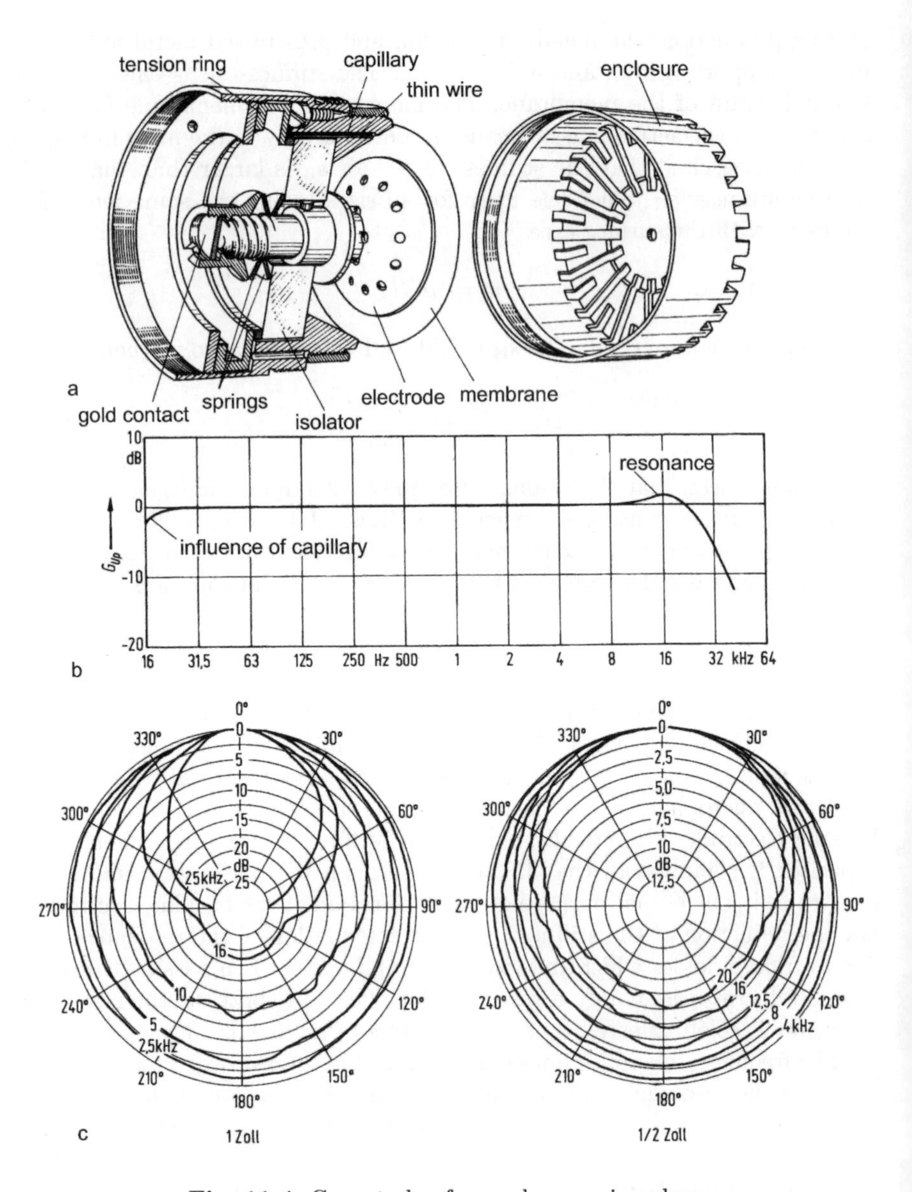

Fig. 11.4. Case study of a condenser microphone

(where S_Q is the cross-sectional area of the capillary). At low frequencies the pressure difference is compensated, $\Delta p \approx 0$. At high frequencies the interior of the capsule is decoupled. At low frequencies, the sensitivity of the microphone is reduced by the pressure compensation. Apparently, the electrical folding frequency of the microphone depicted in Fig. 11.4 is considerably smaller than the folding frequency of the capillary (which represents a mechanical high-pass filter).

The discussion on the dominating stiffness is only of scientific, but not of practical interest, as it can be extracted from Table 11.1: all of the specified quantities cover the frequency range of technical interest (ultrasound is the only exception to this). In practice, the disadvantage of larger microphone membranes compared to smaller ones is not determined by the decreasing frequency range, but rather by the directivity which is already observable in the mid-frequency range. This will be the subject of the next section.

11.2 Microphone directivity

Basically, two principle effects occur which lead to directivity of a microphone at higher frequencies:

1. The microphone represents a reflector for the incident sound field. The geometrical shape of the housing, the optional grip or protective mesh, interfere with the field quantity to be measured. The reflection can differ with the angle of acceleration, but will always result in a higher sound pressure. The resulting sound pressure can have a maximum value of twice that of the incident sound field. Thus, reflections at the microphone appear as an increased sensitivity which is limited to 6 dB at most. Pressure accumulations are therefore secondary phenomena which can be used to make minor corrections to the frequency response by shaping its geometry.
2. Reflections off the body of the microphone do not explain the distinctive, strongly angle-dependent sensitivities at high frequencies. Rapidly changing pick-up characteristics are based on the fact that the sound pressure on the membrane is space-dependent. Only at low frequencies is the force on the membrane the product pS. More generally, and especially at high frequencies it is
$$F = \int_S p \, \mathrm{d}S \, . \tag{11.20}$$
For a corresponding small wavelength and oblique sound incidence, areas of sound pressure with an opposite phase occur. The total force could therefore be zero.

The consequences of the 'pressure-integrating membrane effect' (11.20) can easily be calculated for a circular membrane if the reflection at the membrane itself is neglected. A wave, impinging obliquely on the membrane, is assumed. As shown in Fig. 11.5, the angle between the wave vector and the normal of the membrane surface is denoted by ϑ. The wave in the sketched coordinate system is described by

$$p = p_0 \mathrm{e}^{jkx \sin\vartheta} \mathrm{e}^{jkz \cos\vartheta} \, . \tag{11.21}$$

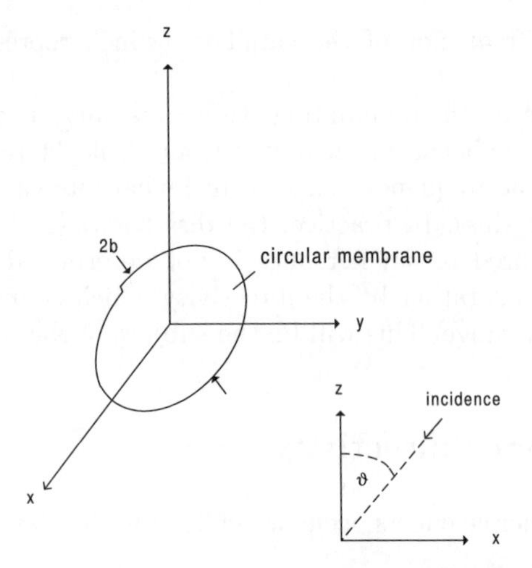

Fig. 11.5. Position of the circular membrane (radius b) in the coordinate system

Using $x = r\cos\varphi$, the pressure integral (11.20) becomes

$$F = p_0 \int_0^b \int_0^{2\pi} \mathrm{e}^{jkr\sin\vartheta\cos\varphi} \mathrm{d}\varphi r \mathrm{d}r = p_0 \int_0^b \int_0^{\pi} \cos\left(kr\sin\vartheta\ \cos\varphi\right) \mathrm{d}\varphi r \mathrm{d}r \ .$$

The inner integral results in the zero-order Bessel function and

$$F = 2p_0 \int_0^b \pi J_0\left(k\sin\vartheta\ r\right) r\mathrm{d}r$$

remains. This integral can also be found in an integral table (see e.g. [?], p. 634, No. 5.52 1) and can be used to calculate the first order Bessel function.

$$F = \pi b^2 \frac{2J_1\left(kb\sin\vartheta\right)}{kb\sin\vartheta} p_0 = Sp_0 \frac{2J_1\left(kb\sin\vartheta\right)}{kb\sin\vartheta} \tag{11.22}$$

In the expression for microphone sensitivity, which is given by (11.13) for the condenser microphone and by (11.29) for the electrodynamic microphone, S has to be replaced by

$$S \rightarrow S\,G\left(u\right) \ , \tag{11.23}$$

using

$$G\left(u\right) = \frac{2J_1\left(u\right)}{u} \tag{11.24}$$

and

$$u = kb\sin\vartheta \ . \tag{11.25}$$

All pick-up characteristics are obtained (in the same way as in Chap. 3 on radiation) by the limited range $u = kb$ of the function $G(u)$. The latter

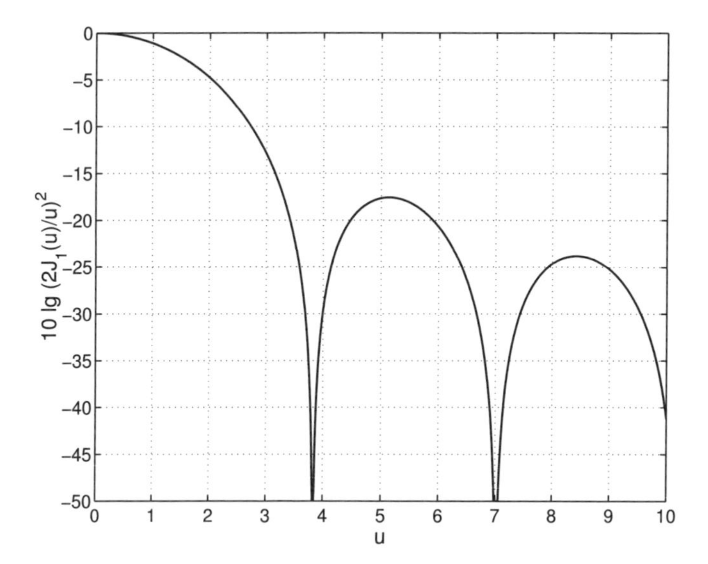

Fig. 11.6. (**a**) Directivity function $G(u) = 2J_1(u)/u$

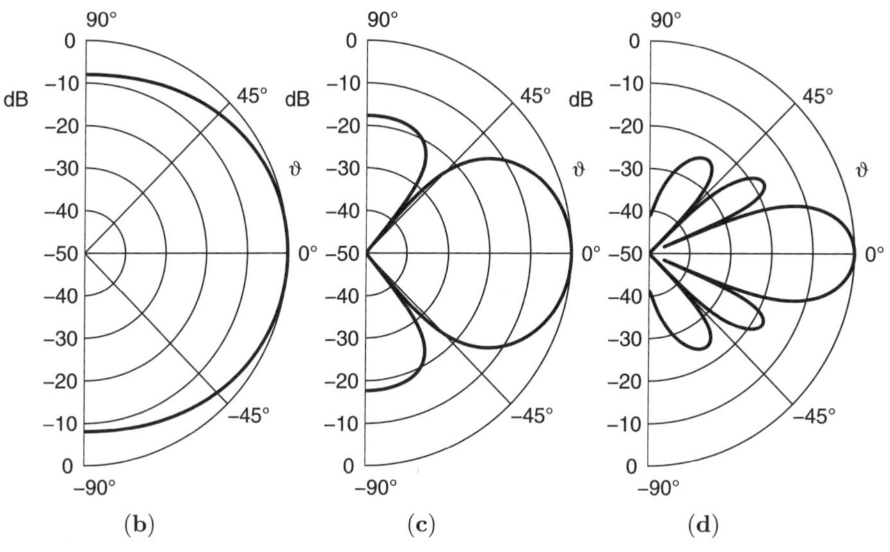

Fig. 11.6. Pick-up characteristic for (**b**) $u_{\max} = kb = 2.5$ (**c**) $u_{\max} = kb = 5$ (**d**) $u_{\max} = kb = 10$

is shown in Fig. 11.6a together with three characteristic examples of pick-up characteristics in Figs. 11.6b,c,d.

At low frequencies the incidence angle is unimportant. All directions have the same microphone sensitivity. If the membrane dimensions and the wave-

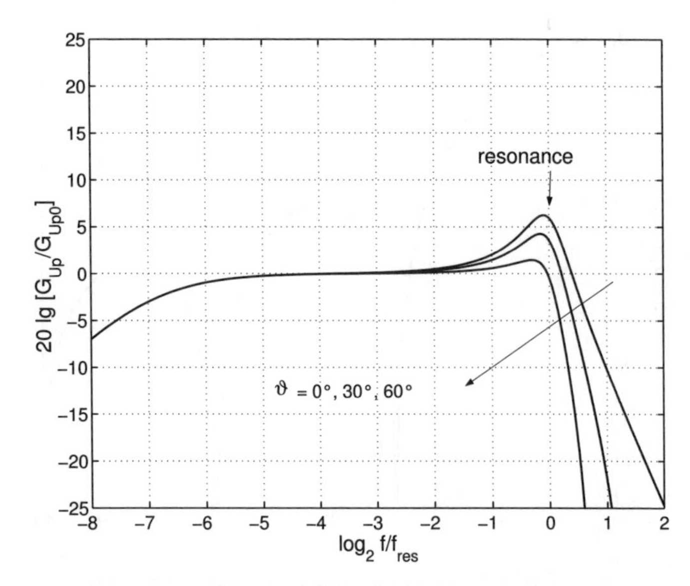

Fig. 11.7. Frequency response function of a condenser microphone for different incidence angles. The electric folding frequency ω_f is 7 octaves below the resonance frequency, here. Calculated for $\eta = 0.5$ and $b/\lambda = 0.4$ at the resonance frequency.

length fall into the same order of magnitude, the angle dependencies are more distinctive. Measured curves are included in Fig. 11.4.

The angle dependence at constant frequencies is drawn in Figs. 11.6b,c,d for three different cases. Likewise, the frequency response for a constant incidence angle could be drawn (examples are shown in Fig. 11.7). At high frequencies, a frequency response function is obtained which varies for each incidence angle. These frequency responses can be influenced up to a few dB by reshaping the microphone (which influences the reflection at the microphone housing) and by choosing a specific microphone sensitivity.

Microphones which have an 'optimized smooth frequency response function' at 0° incidence are called 'free-field microphones'. If, in contrast, the 45° frequency response function is optimized, one speaks of a 'diffuse-field microphone', where it is explicitly assumed that the average incidence angle in a diffuse field is 45°.

11.3 Electrodynamic microphones

For electrodynamic microphones, the physical principle of an AC-generator is used. If loop conductors are placed perpendicular to a the lines in a magnetic field, an electric voltage is induced in them. The most common construction is the voice coil microphone (Fig. 11.10, p. 361), where a voice coil mounted to the membrane moves through a circular gap in a magnet. Electrically, it can be regarded as a real voltage source with the internal resistance $R_\mathrm{i}+j\omega L$ and an ideal voltage source U_i (induction voltage)

$$U_\mathrm{i} = B\ell v\ . \tag{11.26}$$

Here, B is the magnetic induction in the air gap, ℓ is the length of the conductor and v is the velocity of the voice coil and the membrane. The circuit is closed by a load R_a, where the converter voltage U is captured (Fig. 11.8).

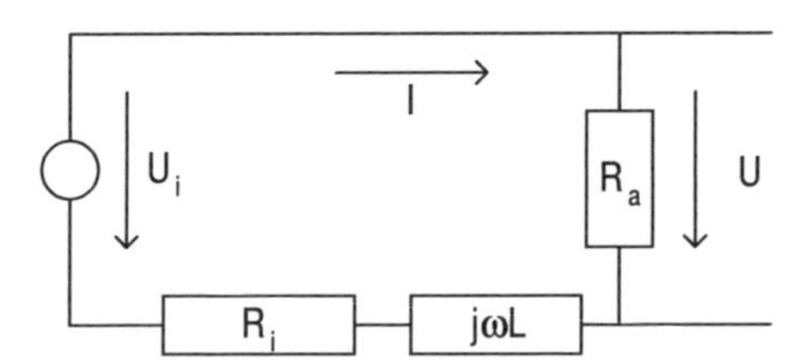

Fig. 11.8. Electrical circuit diagram of the electrodynamic microphone

The converter voltage is given by

$$U = U_\mathrm{i} - (R_\mathrm{i} + j\omega L)\, I \tag{11.27}$$

or, using (11.26) and $I = U/R_\mathrm{a}$,

$$U = \frac{R_\mathrm{a} B\ell}{R_\mathrm{a} + R_\mathrm{i} + j\omega L} v\ . \tag{11.28}$$

Thus, the converter voltage is proportional to the membrane velocity. The converter coefficient U/p is obtained by expressing $v = j\omega x$ by the force $F = pS$, according to (11.2) (and using $R = R_\mathrm{a} + R_\mathrm{i}$):

$$U = \frac{R_\mathrm{a}/R\, B\ell S \frac{j\omega}{s}}{\left(1 + j\omega \frac{L}{R}\right)\left(1 - \frac{\omega^2}{\omega_0^2} + j\frac{\omega}{\omega_0}\eta\right)} p\ . \tag{11.29}$$

Because the combined mass of the membrane and voice coil cannot be reduced indefinitely, tuning the resonance to very high frequencies is impossible. In fact, for all practical purposes, the resonance actually lies below the electric folding frequency

$$\omega_\mathrm{f} = R/L\ . \tag{11.30}$$

At low frequencies $\omega \ll \omega_0$ the frequency response function therefore starts with a gradient ω (corresponding to 6 dB per octave), followed by a weak and broad resonance peak due to the unusually chosen high damping. Between ω_0 and ω_f, the frequency response falls at $1/\omega$ (corresponding to -6 dB per octave, Fig. 11.9). Such a frequency response would be unsuitable for a microphone if there were no way to flatten out the response function by employing

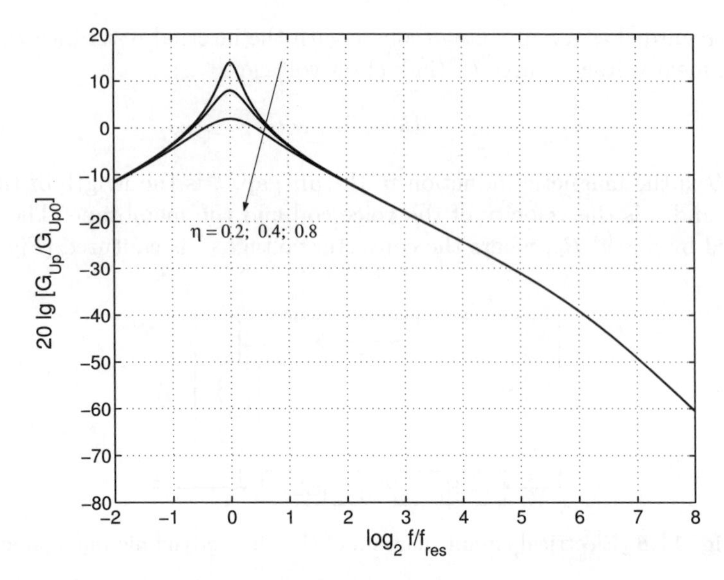

Fig. 11.9. Theoretical frequency response of an electrodynamic microphone. The electrical folding frequency is located 6 octaves above the resonance frequency. Calculated for $\eta = 0.2, 0.4$ and 0.8

certain 'tricks'. By coupling additional mechanical oscillators – for instance, in the form of special cavities behind the membrane – resonance effects of a certain bandwidth tuned to a particular bandwidth are obtained. To effectively compensate the frequency response function along the entire frequency range, several resonators are constructed by subdividing the air space in the microphone housing into several volumes of different sizes. These tubes couple the sub-volumes to the air stiffness which acts directly on the membrane. The air in the tubes acts as the mass and the coupled volume acts as the stiffness. The stiffness is brought about by the simple resonator produced this process. The resonator's resonance can be tuned by way of the tube length and the corresponding volume. The damping is provided by absorbent substances (felt) and by labyrinths with large frictional surface areas (Fig. 11.10). In this way, frequency response functions are obtained which suffice studio requirements. The microphone type in Fig. 11.10 is obviously tuned to an 'average' sound incidence of 45°.

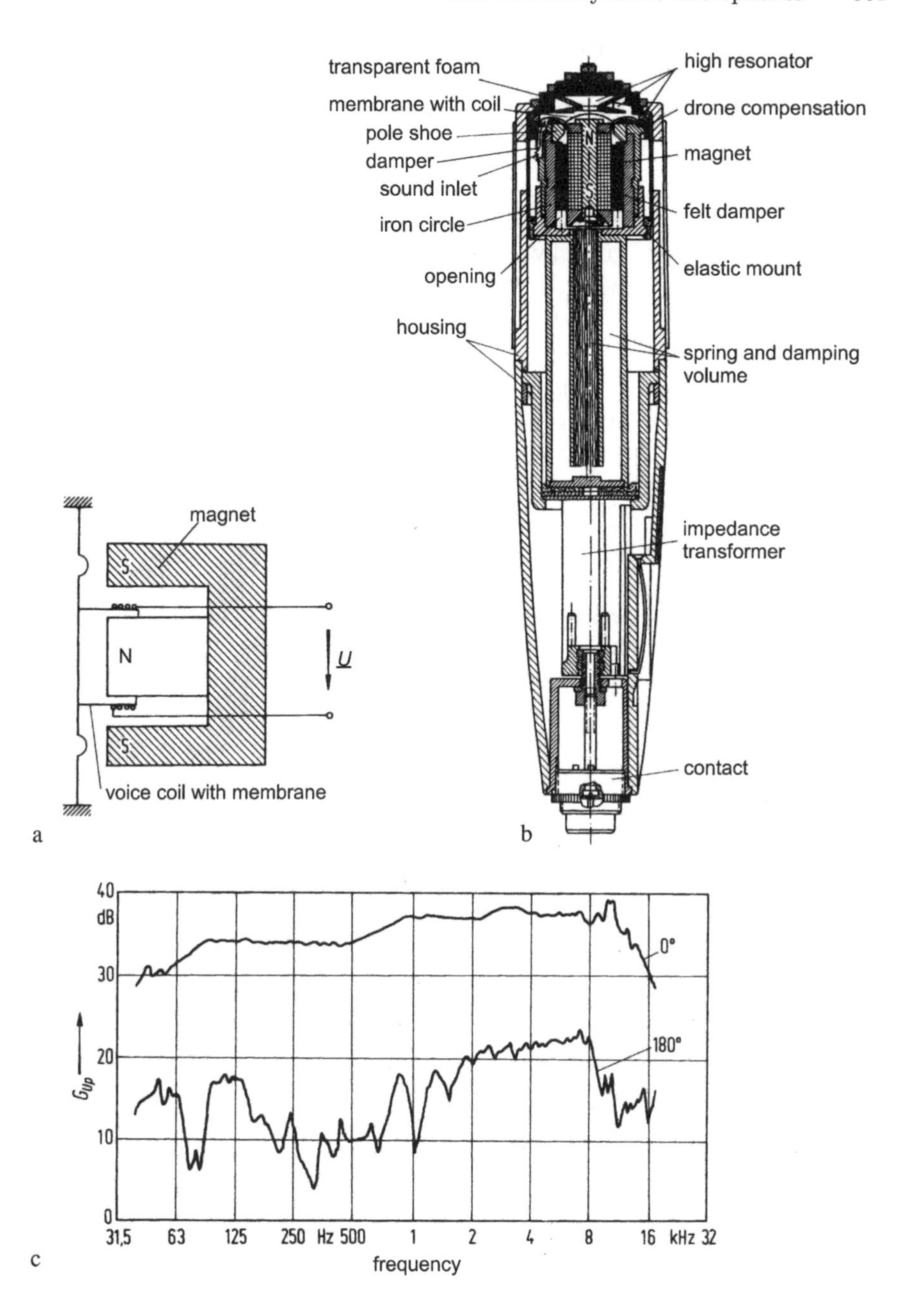

Fig. 11.10. Specific type of an electrodynamic microphone

11.4 Electrodynamic loudspeakers

The electrodynamic loudspeaker is the reversed counterpart to the electrodynamic microphone. An external voltage laid across the voice coil creates an electrical force

$$F = B\ell I \tag{11.31}$$

where the current I is given by

$$I = \frac{U}{R_\mathrm{i} + R_\mathrm{a} + j\omega L}, \tag{11.32}$$

using the impedance of the loudspeaker $R_\mathrm{i} + j\omega L$ and the internal resistance R_a of the voltage source (Fig. 11.11).

According to (11.2), the membrane velocity v is given by

$$v = \frac{j\omega B\ell/s}{(R_\mathrm{i} + R_\mathrm{a} + j\omega L)\left(1 - \frac{\omega^2}{\omega_0^2} + j\frac{\omega}{\omega_0}\eta\right)} U\,. \tag{11.33}$$

The velocity has the same frequency response function in principle as the corresponding microphone (Fig. 11.9). Yet, the membrane velocity is not very important. The sound pressure, produced in a certain distance, is far more important.

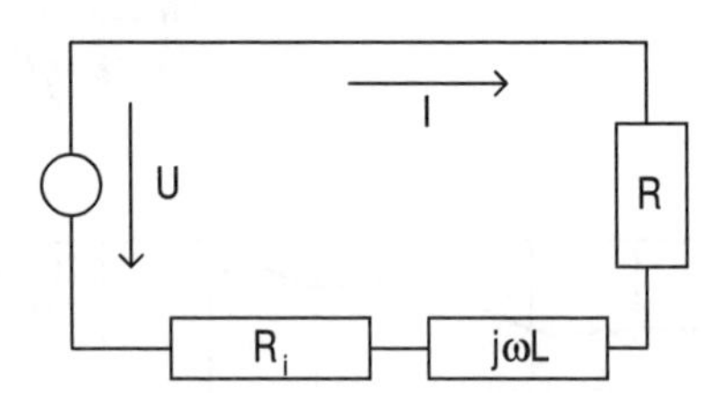

Fig. 11.11. Circuit diagram of the electrodynamic loudspeaker

The radiation of a moving membrane is discussed in Chap. 3. The facts, derived there, are recapitulated here. To suppress the short circuit of the mass at low frequencies (which will make radiation worse), the loudspeakers are mounted in a box (or a 'large' baffle). At low frequencies, they act as a volume velocity source (Sect. 3.3, using (3.13), p. 71) with the resulting far field pressure

$$p \approx \frac{j\omega\varrho Q}{4\pi r}\mathrm{e}^{-jkr} = \frac{j\omega\varrho\pi b^2 v}{4\pi r}\mathrm{e}^{-jkr} \tag{11.34}$$

where b denotes the membrane radius. Thus, for $b \ll \lambda$

$$p = \frac{-\omega^2\varrho\pi b^2 B\ell/s}{(R_\mathrm{i} + R_\mathrm{a} + j\omega L)\left(1 - \frac{\omega^2}{\omega_0^2} + j\frac{\omega}{\omega_0}\eta\right)} U \tag{11.35}$$

is obtained. The electrical folding frequency tor the loudspeaker is also given by

$$\omega_{\mathrm{f}} = \frac{R_{\mathrm{i}} + R_{\mathrm{a}}}{L} \tag{11.36}$$

which is a lot higher than the resonance frequency and is intentionally tuned to lower frequencies. Equation (11.35) shows that the frequency response of the sound pressure is constant in the frequency range $\omega_0 < \omega < \omega_{\mathrm{f}}$ and given by

$$p \approx p_{\mathrm{B}} = \frac{\varrho \pi b^2 B\ell / s}{R_{\mathrm{i}} + R_{\mathrm{a}}} \, .$$

toward the resonance frequency the frequency response function rises at 12 dB per octave, above the resonance it falls at 6 dB per octave. The theoretically resulting frequency response function is depicted in Fig. 11.12. Principally, loudspeaker radiation has band-pass characteristics which are limited by the mechanical resonance at the lower end and by the folding frequency at the upper end.

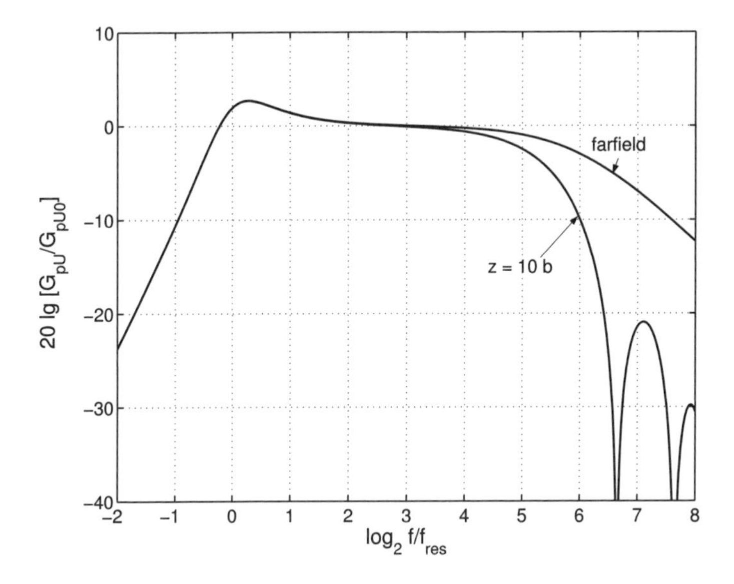

Fig. 11.12. Theoretical frequency response function of an electrodynamic loudspeaker. The electrical folding frequency is 6 octaves below the resonance frequency. Calculated for $\eta = 0.8$ and in the far-field or for a given point at the center axis in front of the loudspeaker

At higher frequencies – as mentioned in Chap. 3 – the directivity of the radiation comes into play. A different frequency response p/U is obtained for each radiation angle. Equation (11.34) is still valid at high frequencies under free-field conditions on the center axis in front of the loudspeaker membrane.

More generally, the sound pressure along the z-axis in front of the membrane, using (3.64), can be described by

$$p = \varrho c v_0 \mathrm{e}^{-jkz} \left\{ 1 - \mathrm{e}^{-jk\left(\sqrt{z^2+b^2}-z\right)} \right\} .$$

The resulting frequency response function of the loudspeaker radiation for a given field point on the z-axis is drawn in Fig. 11.12.

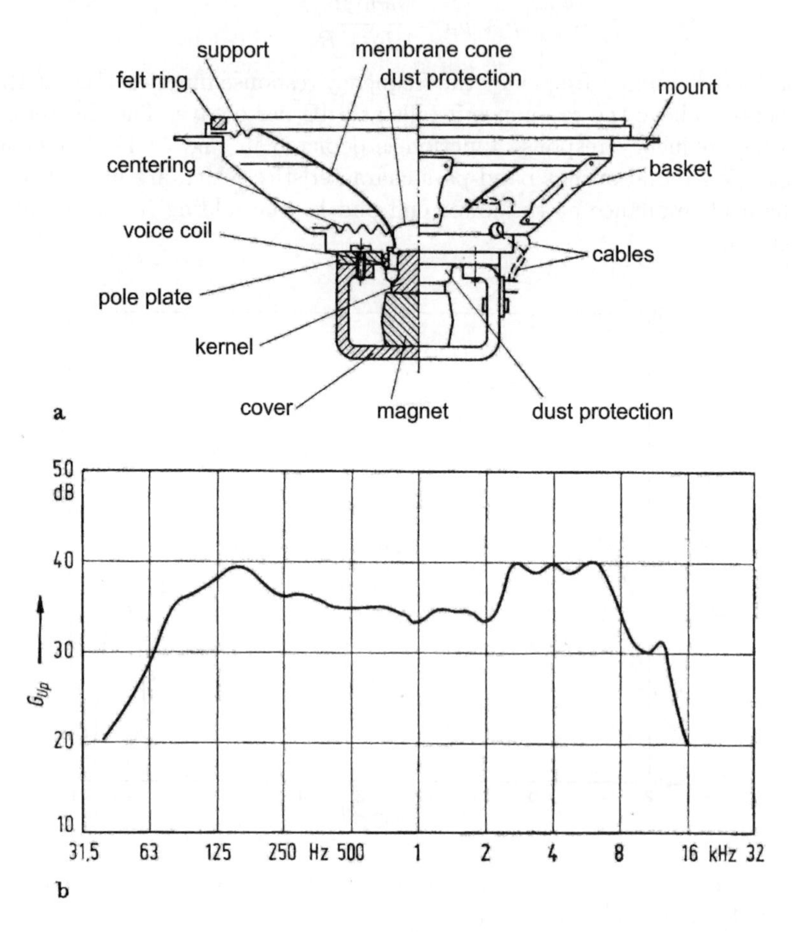

Fig. 11.13. Specific construction of an electrodynamic loudspeaker

In addition, the practically usable frequency range shrinks by the fact that the loudspeaker membrane does not represent a uniformly vibrating surface at higher frequencies. Eigenvibrations occur, similar to bending waves in plates and beams which dynamically deform the membrane. The membrane surface seems to 'fall apart' into oppositely vibrating source elements. The radiation efficiency thus decreases. Since the membrane must be light-weight and thus cannot be very stiff (otherwise the mass impedance, effective above the resonance, would be too large), its stiffness is only increased by shaping the geometry and therefore tuning the first eigenvibration as high as possible. For this reason, so-called NOPRO-membranes ('non-projecting surfaces') are used.

Fig. 11.13 shows a practical construction of an electrodynamic loudspeaker and the corresponding frequency response function. The band-pass characteristics are roughly as predicted. The example is a type that is explicitly sold as a broadband loudspeaker, and which can be implemented between 100 Hz and 8 kHz. It should finally be noted that the efficiency of a loudspeaker is very small and seldom amounts to more than 1%.

11.5 Acoustic antennae

For some microphone-assisted measuring tasks, the spatial distribution of the sound sources become a primary issue. There are plenty of examples of how this is applied. From a layperson's perspective, one might wonder which part of a passing vehicle contributes the most to the overall noise emission: is it the wheels or the engine, or other parts of the car, such as the ventilator, the exhaust, etc.? The use of 'acoustical antennae' (or 'arrays') in measurement techniques can help provide the answers to such questions. Two examples of how this is applied are shown in Figures 11.14 and 11.15.

Obviously, to produce these diagrams, a measuring instrument was required which is not only capable of pivoting in the direction where the measurements are to be taken, but also one that is capable of suppressing undesired directivities. This measuring instrument is referred to as the 'acoustic antenna.' This section will provide further insight into the functionality of this measuring apparatus.

Electronically adjustable and pivotal arrangements for directing acoustic transmission have already been discussed in the context of loudspeaker arrays (see the section on 3.5). Multiple, spatially distributed microphones can be considered an analogous set-up, which can likewise electronically adjust the directivity pattern on the receiver end by means of panning. This electronically adjustable pivoting effect is achieved by setting offset delays between the individual microphones and sensor signals, as described in the following. The similarity between the transmitter and receiver scenario furthermore corresponds to the capability of the 'acoustic antennae' in adjusting receiving signals by suppressing the side lobes, just like loudspeaker arrays.

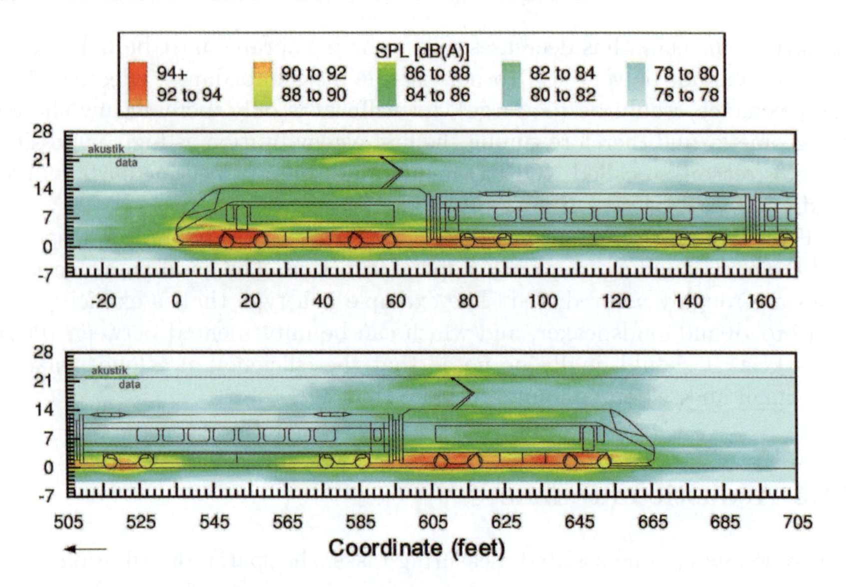

Fig. 11.14. Sound source distribution of a train passing at 240 km/h (measured from 'acoustic-data' using 29 microphones). SPL stands for the sound pressure level. This diagram indicates that the main sound source is the tire-road traction, as well as a significant contribution form the power collector on top of the train.

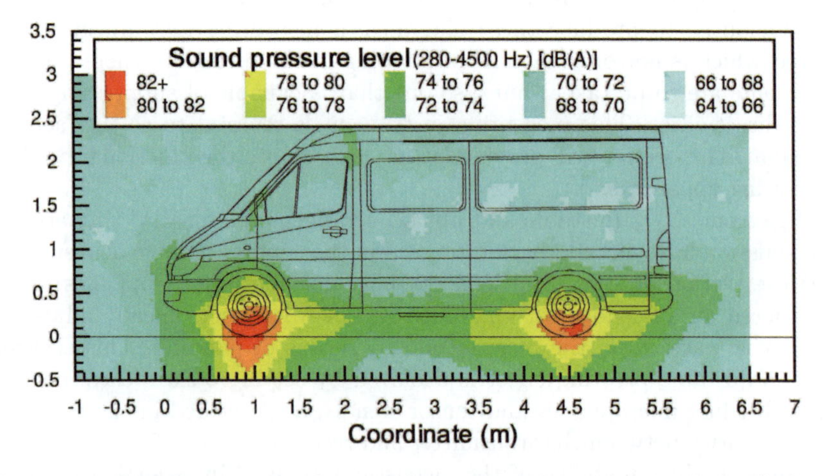

Fig. 11.15. Sound source distribution of a small truck passing at 120 km/h (measured from 'acoustic-data'). It is plain to see the significance of the tire-to-road traction as the primary noise source.

11.5.1 Microphone arrays

The simplest and most transparent arrangement consists of microphones equidistant from one another in a row. The structure and its position in the coordinate system is described in the following Figure11.16. This microphone setup is simply referred to as a 'line array' .

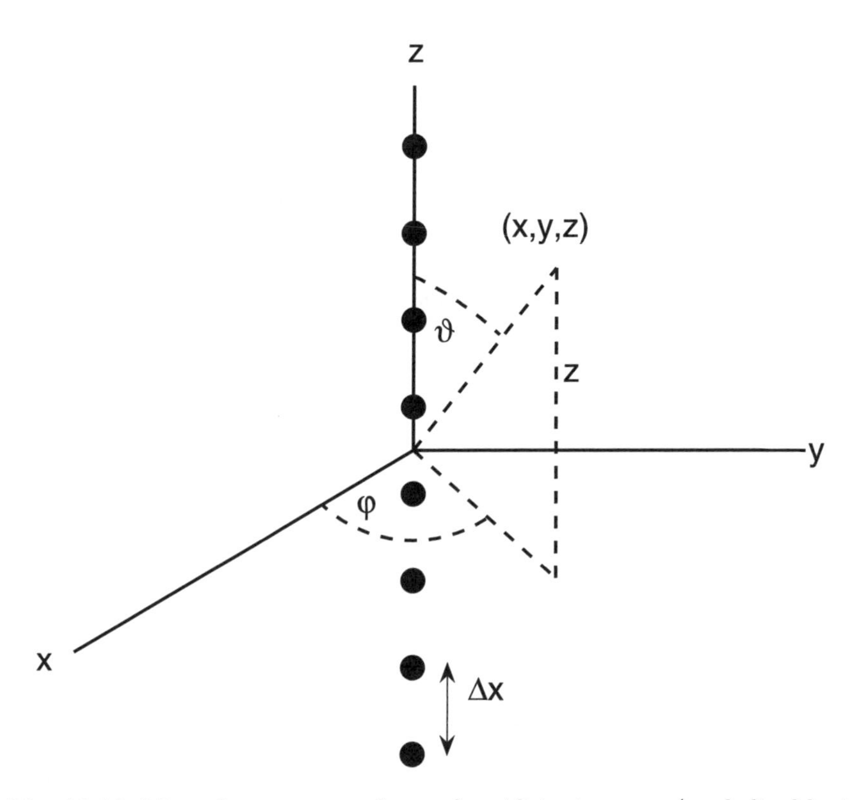

Fig. 11.16. Microphone array made up of equidistant sensors (symbolized by the circles) with corresponding spherical coordinate system

The algorithm used to interpret the microphone signals is based on a self-evident deliberation, namely, that when all microphone signals are in synch, the rms-value S_{eff} of the sum S of all microphone output signals is considerably greater than if there were a delay between the signals. For a signal $r(t)$ with a 'shorter' duration (for instance, a narrow rectangular function) in a synchronized signal system the following applies

$$S^2_{eff}(0) = \frac{1}{T} \int_0^T [Nr(t)]^2 dt = N^2 r^2_{eff} \, ,$$

where r_{eff} represents the rms-value of one signal $r(t)$. If, however, the signals are offset from one another by a mutually varying, non-zero delays time delays $\tau_i \neq 0$, S_{eff} exists in

$$S^2_{eff}(\tau) = \frac{1}{T}\int_0^T [\sum_{i=0}^{N-1} r(t-\tau_i)]^2 dt = Nr^2_{eff} \, .$$

Expressed in levels, the rms-value of the *non-delayed* signals is thus $10lgN$ greater than the rms level of the mutually offset, i.e. *delayed* signals, whereby $10lg[S^2_{eff}(\tau)/S^2_{eff}(0)] = 10lgN$.

This fact can be used to localize sound waves impinging on the microphone array. For skew or oblique sound incidents along the line array at angles ϑ_{in} and φ_{in}, the otherwise synchronized microphone signals are mutually offset by delays in time periods τ_i dependent on ϑ_{in}. The rms-value of the sum of the signals would therefore be comparatively small ($S^2_{eff}(0) = Nr^2_{eff}$ as shown above). However, when the signals are input into the computer (or through delay lines) and appropriately offset in short time intervals, the 'large' value $S^2_{eff} = N^2 r^2_{eff}$ can only be determined in the special case where the delay times produced by the oblique sound incidents are able to be virtually reversed. The 'small' value $S^2_{eff} = Nr^2_{eff}$ applies to all other 'experimentally induced' virtual delays sent through each channel. Plotted over a variable delay time, a maximum of a factor $10lgN$ present in the otherwise relatively constant level gradient would therefore indicate that an oblique impinging wave left precisely these delay times τ_i on the microphones.

Naturally, it still makes sense to express the delay times τ by the direction of incidence. At a given point z on the z axis of the coordinate system shown in Figure 11.16, the difference of delay relative to the origin is

$$\tau = \frac{z}{c}\cos\vartheta \, . \tag{11.37}$$

(c = sound propagation speed). All positive values for τ ahead of the point $z = 0$. The experimental delay times at each microphone position result from varying all possible occurring 'experimental' incidence angles in the interval $0^0 < \vartheta < 180^0$. These microphone positions are used for the summation of these values. The experimental incidence angles are now referred to as 'angles of observance' ϑ. The diagrams shown below are all plotted over this variable angle.

To clearly justify the application of this algorithm for line arrays, the presence of 'short' time-dependent signals is customarily assumed. This procedure characterized in short by the operation 'delay and sum' is a standard procedure in practice.

For longer time signals, the measurement and evaluation results stemming from the algorithm 'delay an sum' is dependent on the structure and duration of the signal, since overlaps can be caused, according to the delay time.

In theory, it is therefore, in light of our discussion regarding general cases to decompose the signal in pure tones (see Chapter 13) and examine each frequency component separately. Therefore, we will consider the effect of the aforementioned algorithm over a variable frequency ω.

For assumed time-dependent sinusoidal signal structures, the delay over time τ is generated by multiplying the complex amplitude with $e^{-j\omega\tau}$. Since, as mentioned before, the signals present on the positive z axis are ahead in time and therefore have to be reversed, the following now applies for the 'delay and sum' algorithm

$$S(\vartheta) = \sum_{i=0}^{N-1} p_i \, e^{-jkz_i \cos\vartheta} \, . \tag{11.38}$$

It is plain to see that $S(\vartheta)$ now represents the sum of the complex (still phase-shifted) series of the microphone amplitudes p_i. For the sake of simplicity, the notation is already in terms of sound pressure (the appropriate calibration has already been made).

The basic shape of the directivity patterns is a simple matter when the impinging sound field is assumed to be composed of an oblique impinging wave at the angles ϑ_{in} and φ_{in} of the form

$$p_{in} = p_0 e^{jkz(z\cos\vartheta_{in} + x\cos\varphi_{in}\sin\vartheta_{in} + y\sin\varphi_{in}\sin\vartheta_{in})} \tag{11.39}$$

The wave produces at the microphone positions z_i the sound pressures

$$p_i = p_0 e^{jkz_i \cos\vartheta_{in}} \, . \tag{11.40}$$

Such cases produce the signal sum

$$S(\vartheta) = p_0 \sum_{i=0}^{N-1} e^{-jkz_i(\cos\vartheta - \cos\vartheta_{in})} \, . \tag{11.41}$$

For microphone line arrays consisting of microphones placed at equal distances to one another z_i,

$$z_i = -l/2 + i\Delta x \tag{11.42}$$

($i = 0, 1, 2, ..., N-1$, l= total length of the line array $= \Delta x(N-1)$) applies. The signal sum then becomes

$$S(\vartheta) = p_0 e^{jkl/2} \sum_{i=0}^{N-1} e^{-jk\Delta x\, i\, (\cos\vartheta - \cos\vartheta_{in})} \, . \tag{11.43}$$

When applying the sum formula for geometric series

$$\sum_{i=0}^{N-1} q^i = \frac{1-q^N}{1-q} \tag{11.44}$$

using (11.43) with $q = e^{-jk\Delta oneobtainsx(\cos\vartheta - \cos\vartheta_{in})}$

$$S(\vartheta) = p_0 e^{jkl/2} \frac{1 - e^{-jk\Delta x\, N\,(\cos\vartheta - \cos\vartheta_{in})}}{1 - e^{-jk\Delta x(\cos\vartheta - \cos\vartheta_{in})}} \,. \tag{11.45}$$

The absolute value allows this function to be expressed in a simpler notation if $e^{-j\frac{k}{2}\Delta x\, N\,(\cos\vartheta - \cos\vartheta_{in})}$ is taken out of the numerator and $e^{-j\frac{k}{2}\Delta x(\cos\vartheta - \cos\vartheta_{in})}$ is taken out of the denominator:

$$S(\vartheta) = p_0 e^{jkl/2} \frac{e^{-j\frac{k}{2}\Delta x\, N\,(\cos\vartheta - \cos\vartheta_{ein})}}{e^{-j\frac{k}{2}\Delta x(\cos\vartheta - \cos\vartheta_{ein})}}$$

$$\frac{e^{j\frac{k}{2}\Delta x\, N\,(\cos\vartheta - \cos\vartheta_{in})} - e^{-j\frac{k}{2}\Delta x\, N\,(\cos\vartheta - \cos\vartheta_{in})}}{e^{j\frac{k}{2}\Delta x(\cos\vartheta - \cos\vartheta_{in})} - e^{-j\frac{k}{2}\Delta x(\cos\vartheta - \cos\vartheta_{in})}} \,.$$

Using $e^{ja} - e^{-ja} = 2j \sin a$ we obtain from the above the absolute value we are primarily interested in

$$|S(\vartheta)| = p_0 \left| \frac{\sin[\frac{k}{2}\Delta x\, N\,(\cos\vartheta - \cos\vartheta_{in})]}{\sin\,[\frac{k}{2}\Delta x(\cos\vartheta - \cos\vartheta_{in})]} \right|$$

$$= p_0 \left| \frac{\sin[\pi N \frac{\Delta x}{\lambda}(\cos\vartheta - \cos\vartheta_{in})]}{\sin[\pi \frac{\Delta x}{\lambda}(\cos\vartheta - \cos\vartheta_{in})]} \right| . \tag{11.46}$$

Before interpreting the solution, it first makes sense to introduce the generalized variable

$$\Omega = \frac{\Delta x}{\lambda}(\cos\vartheta - \cos\vartheta_{ein}). \tag{11.47}$$

The resulting function

$$G(\Omega) = \left| \frac{\sin[\pi N \Omega]}{\sin[\pi \Omega]} \right| \tag{11.48}$$

is quite easily to analyze:

- $G(\Omega)$ is periodic with the period 1: it is $G(\Omega + 1) = G(\Omega)$,
- $G(\Omega)$ possesses roots $\Omega = n/N$ $(n = 1, 2, 3, ...)$ for each period $N - 1$ and
- at the location $\Omega = 0$ $G(0) = N$ (based pm $\sin x = x$ for small x).

A graphical representation of G, over one period, is illustrated in Figure 11.17.

The shape of the directivity patterns $S(\vartheta)$ can be discerned from the $G(\Omega)$ gradient. Because the correlation of the angle of observance. ϑ and the variables Ω is described in eq.(11.47), the intervals in sections of the function G exist in interval

$$\Omega_{min} < \Omega < \Omega_{max} \tag{11.49}$$

with

$$\Omega_{min} = -\frac{\Delta x}{\lambda}(1 + \cos\vartheta_{ein})$$

and

$$\Omega_{max} = \frac{\Delta x}{\lambda}(1 - \cos\vartheta_{ein}) \,.$$

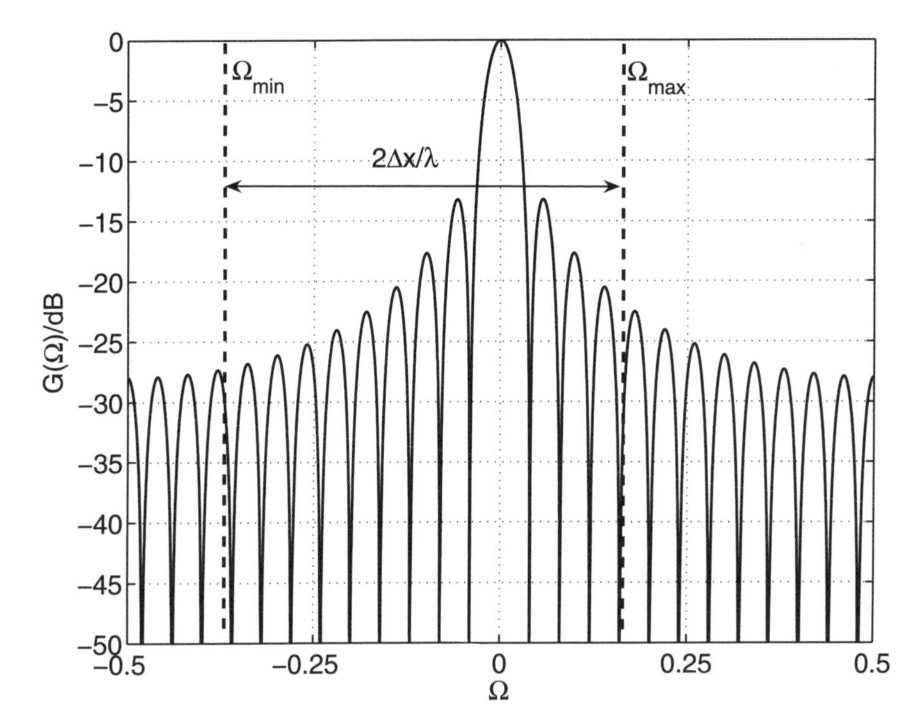

Fig. 11.17. Directivity function $G(\Omega)$ for the generation of directivity patterns for N=25 sensors

This interval cycles through by varying the angle of observance in interval $0 < \vartheta < 180^0$ from top to bottom (that is, beginning at Ω_{max} for $\vartheta = 0^0$). The spanned interval's width is $\Delta\Omega = \Omega_{max} - \Omega_{min} = 2\Delta x/\lambda$ (see Figure 11.17).

The directivity patterns resulting from these discussions are shown in Figures 11.18 to 11.20 for $\Delta x/\lambda = 0.125$, 0.25 and 0.5. As a general rule in respect to directivity patterns, the angles ϑ_n and $\vartheta_{in,n}$ given in these diagrams are accounted for relative to the normal directivity of the microphone line array, resulting in $\vartheta_n = \vartheta - 90^0$ and $\vartheta_{in,n} = \vartheta_{in} - 90^0$.

The contrast and sharpness of each impinging wave can be assessed based on the form of the results obtained. For application in practice, this means that multiple sound sources are simultaneously present at varying signal strengths and angles of incidence. In this case, the output signal generated by the 'delay and sum' algorithm, of course, likewise exists in a sum of the mutually offset functions $G(\Omega)$, depending on the angle of incidence. If pronounced singular maxima are found in the process, this indicates the presence of multiple sources. In such cases, the individual maxima – the so-called main lobes – should be small. If additional 'weak' sources are localized on the basis of the presence of 'stronger' ones, the level intervals between main and side lobes

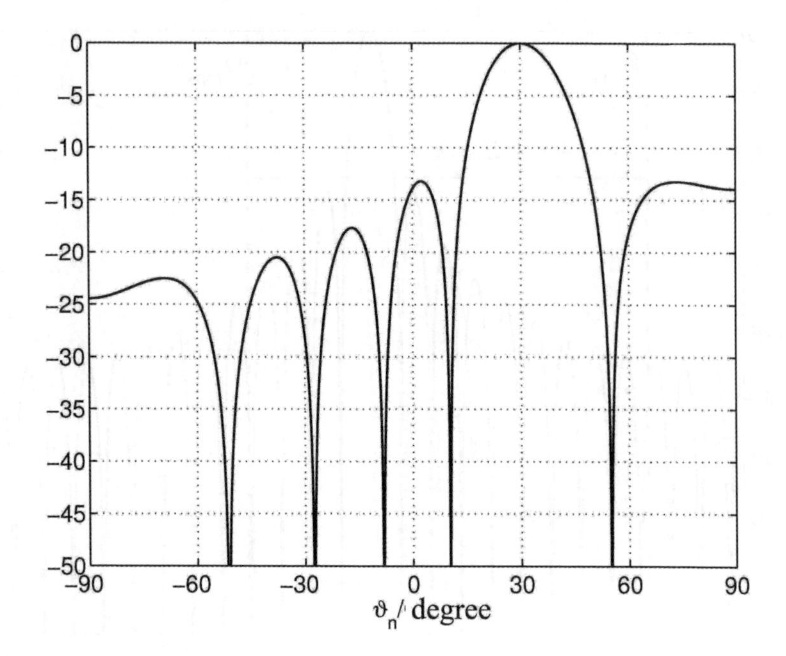

Fig. 11.18. Directivity pattern $D(\vartheta_n)$ over the angle of observance ϑ_n for $\Delta x/\lambda = 0.125$, angle of incidence below $\vartheta_{in,n} = 30°$, calculated for N=25 sensors

should be as great as possible. Otherwise, the side lobes of the stronger source will 'drown out' the weaker source's main lobe. Overall, a significant difference between main and side lobes as well as a narrow main lobe are desirable.

The following principles, based on the directivity patterns and the equations given, guarantee that these two criteria for high-quality measurements are met:

- The lower the frequency, that is, the smaller the ratio between sensor distance and wavelength $\Delta x/\lambda$, the smaller is the width of the interval to be plotted over the angle. Accordingly, the lobes and especially the main lobe are to be as wide as possible. The width decreases as the ratio $\Delta x/\lambda$ increases and therefore, with increasing frequency. At lower frequencies, the resolution is low. The resolution improves at higher frequencies.
- The best resolution in terms of the smallest main lobe width is obtained for $\Delta x/\lambda$=0.5. Larger intervals spanned by varying ϑ should be eliminated, as this would cover more than one period of $G(\Omega)$. This would result in multiple main lobes becoming visible in the directivity pattern, making a unique correlation with each angle of incidence impossible.
- The level interval between main and the smallest side lobe is $20 lg N$ (based on $G(0) = N$ and $G(1) = 0$). The increase of the difference between main

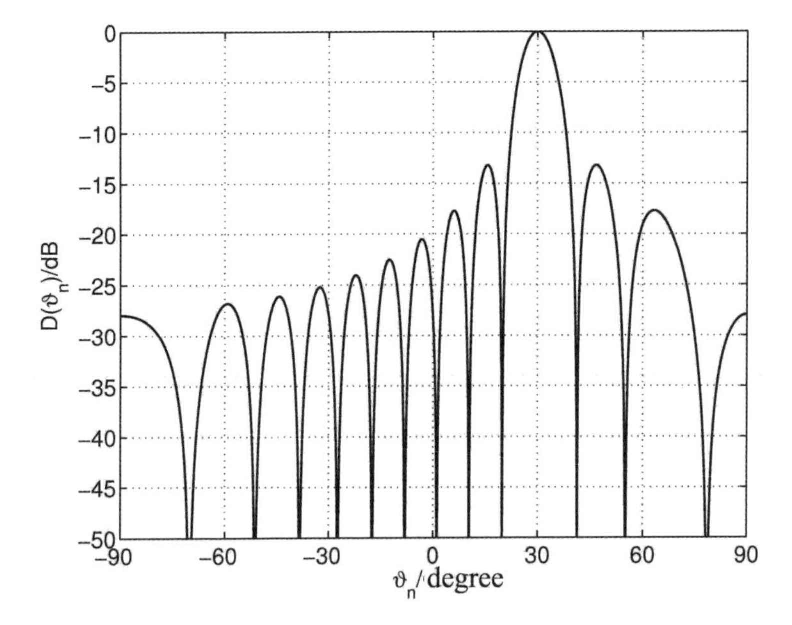

Fig. 11.19. Directivity pattern $D(\vartheta_n)$ over the angle of observance ϑ_n for $\Delta x/\lambda = 0.25$, angle of incidence under $\vartheta_{in,n} = 30°$, calculated for N=25 sensors

and side lobes becomes more gradual for greater numbers of sensors N. To guarantee significant improvements, this number should be doubled.

- The same is also true for the width of the main lobe $\Delta\Omega_{HK} = 2/N$, which only gradually gets smaller as N gets larger.

In conclusion, it should be stressed that microphone signals are independent of φ_{in} (see eq.(11.40)). If the angle of incidence is rotated around the z axis as shown in Figure 11.16, the microphone voltage remains unaffected. If multiple impinging waves occur simultaneously at the same angle ϑ_{in}, but at different angles φ_{in}, the signals present along the z axis exist in the sum of the partial pressures. In terms of the circumferential direction φ, the microphone array adds up to a total of all impinging elements.

Two-dimensional microphone arrangements are required to distinguish sources and/or impinging sound elements from one another in terms of circumferential direction. For the purposes of this text book, it is sufficient to introduce the basic idea of microphone antennae. The following section includes an initial reference to surface arrangements.

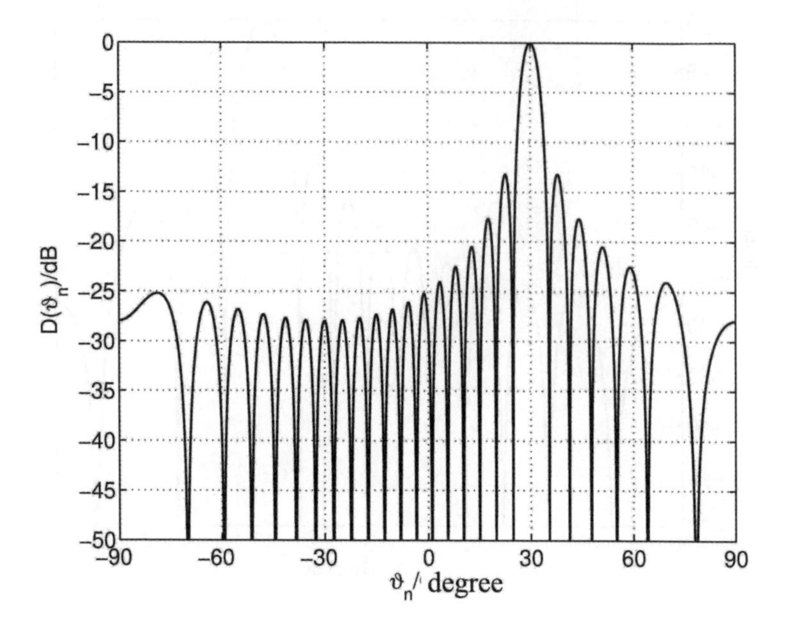

Fig. 11.20. Directivity pattern $D(\vartheta_n)$ over the angle of observance ϑ_n for $\Delta x/\lambda = 0.5$, angle of incidence under $\vartheta_{in,n} = 30°$, calculated for N=25 sensors

11.5.2 Two-dimensional sensor arrangements

The simplest two-dimensional expansion of the microphone array system which allows for localization in the circumferential direction φ consists of an system of two intersecting microphone line arrays at a 90°-angle as shown in Figure (11.21). Such a microphone arrangement is commonly known as a 'cross array'.

The simulation calculations for determining the array characteristics are the same as for the line array: the time-delayed signals are added together. Assuming pure tones, this corresponds, once again, to a phase-shift of the complex amplitudes. The complex amplitude of the array output exists in

$$S(\vartheta) = \sum_{i=0}^{N-1} p_i \, e^{-jk(z_i \cos\vartheta + x_i \cos\varphi \sin\vartheta)} , \tag{11.50}$$

where (x_i, z_i) indicates the microphone positions and p_i, as described above, the complex amplitude of the microphone signal at sensor i. The quantities ϑ and φ provide a clear overview of the desired simulation results of the cross array, which can be virtually reproduced 'photographically'. Do do this, a 'photo plane' parallel to the array is outlined in the quantitative model. A color-coded level value is assigned to every point on the photo plane. Each

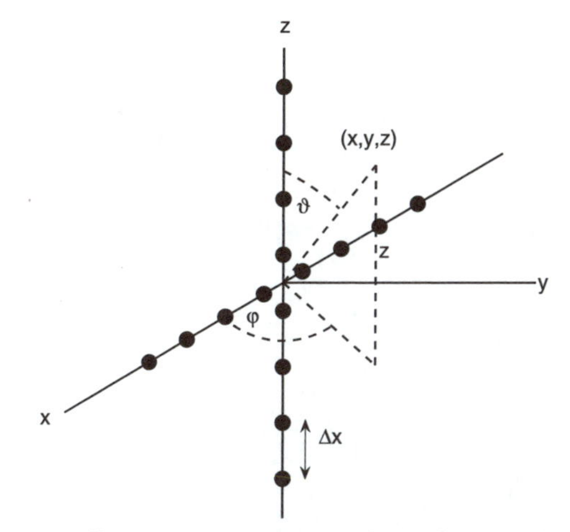

Fig. 11.21. Cross array consisting of equally-placed microphones

level value with its corresponding angle of incidence represented by a photo point has previously been defined using the acoustic antenna. These cross array diagrams are shown in Figures 11.22 and 11.23.

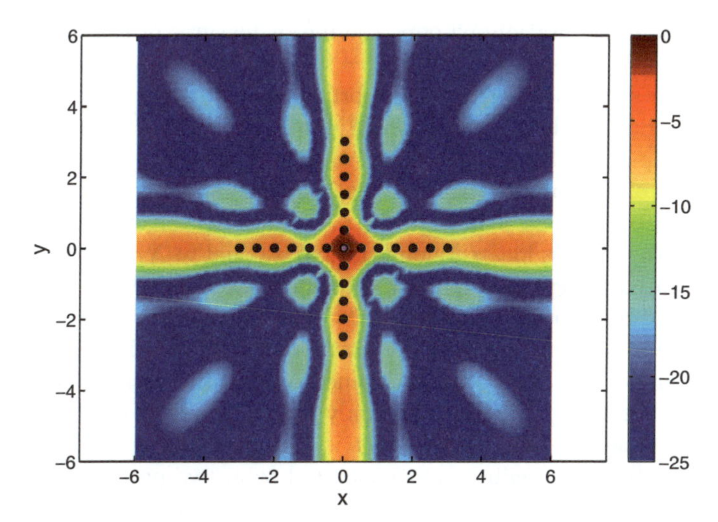

Fig. 11.22. Virtual photographic representation of the level gradient defined by the cross array measurement. The small, light pink dot indicates the actual incidence angle of the impinging plane wave. The black dots symbolize the cross array. The distance between each microphone is $\Delta x/\lambda = 0.5$. The level scale is given on the right.

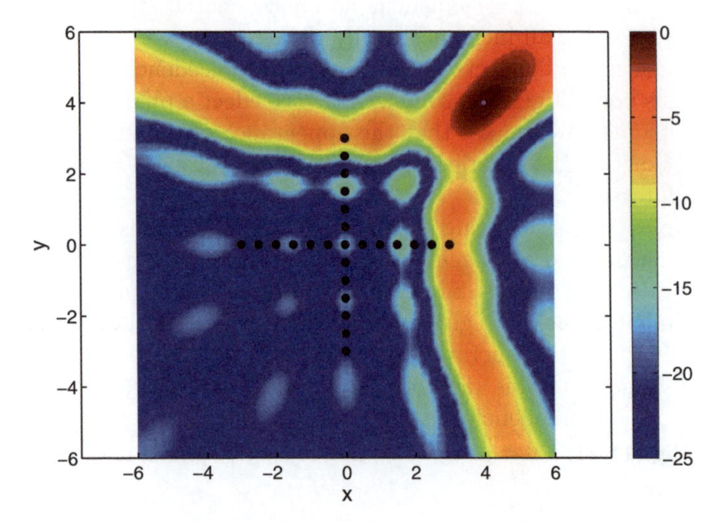

Fig. 11.23. Virtual photographic representation of the level gradient defined by the cross array measurement. The small light pink dot on the upper right indicates the actual incidence angle of the impinging plane wave. The black dots symbolize the cross array. The distance between each microphone is $\Delta x/\lambda = 0.5$. The level scale is given on the right.

A simple consideration explains the major features of the diagrams. The cross array generates signal sums $\varphi = 90^0$ 'without delays' $\vartheta = 90^0$ (from the observer's perspective) for the normal sound incident of a progressive wave impinging in the $-y$-direction (Figure 11.22). These signal sums exist in $2N$-multiples of each signal (identical to one another). Here, N indicates the number of sensors per array. If the incidence angle is rotated $\varphi_{in} \neq 90^0$ $\varphi_{in} = 90^0$ around the z axis from $\vartheta_{in} = 90^0$ in another direction $\vartheta_{in} = 90^0$, the sum of the sensors located along the z axis equal to N, and the sum of the sensors located along the x axis close to zero. Thus the N-multiple signal is now assigned to the observance angle $\vartheta = 90^0$, $\varphi = 90^0$, even though the incidence angle is actually entirely different. The level interval between the 'real' incidence angle (as before, with the value $2N$) and the normal angle of observance with the value N is therefore only $6\,dB$. This level difference between the main and the next higher side lobe is quite small. The diagram in Figure 11.22 shows that the main lobe is generated from two perpendicular 'bright' stripes, the brightness of which is also summed together. The point where the stripes intersect indicates the position of the source. This principle applies, of course, to oblique angles of incidence as well, as shown in Figure 11.23, except for the stripes no longer form a line but instead, take on a rather complex geometric shape. The cross array has two preferred directivities, treating different incidence angles differently.

In addition to the cross array, microphones are often arranged in a ring (Figure 11.24). This arrangement is no longer characterized by a distinctive directivity, treating all incidence angles the same.

Although normally 'facts and figures' take precedence over beautiful pictures in this book, Figure 11.26 illustrates the effect a photographic image can on the outcome. We can see that the appearance of the image differs greatly from the exact same scenario depicted in Figure 11.25. The only change was the level interval.

Examples of cross and ring arrays were examined in this section. Of course, other geometries and arrangements come into play, such as three-lined, spiral and quadratic field arrays. Some of these are also used in practice. In addition, the microphones in a given arrangement do not all have to be the same distance apart. If the microphone distances increase from inward outward, the array increases. This method has the advantage of a better resolution, especially for low frequencies.

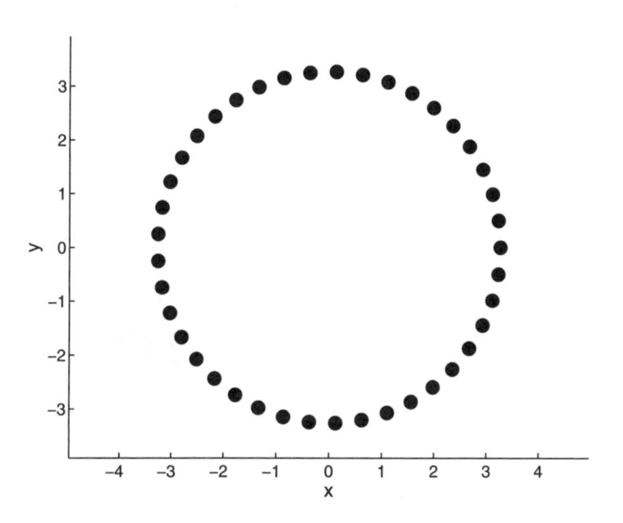

Fig. 11.24. Ring-shaped arrangement of equally-placed microphones

The optimal level distributions for ring arrays are virtually photographed for two examples in Figure 11.25. The diagram shows the area of the source surrounded by a dark ring, providing a far better localization of the source than the cross array.

Note that the visual quality of the image in the virtual photography is strongly dependent on the level range used in imaging. The images appear sharper and the source zones more concentrated if smaller level intervals are used. Such tricks of the trade are understandably frequently implemented in practice.

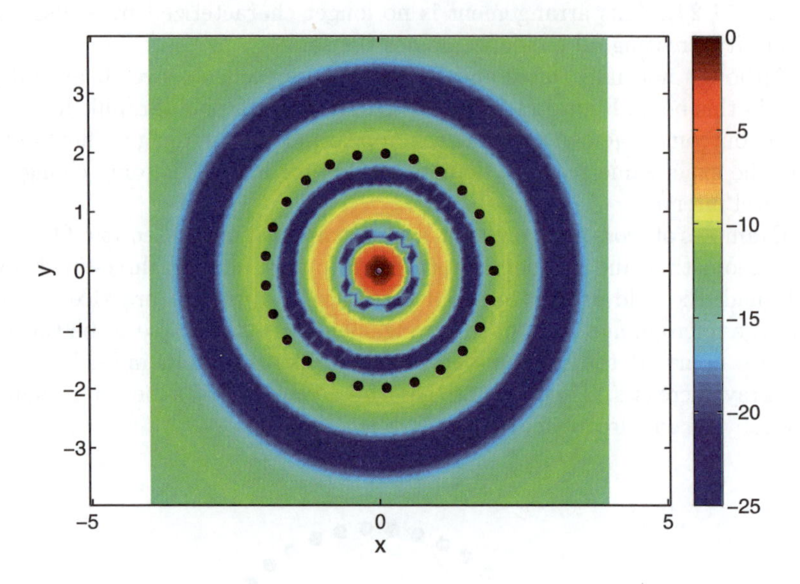

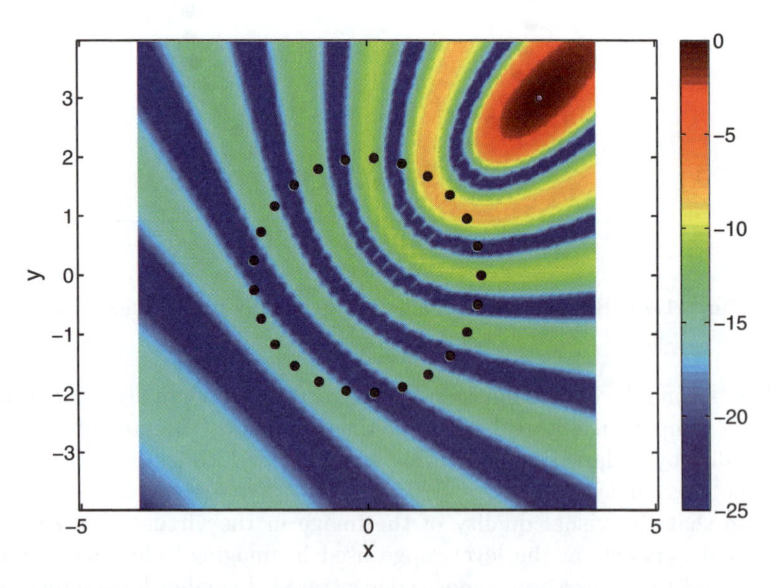

Fig. 11.25. Virtual photographic imaging of the level gradient using a ring array. The small pink dot indicates, here as well, the actual incidence angle. The small dots symbolize the ring array. The distance between each microphone is $\Delta x/\lambda = 0.5$. The level scale is given on the right.

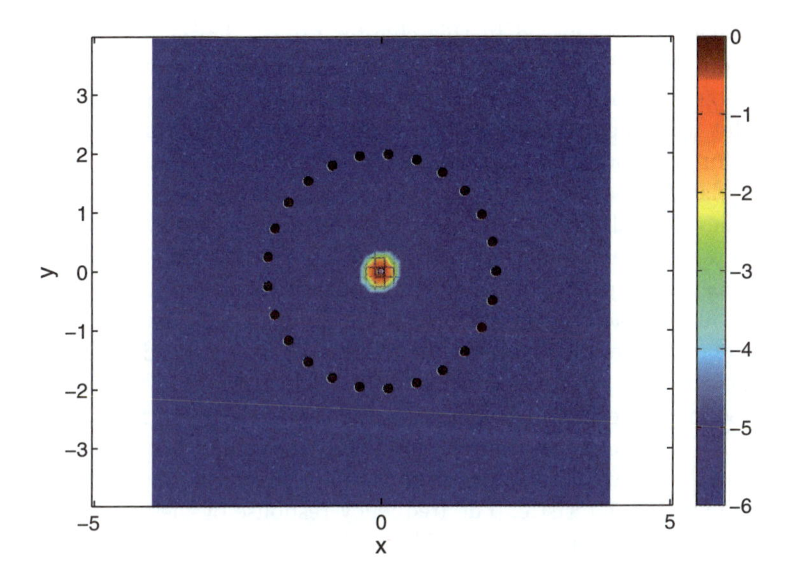

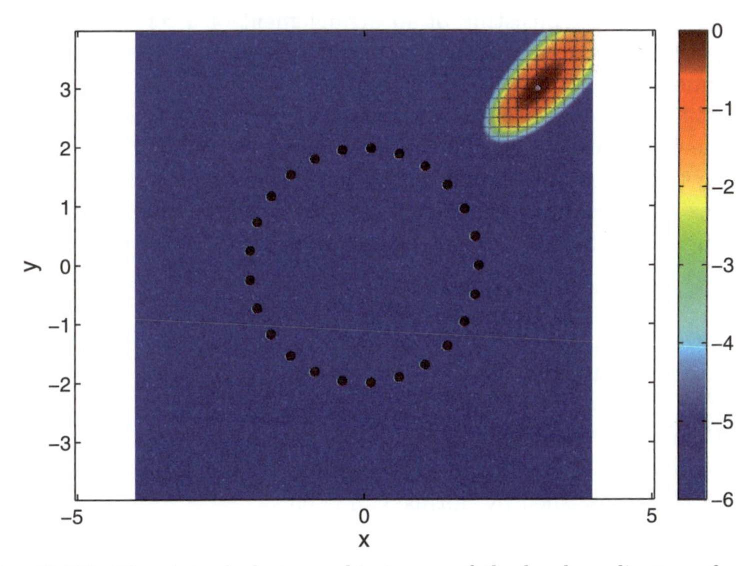

Fig. 11.26. The virtual photographic image of the level gradient produced by the ring array looks like this when the given level gradient interval is decreased.

11.6 Summary

The sound receivers discussed in this chapter are condenser and electrodynamic microphones. The frequency responses of both types of microphones are almost solely defined by their mechanical construction. Due to their characteristic flat frequency response, exclusively condenser microphones are implemented for high-precision measurements. The other types are sufficient for taking relative measurements of level intervals or reverberation times, for example. They may also serve as a more cost-effective option for audio equipment. However, condenser microphones are the only option for measuring intensity, whenever the phase response of microphones is essential, (see Chapter 2). The pick-up characteristic of microphones are primarily derived from the fact that the net force impinging on the membrane is expressed as the sound pressure integral over the membrane's surface. This results in the formation of a unidirectional effect of microphones, which is defined by a main and side lobe structure, depending on the ratio of membrane diameter to wavelength.

The sound pressure frequency response of electrodynamic loudspeakers has a band-pass characteristic. The frequency response of these is somewhat constant in the transmission range. The main reason for this is due to the fact that, at a certain distance to the volume source, the sound pressure is proportional to both the time-dependent change in volume flow as well as the acceleration of the membrane. The transmission range is limited by the mechanical resonance (lower-bound) and by the electrical folding frequency (upper-bound).

'Acoustic antennae' consisting of an arrangement of a certain number of spatially distributed microphones can be used to determine the sound pressure level as a function of directivity. This can be done by summation of transducer's output signals distributed over time. The pick-up characteristic of the measuring device generally constructed in this way is similar to that of loudspeaker arrays and possesses the same tendencies: wide main and side lobes for smaller ratios of total converter dimensions to wavelength, and small-banded lobe structures of high-resolution by small ratios of converter dimensions to wavelength.

11.7 Further reading

The book 'Akustische Messtechnik' (edited by the author M. Möser) provides a valuable supplemental resource for elaboration of these topics. The original in German will be published by Springer in 2009.

11.8 Practice exercises

Problem 1

Which frequencies belong to the pick-up characteristic given in Figure 11.6 if the membrane has a diameter of $25\,mm$?

Problem 2

A recording accelerometer consists of a piezoelectric element which can be mechanically defined as an elastic layer (see the following schema). The underside of the piezo element is rigidly affixed to the object. Let a_m be the object's acceleration. The elastic ceramic part is supporting a mass on top of it. The output voltage U (after selecting an appropriate 'charging amplifier' for a very small charge flow) is proportional to the spring force F_s acting on the piezo element $U = EF_s$, whereby E signifies a converter element independent of frequency. What is the frequency response function of U/a_m?

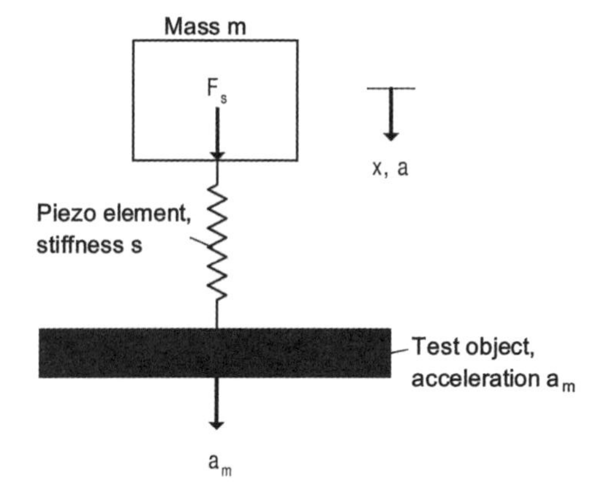

Fig. 11.27. Schema of a recording accelerometer for measuring structure-borne sound

Problem 3

A microphone has a transmission factor of $10\,mV/Pa$ ($1\,Pa = 1\,N/m^2$). An alternating current of $20\,\mu V$ (rms-value) is measured at the output. The voltage has resulted solely from the internal noise of the microphone itself. How high would the sound pressure level be if emitted from a microphone that is ideal and noise-free, but otherwise of the same construction, and producing the same amount of voltage at its output? Such a sound level is otherwise known as the 'equivalent sound pressure level'.

Problem 4

In general, at which angles of incidence ϑ lie breaks in the pick-up characteristic of microphones with output voltage = 0

$$G(\omega,\vartheta) = \frac{2J_1\left(kb\sin\vartheta\right)}{kb\sin\vartheta}$$

(b=membrane radius)? Specify the values for $b/\lambda = 1$; 2 and 3. How great are the frequencies corresponding to each b/λ value for a microphone with a diameter of $2b = 2.5\,cm$ ($2b = 1.25\,cm$)?

12

Fundamentals of Active Noise Control

Methods in eliminating noise 'with noise' were already described in the chapter on sound propagation. Namely, noise power emissions can more or less be reduced by placing two inversely phased sources at a small distance to one another. In other words, one can refer to one of such a source pair as 'an additional sound source introduced for the purpose of noise reduction.' This is the main idea behind what is known as active noise reduction, which is also sometimes referred 'anti-noise.' This term encompasses all major types of noise reduction based on the use of electro-acoustic converters (as described in Chapter 11). The disruptive, somehow unavoidable sound source is referred to here as the 'primary' source; additional electro-acoustic converters which are intended for use as noise reducers are referred to as 'secondary' sources. The example of a noise reduction method outlined extensively in Section (3.4) 'Sound field of two sources' shows, however, that such active methods only work under 'favorable' circumstances. Adding a second source only works in the following special cases:

- An overall effective reduction of sound power emission can only be achieved when the primary noise is composed of low frequencies. To actually implement this reduction method, the frequencies have to be low enough for sources which are placed at a small enough distance to one another, or the primary sound emitter has to be very small in size.
- For higher frequencies, the secondary source may cancel out the primary field at a specific point completely. This effect is, however, extremely localized. Inevitably, the power emission from two sources consists in the sum of the partial performances (in the absence of the other source). For high frequencies, the active method only makes sense if the objective is to insulate a small area.

Such favorable special case conditions, either when there are low pitches or when dealing with very small areas, can nevertheless be practically relevant. Some application areas of active noise reduction are listed below:

M. Möser, *Engineering Acoustics*, DOI 10.1007/978-3-540-92723-5_12,

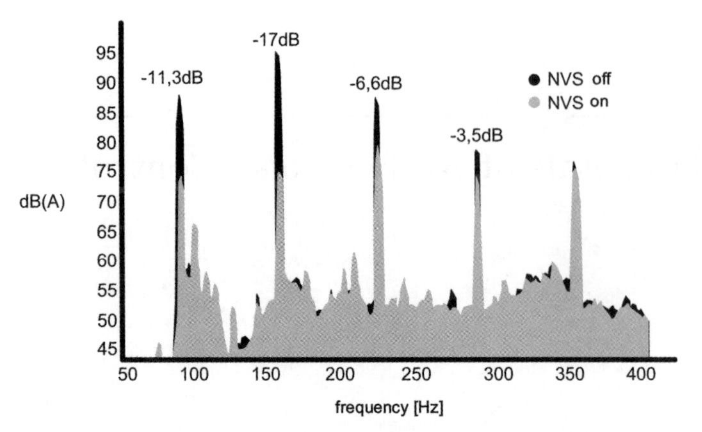

Fig. 12.1. Noise spectra of a turbo-propeller-driven aircraft without (NVS off) and with (NVS on) active sound suppression. (taken from: J. Scheuren 'Aktive Beeinflussung von Schall und Schwingungen', Chapter 13 in G, Müller,G. and M. Möser: Taschenbuch der Technischen Akustik, Springer-Verlag, Berlin 2004)

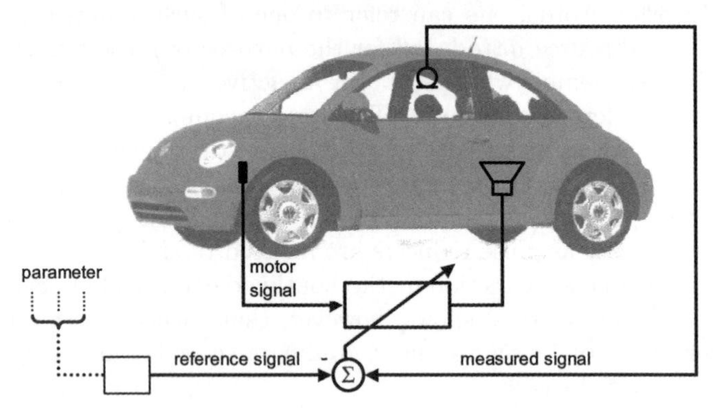

Fig. 12.2. Basic outline of noise control systems implemented in suppressing internal engine noise in automobiles. (taken from: J. Scheuren 'Aktive Beeinflussung von Schall und Schwingungen', Chapter 13 in G. Müller and M. Möser: Taschenbuch der Technischen Akustik, Springer-Verlag, Berlin 2004)

- Only small areas to be insulated at low frequencies exist in environments where it is important to protect the passenger's ear, such as from propeller noise in airplanes. Loudspeakers built into the headrests, for example, can serve to specifically reduce the noise emitted by the propeller (see Figure 12.1).
- Similarly, engine noise in automobiles can be reduced either by mounting similar types of loudspeakers or by introducing additional sources in close proximity to the driver's and passenger's ear (see Figure 12.2).
- An especially effective method is the use of headphones, such as for pilots in the cockpit. There are restrictions on the weight of the headphones of

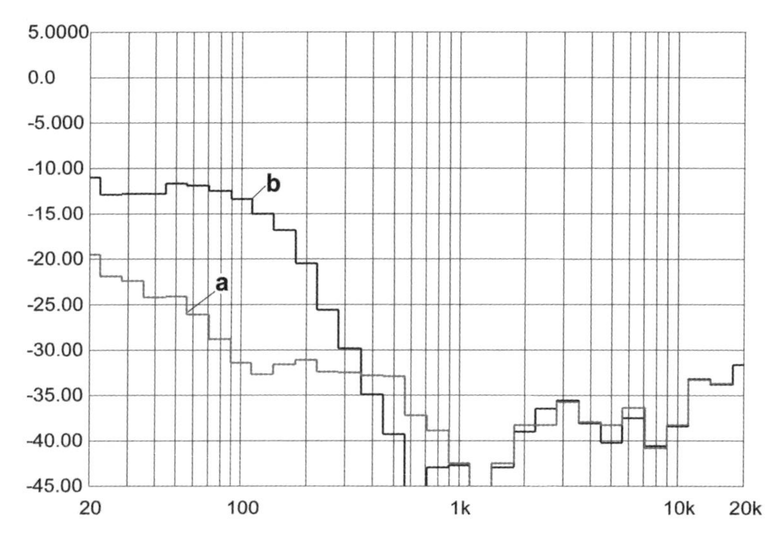

Fig. 12.3. Sound damping curve of a pilot's headset with (a) and without (b) active compensation (taken from: J. Scheuren 'Aktive Beeinflussung von Schall und Schwingungen', Chapter 13 in G. Müller and M. Möser: Taschenbuch der Technischen Akustik, Springer-Verlag, Berlin and Heidelberg 2004)

course, alone due to reasons of the wearer's comfort. This leads to poor sound reduction for low frequencies. By reconstructing the inverse-phase signal, the converse to the signal which is transmitted by the microphone built into the ear cups of the headphones, and by suitable playback through the microphone's membrane, noise reduction can be greatly improved, even for the very low-frequency drone (see Figure12.3). The signal from the transmission tower is also incorporated into the signal.

- Sound propagation in channels also provides favorable conditions for active noise reduction, at least at frequencies below the lowest cut-on frequency. This is because the direction the sound field is travelling is apparent, and the field consists of plane waves. Active noise dampers therefore provide a reasonable alternative to their passive counterparts (Chapter 9).
- And finally, it should be noted that sound damping can be greatly improved in the range of the mass-spring-mass resonance using the proactive method of installing loudspeakers in double windows (see Figures 12.4 and 12.5 for basic construction and effect). The 'most favorable conditions' are due to the fact that passive sound damping is poor near the characteristically low tuned resonance. Active improvements are most effective for small and light-weight windows. As already mentioned, among the areas of application are aircraft engineering and also automotive lightweight construction .

The list of state-of-the-art applications in active noise control goes on. On the other hand, broad-band active sound insulation is certainly not an option

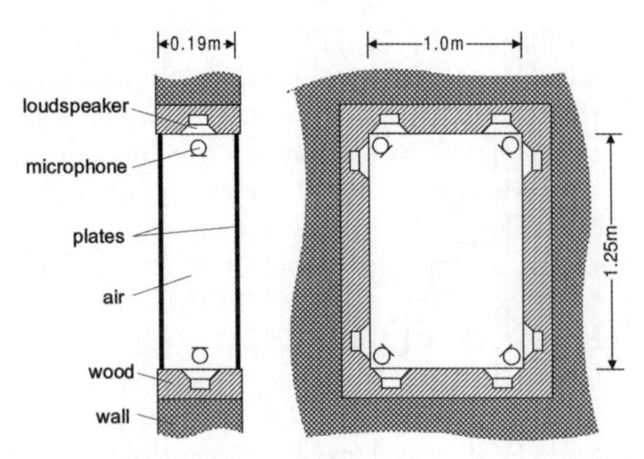

Fig. 12.4. Basic construction of a double window with optimized active noise reduction features. (taken from: A. Jakob and M. Möser: Verbesserung der Schalldämmwirkung von Doppelschalen durch aktive Minderung des Hohlraumfeldes, DAGA 2000)

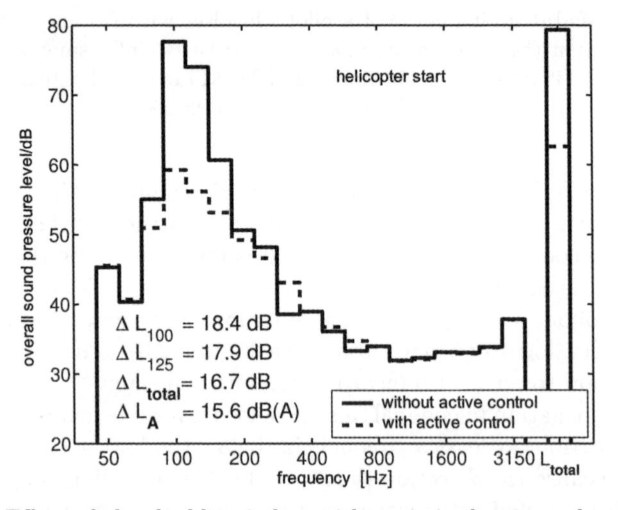

Fig. 12.5. Effect of the double window with optimized noise reduction (average level in emission room); example of a test signal corresponding to the ignition and warm-up of a helicopter. (taken from: A. Jakob, M. Möser, C. Ohly: Ein aktives Doppelglas-Fenster mit geringem Scheibenabstand, DAGA 2002)

for large volumes. This text leaves little room for an extensive treatise on the entire breadth of electronic and technical applications. More information on the specific algorithms required for running processors can be obtained in the supplementary literature recommended at the end of this chapter. The following basic acoustic principles are listed below:

- The practical examples described above lead to the question of which mechanisms limit the dimension of the desired effect. Overall, one can say that the amount of level reduction correlates with the inaccuracies inherent in replicating the primary field in the image of the secondary source. If the primary and secondary sound are not exact inverse opposites, their sum is therefore a non-zero value. The replication error thus defines the dimension of active level reduction. Perpendicularly propagating waves will be used as an illustrative example of the effect of this error. Such waves add up to a sound field sum of zero at one point. The 'error' is simply represented by ears or other 'receivers' present in the field that are not located at precisely the correct spot. This is the central issue in sound insulation for small areas contained by large volumes, such as in the case of airplanes.

- From a basic standpoint, passive and active dampers both possess the same basic characteristics: Both methods allow sound to not only be reflected, but also to be absorbed.
- And finally, it is worth mentioning that secondary sound sources, as in all the applications mentioned above, are not only capable of changing the size of sound fields by means of interference and adding sound to sound, but can also interfere directly in the process of emerging sound. The latter is, however, only possible for a certain class of sound sources, namely, those that are self-excited sources. There are many examples of self-excited vibrations . We will mention just two, the tone of a resonator in contact with air (blowing into a bottle) or flapping vibrations (such as a cloth blowing in a strong wind). The section just preceding the summary of this chapter outlines examples of physics and examples for self-excitations and active methods of prevention.

12.1 The Influence of Replication Errors

The effects of active noise reduction, which are limited by the error in the secondary replication of the primary sound field, can be easily quantified in a simple model. Whatever the actual cause of the error, it can always be attributed to the fact that the signals will slightly vary in amplitude and in phase. The total pressure in complex notation (index p: primary, index s: secondary) consists therefore of

$$p_{total} = p_p - \mathrm{e}^{j\Phi} p_s \,. \tag{12.1}$$

The quantities p_p and p_s describe the amplitudes of the partial sounds. They can be considered here as real and positive. Φ describes the phase error. In the 'ideal case' of complete cancellation due to wave interference, $\Phi = 0$ and $p_s = p_p$. The amplitude-absolute-value-square of the total field is

$$\mid p_{total}^2 \mid = (p_p - \mathrm{e}^{j\Phi} p_s)(p_p - \mathrm{e}^{-j\Phi} p_s) = p_p^2 + p_s^2 - 2 p_s p_p \cos(\Phi). \tag{12.2}$$

The following therefore applies to

$$\Delta L = 10\lg(p_p^2/\mid p_{total}^2\mid) = -10\lg(1 + \frac{p_s^2}{p_p^2} - 2\frac{p_s}{p_p}\cos(\Phi)) \qquad (12.3)$$

the level reduction ΔL induced by the active method. The amplitude ratio of the corresponding level difference between the primary and secondary field levels can be expressed therein by

$$L_{diff} = L_p - L_s = 10\lg(p_p^2/p_s^2) \qquad (12.4)$$

or

$$\frac{p_s^2}{p_p^2} = 10^{-L_{diff}/10}\,. \qquad (12.5)$$

The arrays of curves resulting from variations of L_{diff} and Φ are depicted in Figure 12.6. Only very small margins of tolerance result in significant level reductions ΔL. For instance, one requires an amplitude error of less than $L_{diff} = 0.5\,dB$ and at most 4 degrees phase error to achieve a level reduction of more than 25 dB. For level reductions of over 40 dB, even amplitude errors of $\Phi = 0$ may not exceed $L_{diff} = 0.1\,dB$ (this correlates to a difference of only one percent!) The more effective the desired active noise reduction procedure is, the more precision is required, which is not technologically possible. The limitations of noise reduction effects achieved by active methods is therefore exclusively attributable to the errors always present in measurements of the active replication of the primary field in the secondary field.

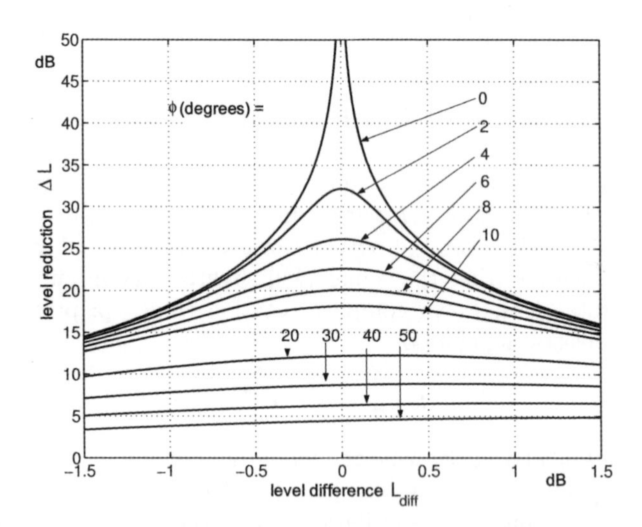

Fig. 12.6. Level reduction achieved in dependency of amplitude and phase errors in the secondary field

In the arrangement described in Chapter 3 of two inverse-phased sources, a finite distance between the two sources can be interpreted as the cause of the error. Only small distances relative to wavelength produce significant level reductions. The effect is also spatially dependent.

12.1.1 Perpendicularly Interfering Waves

Another strongly spatially-dependent level reduction is obtained when restricting the sound damping affect to only small areas, such as in the realm of a passenger in an airplane or automobile. Qualitative statements regarding the dimension of the insulated area can be made by assuming a simple model, one where two sound waves are interfering at a 90-degree angle so that their paths add up to zero at one point. If this main point of effect coincides with the origin of the coordinate system and the directional wave paths encompass the angles 45° to −45° around the y-axis, the sound field composed of both advancing waves where $\vartheta = 45°$ is represented by both

$$p_{total} = p_0[\mathrm{e}^{-jkx\cos(\vartheta)-jky\sin(\vartheta)} - \mathrm{e}^{jkx\cos(\vartheta)-jky\sin(\vartheta)}] \tag{12.6}$$

and

$$p_{total} = -2jp_0 \sin[kx\cos(\vartheta)]\mathrm{e}^{-jky\sin(\vartheta)}. \tag{12.7}$$

This results in the amplitude-absolute-value-square of

$$|p_{total}|^2 = 4p_0^2 \sin^2[kx\cos(\vartheta)]. \tag{12.8}$$

The actively induced insertion loss is therefore

$$\Delta L = 10lg\frac{|p_0|^2}{|p_{total}|^2} = -10\lg[4\sin^2[kx\cos(\vartheta)]], \tag{12.9}$$

increasing with $x \rightarrowtail 0$, because the partial sounds locally cancel each other out without bound. It is evident here that the lines corresponding to the same level reduction are lines $x = const$. A description of the spatial dimensions of the effective areas could possibly be used to describe the boundaries of such areas where a level reduction at least $6\,dB$ to $12\,dB$ – or more general a multiple of $6\,dB$ – has been reached everywhere. For the purposes of this consideration, we set the argument in the logarithm of eq.(12.9) to a power of two with even exponents

$$4\sin^2[kx\cos(\vartheta)] = 2^{-2N}. \tag{12.10}$$

Because $-10\lg(2^{-2N}) = 20N\lg(2) = 6N$, $N = 1$ denotes the $6\,dB$-boundary, $N = 2$ the $12\,dB$-boundary (etc.). Obviously this means

$$\sin[kx\cos(\vartheta)] = 2^{-(N+1)}. \tag{12.11}$$

For $N \geq 1$, the right side is always smaller than 1. This is why we can replace the sin function on the left with its argument

$$kx\cos(\vartheta) = 2^{-(N+1)}. \tag{12.12}$$

Thus, using $\vartheta = 45°$, we finally obtain

$$x/\lambda = \frac{1}{\pi\sqrt{2}}\frac{1}{2^{N+1}} = \frac{0{,}225}{2^{N+1}}. \tag{12.13}$$

The zone characterized by N is sufficient, due to its symmetry of $-x$ to x, resulting in a zone Δx which is twice as wide

$$\Delta x/\lambda = \frac{0,45}{2^{N+1}}. \tag{12.14}$$

Even for the rather lax constraint that the level be reduced by at least $6\,dB$, the width of an area – where this level reduction is to either be attained or surpassed – only amounts to slightly more than one-tenth of a wavelength, so about $34\,cm$ at $100\,Hz$, and only $3.4\,cm$ at $1000\,Hz$. Each time the level reduction requirement is increased by $6\,dB$, the zone is reduced by half. Therefore, only 0.055 wavelengths remain to obtain a reduction of $12\,dB$. This example dramatically underscores the aforementioned fact that active noise control is quite often optimal for low frequency noise. In the case of medium- or high-frequency noise reduction, airplane or automobile passengers would barely be able to move around or turn their head to enjoy sufficient noise reduction at such frequencies.

12.2 Reflection and Absorption

Active noise control methods are not only able to reflect impinging sound waves. Moreover, secondary sources may absorb the primary incident power flow, as in the passive case. This aspect will be elaborated in the following sections. The simplest way to describe the phenomena of simultaneous reflection and absorption is to imagine a one-dimensional continuum, an air-filled channel or duct to model actively-induced noise reduction. The frequency is located below the lowest cut-on frequency of the first non-uniform mode.

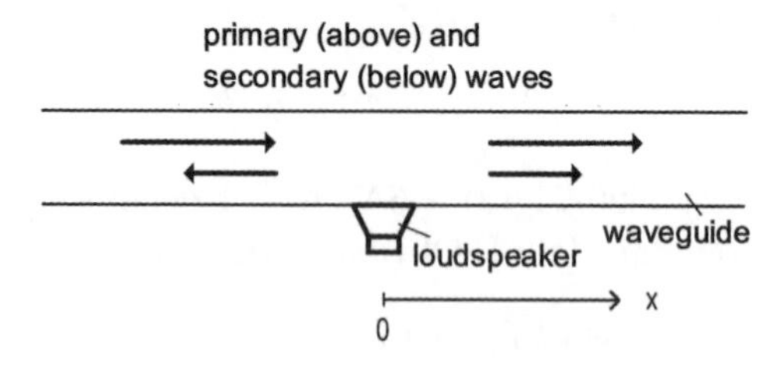

Fig. 12.7. Principle sketch of a wave field made up of primary and secondary components

The pertinent model setup is shown in Figure 12.7. The depicted duct categorically represents all such structures, such as a ventilation duct or a flue, where the active source is installed along the side. The source is assumed to radiate sound equally in all spatial zones to the left and to the right of it. In other words, we assume an omnidirectional source. If its sound field is an r-multiple of the impinging wave, the net field will be assigned spatial zones to the left (zone 1, $x < 0$), and to the right of the source (zone 2, $x > 0$)

$$p_1 = p_0(\mathrm{e}^{-jkx} + r\mathrm{e}^{jkx}) \tag{12.15}$$

and

$$p_2 = p_0(1 + r)\mathrm{e}^{-jkx} \, . \tag{12.16}$$

The 'downstream' zone 2 is precisely that which is completely sound-insulated by $p_2 = 0$ given $r = -1$. In the 'upstream' zone 1, a standing wave develops: $p_1 = -2jp_0 \sin(kx)$. It is evident here that the secondary source functions as a reflector. The dimension r can therefore be denoted as the actively induced reflection factor. The total obliteration of the waves through interference downstream where $r = -1$ is characteristic of "'soft"' reflections, whereby the sound pressure is solely reduced to the pressure of the medium (air) at the point of reflection, with $p_2(0) = 0$. Due to the affect exerted by the secondary source on the net field, no net power is flowing the tube cross-sections to the left and to the right of the source. Therefore, the secondary loudspeaker neither loses nor receives energy to or from the channel in the process.

The power flow emanating from the secondary source does change for other factors r, however. In general, it can be said that the power flowing from the loudspeaker to the channel P_L must equal the difference between both powers to the right and to the left of the loudspeaker, which are transported in the x-direction

$$P_L = P_2 - P_1 \, . \tag{12.17}$$

If, from the perspective of the secondary source, the net power flowing to the right P_2 is greater than the power flowing to the left P_1 through the cross-section, where $P_2 > P_1$, the loudspeaker is losing power to the channel. If, on the contrary, P_2 is less than P_1, that is $P_2 < P_1$, the loudspeaker is absorbing power from the channel and acts as an energy sink. For P_1,

$$P_1 = \frac{Sp_0^2}{2\varrho c}(1 - |r|^2) \tag{12.18}$$

(S=Channel cross-sectional area), and for P_2,

$$P_2 = \frac{Sp_0^2}{2\varrho c}|1 + r|^2 \, . \tag{12.19}$$

It follows that the power introduced into the channel by the secondary source can be described as

$$P_L = \frac{Sp_0^2}{2\varrho c}[|1+r|^2 - (1-|r|^2)]\,. \tag{12.20}$$

Of course, the secondary wave can be modulated according to amplitude or phase, depending on the loudspeaker's input signal. This can be expressed by the complex factor r

$$r = Re^{j\phi} \tag{12.21}$$

R refers to the absolute value, or amplitude, of r. Using

$$|1+r|^2 = (1+Re^{j\phi})(1+Re^{-j\phi}) = 1 + R^2 + 2R\cos(\phi) \tag{12.22}$$

eq.(12.20) becomes

$$P_L = 2P_0(R^2 + R\cos(\phi)) = 2P_0R(R+\cos(\phi)), \tag{12.23}$$

where, for the sake of simplicity, the power of the incident wave alone

$$P_0 = \frac{Sp_0^2}{2\varrho c} \tag{12.24}$$

has been used. Obviously, the phase ϕ can either be adjusted to let the loudspeaker introduce power into the channel or, for negative $\cos\phi$, allows the loudspeaker to function as an energy sink . For $\phi = 180°$,

$$P_L = -2P_0R(1-R) \tag{12.25}$$

is valid. As can be easily illustrated by differentiating the equation by R, the maximum energy loss from the channel to the loudspeaker occurs at $R = 1/2$. In this case,

$$P_{L,max} = -P_0/2 \tag{12.26}$$

is true. Due to r=-1/2, the absorption incurred by the secondary source would cause the sound field in zone 2 downstream of the loudspeaker to be divided in half, leading to a level reduction of 6 dB. Sufficiently phase-modulated loudspeakers, relative to the impinging field, can therefore be implemented as energy sinks. The energy conservation law is not violated here, since loudspeakers represent reversible converters, which either convert electrical into acoustical energy, or vice versa, acoustical into electrical energy. The direction of the power flow is therefore only a matter of the electrical and mechanical conditions driving the loudspeaker. The foreign, or primary, sound field is the one that the loudspeaker must counteract. If the net pressure in front of the membrane is phase-shifted by 180° to the perpendicular velocity at the surface of the membrane, the loudspeaker invariably acts as a sound absorber.

Theoretically, loudspeakers can also be used as generators supplying electrical power to the network. In this case, however, the acoustic power is extremely negligible compared to the loss of electrical power caused by this

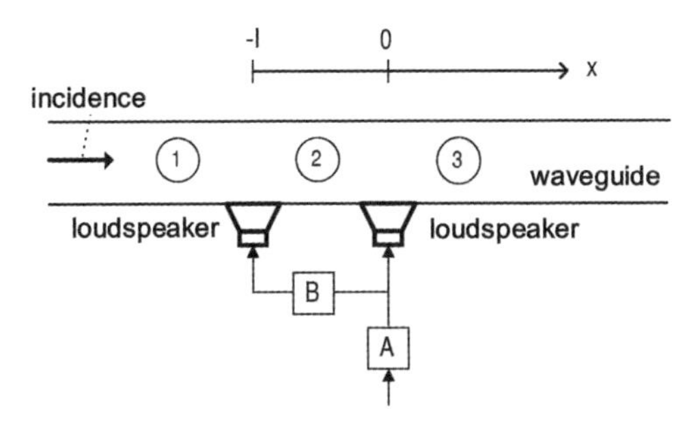

Fig. 12.8. Active noise reduction in a channel with two secondary sources

source's internal electrical resistance. It is precisely this loss of electrical power which initially must be compensated for in the operation of the loudspeaker.

An even more effective form of sound absorption than the method described above is combining a secondary source with a reflection. Such conditions even allow the full absorption of the primary wave. The advantage compared to a pure reflection damper (defined as $r = -1$ in the above sections) is evident. The spatial zone downstream is likewise completely muted, preventing standing waves, potentially susceptible to resonance occurrences, from developing upstream. It is precisely this complete absorption which eliminates such resonances. The reflector can thus be passively – such as through an open duct end – as well as actively induced by a secondary source. The case of setting up a secondary loudspeaker as a reflector will be presented here, as also shown in Figure 12.8.

To this end, the secondary loudspeaker pair to the left of the sum may not produce a secondary sound field. A source combination can be set up which does not emit any sound from one certain direction, using a delay line. The secondary field in zone 1 to the left of both sources exists in

$$p_{1,sec} = Ap_0[\mathrm{e}^{jkx} + B\mathrm{e}^{jk(x+l)}], \tag{12.27}$$

where A and B refer to the complex amplifications depicted in Figure 12.8. The expression e^{jkl} in the wave term for the left source simply states that the sound emitted from this source reaches any location to the left of it if both sources were to simultaneously emit a signal. If the left source is triggered using

$$B = -\mathrm{e}^{-jkl} \tag{12.28}$$

$p_{1,sec} = 0$ will apply to the entire zone 1. In this case, the amplification B represents an interconnection time delay. At the moment that a time delay signal is transmitted from the right secondary source to the left secondary source, the latter will reproduce the right secondary source's field, inverse-phased, so that overall, no secondary sound field can be emitted to the left. Even outdoors, a source pair on a half-axis such as this would not produce any field. An example is shown in Figure 12.9.

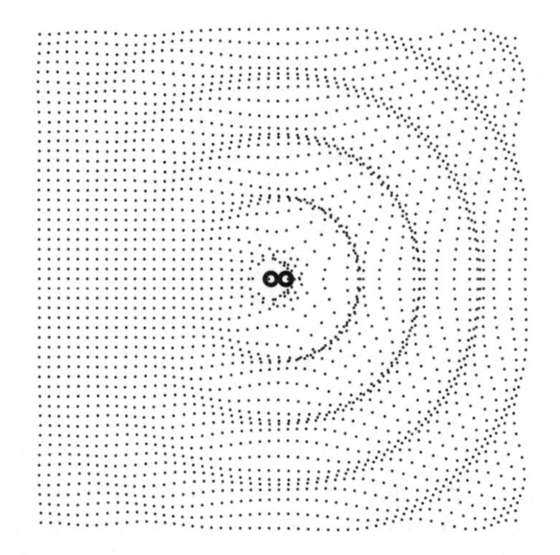

Fig. 12.9. Outdoor sound emission of sound sources which together produce the sound field $p = 0$ via time-delay drive to the right. The source distance in this example is $\lambda/4$.

The only thing left to do is to select an overall drive A so that both sources together compensate for the primary field approaching from the left in zone 3. The secondary field in zone 3 is

$$p_{3,sec} = Ap_0[\mathrm{e}^{-jkx} + B\mathrm{e}^{-jk(x+l)}] . \tag{12.29}$$

Of course at this point, the sound requires more time to reach the receiver at location x from the left source than from the right source, which explains the delay term in parentheses in the second expression. Since $B = -e^{-jkl}$ once again produces a time delay,

$$p_{3,sec} = Ap_0[\mathrm{e}^{-jkx} - \mathrm{e}^{-j2kl}\mathrm{e}^{-jkx}] = Ap_0\mathrm{e}^{-jkx}[1 - \mathrm{e}^{-j2kl}] \tag{12.30}$$

generally applies. The complex amplification A required to compensate the primary wave $p_{prim} = p_0\mathrm{e}^{-jkx}$ is therefore

$$A = -\frac{1}{1 - \mathrm{e}^{-j2kl}}, \tag{12.31}$$

in order to meet the condition $p_{3,sec} = -p_{prim}$. This amplification A obviously cannot be applied to all frequencies, as the above equation requires infinite amplification for $2kl = 2n\pi$, or, expressed in wavelength

$$l = n\lambda/2 . \tag{12.32}$$

This fact provides an immediate explanation for the matter. For the frequencies and wavelengths defined here, both secondary sources are either exactly inversely phased or exactly in phase. Therefore, the fields to the left and to the right of them must be equal in amplitude. If, as intended, the secondary sources combined emit no field to the left, they will not transmit any sound field to the right either. However, in order to initiate a transmittal to the right for the purpose of compensation, infinite amplification would be required. In practice, this means that limits to a usable frequency range must be set.

It is evident that within a usable frequency band, the secondary source structure as described above actually forms an active absorber. The approaching wave impinging on the loudspeaker is not reflected, because the source pair is not transmitting any signal whatsoever to zone 1. The impinging sound power does not arrive in zone 3 either, because here, a complete muting has taken place. This means that the impinging sound energy, as initially mentioned, has been completely absorbed. Simple calculations show that the right sound source downstream takes on the role of a reflector, while the left source functions as the absorber. Since standing waves always develop in front of reflectors and since, on the contrary, standing waves are an indicator for reflectors, the left source has to be the absorber. The total wave field between the sources in zone 2 causes the right source to act as a reflector. Here,

$$p_2 = p_0[\mathrm{e}^{-jkx} + A\mathrm{e}^{jkx} + AB\mathrm{e}^{-jk(x+l)}] = p_0[\mathrm{e}^{-jkx} + A(\mathrm{e}^{jkx} - \mathrm{e}^{-jk(x+2l)})]. \tag{12.33}$$

Following some simple calculations and after substituting for A, one obtains

$$p_2 = p_0 \frac{\mathrm{e}^{-jkx} - \mathrm{e}^{jkx}}{1 - \mathrm{e}^{-j2kl}}. \tag{12.34}$$

The sound field between both secondary sources therefore exists in a standing wave emerging from the soft reflection of the right loudspeaker, resulting in $p_2(x = 0) = 0$.

12.3 Active Stabilization of Self-Induced Vibrations

Under certain conditions, electro-acoustic sources can be used to intervene in sound-producing mechanisms. This is an option for self-induced and self-perpetuating vibrations. Such vibrations can not only be reduced, but even eliminated.

There is a wide variety of types of self-excited vibrations. Sources for such self-induced vibrations are all wind and string instruments, such as flutes, clarinets, saxophones, bassoons, violin, cello and contrabass. In addition, many current-induced types of self-excitation include blowing over a bottle (known as the Helmholtz resonator) or an airfoil, constructions such as smokestacks, bridges, or ducts.

The most important aspect of self-induced vibrations can perhaps be illustrated using the example of aerodynamically induced self-excitation processes. For objects around which air is flowing, the air flow paths form the perimeter of the body, as indicated in Figures 12.10 and 12.11. Symmetrical structures are enveloped by equally dense flow paths above and below. According to the Bernoulli law, the condensed flow paths result in underpressure. If the body is at rest, the underpressures above and below are equal and the net force is thus zero. The situation changes, however, if the body, such as a part of an

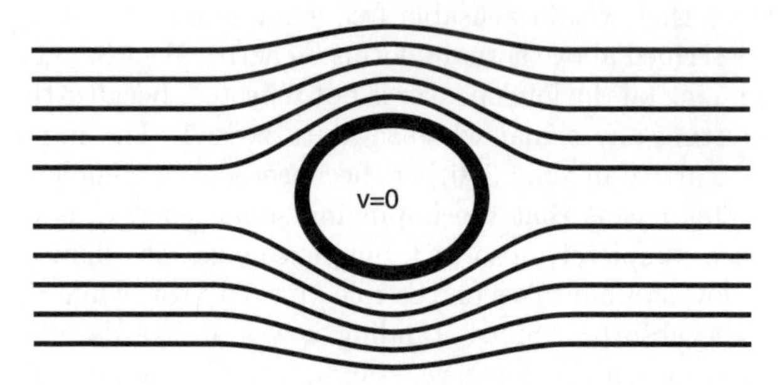

Fig. 12.10. Incident flow of a resting body (Sketch)

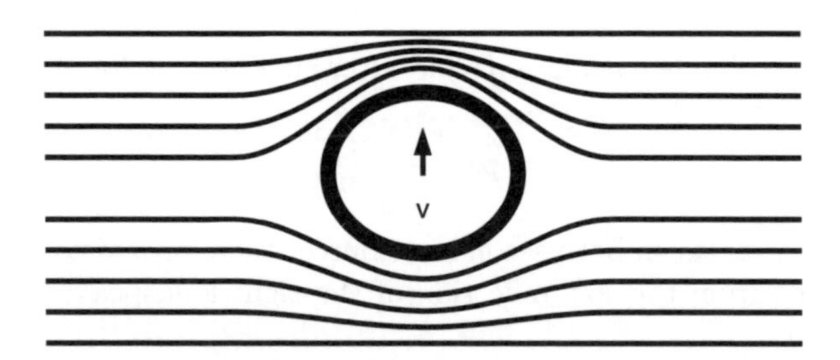

Fig. 12.11. Incident flow of a body travelling upward (Sketch)

oscillator, possesses its own velocity. If the velocity is directed upward, as Figure 12.11 shows, the flow paths become more dense above than those below, causing the body to be pulled upward. If the velocity is directed downward, the net force is likewise directed downward. Thus all external forces induced point toward the same direction as the velocity of the body itself.

At this point, let us assume that the air-cushioned body is resting on a spring bearing, for example, at the end of a beam, as shown in Figure 12.12. The mechanical structure essentially works as a simple resonator. The body makes up the mass m, the elastic beam represents the spring stiffness s. Suppose a small vibration already exists due to contingencies, for example. As soon as the profile at the end of the beam surpasses its resting state and is brought into motion at its momentary maximum velocity, it incurs an external force in the same direction as the velocity. It is as if the body is pushed at just the right moment, an impulse which reinforces the vibration

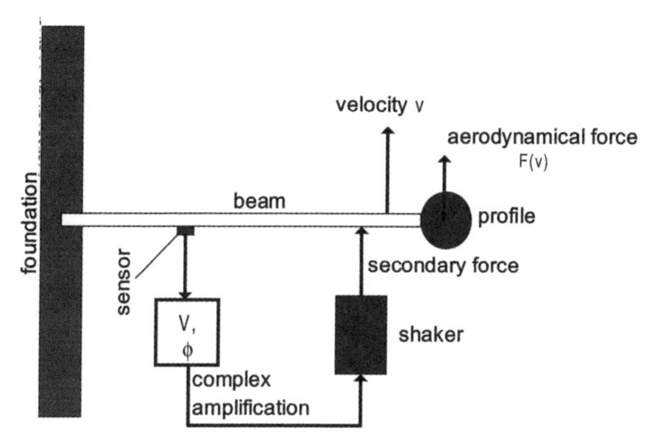

Fig. 12.12. Resonator consisting of a beam (= spring-loaded bearing) and profile (= mass) in the current. The circle representing the secondary force is also included in the graph to serve as an illustration for subsequent discussions in this section.

at each cycle. This causes the amplitudes to increase as if they were 'growing by themselves.' The energy required for this process is obtained from energy the reservoir 'current.' One can use analogy of a child's swingset, to which one gives a push at precisely the right moment. A prototypical time graph in Figure 12.13 depicts the excitation process. The fact that the amplitudes

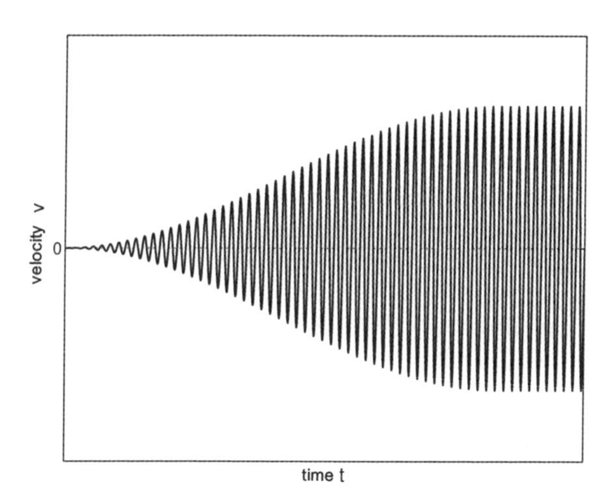

Fig. 12.13. Time graph of a self-induced excitation vibration

do not continue to increase after the onset of the next excitation event can be explained by its non-linear constraints. In the example of the body in the current, the resistance the body encounters increases as the body moves through the current. The inhibiting resistance force grows proportionally to a higher power of the profile's velocity v, compensating the aerodynamic forces at higher velocities. In this manner, the vibration reaches a periodic critical cycle (which is no longer strictly sinusoidal due to its non-linearity).

The basic physical models we considered in the above can be summarized in a characteristic curve showing the basic dependency of aerodynamic force $F(v)$ on the profile velocity (Figure 12.14).

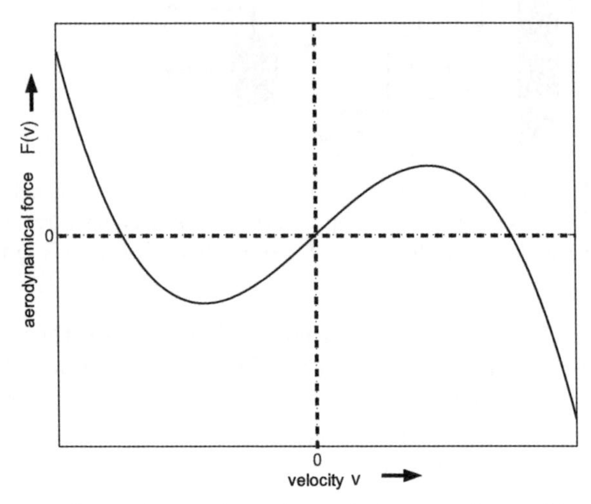

Fig. 12.14. Dependency of aerodynamic force on the profile velocity

Because the product of force and velocity is equal to the vibrational power exerted on the beam, the resonator takes up power, as long as this product is larger than zero. This leads to an increase in the vibration for small amplitudes. Above a certain (non-linear) boundary, the energy supply is shut off completely. At this point, the characteristic curve leaves the linear realm, taking on negative values. No additional power is added to the net energy balance during the course of a complete vibration cycle.

Equations can also easily describe this type of event, at least in its linear portion. The general oscillation equation for resonators is

$$m\frac{d^2x}{dt^2} + r\frac{dx}{dt} + sx = F(v) + F_0. \tag{12.35}$$

F_0 represents the small force induced by contingencies (such as by an eddy passing by. This is the force needed for the onset of the vibration. As usual, m represents the mass, s the stiffness of the oscillator and r the friction constant which accounts for the attenuation.

If you just want to take the excitation event into consideration, the dependency of the aerodynamic force $F(v)$ can be substituted with the first term of the Taylor series:

$$F(v) \simeq r_{aero}v. \tag{12.36}$$

As mentioned in the previous passage, the force and velocity point in the same direction in the range of small vibrations. For this reason, the dimension r_{aero} is therefore a real number and larger than zero. The equation of motion is therefore

$$m\frac{d^2x}{dt^2} + (r - r_{aero})\frac{dx}{dt} + sx = F_0. \tag{12.37}$$

It is evident here that the aerodynamic force acts as a friction term with negative damping. Once again, this only describes the aforementioned energy principle that a negative damper coefficient indicates that external energy flows toward the oscillator. Obviously, an excitation event occurs when the aerodynamic coefficient outweighs

$$\text{instable:} \rightarrowtail r_{aero} > r. \tag{12.38}$$

Because the self-excitation grows by itself, this is described as an 'instable' process. By contrast, there is stability in the absence of vibrations for

$$\text{stable:} \rightarrowtail r_{aero} < r. \tag{12.39}$$

In this case, no excitation occurs in the first place, as the energy loss is greater than the energy gain. Many everyday structures we encounter are potential candidates for self-excited vibrations. These include constructions such as bridges in wind, as well as airfoils in aircrafts. These structures only remain stable due to the existence of sufficient damping effect. Aerodynamic forces may certainly possess more complex properties than in the simplified (laminar) examples described above. Nevertheless, what all current-induced self-excitations have in common is that they tap into the vibrational energy of the current by means of a self-regulating mechanism, as a whole bringing about negative attenuation. Ribbons or cloths flapping in the wind, as well as wind-induced waves on water are all self-excited phenomena, whereby only the forces and displacement are local events.

In the field of acoustical engineering, the Helmholtz resonator perhaps represents one of the most important examples of self-induced sound. The physical processes can be considered akin to a beam with a profile. The air mass in the opening or bottle neck experiences by way of the air current a force pushing it up or down, as long as its velocity is likewise pointing up or down, respectively. The larger the velocity of the air mass, the larger this force. Therefore, the same basic rules governing the effect of aerodynamic force on the profile velocity expressed by the characteristic curve shown in Figure 12.14 apply to the occurrence of sound in Helmholtz resonators. As an aside, the outward transmittal propagation of the sound is simply a byproduct of the excitation event and contributes only negligibly to the damping effect. The level is much greater inside the resonator than outside in the free field.

The idea of attenuating this or other self-excitations using electro-acoustical sources and thereby eliminating them makes sense. The discussions carried out in this chapter prove that this is possible. As shown above, converters can also be used as mechanical energy absorbers, causing electrically-induced attenuation in a self-excited structure. Oscillators are implemented for this purpose in excited beams with profiles (Figure 12.12). To ensure that the right oscillation frequency for the secondary force is met, the exciter is addressed by a sensor signal that has been processed by an amplifier and a phase-shifter. Loudspeakers addressed by a microphone signal serve to extract the sound energy of Helmholtz resonators (Figure 12.15).

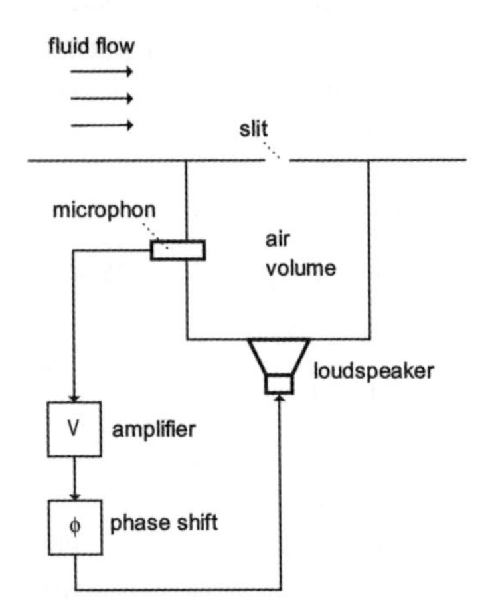

Fig. 12.15. Helmholtz resonator, composed of an air cavity (responsible for the stiffness) and an opening (contains the oscillating mass) with electro-acoustical feedback path consisting of microphone, amplifier, phase-shifter, and loudspeaker for active stabilization

Precision is not the object when extracting vibrational energy in an electro-acoustical absorption circuit. As long as only the energy loss due to the secondary forces is greater than the energy gain invoked by the self-excitation event, the vibration is completely extinguished, preventing the excitation process from happening in the first place. If loudspeakers or oscillators are introduced once the critical cycle has begun, the vibration, as in any attenuation process, gradually subsides, depending on the occurring loss dimensions. The simple calculations utilizing the resonator model in eq. (12.35) show that 'accurate replications' play virtually no role here at all. This time, only the actively induced force is subtracted from the right-hand side (arbitrarily assigned a negative value so that a 0° phase-shift indicates the best case scenario):

$$m\frac{d^2x}{dt^2} + r\frac{dx}{dt} + sx = F(v) - F_{act} + F_0 \tag{12.40}$$

As in the above, once can substitute both forces with a linear approximation if the vibrational amplitudes are small, for $F(v)$ with eq.(12.36), and for the other force F_{act}, with

$$F_{act} \simeq r_{act} v. \tag{12.41}$$

The following applies for the range of small amplitudes, (already notated as ω for pure tones):

$$(j\omega m + \frac{s}{j\omega} + r - r_{aero} + r_{act})v = F_0. \tag{12.42}$$

The main difference between the above and eq.(12.37) is that, while r and r_{aero} are real and positive values, r_{act} is a complex value determined by amplifier and phase-shifter,

$$r_{act} = R_{act} e^{j\Phi}, \tag{12.43}$$

where R_{act}, of course, represents the absolute value of r_{act}. Thus eq.(12.42) - split up into real and imaginary components - exists in

$$([j\omega m + \frac{s}{j\omega} + jR_{act}\sin(\Phi)] + [r - r_{aero} + R_{act}\cos(\Phi)])v = F_0. \tag{12.44}$$

As is generally known, the root of the imaginary component indicates the resonance frequency of the impedance (its value is the entire value given inside the parentheses). The resonance frequency is thus derived from

$$\omega m - \frac{s}{\omega} + R_{act}\sin(\Phi) = 0. \tag{12.45}$$

Obviously, the complete system possesses a different resonance frequency, depending on phase-shift and amplification, than that of just its purely mechanical passive component. By coupling the passive component with an active circuit, one produces a new mechanical-electrical hybrid , a mechanical-electrical hermaphrodite structure, whose properties are determined by mechanical as well as electrical parameters. The stability properties of the hybrid oscillator can be derived from the positive or negative sign of the impedance-real component. If this is larger than zero, it becomes stable. This is the case for

$$R_{act}\cos(\Phi) > r_{aero} - r\ , \tag{12.46}$$

or, expressed without dimensions, for

$$R_{act}/r > \frac{r_{aero}/r - 1}{\cos(\Phi)}\ . \tag{12.47}$$

If eq.(12.47) is not satisfied, instability ensues, with its characteristic self-perpetuating accumulating process. The stability boundary indicates the boundary between the zones of stability and instability in the $(R_{act}/r, \Phi)$ level. The borderline is depicted in Figure 12.16 for some values of r_{aero}/r. As mentioned before, errors only arise if the stability boundary is crossed. Changes in parameters that occur within the stable zone are irrelevant. This lends to some flexibility in choosing the phase and amplification to eliminate the oscillations. The effects and principles described above have been shown

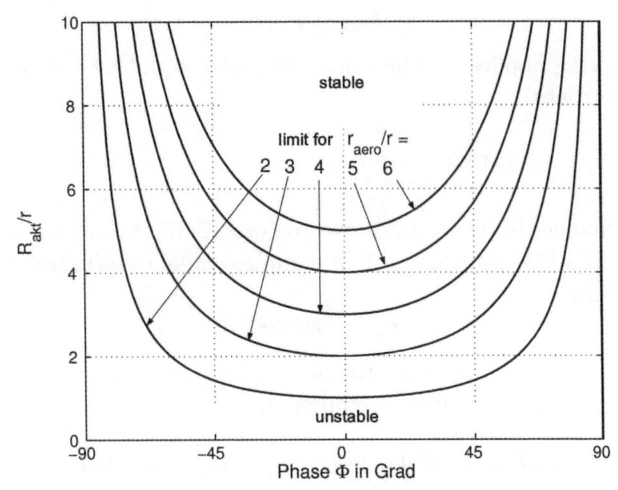

Fig. 12.16. Stability chart of actively induced vibrational obliteration

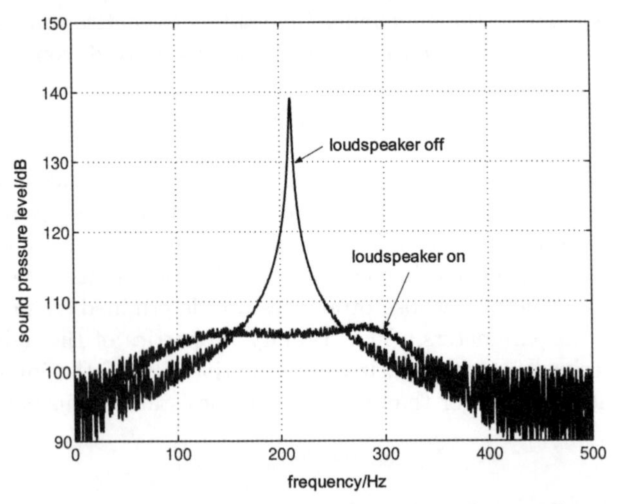

Fig. 12.17. Amplitude spectra on internal microphone

in experiments with Helmholtz resonators. Figure 12.17 shows two typical spectra using the structure shown in Figure 12.15. Shown here are the amplitude spectra on the microphone inside the resonator. The level here can be as high as 140 dB! This results in a wide scope of the parameters for the amplification of the electro-acoustical circuit, as shown in Figure 12.18. The level reductions attained for each resonance frequency are shown here. These are given relative to a loudspeaker that has been shut off.

The amplitude spectra in Figure 12.17 show technically and perceptually effective active attenuation obtained using the predetermined parameters. At

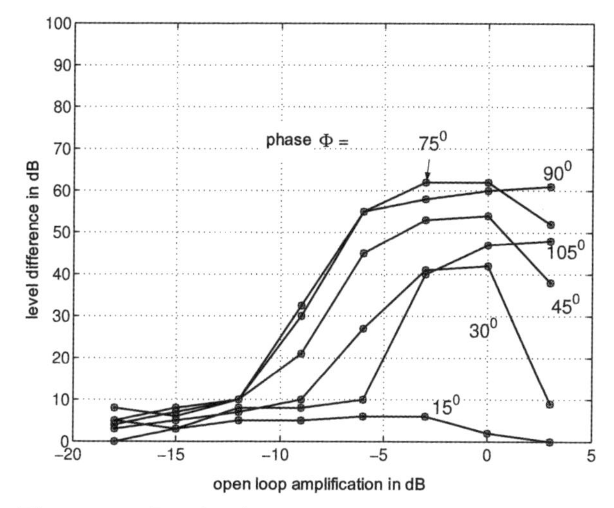

Fig. 12.18. Level reduction in the resonance frequency

the same time, an accentuation of the existing wide-banded noise can occur in another frequency range as well. Therefore, parameters which provide a 'favorable environment' in terms of stabilization for the self-excited vibration can certainly have negative effects on other sound components. Problems can particularly arise if two different oscillation modes with different resonance frequencies are involved in the self-excitation process. Under certain conditions, controlling one mode can actually bring out the destabilization of an other. As seen in the spectrum in such a case, one peak is reduced while another grows above the noise.

12.4 Summary

Active methods which utilize the principle of interference are only suitable for application in noise control if the space and/or time-dependent properties of the primary noise field to be reduced have a simple structure. A typical scenario calling for the application of such active noise control methods is engine noise in small areas to be protected (such as a motorist's ear or a pilot's noise-deterrent head phones). Sound reduction provisions can entail either the reflection or the absorption of the sound field incidence present at the relevant point. Replicating the secondary sound field as accurately as possible is essential in producing significant sound level reductions. If, for example, the direction of propagation cannot be determined, the noise reduction will be severely limited at some points in space, depending on the frequency.

Since electro-acoustic sources can also be utilized as sound energy sinks, such sources can be implemented in order to control accumulating, self-induced sound and reverberation processes. Stabilizing the otherwise instable process in this manner eliminates the problem entirely. Errors in the parameters of the secondary circuit are only detrimental when they cross the bounds of the broad stability area.

12.5 Further Reading

The chapter 'Aktive Beeinflussung von Schall und Schwingungen' by Joachim Scheuren (Chapter 13 in G. Müller and M. Möser: Handbuch der Technischen Akustik, Springer, Berlin 2004) is recommended for an informative overview of this topic. A supplemental and in-depth exploration of the algorithms and control strategies is offered in Hansen and Snyder's book 'Active Control of Noise and Vibration', E and FN SPON, London 1997.

12.6 Practice Exercises

Aufgabe 1

Two waves are propagating opposite to one another in a one-dimensional wave guide, a rigid tube. Show the general condition that the net power inside the tube consists solely of the power difference between the waves propagating in the $+x$-direction and in the $-x$-direction. Account for all powers in the positive x-direction in terms of the time-averaged power.

Problem 2

A single loudspeaker mounted on the side of a one-dimensional wave guide, as shown in Figure 12.7, can be used as a sound absorber. How large is the resulting maximum absorption coefficient? How does the secondary source have to accordingly be operated? In the optimal case, how great is the net power flow to the right and to the left from the secondary source, and into the secondary source itself?

Problem 3

As shown in the following outline, the sound field of the primary volume velocity source Q_0 is to be actively reduced by introducing two large secondary volume sources with the volume velocity flows $-\beta Q_0$, each at a distance of h to the primary source. How great is the total sound power emitted from all three sources in dependence of both β (a real value) and kh? In addition, find the frequency response function of the sound power, if the net volume velocity flow of all three sources together amounts to zero ($\beta = 1/2$).

Finally, determine the amplification factor β for the case of the smallest possible power resulting from all three sound sources together.

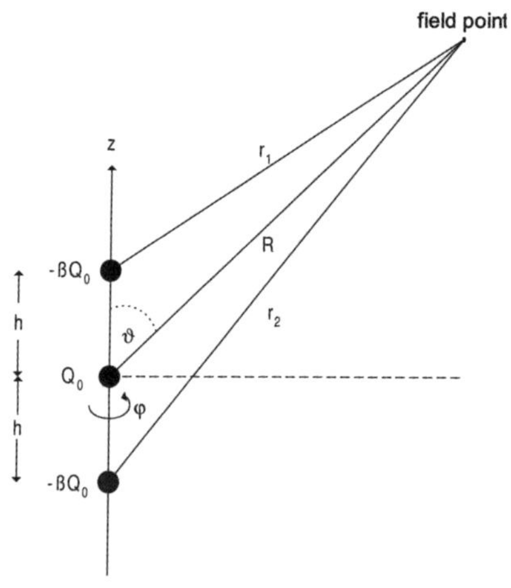

Fig. 12.19. Arrangement of primary source Q_0 and both secondary opposing sources

13

Aspects and Properties of Transmitters

It actually goes without saying that an essential cornerstone of a book on acoustical engineering would be a section devoted to the basic properties of transmitters. The previous chapters served as an extensive treatise on the entire scope of transmitters, whether they are microphones, loudspeakers, sound damper channels, walls, elastic bearings. All of these constructions transmit a time-dependent stimulus signal and in the process, change the signal. As transmitters are often referred to as 'systems,' this chapter could otherwise be titled 'The Basics of System Theory'.

In general, a transmitter, or system, can be understood as any mechanism which can create an output signal out of the distortion of the original time-dependent input signal. The excitation signal $x(t)$ already is referred to as 'input signal' and can be considered the cause of a time-dependent process producing the output signal from the input signal. The output signal $y(t)$ for a given input signal can therefore be understood as the 'effect'. The specific selection of input $x(t)$ and output $y(t)$ depends on the circumstances, the intention of the analysis, and - last but not least - on the observer. For example, it makes most sense to consider the sound pressure time-dependency at the line-in of a microphone as the input signal and the microphone voltage as the output signal; but it is perfectly conceivable to use the electrical time-dependent current in the circuit as the output, for example. If one carries out a local analysis of a loudspeaker, the voltage supply going into the loudspeaker is typically defined as the input and the local membrane velocity or membrane acceleration is usually defined as the output. If one is looking for the radiated signal, one might then consider the sound pressure signal at a given point as the output. If, for whatever reason, the air-borne particle velocity should be of interest, this could also be chosen as the output signal.

The operation L will henceforth describe the exact manner in which the transmitter transforms the input into the output signal:

$$y(t) = L[x(t)] \, . \tag{13.1}$$

M. Möser, *Engineering Acoustics*, DOI 10.1007/978-3-540-92723-5_13,

This allows the assignment of any distortion of any input signal $x(t)$ to any output $y(t)$ brought about by any constraint L. As an example, there is a given filter which distorts any periodic triangular signal to its corresponding sine function (of the same frequency).

Many transmitters are mentioned in this book, an additional example is a rectifier. Every one of us who has a mains connection at home has more or less as many rectifiers as electrical devices. A rectifier is a type of signal filter which transforms an alternating current into direct current voltage. A similar transmitter is a squaring device, which will be used as an example later in this chapter.

13.1 Properties of Transmitters

13.1.1 Linearity

Transmitters can be linear or non-linear. Systems are referred to as linear when the principle of superposition holds. That means,

$$L[c_1 x_1(t)] + L[c_2 x_2(t)] = c_1 L[x_1(t)] + c_2 L[x_2(t)] \tag{13.2}$$

is true for any signal $x_1(t)$ and $x_2(t)$ and corresponding constants c_1 and c_2. The reaction of the transmitter to a linear combination of signals is equal to the sum of the partial reactions.

Nearly all transmitters mentioned in this book behave as linear systems, given that the input signals remain within the same limits. For instance, air-borne sound transmission is linear below approximately 130 dB. For very high sound levels beyond 140 dB, however, sound propagation in air becomes non-linear. Also electro-acoustic converters are linear at low excitation levels. Only at high sound pressures non-linearities occur in microphones. Non-linearities occur in loudspeakers somewhat more often when their input voltage is set at high levels for the sake of high-volume output, especially in more economical loudspeaker systems.

A simple example of a non-linear transmitter is the squaring device $y(t) = x^2(t)$. This example already points out that non-linear systems change the signal frequency of a harmonic input signal. For $x(t) = x_0 \cos \omega t$ is

$$y(t) = x^2(t) = x_0^2 \cos^2 \omega t = \frac{x_0^2}{2}(1 + \cos 2\omega t)\,. \tag{13.3}$$

The output signal thus consists of a constant component as well as an output at double the input frequency. Overall, this entails the general consequence of a non-linearity. That is, each input frequency results new frequencies in the output. Even though this is often the case, new resulting frequencies do not always have to be multiples of the input frequencies, as is the case with squaring devices. As a matter of fact, frequencies can be halved or result in fractions of the input signal frequency, for example.

As demonstrated above, the transition between the linear and non-linear realm of a transmitter is fluid as opposed to abrupt. For this reason, specification measurements are always given for the degree of non-linearity. The best known type of non-linearity is the so-called distortion factor. This consists of the amplitude ratio of the sum of the amplitudes of all the harmonic frequencies not in the fundamental frequency (all the frequencies not existing in the input signal) to the sum of the amplitudes contained of the input frequency. Rather than simply categorizing a transmitter as 'linear,' specifying the distortion factor as a function of input amplitude and input frequency is much more precise. The distortion factor is often quite small, however. The example of air-borne sound transmission below $100\,dB$ is just one of many such examples addressed in this book.

Studio technicians who used the old-fashioned tape recorders will remember adjusting the amplification before recording in order to set the distortion factor to approximately 0.03.

13.1.2 Time Invariance

A transmitter is referred to as time-invariant when any time delay τ produces an equal delay in the system reaction $L[x(t)]$. For any $x(t)$ and τ

$$y(t-\tau) = L[x(t-\tau)] \tag{13.4}$$

applies, presuming $y(t) = L[x(t)]$ for the non-delayed system.

Almost all transmitters mentioned in this book not only exhibit linear behavior, but also can be assumed to be time-invariant. Time-variant systems are systems where their parameters undergo noticeable changes after a certain amount of time. For instance, the temperature can change in a sound studio, and therefore the speed of sound in the room changes too; sound propagation in rooms is thereby a time-variant transmission. However, the rise in temperature usually takes place so slowly that time-invariance is presumed for short time intervals of just a few minutes each for measurements of such sound events.

The only time-variant system actually addressed in this book is sound transmission where the transmitter and receiver are moving relative to one another (refer to the section on sound propagation in moving medium). This is because the duration of the sound event's travel from the location of the sound source and the microphone is time-dependent itself. This time-variant effect is known as the Doppler shift in the signal frequency. Not only non-linear systems, but also time-variant systems exert changes on the input signal frequency. Such transmissions are thus characterized by a change in the input frequency.

On the other hand, experience seems to show that linear and time-invariant transmitters have a consistent frequency. If the signal input of a linear and time-invariant system consists of a harmonic signal (a cosine function) at a given frequency, the output produces a likewise harmonic signal at the same frequency. This resulting signal may differ from the input solely in amplitude and phase. This quality is implicit in nearly every example offered in this book. Indeed, it can be shown that the law of frequency consistency described here applies to all linear and time-invariant transmitters. This will be proven in the following section.

13.2 Description using the Impulse Response

The simplest way to describe the effect of an inhomogeneous system is to theoretically divide it into many (in the extreme case, into infinitely many) indivisible parts, analyzing the effect of each of these minuscule components separately. For example, this analytical method is used to examine the gravity field of an inhomogeneous body. The object is divided into infinitely small cubes with constant density. The gravity field can thus be determined for each of these small cubes, whereby the whole is of course the sum of these parts (this comes down to describe the total field using an integral).

Describing the transmission of a linear and time-invariant system is done the same way. The input signal is first divided up into its most basic and indivisible components. Then, one examines the transmission of these parts and finally constructs the output signal by way summation these components together. This last step utilizes the prerequisites of systematic linearity and time-invariance.

For didactical reasons, we will use finite components in the following analysis first; arbitrarily narrow components will be considered a little later.

First one must identify the component itself, consisting in the rectangular function $r_{\Delta T}(t)$ shown in Figure 13.1 , whose value is $1/\Delta T$ inside the interval $-\Delta T/2 < t < \Delta T/2$. Outside of this interval, this function's value is $r_{\Delta T}(t) = 0$. The integral over $r_{\Delta T}(t)$ is thereby independent of ΔT equal to 1, as long as the integration area fully contains the interval $-\Delta T/2 < t < \Delta T/2$ $(a, b > \Delta T/2)$:

$$\int_{-a}^{b} r_{\Delta T}(t)dt = 1. \tag{13.5}$$

The decomposition of the input signal $x(t)$ into a step function consisting of a series of shifted components results in the approximation function

$$x_{\Delta T}(t) = \sum_{n=-\infty}^{\infty} x(n\Delta T) r_{\Delta T}(t - n\Delta T) \tag{13.6}$$

(see Figure 13.2), which, of course, is only a rough estimate of the original function $x(t)$ for finite ΔT. Only after considering the bound $\Delta T \to 0$ can we approach an accurate model, because only then the definition of 'indivisible' components actually has been made.

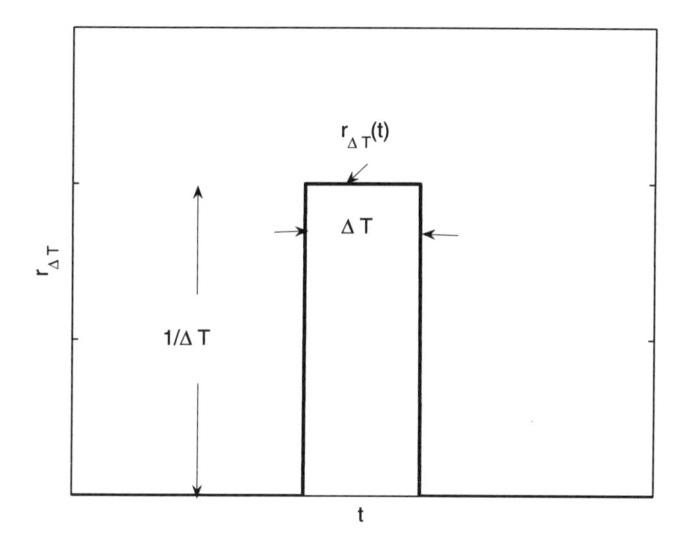

Fig. 13.1. Step function $r_{\Delta T}(t)$

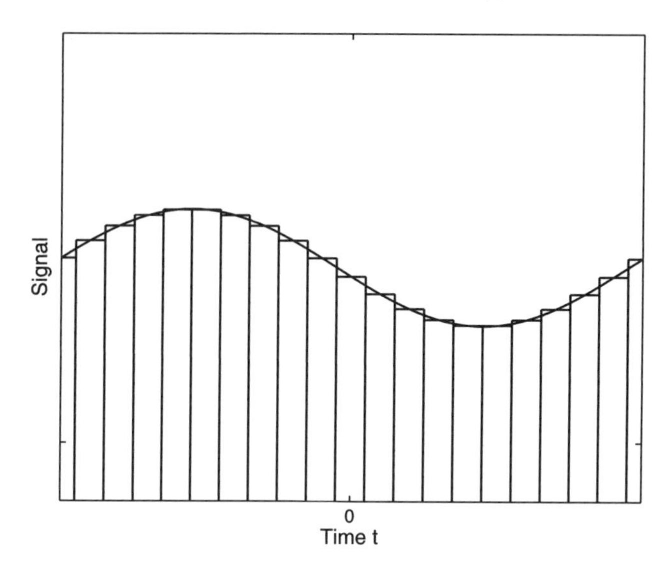

Fig. 13.2. Original signal $x(t)$ and its reconstruction $x_{\Delta T}(t)$ using step functions

If the necessary prerequisite condition of a linear and time-invariant transmitter is to be fulfilled, calculating the transmission of $x_{\Delta T}(t)$ is simple, as long as the reaction of the system to a single rectangular function $r_{\Delta T}(t)$ is known. This information must be assumed to be given, therefore

$$h_{\Delta T}(t) = L[r_{\Delta T}(t)] \tag{13.7}$$

is known. Due to linearity and time-invariance

$$y_{\Delta T}(t) = L[\sum_{n=-\infty}^{\infty} x(n\Delta T) r_{\Delta T}(t - n\Delta T)] = \sum_{n=-\infty}^{\infty} x(n\Delta T) L[r_{\Delta T}(t - n\Delta T)]$$

$$= \sum_{n=-\infty}^{\infty} x(n\Delta T) h_{\Delta T}(t - n\Delta T) \tag{13.8}$$

apply for the system response to the input $x_{\Delta T}(t)$.

When $\Delta T \to 0$ the rectangular function approaches the infinitely narrow Dirac delta function:

$$\lim_{\Delta T \to 0} r_{\Delta T}(t) = \delta(t) \tag{13.9}$$

which only has an infinitely large value other than zero at $t = 0$. This delta function meets the criteria of not being able to be divided further. The integral over the delta function exists and is equal to 1 $(a, b > 0)$:

$$\int_{-a}^{b} \delta(t) dt = 1. \tag{13.10}$$

The delta function is also referred to as a special function due to its unusual appearance.

As ΔT gets smaller, the individual rectangular functions $r_{\Delta T}(t - n\Delta T)$ come closer together, and thereby increasing in density. In the limiting case $\Delta T \to 0$ the components $r_{\Delta T}(t - n\Delta T)$ are infinitely close to one another, causing the discrete delay time $n\Delta T$ to cross over into the realm of the continuous variable τ: $n\Delta T \to \tau$. Summation in the equation for $y_{\Delta T}(t)$ results in an integration, the discrete distance ΔT between two rectangular functions then becomes the infinitely small element $d\tau$. This leads to the exact description of the system output:

$$y(t) = \int_{-\infty}^{\infty} x(\tau) h(t - \tau) d\tau \,. \tag{13.11}$$

Obviously, $h(t)$ denotes the response of the transmitter to the delta-shaped input,

$$h(t) = \lim_{\Delta T \to 0} h_{\Delta T}(t) = \lim_{\Delta T \to 0} L[r_{\Delta T}(t)] = L[\lim_{\Delta T \to 0} r_{\Delta T}(t)] = L[\delta(t)] \,. \tag{13.12}$$

The system response $h(t)$ to the delta impulse at the input is known as the impulse response.

The integral on the right side of eq.(13.11) is referred to as the convolution integral, because the integration variable appears in $h(t - \tau)$ with a negative sign. This inversion of the order can be imagined by folding a sheet of paper at $\tau = t$. The name 'convolution' is simply an allusion to this technique of 'folding' the function. It's purpose on the other hand is to decompose the input signal into indivisible delta-function parts

$$x(t) = \int_{-\infty}^{\infty} x(\tau)\delta(t-\tau)d\tau\ , \tag{13.13}$$

which enables the calculation of the transmitter output as shown above. The operation on the right-hand side of eq.(13.11), which is applied to the input signal $x(t)$ and the impulse response $h(t)$, is known as 'convolution.' The output signal of a linear and time-invariant transmitter is equal to the convolution of the impulse response and the input signal. The input signal and the impulse response are mutually interchangeable, meaning that

$$y(t) = \int_{-\infty}^{\infty} x(\tau)h(t-\tau)d\tau = \int_{-\infty}^{\infty} h(\tau)x(t-\tau)d\tau\ , \tag{13.14}$$

can also be used, as can be shown by substituting the variables in eq.(13.11) (one can replace $u = t - \tau$ with $du = -d\tau$ and again, just use τ to express u). The convolution is invariant against the interchange of the signals to be convolved. Therefore, a transmitter's input signal and the impulse response are also interchangeable without affecting the output.

The crux of the preceding discussions is the decomposition of signals in maximally dense delta impulses which have infinitely narrow intervals. In order for its integral to be other than zero, its value must be infinitely large in the middle of the function. This way of thinking can be compared to decomposing any type of series function. The only difference is that in the case of Convolution, integration instead takes place over infinitely narrow elements, as opposed to summation of discrete elements. The purpose of representing signals by use of a delta comb function as also described in eq.(13.13) is that the transmitter output can now be directly derived from such input calculations. This is of course done directly by way of decomposition (13.13) without the intermediate step of using the step function, as described for better understanding at the beginning of this section:

$$y(t) = L[\int_{-\infty}^{\infty} x(\tau)\delta(t-\tau)d\tau]\ . \tag{13.15}$$

Due to the required linearity, the sequential order of integration and the L operation can be reversed:

$$y(t) = \int_{-\infty}^{\infty} L[x(\tau)\delta(t-\tau)]d\tau = \int_{-\infty}^{\infty} x(\tau)L[\delta(t-\tau)]d\tau\ . \tag{13.16}$$

Due to assumed time invariance, this once again results, of course, in

$$y(t) = \int_{-\infty}^{\infty} x(\tau)h(t-\tau)d\tau \, .$$

Convolution integral and impulse response $h(t)$ both describe the transmission in the time domain, where the effect of many input impulses 'squeezed together' is summed up at the output. The main disadvantage of this method is the overall complexity of the impulse response in terms of characterizing the transmission. For instance, the reaction of a room receiving the signal of an explosion coming from an emitter which has been transmitted via a single-layered thin wall is a type of 'extended' impulse; in actuality, the impulse response exists in a very rapidly diminishing exponential function ($h(t) \sim e^{-t/T}$ with $T = m''/\varrho c$ for perpendicular sound incidence and for $t \geq 0$, for $t < 0$ is of course $h(t) = 0$ (for further explanation please see Chapter 8). Although the sound transmission can certainly be correctly ascertained through mathematical calculations, the assessment of this physical phenomenon is in fact quite difficult to attain based on the impulse response.

Describing the transmission with the help of the frequency response functions to be discussed in the following sections provide a conclusive means of dealing with signal tone distortions in the transmission.

13.3 The Invariance Principle

We already presumed in the last section that linear and time-invariant transmitters will transmit a harmonic (or sinusoidal) input signal without distorting it. The output signal is comprised of a harmonic signal of the same frequency. Only its amplitude and phase have been changed by the transmitter, while signal structure remains the same after the transmission.

The convolution integral shows that this presumed principle is universal (13.14). This assumes an input signal

$$x(t) = \mathrm{Re}\{x_0 e^{j\omega t}\} \tag{13.17}$$

which has the complex amplitude x_0. The corresponding output that results is according to eq.(13.14)

$$y(t) = \int_{-\infty}^{\infty} h(\tau)\mathrm{Re}\{x_0 e^{j\omega(t-\tau)}\}d\tau = \mathrm{Re}\{x_0 e^{j\omega t} \int_{-\infty}^{\infty} h(\tau)e^{-j\omega\tau}d\tau\} \, . \tag{13.18}$$

The last integral is only dependent on the signal frequency ω, and in particular not dependent on t. Defined by

$$H(\omega) = \int_{-\infty}^{\infty} h(\tau)e^{-j\omega\tau}d\tau\,, \tag{13.19}$$

it can be abbreviated to

$$y(t) = \mathrm{Re}\{H(\omega)x_0 e^{j\omega t}\}\,, \tag{13.20}$$

which proves the previous assumption of the invariance principle: if the input is harmonic with the frequency ω, the output will also be harmonic with the same frequency. The transmission is thus completely defined by describing its change in amplitude $|H|$ and phase φ, which are denoted in the complex transmission factor

$$H(\omega) = |H|e^{j\varphi}\,. \tag{13.21}$$

13.4 Fourier Decomposition

It is an attractively simple and intuitive idea to reduce the transmission of signals of other forms to its closest harmonic proximity. To do this, one can express the given signal function of a system, such as the input $x(t)$, using a function series of the form Form:

$$x(t) = \sum_n x_n e^{j\omega_n t} \tag{13.22}$$

If this is possible, meaning, that its contained frequencies ω_n and their corresponding complex amplitudes x_n have been defined, describing such transmission processes becomes quite simple. According to the invariance principle, and its precondition of linearity, the output must be composed of the same frequencies as the input, whereby only their respective amplitudes are altered during the transmission:

$$y(t) = \sum_n H(\omega_n) x_n e^{j\omega_n t} \tag{13.23}$$

Depending on the input signal, different frequencies with their corresponding amplitudes occur, each unique to their given signal. If several should be defined at once, then the complex-valued transmission factor H will have to be known for all those frequencies. To this end, we use the frequency response $H(\omega)$. $H(\omega)$ is referred to as the transmission function, in order to denote that the frequency is a continuous variable used to describe the transmission over its entire spectrum. Using the frequency response to describe the transmission function can be easily interpreted as follows. If one imagines the input broken up into frequencies - or qualitatively described as tone colors - one can simply interpret the transmission as tone color distortion. For instance, a sound signal transmitted over a wall sounds quieter and more muffled in the receiver space because the transmission function is small at higher frequencies.

Interpreting transmissions as tone color distortions presupposes that arbitrary signals can actually be described using a function series containing harmonic components, as is the prerequisite for eq.(13.22).

The following section will focus on the particulars of such signals which can be described by their Fourier decomposition, as will as discuss the properties of these types of signals. To reiterate the basic idea, eq.(13.22) is an expression of the ability to approximate a given, known signal $x(t)$ by way of a function series. The elements of such function series exist harmonic, sinusoidal signals containing (many) different frequencies. The invariance principle of transmission lets us use simple methods to describe what is already given.

13.4.1 Fourier Series

The discussion of Fourier decomposition can begin with the simplest case, where the function to be decomposed is periodic itself with the period T. The advantage to this assumption is that the frequencies contained therein are already given. The still unspecified frequencies ω_i in eq.(13.22) are, in this case, known from the beginning. The components occurring in eq.(13.22) contain the periods T_n

$$\omega_n = \frac{2\pi}{T_n} \,. \tag{13.24}$$

A series expansion of a periodic function T can only contain components which are characterized by periods T_n as multiples within T. All periods of any component which occurs within the equation can be defined as

$$T_n = \frac{T}{n} \,. \tag{13.25}$$

For the purposes of the discussion, we will define a 'model function' which only consists of the components into which the function will ultimately be decomposed. Thus, we specify:

$$x_M(t) = \sum_{n=-N}^{N} A_n e^{j2\pi n \frac{t}{T}} \tag{13.26}$$

As can be seen, initially N positive as well as negative frequencies ω_n have been permitted in this function.

It only remains to specify the coefficients A_n in such a way that the model function $x_M(t)$ 'matches' the given signal $x(t)$. There are at least two different ways to proceed accordingly, in specifying the unknown amplitudes A_n in the given signal $x(t)$, so that the error diminishes as N increases in $x_M(t)$. The standard quadratic deviation used to minimize this error is usually defined as

$$E = \frac{1}{T} \int_0^T |x(t) - x_M(t)|^2 dt \,. \tag{13.27}$$

This method is known as the method of least squares. This more formal procedure will not be discussed here, the method is described in detail in many works which serve as treatises on Fourier sums. Instead, we will proceed to introduce a much more handy method to specify the unknown amplitudes A_n. They now are calculated in such a way, that the model $x_M(t)$ and the original $x(t)$ coincide completely in $2N+1$ points inside a period T:

$$x_M(i\Delta t) = x(i\Delta t) \tag{13.28}$$

for

$$i = 0, 1, 2, 3, ...2N$$

The increment Δt used here is defined by

$$\Delta t = \frac{T}{2N+1} \,. \tag{13.29}$$

Figure 13.3 serves as a visual aid to better outline this method.

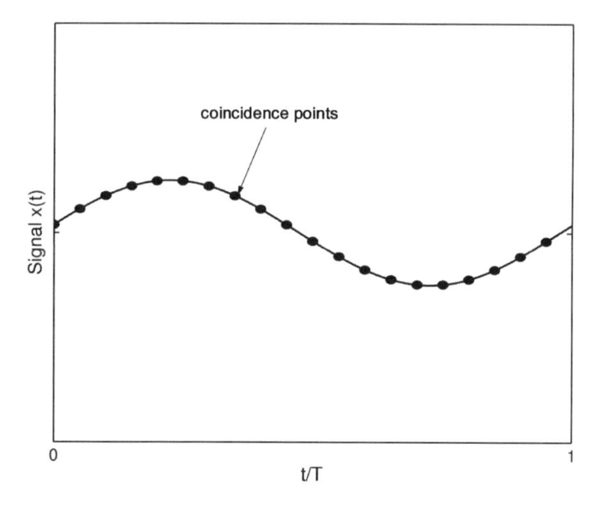

Fig. 13.3. Alignment of $x_M(t)$ along $x(t)$ at the sampling points $t = i\Delta t$, where $x_M(i\Delta t) = x(i\Delta t)$.

The $2N+1$ conditional equations (13.28) now serve as the system of linear equations derived from the model definition (13.26) to find the coefficients:

$$\sum_{n=-N}^{N} A_n e^{j2\pi \frac{ni}{2N+1}} = x(i\Delta t) \,. \tag{13.30}$$

As just mentioned, eq.(13.30) specifies the system of equations in order to find the unknown coefficients A_n. Inserting values for the index $i = 0, 1, 2, 3, ..., 2N$ (13.30) produces $2N+1$ equations.

This system of equations can now easily be solved for any chosen unknown A_m as outlined in the following. To do this, we multiply the i-th equation of the system (13.30) with $e^{-j2\pi \frac{mi}{2N+1}}$, giving

$$\sum_{n=-N}^{N} A_n e^{j2\pi \frac{(n-m)i}{2N+1}} = x(i\Delta t) e^{-j2\pi \frac{mi}{2N+1}} . \tag{13.31}$$

Next, we add all the $2N+1$ equations of this system:

$$\sum_{i=0}^{2N} \sum_{n=-N}^{N} A_n e^{j2\pi \frac{(n-m)i}{2N+1}} = \sum_{i=0}^{2N} x(i\Delta t) e^{-j2\pi \frac{mi}{2N+1}} , \tag{13.32}$$

or

$$\sum_{n=-N}^{N} A_n \sum_{i=0}^{2N} e^{j2\pi \frac{(n-m)i}{2N+1}} = \sum_{i=0}^{2N} x(i\Delta t) e^{-j2\pi \frac{mi}{2N+1}} . \tag{13.33}$$

The inner sum on the left side constitutes a geometrical series with

$$\sum_{i=0}^{2N} e^{j2\pi \frac{(n-m)i}{2N+1}} = 0 , \tag{13.34}$$

when $n - m \neq 0$. For $n = m$ is

$$\sum_{i=0}^{2N} e^{j2\pi \frac{(n-m)i}{2N+1}} = \sum_{i=0}^{2N} 1 = 1 + 1 + 1 + ... = 2N + 1 , \tag{13.35}$$

All elements in the sum with the running index n on the left side of (13.33) are therefore equal to zero, with the exception of the sole summand with $n = m$. We can thus solve (13.33) for the unknown A_m accordingly, using the calculation steps described above, for which

$$A_m = \frac{1}{2N+1} \sum_{i=0}^{2N} x(i\Delta t) e^{-j2\pi \frac{mi}{2N+1}} . \tag{13.36}$$

applies. Eq.(13.36) defines the solution to the system of equations (13.30) for the uniquely specified unknown A_m. As it is completely arbitrary, which specific unknown A_m is selected here, eq.(13.36) universally applies for all unknowns A_m. All amplitudes A_n of the model function eq.(13.26) are thus derived from the original signal $x(t)$ using eq.(13.36).

The above has shown that a given signal can be exactly replicated by decomposing it into a finite but arbitrarily high amount of points by way of the function series (13.26). The problem posed in this section is therefore reasonable and solvable.

How well or badly a given signal $x(t)$ can be replicated by its model $x_M(t)$ can subsequently be shown in the following examples. The figures 13.4 through

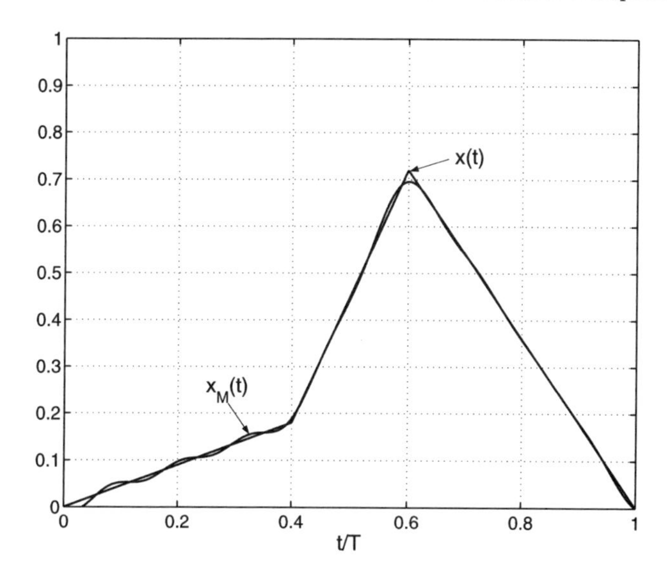

Fig. 13.4. Model of a deviated signal using N=8. Only one period of the periodic signals x and x_M is represented here.

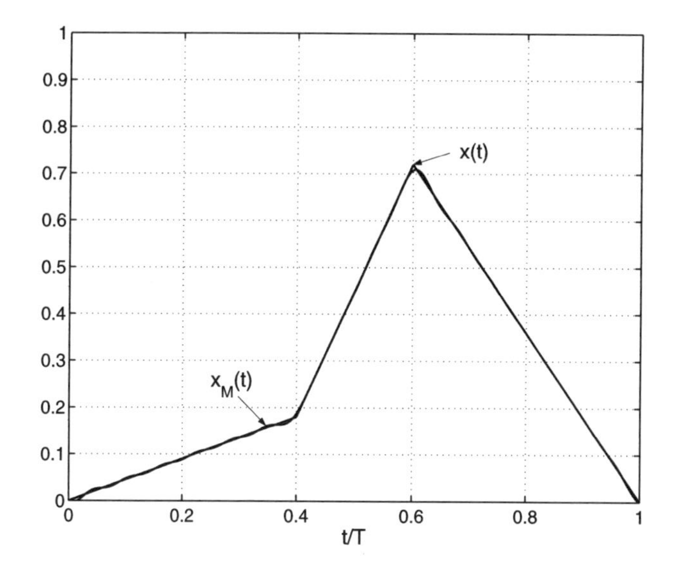

Fig. 13.5. Model of a deviated signal using N=16. Only one period of the periodic signals x and x_M is represented here.

13.6 demonstrate the quality of the replicas using the example of a specific signal consisting of three line segments, showing the difference between the original $x(t)$ and the model $x_M(t)$ for $N = 8$, $N = 16$ and $N = 32$. As these graphs shown, the model is already quite accurate with 17 sampling points

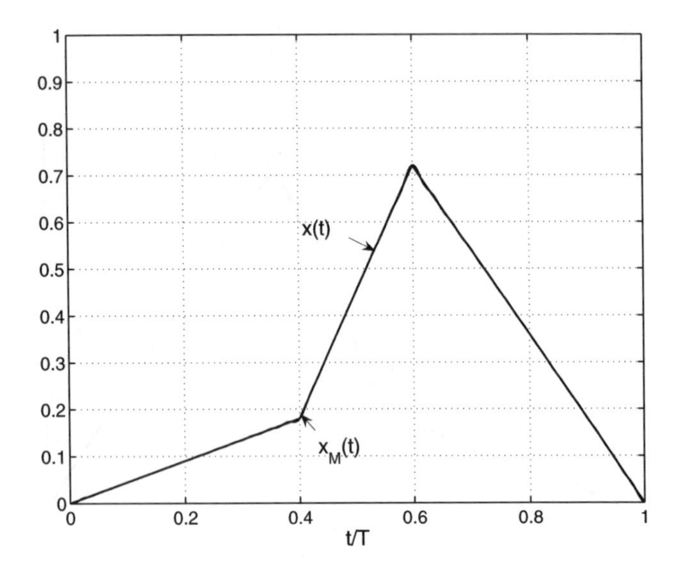

Fig. 13.6. Model of a deviated signal using N=32. Only one period of the periodic signals x and x_M is represented here. The differences between signal x and replicate signal x_M have been reduced to within the line width.

(N=8). For 65 points (M=32) the differences between x and x_M are quite difficult to discern. Apart from the exact likeness at the discrete sampling points (where x and x_M are always the same), the series also converges very rapidly between the sampling points toward the signal representing its expansion. There is an easy explanation for this. There are no significant changes between the discrete points (the points $t = i\Delta t$) in the signal to be expanded. The signal gradient is therefore 'flat.' In mathematical terminology, such a signal is referred to as continuous. For continuous signals, as this example shows, only a relatively small number of frequencies and sampling points 2N+1 are required to sufficiently model a signal. Overall, the series converges 'rapidly' toward the representative signal as N increases.

Conversely, signals which fluctuate dramatically between two signal points are characterized by slow convergence of the series. The worst case scenario is a function which has a break, in other words, is discontinuous. Examples of such signals and their series expansions with different amounts of sampling points are shown in Figures 13.7, 13.8 and 13.9. It is clear to see in these examples that such signals require a lot more sampling points and expansion terms to ensure a 'good' image of x through the model function x_M. The reasons for this slow convergence with increasing N are given below:

- There are no sampling points for a finite N in the extreme case of an infinitely rising curve, leading to a 'bad' model of this section of the curve. Ideally, only an infinite number of sampling points can replicate the section of the curve accurately.

- The function series (13.26) consists of element functions which in themselves constitute universally continuous functions. A signal obviously cannot actually be modelled at points of discontinuity using a finite number of continuous functions, rather theoretically, can only be accurately represented through an infinite number of series terms.

It is therefore evident in Figures 13.7 through 13.9 that the principle problem of modelling signals at their points of discontinuity persists, even with increasing N. Increasing N improves the modelling of the continuous neighboring branches of the function, but the fact remains that x_M overshoots and deviates at the mean value of the left- and right-hand threshold at the point of discontinuity. (at the discontinuity t_0 is $x_M(t_0) = (x(t_0 - \varepsilon) + x(t_0 + \varepsilon))/2$, as is clearly represented in Figure 13.7). Here, it can be seen that the series converges at every point toward the signal, but the only improvement which results is the narrowing of the 'problem zone' where the discontinuity lies. Only in the case of infinite summands in the series does this zone disappear altogether.

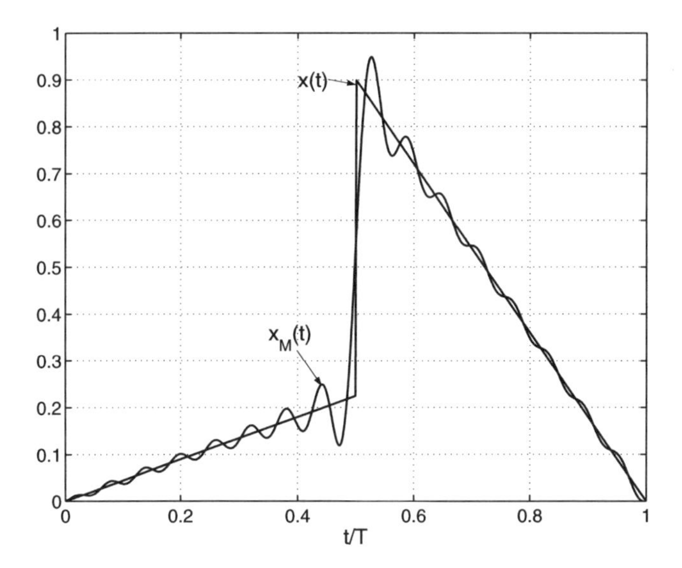

Fig. 13.7. Modelling a discontinuous signal with N=16. Only one period of the periodic signals x and x_M is represented here.

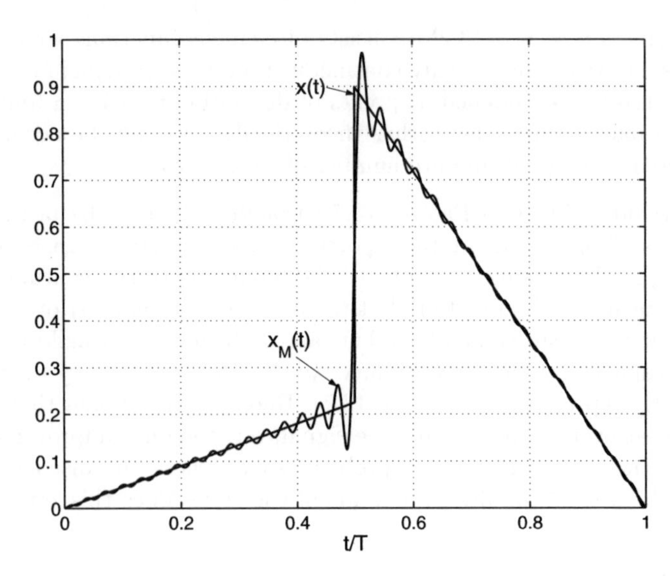

Fig. 13.8. Modelling a discontinuous signal with N=32. Only one period of the periodic signals x and x_M is represented here.

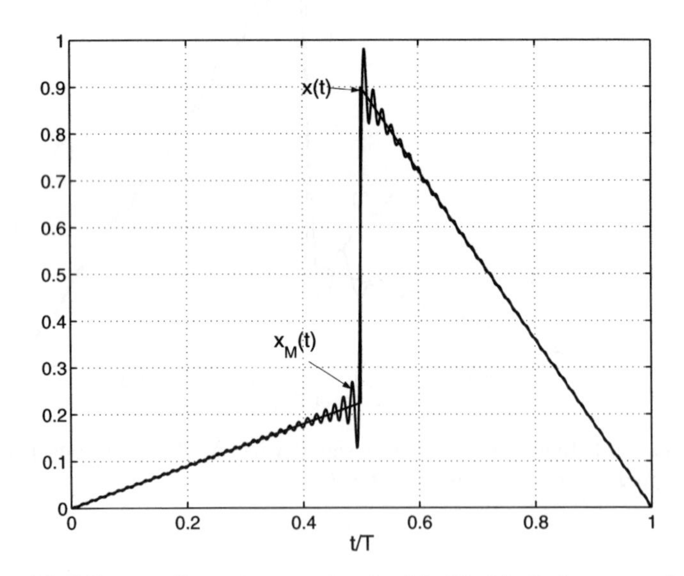

Fig. 13.9. Modelling a discontinuous signal with N=64. Only one period of the periodic signals x and x_M is represented here.

Just before and after the point of discontinuity, which only occurs in a narrow time interval when using an increasing finite N, results in overshooting curves. This effect is known as Gibb's phenomenon. As can also be discerned in Figures 13.7 through 13.9, the number of N therefore does not influence the maximum height of the overshot of the function whatsoever.

The question remains as to which form the eq.(13.36) takes on for the amplitudes when the number of sampling points increases and eventually surpasses all bounds. At this point, the increment Δt - the distance between two sampling points - gets smaller and converges to zero. The discrete points $i\Delta t$ then become the continuous time t. The expression $1/(2N+1)$ denotes the ratio of the increment Δt to the period duration T, $1/(2N+1) = \Delta t/T$. The increment Δt becomes the infinitesimally small element of distance dt, and the sum becomes an integral. This results in the limiting case of

$$A_n = \frac{1}{T}\int_0^T x(t)e^{-j2\pi n\frac{t}{T}}dt\,. \tag{13.37}$$

This equation shows how the coefficients of the Fourier series representation of a periodic signal $x(t)$s can be obtained. If an infinite amount of summands are reflected in the series expansion of the signal, the model function x_M converges at every point t toward the given signal x. Therefore, the equation above can be shortened to

$$x(t) = \sum_{n=-\infty}^{\infty} A_n e^{j2\pi n\frac{t}{T}} \tag{13.38}$$

The integrand in eq.(13.37) is periodic with T. This is why the integration boundaries can be shifted arbitrarily, as long as the width of the interval remains T. This can be particularly be expressed in this form,

$$A_n = \frac{1}{T}\int_{-T/2}^{T/2} x(t)e^{-j2\pi\frac{nt}{T}}\,. \tag{13.39}$$

which will be used for the discussion in the following section.

Eq.(13.37) (or (13.39)) can also be referred to as the 'transform' or 'mapping function'. The signal x is thereby mapped in the sequence A_n. Eq.(13.38) therefore shows how the original x can be obtained form the map A_n; thus lending to another name for this equation, the 'inverse transform'. An non-mathematical analogy to this process of reverse mapping might be the relationship between a photo's positive and its negative image. Naturally, by using a good camera, the positive and negative pictures can be used interchangeably to produce the one or the other, and both contain the same information. Fourier sum transformation works exactly the same way. The signal x is represented in A_n 'by other means'. No information is gained or lost. Nevertheless, the use of transformation makes sense. It enables the simple description of transmission processes.

As is the case with our photo analogy, Fourier sum transformations are mutually unique. Every periodic signal has exactly one representation A_n.

13.4.2 Fourier Transform

Of course, it should not only be possible to decompose periodic signals into their frequency components, but also any other non-periodic signal as well.

Since practically all real signals have an end and a beginning, we are only interested in non-recurring processes which are set to zero before and after a certain given time interval. The principles discusses in the last section can be applied when an arbitrary segment is taken out of a time signal (which does not have to equal the duration of the entire signal) and be considered as the specified period duration. This is shown in the Sketch 13.10. This signal segment is virtually periodically repeated, so that it can be assigned an amplitude spectrum as discussed in the previous section. Subsequently, the period duration T is allowed to grow. That means that both dotted lines to the right and to the of the middle line migrate outward, as shown in Figure 13.10. The next period comes increasingly later, and the preceding period moves back further and further into the past. The most extreme case of an infinite period duration is therefore the most accurate description of a signal of finite length.

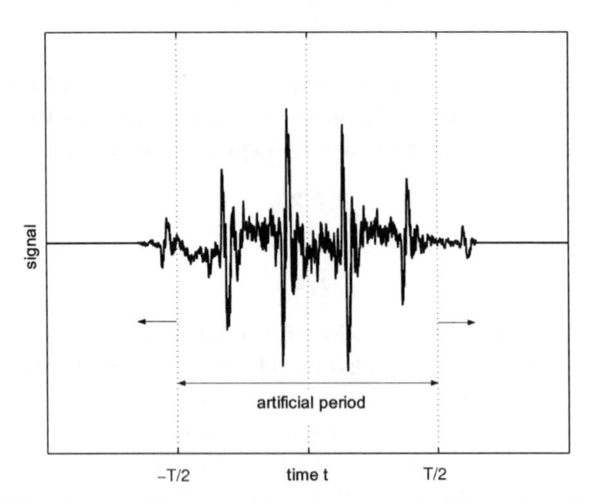

Fig. 13.10. 'Non-recurring' time signal of finite length

When approaching $T \to 0$ it must be noted that the distance $\Delta f = 1/T$ of the frequencies n/T keeps getting smaller. The discrete frequencies then become a continuous frequency variable: $n/T \rightarrow f$. The fact that all frequencies must be permissible to describe any signal is self-explanatory. As opposed to periodic signals, it is no longer necessary to emphasize certain frequencies over others. The inclusion of all frequencies in the definition necessitates the specification of a va continuous frequency variable f. As a short-cut we will use the angular frequency $\omega = 2\pi f$.

Using (13.39) this results in

$$\lim_{T\rightarrow\infty} A_n = \lim_{T\rightarrow\infty} \frac{1}{T} \int\limits_{-T/2}^{T/2} x(t)e^{-j\omega t}dt \,. \tag{13.40}$$

The limit value on the left side is in any case zero. Thus the integral contained in a specified frequency ω converges toward a specific value. If the virtually assigned period duration exceeds the beginning and end of the signal itself, the value of the integral stops changing for values of T beyond this point. Since this value is nevertheless divided by T, the total limit value approaches zero. For this reason, the following discussions are not able to be applied to the limiting case of A_n. It makes sense to use the product TA_n, since this value approaches a limit value but one that does not consistently result in zero. The spectrum $X(\omega)$ of $x(t)$ is therefore defined as

$$X(\omega) = \lim_{T\rightarrow\infty} TA_n = \int\limits_{-\infty}^{\infty} x(t)e^{-j\omega t}dt \,. \tag{13.41}$$

$X(\omega)$ is also known as the Fourier transform of $x(t)$.

The next question to be examined is what constitutes the inverse transformation rule. To examine this, we will apply the boundary to eq.(13.38):

$$x(t) = \lim_{T\rightarrow\infty} \frac{1}{T} \sum_{n=-\infty}^{\infty} TA_n e^{j2\pi n\frac{t}{T}} \,. \tag{13.42}$$

The following variables are redefined:

- $2\pi n/T$ becomes the continuous frequency variable ω,
- TA_n becomes $X(\omega)$,
- the frequency increment $1/T$ becomes the infinitesimally short distance *df* using $(1/T \rightarrow df = d\omega/2\pi)$ and
- the summation becomes an integration.

This results in the universal inverse transformation rule:

$$x(t) = \frac{1}{2\pi} \int\limits_{-\infty}^{\infty} X(\omega)e^{j\omega t}d\omega \,. \tag{13.43}$$

Eq.(13.43) is also referred to as inverse Fourier transform.

The basic concept and principles already discussed in the last section essentially apply here, other than the fact that signals cannot be analyzed based on their harmonic components by summing them over their discrete elements, rather, the inverse analysis takes place using integration instead of summation. For this reason, Fourier transformation is also referred to as an integral transformation. As is the case with Fourier sums, Fourier transformation likewise constitutes a unique and reversable mapping of a signal, in order to describe a signal through integration, as with summation, based on pure tones of the form $e^{j\omega t}$. Likewise, we can use the photo analogy described above to also better understand Fourier transformation.

For the sake of convenience, we will use short-cuts in the following discussions. For example, in order to reiterate that $X(\omega)$ refers to the Fourier transform of $x(t)$, from here on we will use the following notation:

$$X(\omega) = F\{x(t)\} = \int_{-\infty}^{\infty} x(t)e^{-j\omega t}dt \tag{13.44}$$

Likewise

$$x(t) = F^{-1}\{X(\omega)\} = \frac{1}{2\pi}\int_{-\infty}^{\infty} X(\omega)e^{j\omega t}d\omega\ , \tag{13.45}$$

means that $x(t)$ is the inverse transform of $X(\omega)$. Due to the unique mutual reversability of such functions, the operations F and F^{-1} cancel each other out, meaning

$$F^{-1}\{F\{x(t)\})\} = x(t) \tag{13.46}$$

and likewise

$$F\{F^{-1}\{X(\omega)\}\} = X(\omega)\ . \tag{13.47}$$

In conclusion, it is to be noted that the signal $x(t)$ and the spectrum $X(\omega)$ do not possess the same physical dimensions (the units). It is plain to see that the following statement is true for the units:

$$Dim[X(\omega)] = Dim[x(t)]s = \frac{Dim[x(t)]}{Hz}\ . \tag{13.48}$$

For this reason, $X(\omega)$ is sometimes referred to as the amplitude density function.

13.4.3 The Transmission Function and the Convolution Law

The reason for the previous introduction to Fourier transformation was, as mentioned, to construct an effective tool for describing the transmission of linear and time-invariant systems through multiplication with the complex-valued transmission function. Based on the principle of invariance, the Fourier representation of the input signal (13.43) implies that the output must always have the form

$$y(t) = \frac{1}{2\pi} \int_{-\infty}^{\infty} H(\omega)X(\omega)e^{j\omega t} d\omega \,. \tag{13.49}$$

In the frequency domain, the transmission is described by the product of the Fourier transform of the input and a transmission function $H(\omega)$ which characterizes the transmitter. The Fourier transform $Y(\omega)$ of the output $y(t)$ is

$$Y(\omega) = H(\omega)X(\omega) \,. \tag{13.50}$$

The transfer function always allows the calculation of the output as long as the input is known. In addition, it has already been discussed in the previous sections that the impulse response of the the transmitter likewise provides a complete description of the transmitter. The impulse response as well enables the specification of the output based on the input and thus also provides a complete description of the transmitter such as $H(\omega)$. The transmission function along with the impulse response represent the very same thing and therefore cannot be considered independent of one another. Rather, they have a specific relationship to one another.

This relationship can be easily derived from the convolution integral eq.(13.11). To do this, the integral is first Fourier transformed:

$$Y(\omega) = F\{y(t)\} = \int_{-\infty}^{\infty} \int_{-\infty}^{\infty} x(\tau)h(t-\tau)d\tau \, e^{-j\omega t} dt \,. \tag{13.51}$$

The sequence of the component integrations is interchangeable, meaning that

$$Y(\omega) = \int_{-\infty}^{\infty} x(\tau) \int_{-\infty}^{\infty} h(t-\tau) \, e^{-j\omega t} \, dt \, d\tau \,, \tag{13.52}$$

or

$$Y(\omega) = \int_{-\infty}^{\infty} x(\tau)e^{-j\omega\tau} \int_{-\infty}^{\infty} h(t-\tau) \, e^{-j\omega(t-\tau)} \, dt \, d\tau \,, \tag{13.53}$$

also apply. The inner integral exists in the Fourier transform of the impulse response (a formal proof explaining this can be obtained using variable substitution $u = t - \tau$), the remaining integral represents the transform $X(\omega)$ of $x(t)$. Therefore

$$Y(\omega) = F\{h(t)\}X(\omega) \,. \tag{13.54}$$

Using (13.50), we find the relationship between impulse response $h(t)$ and transmission function $H(\omega)$:

$$H(\omega) = F\{h(t)\} \,. \tag{13.55}$$

The transmission function is therefore the Fourier transform of the impulse response.

From a mathematical standpoint, we showed in the above that convolution in the time domain corresponds to multiplication in the frequency domain:

$$X(\omega)H(\omega) = F\{ \int_{-\infty}^{\infty} x(\tau)h(t-\tau)d\tau \} , \qquad (13.56)$$

where, of course, $X(\omega) = F\{x(t)\}$ and $H(\omega) = F\{h(t)\}$ are Fourier pairs. This relationship is known as the 'convolution law.'

The convolution law, in a slightly different form, also basically applies to the product of two time signals. As the reader can easily see by inverse-transforming the right-side of the 'convolution integral' in the frequency domain, as outlined above, the following applies

$$x(t)g(t) = F^{-1}\{ \frac{1}{2\pi} \int_{-\infty}^{\infty} X(\nu)G(\omega-\nu)d\nu\} , \qquad (13.57)$$

where likewise x, X and g, G also represent Fourier pairs. Convolution in the frequency domain corresponds to multiplication in the time domain. The only difference here is that the factor $1/2\pi$ appears as an element in the convolution integral.

13.4.4 Symmetries

The symmetrical properties of certain signals also belong to the body of basic knowledge of Fourier transforms.

Real-valued Signals

The spectrum of a real-valued function $x(t)$

$$X(\omega) = \int_{-\infty}^{\infty} x(t)e^{-j\omega t}dt .$$

converts into itself when ω is replaced by the value $-\omega$ and both sides of the equation are considered conjugated complex (*):

$$X^*(-\omega) = \int_{-\infty}^{\infty} x(t)e^{-j\omega t}dt .$$

This obviously results in

$$X^*(-\omega) = X(\omega) . \qquad (13.58)$$

The absolute spectrum is therefore axis-symmetric

$$|X(-\omega)|^2 = |X(\omega)|^2 \, , \tag{13.59}$$

the real component of the spectrum is likewise axis-symmetric

$$Re\{X(-\omega)\} = Re\{X(-\omega)\} :, \tag{13.60}$$

while the imaginary part is point-symmetric

$$Im\{X(-\omega)\} = -Im\{X(-\omega)\} :, \tag{13.61}$$

This is true for any real signal regardless of its shape.

Real-valued and Axis-symmetric Signals

A signal which is axis-symmetric with $x_g(-t) = x_g(t)$ has a real-valued spectrum, as is illustrated in the following simple discussions. In the integral

$$X(\omega) = \int_{-\infty}^{\infty} x(t)e^{-j\omega t}dt = \int_{-\infty}^{\infty} x(t)[cos(\omega t) - j\ sin(\omega t)]dt$$

the product $x(t)sin(\omega t)$ is a point-symmetric function, the integral over this portion of the integrand is therefore zero . As a result, only the real-valued transform remains

$$X(\omega) = \int_{-\infty}^{\infty} x(t)cos(\omega t)dt$$

The imaginary part is equal to zero $Im\{X(\omega)\} = 0$, the real part is of course still axis-symmetric: $Re\{X(-\omega)\} = Re\{X(\omega)\}$

Real-valued and Point-symmetric Signals

A signal which is point-symmetric with $x_u(-t) = -x_u(t)$ has a pure imaginary spectrum. In the integral

$$X(\omega) = \int_{-\infty}^{\infty} x(t)[cos(\omega t) - j\ sin(\omega t)]dt$$

the product $x(t)cos(\omega t)$ is a point-symmetric function, the integral over this portion of the integrand is therefore equal to zero. As a result, only the imaginary transform remains

$$X(\omega) = -j \int_{-\infty}^{\infty} x(t)sin(\omega t)dt$$

The real part is equal to zero, $Re\{X(\omega)\} = 0$, the imaginary part is of course still point-symmetric: $Im\{X(-\omega)\} = -Im\{X(\omega)\}$

Decomposition into Axis-symmetric and Point-symmetric Components

In general, all real signals without specific symmetry properties can be decomposed into their axis- and point-symmetric counterparts using the following method:

$$x(t) = \frac{1}{2}[x(t) + x(-t)] + \frac{1}{2}[x(t) - x(-t)] = x_g(t) + x_u(t)\,. \tag{13.62}$$

Here

$$x_g(t) = \frac{1}{2}[x(t) + x(-t)]$$

is of course the axis-symmetric and

$$x_u(t) = \frac{1}{2}[x(t) - x(-t)]$$

the point-symmetric part.

The spectrum of $x_g(t)$ is real, the spectrum of $x_u(t)$ is pure imaginary. Overall, one can ascertain that the Fourier transform of the axis-symmetric signal component is equal to the real component of the entire spectrum of $x(t)$:

$$Re\{X(\omega)\} = F\{x_g(t)\}\,. \tag{13.63}$$

Likewise the point-symmetric signal component corresponds with the imaginary part of the entire spectrum:

$$jIm\{X(\omega)\} = F\{x_u(t)\} \tag{13.64}$$

(with z=Re{z}+ j Im{z} for every complex value z).

13.4.5 Impulse Responses and Hilbert Transformation

The aforementioned fact that axis- and point-symmetric signal components pertain to the real and imaginary parts of the spectrum has intriguing implications for impulse response transforms. The latter consist of the reaction of a transmitter to the delta function $\delta(t)$. The signal excitation begins therefore at the time point $t = 0$. For this reason, the system response $h(t)$ can only be zero for negative time points as well. Based on the principle of causality ('nothing comes from nothing') the impulse response can likewise only begin at $t = 0$ at the earliest. Consequently, for causal impulse responses

$$h(t < 0) = 0$$

is true. Now the signal, as with any signal, can be decomposed into its axis- and point-symmetric components at either side of the impulse response:

$$h(t) = h_g(t) + h_u(t)\,.$$

Naturally this necessitates that h_g and h_u have to be equal also in the negative for all time points $t < 0$ as well. Only then is the impulse response likewise causal. Therefore

$$h_u(t < 0) = -h_g(t < 0)$$

must apply. Due to the required symmetry properties of h_g and h_u the following applies for positive time points

$$h_u(t > 0) = h_g(t > 0) ,$$

or, summarized

$$h_u(t) = sign(t)\, h_g(t)$$

$(sign(t > 0) = 1,\ sign(t < 0) = -1)$. Figure 13.11 underscores this fact with a simple illustration. For negative time points, h_u and h_g are equal value, one positive and one negative. Consequently, these values are identical for positive time points. It follows that, for $t > 0$, $h_u = h_g = h/2$.

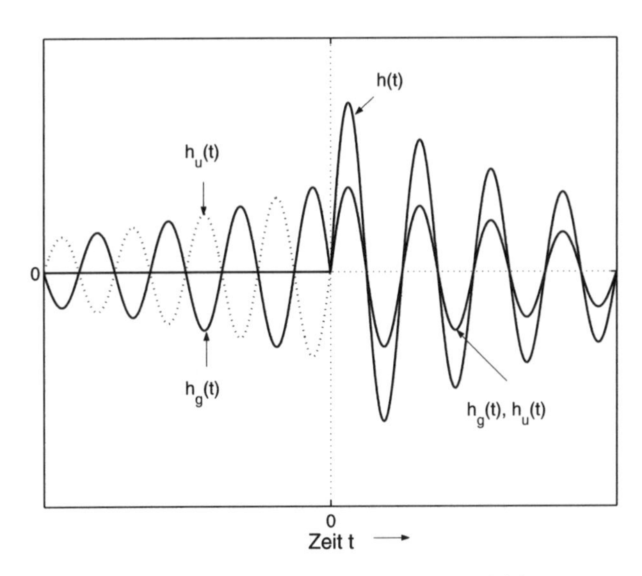

Fig. 13.11. Decomposition of the causal impulse response $h(t)$ in axis-symmetric part $h_g(t)$ and point-symmetric part $h_u(t)$ as exemplified in a swing-out transient

In sum, it has been shown that point-symmetric signal components, as shown, can be ascertained from the axis-symmetric signal component, as long as we are dealing with a causal impulse response. Because the transform of the axis-symmetric part h_g is equal to the real part $Re\{H(\omega)\}$ of the transmission function $H(\omega)$ and the point-symmetric part h_u corresponds to the imaginary part $Im\{H(\omega)\}$ of the transmission function $H(\omega)$, the real and imaginary part of H are dependent on one another. Indeed, the relationship between $Re\{H(\omega)\}$ and $Im\{H(\omega)\}$ can be easily discerned by first transforming $h_u(t) = sign(t)\, h_g(t)$

$$jIm\{H(\omega)\} = F\{h_u(t)\} = F\{sign(t)\, h_g(t)\}$$

then expressing $h_g(t)$ in terms of $h_g(t) = F^{-1}\{Re\{H(\omega)\}\}$:

$$Im\{H(\omega)\} = -jF\{h_u(t)\} = -jF\{sign(t)F^{-1}\{Re\{H(\omega)\}\}\} \,. \qquad (13.65)$$

Eq.(13.65) states the relationship that the real and imaginary parts of the transmission function must have toward one another for every causal (meaning linear and time-invariant) transmitter. In measurement practice, it suffices simply to compute the real part of the transmission function; the imaginary part can be calculated based on the real part.

The order of operations on the right side of eq.(13.65) is known as the 'Hilbert transformation' . Of course, this constitutes a transformation insofar that a spectrum is calculated out of another. Hilbert transformations, however, do not serve to represent a signal in a series of many other signals, as is the case with Fourier transformations. On the contrary, Hilbert transformations state a relationship $Re\{H(\omega)\}$ $Im\{H(\omega)\}$ in such a way so that the transmitter described behaves in a causal manner.

Fourier analysis of measurement signals in the form of transmission functions by nature define complex transmission functions and thereby entail a certain redundancy. An interesting question could be if this redundancy can be utilized to judge the results measuring a system's effectiveness or other factors.

13.5 Fourier Acoustics: Wavelength Decomposition

The use of Fourier transformation is not just limited to time dependencies and their frequency decomposition. The labelling of the time variable with t and the frequency variable with ω in equations (13.44) and (13.45) is just an arbitrary decision to differentiate certain physical quantities. One could just as well assign t a localized variable, a coordinate direction with the unit m (meter) and ω could be designated as a variable defining wave numbers (with the unit $1/m$). In short, it doesn't matter whether decomposition into harmonics is applied to time functions or spatial functions. Only the significance underlying the variable has been changed.

On the other hand, changing the meaning behind the label might cause some lack of clarity. To avoid confusion, from here on the one-dimensional coordinate axis will be referred to as x and the wave number as k. For the Fourier transformation of a spatial function $g(x)$ and its wave number spectrum $G(k)$ instead of (13.44) and (13.45)

$$G(k) = F\{g(x)\} = \int_{-\infty}^{\infty} g(x)e^{-jkx}dx \tag{13.66}$$

and

$$g(x) = F^{-1}\{G(k)\} = \frac{1}{2\pi}\int_{-\infty}^{\infty} G(k)e^{jkx}dk\,. \tag{13.67}$$

will be used. Eq.(13.67) interprets the spatial function $g(x)$ as consisting of several wave functions $G(k)e^{jkx}$; Eq.(13.66) specifies how the corresponding amplitude density function $G(k)$ can be derived from the space dependency $g(x)$.

The advantage gained by Fourier transformation of spatial functions is that it can be used to directly express the universal solution for the wave equation, also expressed in terms of a Fourier transform. To illustrate this fact and the resulting implications, for the sake of simplicity we will first examine two-dimensional fields in the following sections ($\partial/\partial z = 0$). The more universal, three-dimensional case will be subsequently discussed. Furthermore, we will presuppose the Helmholtz equation

$$\frac{\partial^2 p}{\partial x^2} + \frac{\partial^2 p}{\partial y^2} + k_0^2 p = 0 \tag{13.68}$$

which results from the wave equation (2.55), when we are assuming either complex amplitudes arising from sound field excitation with pure tones or amplitude densities arising from time-dependent Fourier transformation for all field quantities from this point on. The wave number of free waves will hereto be referred to as k_0 (with $k_0 = \omega/c$). A distinction has to be made, however, between k_0 and the wave number variable k in localized Fourier transformations.

According to the analogy of (13.67), the following ansatz applies for the case of sound pressure

$$p(x,y) = F^{-1}\{P(k,y)\} = \frac{1}{2\pi}\int_{-\infty}^{\infty} P(k,y)e^{jkx}dk\,, \tag{13.69}$$

Thus, one expresses on every plane $y = const.$ the localized sound pressure through its wave number decomposition $P(k,y)$, which of course may take on a different gradient on every plane $y = const.$, that is, be dependent on y. Using the Helmholtz equation (13.68) one obtains

$$\frac{\partial^2 P(k,y)}{\partial y^2} + (k_0^2 - k^2)P(k,y) = 0\,. \tag{13.70}$$

The solution of this typical differential can now quite easily be obtained:

$$P(k,y) = P_+(k)e^{-jk_y y} + P_-(k)e^{jk_y y}\,. \tag{13.71}$$

In so doing, the wave number k_y is defined as follows:

$$k_y = \begin{cases} +\sqrt{k_0^2 - k^2}\,, & k_0^2 > k^2 \\ -j\sqrt{k^2 - k_0^2}\,, & k^2 \geq k_0^2\,. \end{cases} \tag{13.72}$$

As in many examples of this book, the sign after taking the square root has been specified so that the sound field $e^{-jk_y y}$ means either

- a wave travelling in the positive y-direction ($k_0^2 > k^2$) or
- a near field subsiding in the y-direction, exponentially decreasing ($k^2 > k_0^2$).

The overall solution of the Helmholtz equation is, according to this definition

$$p(x,y) = \frac{1}{2\pi}\int_{-\infty}^{\infty} [P_+(k)e^{-jk_y y} + P_-(k)e^{jk_y y})]e^{jkx}dk\,. \tag{13.73}$$

Moreover, it is to be assumed that neither sources nor reflectors exist in the partial space $y > 0$; in this case, the wave can neither occur in the negative y-direction nor in near-fields growing in the positive y-direction, leaving us only with

$$p(x,y) = \frac{1}{2\pi}\int_{-\infty}^{\infty} P_+(k)e^{-jk_y y}e^{jkx}dk \tag{13.74}$$

for the sum field.

Eq.(13.74) primarily opens two kinds of areas of application which will be discussed in the following.

13.5.1 Radiation from Planes

The 'classic' application of wave decomposition of a spatial function described in the following is in any case the measuring of sound emission of vibrating surfaces or planes. The assumption when doing this is that the y components of the local velocity $v_y(x)$ on a plane are known either through assuming or measuring the absolute value and phase. For the sake of simplicity we will suppose that the source element is located on the plane $y = 0$. The only thing left to do now is to establish the relationship between the still unknown amplitude density $P_+(k)$ in (13.74) and the source element velocity $v_y(x)$. This can be done quite easily. First, the following velocity is construed using the ansatz eq.(13.74). This is

$$v_y(x,y) = \frac{j}{\omega\varrho}\frac{\partial p}{\partial y} = \frac{1}{2\pi}\frac{1}{\varrho c}\int_{-\infty}^{\infty} \frac{k_y}{k_0}P_+(k)e^{-jk_y y}e^{jkx}dk\,. \tag{13.75}$$

For $y = 0$ the Fourier transform $V_y(k)$ of the source velocity $v_y(x)$ must appear as a term on the left-hand side, resulting in

$$V_y(k) = \frac{1}{\varrho c}\frac{k_y}{k_0}P_+(k) , \tag{13.76}$$

whereby the amplitude density $P_+(k)$ is derived from the source velocity. Overall,

$$p(x, y) = \varrho c \frac{1}{2\pi}\int_{-\infty}^{\infty}\frac{k_0}{k_y}V_y(k)e^{-jk_y y}e^{jkx}dk , \tag{13.77}$$

applies, whereby $V_y(k)$, as previously mentioned, represents the Fourier transform of $v_y(x)$:

$$V_y(k) = \int_{-\infty}^{\infty} v_y(x)e^{-jkx}dx . \tag{13.78}$$

For the complete interpretation of eq.(13.77), the wave number variable k is also expressed through the designated wavelength variable λ with $k = 2\pi/\lambda$. With this, we will reiterate the principles based on eq.(13.77) introduced in Chapter 3 here below:

- Only long-waved source elements (with $\lambda > \lambda_0$ (λ_0=length of sound wave in air) and therefore $|k| < k_0$ have a real-valued wave number k_y in the y-direction according to eq.(13.72). Only such long-waved source elements are emitted as skew waves and are thus apparent even at large distances from the source.
- Short-waved source elements with $\lambda < \lambda_0$ and $|k| > k_0$ on the other hand possess an imaginary wave number. The corresponding sound field element possesses a near-field characteristic, and is only noticeable in the vicinity of the source in $y = 0$, gradually diminishing outward until it ceases to influence the sound field altogether at large distances from the source.

A long-waved source element $\lambda > \lambda_0$ is emitted as a plane wave at a skew angle

$$\sin\vartheta = -\frac{\lambda_0}{\lambda} \tag{13.79}$$

as can be seen when comparing the source element $e^{-jk_y y}e^{jkx}$ with the overall form of such a wave

$$p = p_0 e^{-jk_0 x sin\vartheta}e^{-jk_0 y cos\vartheta} \tag{13.80}$$

The angle ϑ counts relative to the y-axis 'upward', in other words, the wave travels 'downward,' as expressed trough the negative sign.

In the far-field (already elaborated in Chapter 3) only the long-waved wave number elements $|k| < k_0$ of the source are present, due to the fact that the short-waved elements are solely concentrated in the source's proximity and then rapidly diminish. Only the 'visible' portion corresponding to the long source wavelengths $\lambda > \lambda_0$ exerts any influence at all on the emission at large distances $|k| < k_0$. A specific directivity is attributed to each wave number component within the visible, long-waved portion segment. For this reason, a direct relationship exists between the directivity pattern of a source and Fourier transformation of the source velocity. In the eq.(3.37) introduced in Chapter 3 for narrow source bands with the width b, the integral on the right-hand side can then be expressed through the Fourier transform of the source velocity:

$$p_{\text{fern}} = \frac{j\omega\varrho\, b}{4\pi R} e^{-jk_0R} \int_{-l/2}^{l/2} v_y(x) e^{jk_0 x \sin\vartheta} \mathrm{d}x = \frac{j\omega\varrho\, b}{4\pi R} e^{-jk_0R}\, V_y(k = -k_0 \sin\vartheta)\;, \tag{13.81}$$

where V signifies the sound velocity defined in eq.(13.78). The circumferential distribution of the sound field is equal to the velocity transform, the emission at large distances clearly constituting a physical 'Fourier transformer'. The fundamental framework exists in the fact that a specific direction of travel is assigned to every source wave number.

13.5.2 Emission of Bending Waves

First, we will omit the factors of a finite plate dimension and other effects such as the amplitude decay at the point of induction from our considerations of the fundamental aspects of sound emitting from a plate vibrating with bending waves. The local sound velocity is described on the whole plane $y = 0$ as

$$v_y(x) = v_0 e^{-jk_B x} \tag{13.82}$$

whereby $k_B = 2\pi/\lambda_B$ refers to the bending wave number (λ_B = bending wavelength) (See Chapter 4 for more information on the fundamental properties of bending waves and their transmission along plates). The Fourier transform of this monochromatic event (monochromatic: consisting of only one wave component) must possess the delta form. It is

$$V_y(k) = 2\pi v_0 \delta(k + k_B)\;, \tag{13.83}$$

as can easily be discerned by substituting (13.83) in eq.(13.67). For the sound pressure emanating from this source, using eq.(13.77) one obtains

$$p(x,y) = \varrho c v_0 \frac{k_0}{k_y} e^{-jk_y y} e^{-jk_B x}\;, \tag{13.84}$$

where the wave number k_y has been shortened to

$$k_y = \begin{cases} +\sqrt{k_0^2 - k_B^2} & k_0^2 > k_B^2 \\ -j\sqrt{k_B^2 - k_0^2} & k_B^2 \geq k_0^2\,. \end{cases} \tag{13.85}$$

The basic qualities of this sound field in air produced by the monochromatic bending wave can generally be summarized in the following:

- if the source wavelength (in this example: λ_B) is greater than the sound wavelength in air λ_0, a wave at the skew angle ϑ with $\sin\vartheta = -\lambda_0/\lambda_B$ will be emitted,
- for a source wavelength λ_B, which is shorter than the sound wavelength in air λ_0, is only present in a diminishing near field relative to the source surface. This diminishing near field does not transport any power in the time average.

That even in the presence of short bending waves below the coincidence critical frequency many sound fields that are not equal to zero are visible at larger distances from the emitting surface lends itself to the fact that in reality, source surfaces have only finite dimensions.

The fact that a weak residual emission results from short-waved sources of finite dimensions can be drawn from the simple concept which is outlined in Figure 13.12. To provide a simpler example, short-waved source velocities are depicted in the form of standing waves (standing waves result from the sum of two waves travelling in opposite direction; a progressive wave can also be understood as the sum of two standing waves).

If the distances of the inversely phased reverberating zones are small as compared to the sound wavelength in air (i.e. when $\lambda_B \ll \lambda_0$), this gives rise to a pattern of de-facto sources with rapidly changing signs for both cases depicted. "'Almost all"' sound sources cancel out each other's effects. The point of each pair of opposing signs can be understood as 'at the same place at the same time.' The net volume flow emanating from such pairs is therefore equal to zero. The motion field induced by the pair in the surrounding air consists in mere shifts of mass. The shifting air mass lifted along with the rising source zone is shoved to the side of the sinking source zone.

The residual sound emission for short-waved sources that ensues obviously depends primarily on whether each local source contacts a neighboring source, obliterating each other to zero. This already occurs in the middle of the source element, but along the margins it is possible that some sources remain "without a partner." As shown in Figure 13.12, the short-circuit of pairs in vibrating sources "'with antinodes at the margins" is complete and the resulting sound emission therefore very negligible. In the case of "'nodes at the margins"', source zones remain at the external regions of the source margin. These behave like volume sources. Compared to long-waved sources (where the entire source surface plays a role in the emission event), the emission is still quite negligible because "'most"' source elements do not actually contribute to the emission, as compared with the case of "'antinodes at the margins"' the sound field is however considerably greater, because this time, a net volume flow is left over.

The sound field of a bending wave guide which has vibrational nodes at the upper end and a vibrational antinode at the lower end is depicted in Figure 13.13. The particle displacement was derived from the specified local source velocity using the methods described in Fourier acoustics. The upper margin

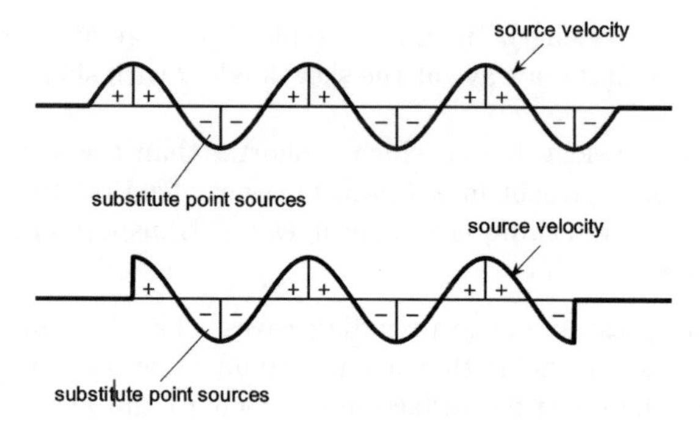

Fig. 13.12. De-facto sources for short-waved source elements with vibrational antinode or nodes at the margins.

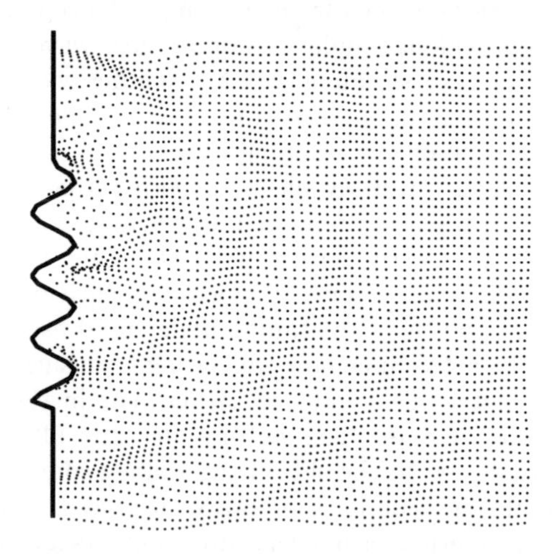

Fig. 13.13. Sound field of a short-waved source with vibrational nodes at the upper, vibrational antinodes at the lower end. The vibrational node at the upper end constitutes a volume source. Sound emanates with circular wave fronts surrounding it. The antinode emits virtually nothing.

is an easily identifiable element of the sound field with its vibrational nodes while the vibrational antinode emits much less sound.

13.5.3 Acoustic Holography

Another interesting application of sound field decomposition into wave functions along a coordinate axis is what is known as 'acoustic holography', which we will describe below.

The concept can be described as follows. If one measures the space-dependency of the sound pressure $p(x, d)$ in terms of absolute value and phase along an entire plane, such as in the $y = d$ plane just before the source element, one can also ascertain its wave number spectrum:

$$P_+(k,d) = F\{p(x,d)\} = \int_{-\infty}^{\infty} p(x,d)e^{-jkx}dx\,. \tag{13.86}$$

After comparing eq.(13.86) with eq.(13.74) we obtain:

$$P_+(k) = P_+(k,d)e^{jk_yd}\,. \tag{13.87}$$

By these means and incorporating eq. (13.74) we can compute the sound field in the entire space, as long as the constraints are met: no reflectors or sources in the reconstructive space existing in $y > 0$. Basically, it is possible to take the measurements along just one plane and reconstruct the results for all other planes, a concept akin to a hologram. Solely a numerical procedure is required to do this. This procedure utilizes Fourier transformation and its inverse process in numerical terms. The principle methods are not only limited to the sound pressure, but can also be expanded to include the measurement of pressure along one plane and sound velocity components in the entire room, utilizing eq.(13.75).

The main problem in respect to acoustic holography constitutes those source elements which have shorter wavelengths than those of the surrounding medium. These source elements are characterized by an imaginary wave number k_y in the y-direction, $k_y = -j \mid k_y \mid$, based on eq.(13.72). If the measurements $p(x,d)$ and likewise $P_+(k,d)$ contain small inconsistencies or errors, these will weigh heavily in the short-waved source domain in eq.(13.87) due to its multiplication with the exponential function $e^{|k_y|d}$. The space-dependencies measured as such then contain intense, localized noise, limiting the quality of the outcome considerably. This effect can be mitigated or avoided by taking measurements in as small distances from the source's surface d as possible, or by omitting extremely short source wavelengths from the calculations.

13.5.4 Three-dimensional Sound Fields

All the aspects and procedures described above in the previous sections in Fourier acoustics can easily be translated to the more realistic, three-dimensional case of emissions originating from plane surfaces, if the two-dimensional Fourier transform is applied to all occurring spatial functions. The transformation now becomes

$$G(k_x,k_y) = \int_{-\infty}^{\infty}\int_{-\infty}^{\infty} g(x,y)e^{-jk_xx}e^{-jk_yy}dxdy \tag{13.88}$$

and the inverse transformation is

$$g(x,y) = \frac{1}{4\pi^2} \int_{-\infty}^{\infty} \int_{-\infty}^{\infty} G(k_x, k_y) e^{jk_x x} e^{jk_y y} dk_x dk_y \,. \tag{13.89}$$

For the sake of simplicity, we will assume as we did in Chapter 3 that the source lies on the plane $z = 0$ (x, y-plane) and has a known local velocity $v_z(x, y)$ in the z-direction.

The sound field in the entire room $z > 0$ (where we once again presume the absence of a reflector or any other sound source) is then ascertained similarly as in (13.74) on the basis of

$$p(x,y,z) = \frac{1}{4\pi^2} \int_{-\infty}^{\infty} \int_{-\infty}^{\infty} P_+(k_x, k_y) e^{-jk_z z} e^{jk_x x} e^{jk_y y} dk_x dk_y \,. \tag{13.90}$$

Thus, the double Fourier transform $P_+(k_x, k_y)$ of the sound pressure $p(x, y, 0)$ in the plane $z = 0$ related to the double Fourier transform $V_z(k_x, k_y)$ of the source velocity $v_z(x, y)$ (see eq.13.88 for a definition of transform) by way of

$$P_+(k_x, k_y) = \varrho c \frac{k_0}{k_z} V_z(k_x, k_y) \tag{13.91}$$

Using the wave equation, the wave number k_z in the z-direction becomes

$$k_z = \begin{cases} +\sqrt{k_0^2 - (k_x^2 + k_y^2)}\,, & k_0^2 > k_x^2 + k_y^2 \\ -j\sqrt{(k_x^2 + k_y^2) - k_0^2}\,, & k_x^2 + k_y^2 \geq k_0^2 \,. \end{cases} \tag{13.92}$$

Furthermore, eq.(13.92) indicates that short-waved source elements $k_x^2 + k_y^2 \geq k_0^2$ only attract near field, while long-waved source components $k_x^2 + k_y^2 < k_0^2$ are emitted in the form of a progressive wave travelling at a skew angle.

It is additionally possible to ascertain the sound velocity components by differentiating eq.(13.90).

If one considers the source velocity $v_z(x, y)$ as given, the entire set of formulas can then be used to measure the sound field in the half-space $z > 0$ before the source. The measurements of the sound pressure, according to absolute value and phase, in a whole plane $z = d$, enables once again the implementation of acoustic holography.

The Fourier acoustics discussed in this chapter are drawn from the fact that arbitrary spatial functions can be mapped using a wave sum in the form of the Fourier integral eq.(13.66). When applying the decomposition into source wavelengths to the source velocity, the linearity of the emission process allows for the analysis of each sound field brought about by each source wave number separately. The sum field subsequently results simply from the 'sum of parts' – precisely derived from the integral over the corresponding wave number spectra. This method can be generally described as the 'method of decomposition into wavelengths'.

This procedure can be compared to the description of a sound field using the Rayleigh integral introduced in Chapter 3, the 'method of decomposition into point sources,'

$$p(x,y,z) = \frac{j\omega\varrho}{2\pi} \int_{-\infty}^{\infty}\int_{-\infty}^{\infty} v_z(x_Q, y_Q) \frac{e^{-jkr}}{r} \mathrm{d}x_Q \mathrm{d}y_Q \tag{13.93}$$

where r denotes the distance form source point to field point

$$r = \sqrt{(x-x_Q)^2 + (y-y_Q)^2 + z^2} \tag{13.94}$$

It goes without saying that the methods of both 'decomposition into waves' and 'decomposition into sources' lead to the same results. Indeed, this fact can be proven with the help of the convolution law. In the expression for the sound pressure

$$p(x,y,z) = \frac{1}{4\pi^2} \int_{-\infty}^{\infty}\int_{-\infty}^{\infty} \varrho c \frac{k_0}{k_z} e^{-jk_z z} V_z(k_x, k_y) e^{jk_x x} e^{jk_y y} dk_x dk_y \tag{13.95}$$

which results from substituting eq.(13.91) into eq.(13.90), the wave number spectrum to be inverse transformed can be seen as the product of a 'typical emission' function $H(k_x, k_y, z)$ with

$$H(k_x, k_y, z) = \varrho c \frac{k_0}{k_z} e^{-jk_z z} \tag{13.96}$$

and a 'typical source' function $V_z(k_x, k_y)$:

$$p(x,y,z) = \frac{1}{4\pi^2} \int_{-\infty}^{\infty}\int_{-\infty}^{\infty} H(k_x, k_y, z) V_z(k_x, k_y) e^{jk_x x} e^{jk_y y} dk_x dk_y \, . \tag{13.97}$$

The 'typical emission function' $H(k_x, k_y, z)$ is nothing more than the (localized) transmission function which defines how heavily a wavelength segment of the source weighs into the overall emission. It is therefore evident that $V_z(k_x, k_y)$ describes the source itself by specifying its wavelength decomposition.

Sound pressure in space accordingly results from the inverse transformation of the product $H(k_x, k_y, z)V_z(k_x, k_y)$. For this reason, $p(x, y, z)$ can be derived from convoluting the corresponding inverse transforms. If the inverse transform of $H(k_x, k_y, z)$ can be described as $h(x, y, z)$ then

$$h(x,y,z) = \frac{1}{4\pi^2} \int_{-\infty}^{\infty}\int_{-\infty}^{\infty} H(k_x, k_y, z) e^{jk_x x} e^{jk_y y} dk_x dk_y \, . \tag{13.98}$$

is also true. The obvious definition of $h(x,y,z)$ can be therefore construed from the following discussion. For a delta shaped source $v_z(x,y) = \delta(x)\delta(y)$ the Fourier transform of the source velocity exists in $V_z(k_x,k_y) = 1$. One can assume $p(x,y,z) = h(x,y,z)$ based on this specific case by comparing eq.(13.97) and eq.(13.98). Obviously, following this logic, $h(x,y,z)$ describes the 'localized impulse response', that is, the spatial reaction to a delta-shaped source at the origin.

Finally, the spatial sound pressure must be obtainable by convoluting the source velocity with the localized impulse response:

$$p(x,y,z) = \int_{-\infty}^{\infty}\int_{-\infty}^{\infty} v_z(x_Q,y_Q)h(x-x_Q,y-y_Q,z)dx_Q dy_Q. \qquad (13.99)$$

In any case, the localized impulse response $h(x,y,z)$ can be quite easily discovered based on the discussions in Chapter 3. Since we are dealing with a volume source $v_z(x,y) = \delta(x)\delta(y)$ emitting in a half-space $z > 0$ with the volume flow $Q = 1$, using eq.(3.13)

$$h(x,y,z) = \frac{j\omega\varrho}{2\pi\sqrt{x^2+y^2+z^2}} e^{-jk\sqrt{x^2+y^2+z^2}}\,, \qquad (13.100)$$

is true, whereby eq.(13.99) ultimately converges into the Rayleigh integral (13.93). The Rayleigh integral, which precisely comprises the 'method of source decomposition' , can therefore be justified by means of the convolution law pertaining to the 'method of wave decomposition'. Both methods give one and the same result.

In conclusion, we would like to remark that the far field approximation discussed in Chapter 3.6 contains the Fourier transform of the source velocity, as is the case with 'one-dimensional' sources. When comparing eq.(3.59) to eq.(13.88) and applying them to the source velocity, we obtain namely

$$p_{\text{fern}}(R,\vartheta,\varphi) = \frac{j\omega\varrho}{2\pi R} e^{-jkR} V_z(k_x = -\sin\vartheta\cos\varphi,\ k_y = -k\sin\vartheta\sin\varphi)\,, \qquad (13.101)$$

where (R,ϑ,φ) specifies the coordinate system of the spherical coordinates. Naturally, this again only states that every combination of long wavelengths (λ_x,λ_y) emitting from the source (with $k_x = 2\pi/\lambda_x$ and $k_y = 2\pi/\lambda_y$) is designated a specific trajectory which can be found in eq.(13.101).

13.6 Summary

Linear and time-invariant transmitters (LTI) follow the invariance law: in the case of an input signal consisting of a pure tone, meaning a sinusoid time-dependency, the output signal will always consist of a pure tone of the same frequency. The transmission of pure tones

- occurs therefore without distortion and
- can be fully described by the changes in amplitude and phase of the input signal as a result of the transmission.

The signal-series scheme of pure tones of varying frequencies and complex amplitudes provides a collection of simple and easy-to-understand tools for describing the LTI transmission of any input signal, be it speech, engine noise, music, etc., by integration of the tones at varying frequency. The mathematical decomposition and composition procedures into pure tones and from pure tones is known as the Fourier series expansion (for periodic functions), and Fourier transformation for the non-periodic case. For both forms, a respective clear inverse exists; that is, for each time-dependency there exists exactly one corresponding transformed function, and vice versa.

In the frequency domain, every LTI transmission can be described by the product of the input signal's spectrum $X(\omega)$ (denoting the Fourier transform's input signal) and the transmission function $H(\omega)$: $Y(\omega) = H(\omega)X(\omega)$ is true for the spectrum of the output signal $Y(\omega)$. The resulting shape of the transmission function can thus be ascribed both the tone color characteristics of the input signal and the filtering affects of the transmitter.

Fourier transform proves useful in examining sound emission from planes. The source can thus be decomposed into multiple components of varying wavelengths. This composition is characteristic of the given source. Considering the emission of individual components leads to the basic principle of sound emission:

- short source wavelengths, smaller than the wavelengths of the surrounding medium, lead solely to near fields on the source's surface,
- long-rippling source components are, at a certain angle dependent on both wavelengths involved, emitted diagonally to the surface in the form of a plane wave, which can therefore also be perceived in the far field.

Because exactly one well-defined angle of emission can be attributed to every source wavelength, there is a direct correlation between the beam pattern in the far field and the long-rippling range within the source wave number spectrum. Above and beyond this very important discovery in Fourier acoustics, the latter has made acoustic holography – the method of mapping a sound field, based on its measurements in a room, onto a surface (according to amplitude and phase) – possible.

13.7 Further Reading

The author owes his knowledge of system theoretical fundamentals to a considerable work by Rolf Unbehauen (Unbehauen, R.: 'Systemtheorie', R. Oldenbourg Verlag, München 1971), which also includes a treatment of Fourier transformation.

The journal article by Manfred Heckl (Heckl,M.: 'Abstrahlung von ebenen Schallquellen', ACUSTICA 37 (1977), S. 155 - 166) represents the first basic step in the development of Fourier acoustics.

According to the author, the work Papoulis, A.:'The Fourier Integral and Its Applications', McGraw-Hill, New York 1963 provides the groundwork for understanding Fourier transformation.

13.8 Practice Exercises

Problem 1

Calculate the amplitude spectrum A_n of the rectangular signal with period T, which is defined within the period of $-T/2 < t < T/2$ by

$$x(t) = \begin{cases} 1, & |t| < T_D/2 \\ 0, & else \end{cases}$$

T_D ($T_D < T$) is the duration of signal excitation within the period. How does A_n converge?

Problem 2

One can generally assume that the principle convergence characteristics of the amplitudes A_n behave like the rectangular signal in Problem 1 for all discontinuous functions. Based on this assumption:

- How do A_n in a continuous signal – whose first derivative is discontinuous – converge?
- How do A_n of a continuous signal with a continuous first derivative and a discontinuous second derivative converge?
- How do A_n of a continuous signal converge with the first m continuous derivatives, but with a discontinuous $m+1$-th derivative?

Comment on the results in respect to the number of numerically genuine model x_M for x.

Problem 3

Which physical unit characterizes the rectangular function $r_{\Delta T}(t)$ described in Figure 13.1? Which physical unit characterizes the delta function $\delta(t)$?

Problem 4

Prove the 'energy law', according to which

$$\int_{-\infty}^{\infty} |x(t)|^2 dt = \frac{1}{2\pi} \int_{-\infty}^{\infty} |X(\omega)|^2 d\omega$$

is true. Proceed by using the convolution principle which applies to the product of two time signals.

The left integral can be defined as the 'signal energy from time-function', and the right integral, the 'spectral signal energy'. Viewed in this way, the energy law establishes a conservation principle.

Problem 5

Prove that the so-called auto-correlation function

$$a(t) = \int_{-\infty}^{\infty} x(t+\tau)x^*(\tau)dt$$

is the inverse transform of $|X(\omega)|^2$:

$$a(t) = F^{-1}\{|X(\omega)|^2\}$$

Use the convolution principle which applies to the product of two spectra.

For the sake of universality, allow complex time functions. The conjugate annotation * is omitted for real-valued time functions.

Problem 6

Show that for every linear and time-invariant transmitter, the relationships

- between the complex amplitudes $\underline{x}$ and $\underline{y}$ of the input and output
- and between the Fourier transformed $X(\omega)$ and $Y(\omega)$ of the input and outputs

are identical.

Problem 7

Calculate the impulse response for the particle displacement x of the mass for a simple resonator (see Chapter 5) based on the transmission function

$$H(\omega) = \frac{X(\omega)}{F(\omega)} = \frac{\frac{1}{s}}{1 - \frac{\omega^2}{\omega_0^2} + j\eta\frac{\omega}{\omega_0}} .$$

Here, $\omega_0 = \sqrt{s/m}$ is the resonance frequency of the oscillator, η describes the loss factor.

Tip: The simplest method for solving this problem is to use the residue principle. Otherwise, use partial fraction expansion.

Problem 8

Calculate the impulse response of the speed of a sound of a beam excited to bending waves by a point force $F'_a = F_0\delta(x)\delta(t)$, in dependency of x on the beam.

Note: F_0 denotes the total impulse contained in the force function. Namely, the unit of F_0 is $dim(F_0) = Ns$.

Tip: This problem is asking for the inverse transform of the transmission function

$$V(\omega) = \frac{F_0}{4k_B\sqrt{m'B}}(e^{-jk_Bx} - je^{-k_Bx})$$

(only true for $\omega > 0$), which is a direct result of the solution of the last problem in Chapter 4, based on the aspects mentioned in Problem 6. Keep in mind, that $V(-\omega) = V^*(\omega)$ must apply; otherwise, the impulse response asked for would not be real. Refer to Chapter 4 to review the significance of the dimensions as well as the wave number k_B of bending waves.

Problem 9

Find the Fourier transform of the Gauss function

$$f(t) = f_0 e^{-\gamma t^2} .$$

Problem 10

Find the far field beam pattern of a radiating strip with width b ($b << \lambda_0$), with the oscillation course

$$v_y(x) = v_0 e^{-|x|/x_0} .$$

Problem 11

An oscillation of a radiator strip b wide ($b << \lambda_0$) is given below:

$$v_y(x) = \frac{v_0}{2}[e^{j2\pi nx/l} + \varepsilon e^{-j2\pi nx/l}]$$

($0 < x < l$, l= length). Show, that for $\varepsilon = 1$ the oscillation has antinodes at the edges $x = 0$ and $x = l$, and for $\varepsilon = -1$ the oscillation provides nodes at the edges. The Fourier-transformed oscillation should serve as a tool for qualitatively estimating the difference of the powers radiated for $\varepsilon = 1$ and $\varepsilon = -1$.

Problem 12

Find the Fourier-transformed amplitude-modulated signals

$$f_1(t) = g(t)e^{j\omega_0 t}$$

and

$$f_2(t) = g(t)\ \cos\omega_0 t\ :$$

Here, $g(t)$ is an enveloping function, as a Gauss function, over the carrier signal with frequency ω_0.

Problem 13

This problem examines a one-dimensional, anechoic wave guide, on which a sound and reverberation field $v(x,t)$ with the (real) wave number k disperses, $k = k(\omega)$ can be characterized by an arbitrary frequency dependency. For example, it can defined as $k \sim \sqrt{\omega}$ for bending waves. An amplitude-modulated oscillation gradient of

$$v(0,t) = g(t)\ \cos\omega_0 t$$

occurs on the wave guide at the point $x = 0$. As in Problem 11, assume $g(t)$ is an enveloping function (e.g. a Gauss function), which, in addition, can be assumed here to have a small-banded character. Describe approximately how this signal propagates along the wave guide.

A

Level Arithmetics

A.1 Decadic Logarithm

The decadic logarithm is defined as the inversion of the calculation of a power to the base 10. If the relation between two numbers x and y is given as

$$x = 10^y \tag{A.1}$$

then y is denoted as the decadic logarithm of x:

$$y = \lg x \, . \tag{A.2}$$

The facts expressed by (A.1) and (A.2) can also be described by the following problem: for a given number x a second number with the name 'logarithm of x' is searched for in such a way, that 10 to the power of the second number results in x ($x = 10^{\lg(x)}$). Basically, nothing else is expressed except the fact that taking the logarithm and '10 to the power of' are operations which cancel each other out.

Some values, directly following from (A.2), like

$$\begin{aligned} \lg(10) &= 1 \\ \lg(100) &= 2 \\ \lg(10^n) &= n \end{aligned}$$

also show that the logarithmic function displays the principal characteristics of the sensitivity relation shown in Fig. 1.2.

Some simple calculation rules directly follow from the definition. For example, a product of two numbers ab can be written as

$$ab = 10^{\lg(ab)}$$

which is equivalent to

$$10^{\lg(a)}10^{\lg(b)} = 10^{(\lg(a)+\lg(b))} = 10^{\lg(ab)}$$

because $a = 10^{\lg(a)}$ and $b = 10^{\lg(b)}$ and thus the rule for the product results in

$$\lg(ab) = \lg(a) + \lg(b) . \tag{A.3}$$

Likewise,

$$\lg(a/b) = \lg(a) - \lg(b) \tag{A.4}$$

and also

$$\lg(a^b) = b \lg(a) . \tag{A.5}$$

When migrating to another base, the 10 used above can be exchanged arbitrarily with a different number. The logarithm of x to the base a is therefore defined as

$$x = a^{\log_a(x)} . \tag{A.6}$$

The relation between two logarithms of different bases can be produced as follows. For two bases a and b, according to (A.6) it is

$$a^{\log_a(x)} = b^{\log_b(x)} .$$

Using the operation $\log_a$ it is

$$\log_a(a^{\log_a(x)}) = \log_a(x) = \log_a(b^{\log_b(x)}) = \log_b(x) \log_a(b) ,$$

hence

$$\log_a(x) = \log_b(x) \log_a(b) . \tag{A.7}$$

Thus, all logarithmic functions have the same shape, independent of their base (apart from the scaling on the abscissa).

The future acoustician should learn the value $\lg 2 = 0.3$ by heart; it will be useful very often.

A.2 Level Inversion

The level definition

$$L = 10 \lg(p/p_0)^2$$

(where $p_0 = 2\,10^{-5}\,\mathrm{N/m^2}$ is the reference sound pressure) can be solved for the square of the sound pressure by calculating $10^{L/10}$ and, due to $10^{\lg(x)} = x$:

$$\left(\frac{p}{p_0}\right)^2 = 10^{\frac{L}{10}} . \tag{A.8}$$

Certainly, the physical sound pressure can be regained from the level.

A.3 Level Summation

Often, two levels have to be summarised to one. One simple example: two vehicles produce the individual level L_1 and L_2 at a given position. How large is the total sound level, if both vehicles are operated at the same time? Similar problems are very often found. The vehicles substitutionally stand for the radiation of two so-called incoherent signals. This means signals which do not contain the same frequencies. It would be a great coincidence, if the engines of two vehicles ran at exactly the same speed. It is nearly always justified to assume incoherent signals. Sound incidents, originating from independently operating technical devices and machines, and also speech signals are of course incoherent amongst each other. This is clearly not the case if the same cause is hidden behind the sound sources: electrical machines, for instance, which are supplied from the same network, are of course coherent and contain the same frequencies.

The simplest model for a signal containing two incoherent parts is given by a total signal, composed of two different frequencies:

$$p = p_1 \cos \omega_1 t + p_2 \cos \omega_2 t \, .$$

The squared root-mean-square which is generally given by

$$p_{\mathrm{eff}}^2 = \frac{1}{T} \int_0^T p^2(t) \mathrm{d}t \, , \tag{A.9}$$

thus results in

$$p_{\mathrm{eff}}^2 = \frac{1}{T} \int_0^T (p_1^2 \cos^2 \omega_1 t + p_2^2 \cos^2 \omega_2 t + 2 p_1 p_2 \cos \omega_1 t \cos \omega_2 t) \, \mathrm{d}t \, .$$

If, as already assumed, the frequencies ω_1 and ω_2 are not equal, the latter integral is much smaller than the two first parts, due to $\cos \omega_1 t \cos \omega_2 t = (\cos(\omega_1 - \omega_2)t + \cos(\omega_1 + \omega_2)t)/2$, and the remainder is

$$p_{\mathrm{eff}}^2 = \frac{1}{2} \left(p_1^2 + p_2^2 \right) = p_{\mathrm{eff},1}^2 + p_{\mathrm{eff},2}^2 \, . \tag{A.10}$$

The squared rms-value of the total signal is the sum of the individual squared rms-values.

More generally, the squared rms-value of a signal composed of N different frequencies is given by

$$p_{\mathrm{eff}}^2 = \sum_{i=1}^{N} p_{\mathrm{eff,i}}^2 \, . \tag{A.11}$$

The level summation is gained, using (A.11), by expressing all rms-values by levels (and by using the reference sound pressure $p_0 = 2\,10^{-5}\,\mathrm{N/m^2}$)

$$L_{\text{ges}} = 10 \lg p_{\text{eff}}^2 / p_0^2 = 10 \lg \sum_{i=1}^{N} p_{\text{eff,i}}^2 / p_0^2 = 10 \lg \sum_{i=1}^{N} 10^{L_i/10} \,. \tag{A.12}$$

Equation (A.12) is called 'law of level summation'. It states that the levels are in fact *not* summed, but the individual levels must be transformed to squared rms-pressure values, before they are added together to yield the total squared rms-value.

B

Complex Pointers

Appendix B serves two purposes:

- a short introduction into the definition of complex numbers and their calculation rules and
- an explanation how and why complex numbers are used to describe acoustic processes.

B.1 Introduction to Complex Pointer Arithmetics

Complex numbers can be regarded as points in a plane, where one of the two axes represents the straight line of real numbers (denoted by the x-axis, here). Complex numbers are usually depicted graphically (as in Fig. B.1) as a connecting line between the origin and the point; this line is called a 'pointer'.

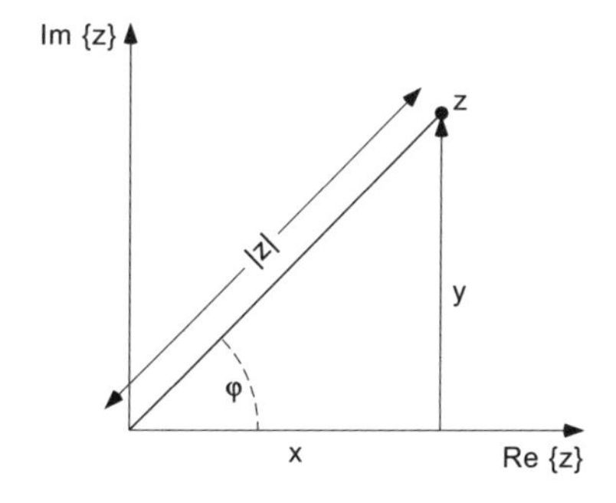

Fig. B.1. Representation of a complex number z in the complex plane

As in all number calculations the reason for their definition is given by the resulting operational possibilities. All calculation rules and operations for complex numbers can be derived by the purpose that an element, called 'j', should rotate an arbitrary pointer by 90° in a mathematically positive sense. Thus, the y-axis, normal to the x-axis, evolves from the real number line by a multiplication with j. A complex number is therefore composed of an 'unrotated' and a 'rotated' part:

$$z = x + jy \tag{B.1}$$

where x and y are real numbers.

The addition of complex numbers is performed in algebra as usual. In mathematics apples are added to apples and pears to pears, so 'unrotated' and 'rotated' elements are summed individually. Using $z_1 = x_1 + jy_1$ and $z_2 = x_2 + jy_2$, a sum of complex numbers results in

$$z_1 + z_2 = (x_1 + x_2) + j(y_1 + y_2) . \tag{B.2}$$

All real quantities can also have negative parts, thus (B.2) also contains the subtraction.

It follows from the definition 'multiplication with j rotates by 90°' that j multiplied by j results in the number -1:

$$j\,j = j^2 = -1 . \tag{B.3}$$

In the same way $j^3 = -j$, $j^4 = 1$, etc. The facts stated in (B.3) can also be written as

$$j = \sqrt{-1} . \tag{B.4}$$

Thus, j is also called the 'imaginary' unit. The interval on the x-axis of the complex number z (B.1) is called the *real part* of z, abbreviated by

$$x = \mathrm{Re}\{z\} . \tag{B.5}$$

The interval on the y-axis is called the *imaginary part* of z, abbreviated by

$$y = \mathrm{Im}\{z\} \tag{B.6}$$

which also yields

$$z = x + jy = \mathrm{Re}\{z\} + j\mathrm{Im}\{z\} . \tag{B.7}$$

It is necessary to keep in mind that y denotes a real number. The absolute value $|z|$ of a complex number is the length of the pointer. According to Pythagoras it follows from Fig. B.1 that

$$|z| = \sqrt{x^2 + y^2} . \tag{B.8}$$

The pointer can also be described by the absolute value and the angle, which is enclosed with the real axis. Due to

$$x = |z| \cos\varphi \tag{B.9}$$

and

$$y = |z| \sin \varphi \tag{B.10}$$

it is

$$z = |z|(\cos \varphi + j \sin \varphi) \; . \tag{B.11}$$

The absolute value of the pointer $\cos \varphi + j \sin \varphi$ is unity.

Using the polynomial series expansion

$$\cos \varphi = \sum_{n=0}^{\infty} (-1)^n \frac{\varphi^{2n}}{(2n)!} \quad \text{and} \quad \sin \varphi = \sum_{n=0}^{\infty} (-1)^n \frac{\varphi^{2n+1}}{(2n+1)!}$$

and

$$\mathrm{e}^{j\varphi} = \sum_{n=0}^{\infty} \frac{j^{2n} \varphi^{2n}}{(2n)!} + \sum_{n=0}^{\infty} \frac{j^{2n+1} \varphi^{2n+1}}{(2n+1)!} = \sum_{n=0}^{\infty} (-1)^n \frac{\varphi^{2n}}{(2n)!} + j \sum_{n=0}^{\infty} \frac{(-1)^n \varphi^{2n+1}}{(2n+1)!}$$

it can be stated that

$$\cos \varphi + j \sin \varphi = \mathrm{e}^{j\varphi} \tag{B.12}$$

is valid. Equation (B.12) is very useful, if multiplications and divisions are performed. It is

$$z_1 = |z_1| \mathrm{e}^{j\varphi_1}$$

and

$$z_2 = |z_2| \mathrm{e}^{j\varphi_2} \; .$$

It then follows that

$$z_1 z_2 = |z_1||z_2| \mathrm{e}^{j(\varphi_1 + \varphi_2)} \tag{B.13}$$

and likewise

$$z_1 / z_2 = \frac{|z_1|}{|z_2|} \mathrm{e}^{j(\varphi_1 - \varphi_2)} \; . \tag{B.14}$$

According to (B.13), the complex multiplication of z_1 with z_2 results in a rotation of z_1 by φ_2 and an ‘elongation’ of z_1 by the factor $|z2|$.

The square root of a complex number $z = |z| \mathrm{e}^{j\varphi}$ is given by

$$\sqrt{z} = \pm \sqrt{|z|} \mathrm{e}^{j\varphi/2} \; . \tag{B.15}$$

B.2 Using Complex Pointers in Acoustics

All transmission processes treated in this book have two basic attributes.

1. They are linear (for sufficiently small amplitudes); the principle of superposition can be utilised.
2. The structures are time invariant.

It can, for instance, be assumed that the speed of sound does not change for sound radiation into the free field or that a total sound field is the sum of the individual sound fields which in turn are caused by individual elements of a loudspeaker voltage. For walls it can reasonably enough be assumed that mass and bending stiffness do not change with time and that they react to sums of forces, according to the principle of superposition. Similar considerations can be made for all acoustic transducers treated in this book.

All linear and time-invariant transducers have in common that they always react to a sinusoidal excitation with a sine of the same frequency. If, for example, a wall is excited with a sine wave, the same tone is audible on the receiving side, which is attenuated in its amplitude and shifted in phase. A microphone, assuming a pressure of the form $p_0 \sin \omega_0 t$, yields an output voltage with the same signal form and the same frequency; and a sound pressure in an excited duct – to mention just a third example – always shows the same signal shape of the sine with space-dependent amplitude and phase.

For all discussed transducers, their 'output' $y(t)$

$$y(t) = |H(\omega)| x_0 \cos(\omega t + \varphi_{\mathrm{H}} + \varphi_{\mathrm{x}}) \tag{B.16}$$

is a version of the 'input'

$$x(t) = x_0 \cos(\omega t + \varphi_{\mathrm{x}}) \ , \tag{B.17}$$

which is phase-shifted by φ_{H} and 'amplified' by $|H|$. The output y denotes the vibration reaction (the membrane displacement, the output voltage, the sound pressure in the duct, etc.) and the input x denotes the excitation (the loudspeaker voltage, the force, etc). The fact that the signal shape is not changed during the transmission of pure tones is a special case only and by no means an obvious property of linear, time-invariant structures and the sine-shape. Other signal forms (triangle or square waves) are not transmitted with the same shape; even the simple time derivation in the radiation of volume velocity sources in Chap. 3.3 according to (3.14) results in a signal deforming, where, for instance, a triangle shape is transformed to a square.

The special case that a sine wave is always transmitted unchanged, leads to a very simple description: for pure tones the transmission is completely described by an 'amplification factor' $|H(\omega)|$ (which can also be smaller than unity or has a dimension) and by a phase shift φ_{H}.

It is suggested to describe the effects of transducers on their input signal by a complex multiplication. To make this possible, the real time signals of input and output have to be assigned complex amplitudes. This is done with the aid of the so-called time convention

$$f(t) = \mathrm{Re}\{\underline{f} e^{j\omega t}\} \tag{B.18}$$

where $\underline{f}$ is a complex amplitude, which is used for the description of a real process $f(t)$. Using

$$\underline{f} = |\underline{f}|e^{j\varphi_f} , \quad (B.19)$$

the time convention (B.18) maps the complex amplitude $\underline{f}$ to the real and observable reality

$$\underline{f}(t) = |\underline{f}| \cos(\omega t + \varphi_f) . \quad (B.20)$$

Equation (B.18) allows the description of sinusoidal signals by complex amplitudes, where, according to (B.19), the signal amplitudes are made equal to the absolute value of the complex amplitudes and the angle φ_f is made equal to the phase of the signal.

This enables the description of the transmission by a complex multiplication. The operation

$$\underline{y} = \underline{H}\, \underline{x} \quad (B.21)$$

includes the complete description of the transmission. It contains the real amplitude amplification $|H|$ as well as the phase shift φ_H. As a matter of fact, a check performed with

$$\underline{x} = |\underline{x}|e^{j\varphi_x}$$
$$\underline{y} = |\underline{y}|e^{j\varphi_y}$$
$$\underline{H} = |\underline{H}|e^{j\varphi_H}$$

for the time signals

$$y(t) = |x||H| \cos(\omega t + \varphi_x + \varphi_H)$$

shows that $y(t)$ is a version of the input which is amplified by $|H|$ and phase-shifted by φ_H.

The main advantage when using complex numbers is the use of clearer and simpler calculation rules. A signal sum, for example, given by two pure tones

$$x(t) = x_1 \cos(\omega t + \varphi_1) + x_2 \cos(\omega t + \varphi_2) \quad (B.22)$$

is a pure tone itself, with the total amplitude x_{tot} and a total phase ϕ_{tot}

$$x(t) = x_{tot} \cos(\omega t + \phi_{tot}) .$$

The actual values of x_{tot} and ϕ_{tot} would not be easy to calculate without complex numbers; the realization of the calculation needs certain skills and some experience using addition theorems. In contrast, using pointers, it is a breeze:

$$\underline{x}_{tot} = \underline{x}_1 + \underline{x}_2 \quad (B.23)$$

with

$$\underline{x}_1 = x_1 e^{j\varphi_1} \quad (B.24)$$

and

$$\underline{x}_2 = x_2 e^{j\varphi_2} . \quad (B.25)$$

Even the description of waves becomes very simple with the aid of the definitions of complex amplitudes mentioned earlier. The wave, travelling in positive direction, is described by

$$\underline{p} = p_0 \mathrm{e}^{-jkx} . \tag{B.26}$$

The only measurable quantity with a real-value is the time-space-characteristics of the pressure, given by

$$p(x,t) = \mathrm{Re}\{\underline{p}\mathrm{e}^{j\omega t}\} = \mathrm{Re}\{p_0 \mathrm{e}^{j(\omega t - kx)}\} = p_0 \cos(\omega t - kx) \tag{B.27}$$

(where $k = \omega/c$ is the wavenumber).

Generally, sound fields composed of pure tones can be described by complex spatial functions.

C

Solutions to the Practice Exercises

C.1 Practice Exercises Chapter 1

Problem 1

The pump level we are looking at is designated with L_P. The following applies to the total level L_{tot}

$$10^{L_{tot}/10} = 10^{5,5} = 10^{L_P/10} + 10^5 .$$

resulting in

$$10^{L_P/10} = 10^{5,5} - 10^5$$

or

$$L_P = 10lg(10^{5,5} - 10^5) = 53,3\,dB(A) .$$

Problem 2

Both octave levels given are $L(500\,Hz) = 81.1\,dB$ and $L(1000\,Hz) = 78,8\,dB$. For the non-weighted total level, $L(lin) = 83.1\,dB$ is true, and the A-weighted total level comes out to be $L(A) = 81.2\,dB$.

Problem 3

The octave levels increase by $3\,dB$ each time the center-frequency doubles. The non-weighted total level is found by

$$L_{tot} = 10lg(\sum_{i=0}^{N-1} 10^{(L+i)/10}) ,$$

where L refers to the third-octave level of the lowest third-octave. Using the sum formula for geometric series, we obtain

$$\sum_{i=0}^{N-1} 10^{i/10} = \frac{10^{N/10}-1}{10^{1/10}-1} \,.$$

The total level is

$$\Delta L = 10lg\frac{10^{N/10}-1}{10^{1/10}-1}$$

above the third-octave level L of the lowest third-octave. For N=10, this results in $\Delta L = 15.4\,dB$.

Problem 4

Octave levels are likewise equal to one another. They exceed the third-octave levels by $4.8\,dB$. The total level is $\Delta L = 10\,lg\,N$ above the third-octave level. For N=10, the total level exceeds the third-octave level by $10\,dB$.

Problem 5

The energy-equivalent continuous noise level pertaining to a lengthy time period of 16 hours of the train tracks alone is

$$L_{eq}(train) = L_{eq}(train, 2\,min) - 10\,lg\frac{120\,min}{2\,min} =$$

$$L_{eq}(train, 2\,min) - 17.8 = 57.2\,dB(A).$$

According to the law of level addition, the long-term level of the street and tracks together then amounts to $L_{eq}(total) = 59.2\,dB(A)$.

Problem 6

The energy-equivalent continuous noise level for the reference time period "daytime" can be found in

$$L_{eq}(day) = L_{eq}(30s) - 10lg\frac{5min}{30s} = 78 - 10 = 68\,dB(A)\,.$$

The energy-equivalent continuous noise level for the reference time period "nighttime" can be found in

$$L_{eq}(night) = L_{eq}(30s) - 10lg\frac{20min}{30s} - 10lg\frac{8hours}{4hours} =$$

$$= 78 - 16 - 3 = 59\,dB(A)\,.$$

Problem 7

The actual measured level L_m results in

$$L_m = 10lg(10^{L/10} + 10^{L_H/10}) ,$$

whereby L is the actual level exclusively of the event of interest in this case. L_H is the level which solely pertains to the background noise. If L_H is lower than L by ΔL,

$$L_m = 10lg(10^{L/10} + 10^{(L-\Delta L)/10}) = 10lg(10^{L/10}(1 + 10^{-\Delta L/10})) =$$

$$= 10lg(10^{L/10}) + 10lg(1 + 10^{-\Delta L/10}) = L + \Delta L_F$$

applies to $L_H = L - \Delta L$, whereby L_F refers to the margin of error in the measurement caused by the background noise level. This margin of error is defined by

$$\Delta L_F = 10lg(1 + 10^{-\Delta L/10})$$

The margin of error is

- for $\Delta L = 6\,dB$, $\Delta L_F = 1\,dB$,
- for $\Delta L = 10\,dB$, $\Delta L_F = 0.4\,dB$ and
- for $\Delta L = 20\,dB$, $\Delta L_F = 0.04\,dB$.

Problem 8

The last result in Problem 7 is solved for $10^{-\Delta L/10}$:

$$10^{-\Delta L/10} = 10^{\Delta L_F/10} - 1 .$$

By logging both sides and multiplying them by 10, we obtain

$$\Delta L = -10lg(10^{\Delta L_F/10} - 1) .$$

This results in the necessary intervals between noise disturbances for $\Delta L_F = 0.1\,dB$, resulting in $\Delta L = 16.3\,dB$.

An error of measurement of $1\,dB$ therefore requires a noise interval of $6\,dB$ (see Problem 7). A noise interval of $16.3\,dB$ becomes necessary for measurement errors of only $0.1\,dB$.

Problem 9

$$f_o = \sqrt[6]{2} f_u$$

applies to out-of-band frequencies f_o and f_u for sixth-octaves, since 6 sixth-octaves amount to an octave. For the center-frequencies,

$$f_m = \sqrt[2]{f_o f_u} = \sqrt[12]{2} f_u$$

applies, resulting in

$$\Delta f = f_o - f_u = (\sqrt[6]{2} - 1) f_u \, .$$

for the bandwidth of the pass band. The center-frequencies therefore adhere to the law

$$f_m^{(n+1)} = \sqrt[6]{2} f_m^{(n)} \, ,$$

whereby $f_m^{(n)}$ refers to the center-frequency of the n-th filter.

Problem 10

The octave level L_{oct} can be ascertained from the three third-octaves L_1, L_2 and L_3 based on the law of level addition

$$L_{oct} = 10 lg(10^{L_1/10} + 10^{L_2/10} + 10^{L_3/10}) \, ,$$

or

$$10^{L_{oct}/10} = 10^{L_1/10} + 10^{L_2/10} + 10^{L_3/10} \, .$$

This leads to

$$10^{L_3/10} = 10^{L_{okt}/10} - 10^{L_1/10} - 10^{L_2/10} \, ,$$

and consequently

$$L_3 = 10 lg(10^{L_{oct}/10} - 10^{L_1/10} - 10^{L_2/10}) \, ,$$

which can be used to check the 'uncertain' third-octave level L_3.

C.2 Practice Exercises Chapter 2

Problem 1

The functions f_1, f_2 and f_3 fulfill the wave equation, whereas f_4 does not constitute a solution to the given partial differential equation.

The following example can be used to prove this:

$$\frac{\partial^2 f_1}{\partial t^2} = -\frac{1}{(t+x/c)^2} ,$$

$$\frac{\partial^2 f_1}{\partial x^2} = -\frac{1}{c^2}\frac{1}{(t+x/c)^2} ,$$

and therefore

$$\frac{\partial^2 f_1}{\partial x^2} = \frac{1}{c^2}\frac{\partial^2 f_1}{\partial t^2}$$

the wave-equation is satisfied.

Problem 2

At equal pressure and an equal specific heat ration, the squares of the sound velocities of various gases have an inverse relationship to their densities:

$$\frac{c^2(gas)}{c^2(air)} = \frac{\varrho(air)}{\varrho(gas)}$$

Therefore, we obtain the sound velocity for hydrogen $c = 1290\,m/s$, for oxygen $c = 323\,m/s$, and for carbon dioxide $c = 275\,m/s$.

The elasticity modules are all identical based on $E = \varrho\, c^2 = \kappa p_0$ ($p_0 =$ static pressure) and result in $E = 1.4\ 10^5\,kg/ms^2 = 1.4\ 10^5\,N/m^2$.

The wavelengths for a frequency of $1000\,Hz$ are

- $\lambda = 1.29\,m$ in hydrogen
- $\lambda = 0.323\,m$ in oxygen
- $\lambda = 0.275\,m$ in carbon dioxide and
- $\lambda = 0.34\,m$ in air.

Problem 3

- Speed of sound $= 10^{-4}\,m/s = 0.1\,mm/s$.
- Particle displacement $= 0.16\ 10^{-6} m$ at $100\,Hz$,
- Particle displacement $= 0.016\,10^{-6} m$ at $1000\,Hz$.
- Sound intensity $= 4\ 10^{-6}\,W/m^2$, sound power $= 16\ 10^{-6}\,W$.
- Sound pressure level $= 20\,lg(2\ 10^3) = 66\,dB =$ intensity level.
- Power level $=$ intensity level $+\ 10\,lg(S/1m^2) = 72\,dB$.

Problem 4

Based on equations(2.75) through (2.79), the following is true for N number of equal surface sections S_i:

$$\frac{P}{P_0} = \frac{S_i}{1m^2} \sum_{i=1}^{N} \frac{p_{eff,i}^2}{p_0^2} = \frac{S_i}{1m^2} \sum_{i=1}^{N} 10^{L_i/10}$$

Consequently, the following is true for the power level

$$L_w = 10 \lg \left(\frac{S_i}{1m^2} \sum_{i=1}^{N} 10^{L_i/10}\right) = 10 \lg \left(\sum_{i=1}^{N} 10^{L_i/10}\right) + 10 \lg \left(\frac{S_i}{1m^2}\right).$$

Therefore, we obtain the solution $L_w = 96.7dB(A)$.

Problem 5

The Mach numbers are 0.0408 ($50\,km/h$); 0.0817 ($100\,km/h$) and 0.1225 ($150\,km/h$), resulting in the following receiver frequencies for source and receiver moving away from one another:

Receiver at rest in fluid	Source at rest in fluid
$960.8\,Hz$	$959.2\,Hz$
$924.5\,Hz$	$918.3\,Hz$
$890.1\,Hz$	$877.5\,Hz$

If the source and receiver are moving toward one another, we obtain the following receiver frequencies:

Receiver at rest in fluid	Source at rest in fluid
$1042.5\,Hz$	$1040.8\,Hz$
$1089.0\,Hz$	$1081.7\,Hz$
$1139.6\,Hz$	$1122.5\,Hz$

Problem 6

The sound velocity of nitrogen (N_2) is c $= 349\,m/s$ at $293\,K$ ($= 20^0\,C$), in oxygen, c $= 326.5\,m/s$. Therefore, the sound travels slightly faster in 'stale air.'

Problem 7

Pressure-space dependency:

$$p = p_0 \sin kx$$

Speed space dependency:

$$v = \frac{jp_0}{\varrho c} \cos kx$$

Resonance equation:

$$\cos kl = 0$$

(l=length) or $kl = \pi/2 + n\pi$ with $n = 0; 1; 2...$ and

$$f = (\frac{1}{4} + \frac{n}{2})\frac{c}{l}\,.$$

The first three resonance frequencies are $340\,Hz$, $1020\,Hz$ and $1700\,Hz$.

Problem 8

The wavelengths in water are at $500\,Hz$: $\lambda = 2.4\,m$, at $1000\,Hz$: $\lambda = 1.2\,m$, at $2000\,Hz$: $\lambda = 0.6\,m$ and at $4000Hz$: $\lambda = 0.3\,m$.

Problem 9

Field and energy dimensions in the interval $0 < t - x/c < T$ (otherwise, outside of this interval for $t - x/c < 0$ as well as for $t - x/c > T$ all dimensions are equal to zero):

$$v(x,t) = v_0 \sin \frac{\pi(t - x/c)}{T}$$

$$p(x,t) = \varrho_0 c\, v_0 \sin \frac{\pi(t - x/c)}{T}$$

$$I(x,t) = \frac{p^2(x,t)}{\varrho_0 c} = \varrho_0 c\, v_0^2 \sin^2 \frac{\pi(t - x/c)}{T}$$

$$E(x,t) = \frac{p^2(x,t)}{\varrho_0 c^2} = \varrho_0\, v_0^2 \sin^2 \frac{\pi(t - x/c)}{T}$$

The energy produced by the source E_Q is – according to the amount of energy exerted – is completely absorbed by the field. Therefore,

$$E_Q = S \int_0^\infty E(x,t)\, dx = \varrho_0\, v_0^2 S \int_{ct}^{c(t+T)} \sin^2 \frac{\pi(t - x/c)}{T}\, dx.$$

Using $\sin^2 x = (1 - \cos 2x)/2$ this becomes

$$E_Q = \frac{\varrho_0\, v_0^2}{2} S \int_{ct}^{c(t+T)} 1 - \cos\frac{2\pi(t - x/c)}{T}\, dx = \frac{\varrho_0\, v_0^2}{2} ScT$$

The energy produced by the source, at a speed of $v_0 = 0.01\, m/s = 1\, cm/s$, with a wave guide diameter of $10\, cm$ (with cylindrical surface area $S = \pi\, 0.05^2\, m^2$ and signal length of $T = 0.01s$) is thus $E_Q = 1.6\, 10^{-6}\, Ws$.

Problem 10

The following applies to the relationship between measured intensity I_M and actual intensity I according to the given problem

$$10lg\frac{I_M}{I} = -2 \;\; (-3)\,,$$

and therefore

$$\frac{\sin k\Delta x}{k\Delta x} = 10^{-0,2} \;\; (10^{-0,3}) = 0.63 \;\; (0,5)\,.$$

One obtains from the resulting table for the given sinc function $\sin k\Delta x/k\Delta x$ the values $k\Delta x = 0.5\pi$ ($k\Delta x = 0.6\pi$). Based on these values, the function always complies with $\Delta x/\lambda < 1/4 = 0.25$ ($\Delta x/\lambda < 0.3$). Using $\Delta x = 2.5\, cm$ we obtain the frequency threshold for these measurements of $f = 3.4\, kHz$ ($f = 4.1\, kHz$).

Problem 11

The phase tolerance φ must comply with

$$\frac{\varphi}{2\pi} < \frac{f\Delta x}{5c}\frac{p_p}{p_s}\,.$$

A reasonable phase tolerance within these bounds is obtained using $f = 100\, Hz$, $\Delta x = 5\, cm$ and $p_s/p_p = 10$

$$\frac{\varphi}{2\pi} < 0,3\, 10^{-3} \;\; (0,3\, 10^{-4})\,.$$

This corresponds to the phase error 0.11° (0.011°) – expressed in degrees.

Problem 12

In the absence of pre-amplification using a microphone, a police car approaches a receiver. The relationship between transmission frequency f_Q and receiver frequency f_{E1} can therefore be described by

$$f_{E1} = \frac{f_Q}{1 - |M|}$$

(M=Mach number). Subsequently, the source moves away from the receiver

$$f_{E2} = \frac{f_Q}{1 + |M|} \,.$$

The result of both equations is

$$\frac{f_{E2}}{f_{E1}} = \frac{1 - |M|}{1 + |M|} \,,$$

or, expressed as a Mach number,

$$|M| = \frac{1 - \frac{f_{E2}}{f_{E1}}}{1 + \frac{f_{E2}}{f_{E1}}} \,.$$

Using $f_{E1} = 555.6\,Hz$ and $f_{E2} = 454.6\,Hz$, one obtains $M = 0.1$, corresponding to a velocity of $U = |M|c = 34\,m/s = 122.4\,km/h$. We see that at the source frequency f_Q and based on

$$f_Q = f_{E1}(1 - |M|)$$

and

$$f_Q = f_{E2}(1 + |M|),$$

$f_Q = 500\,Hz$ is true.

C.3 Practice Exercises Chapter 3

Problem 1

The A-weighted sound power level may not exceed a maximum possible value of 95.3 $dB(A)$ (pump on solid, reflective platform).

Problem 2

The change of the volume flow over time only has a non-zero value in the time intervals $2 < t/T_F < 3$ and $7 < t/T_F < 8$, at which $dQ/dt = const = Q_0/T_F$. The resulting gradient of the sound pressure square is shown in Fig. C.1. For the value given in the graph p_A,

$$p_A = \frac{\varrho Q_0}{4\pi r T_F}$$

applies. For $Q_0 = 1\,m^3/s$ and $T_F = 0,01\,s$, we obtain $p_A = 0.95\ N/m^2$ ($p_A = 0,3\ N/m^2$ for $T_F = 0.0316\,s$ and $p_A = 0.095\ N/m^2$ for $T_F = 0.1\,s$), using $\varrho = 1.2\,kg/m^3$. When defining the sound pressure level

$$L = 20\,lg\,p_A/p_0\;,$$

using $p_0 = 2\ 10^{-5}\,N/m^2$, the level corresponds to $L(0.01\,s) = 93.6\,dB$ within both non-zero-sound-pressure time intervals ($L(0.0316\,s) = 83.6\,dB$ and $L(0,1s) = 73.6\,dB$).

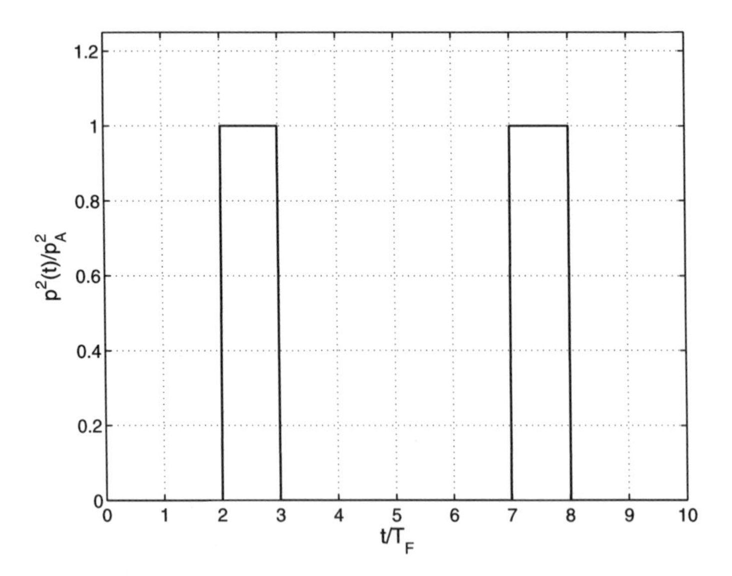

Fig. C.1. Time dependency of the sound pressure square

Problem 3

The power level is $117\,dB$. The power is therefore equal to $P = 10^{11.7}\,P_0 = 0.5\,W$. The efficiency coefficient therefore has a value of 0.01.

Problem 4

The sound pressure in the far-field is

$$p_{fern} = \frac{j\omega\varrho\, b\, l\, v_0}{2\pi R} e^{-jkR} \frac{\sin^2\left(\frac{kl}{4} sin\vartheta_N\right)}{\left(\frac{kl}{4} sin\vartheta_N\right)^2} = \frac{j\omega\varrho\, b\, l\, v_0}{2\pi R} e^{-jkR} \frac{\sin^2\left(\pi\frac{l}{2\lambda} sin\vartheta_N\right)}{\left(\pi\frac{l}{2\lambda} sin\vartheta_N\right)^2} .$$

The directivity pattern exists in segments of the $sin^2(\pi u)/(\pi u)^2$ function, which is constrained by $u = \pm l/(2\lambda)$.

Problem 5

Based on $R = 5\,l$, $5 >> l/\lambda$ and $5\,l >> \lambda$ apply to both remaining far-field conditions

- a) $\lambda > l$ results from the first condition, $\lambda < l$ from the second. Values can therefore only be measured for $f = 340\,Hz$ ($680\,Hz$, $170\,Hz$).
- b) $2.5 > l/\lambda$ or $\lambda > l/2.5$ results based on the first condition. $\lambda < 2.5\,l$ is obtained based on the second condition. Measurable values can therefore be obtained with the frequency interval $f = 136\,Hz$ to $f = 850\,Hz$ (in the interval $272 Hz$ to $1700\,Hz$; in the interval $68\,Hz$ to $425\,Hz$).

Problem 6

The level at a distance of $20\,m$ is measured at $3\,dB$ less than that which is measured at a distance of $10\,m$. It is therefore $81\,dB(A)$.

To calculate the level at a distances of $200\,m$ and $400\,m$ respectively, one must first find the power level of the line source (mounted on a non-reflective platform). It is $L_w = 119\,dB(A)$. Next, we find the pressure level of the point sound source at a distance of $200\,m$, which is $65\,dB(A)$, leaving a remaining level of $59\,dB(A)$ measured at a distance of $400\,m$.

Problem 7

The basic idea here is that the continuous phase shift from source to source produces a virtual rotating source in a given period, emanating in a spiral wave at lower frequencies, (see the following images which depict the solution to this problem).

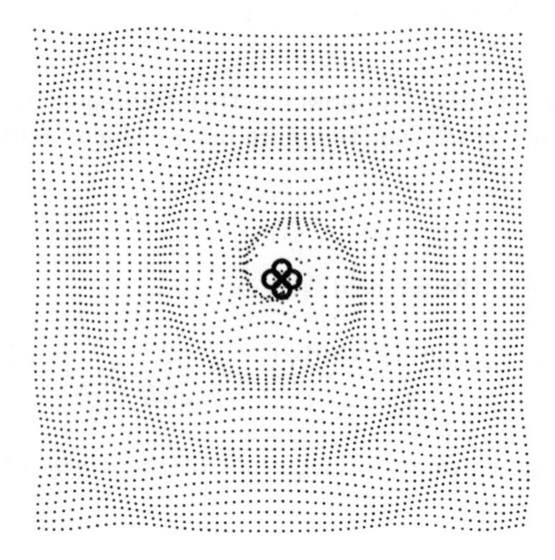

Fig. C.2. Sound field of the sources for $2h/\lambda = 0.25$

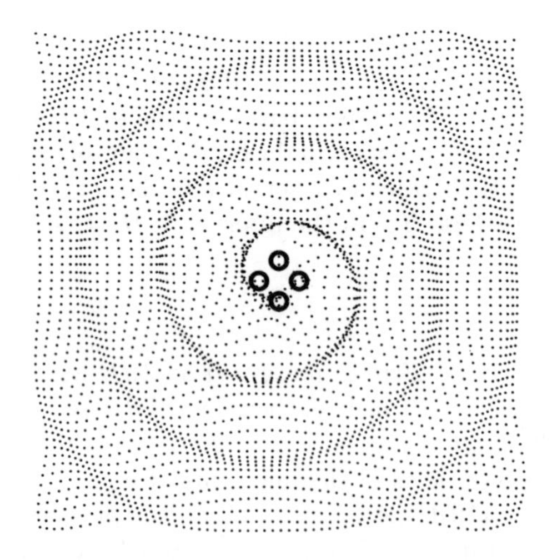

Fig. C.3. Sound field of the sources for $2h/\lambda = 0.5$

Locally distributed interferences occur at higher frequencies.

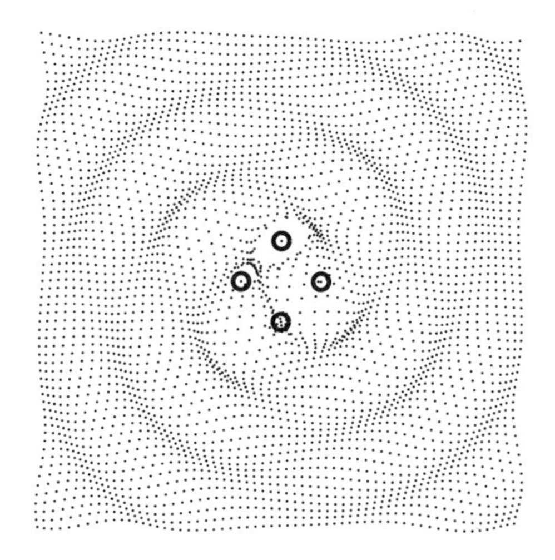

Fig. C.4. Sound field of the sources for $2h/\lambda = 1$

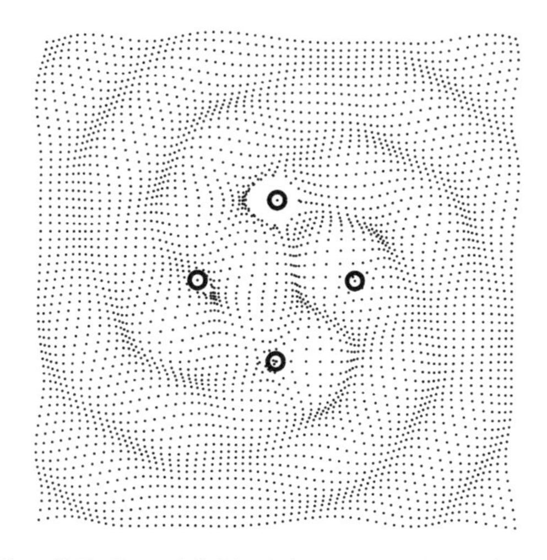

Fig. C.5. Sound field of the sources for $2h/\lambda = 2$

A routine written in Matlab, which can produce a short film of such an event, is given below:

```
clear all
xmax=3.;
abstand=0.5;
dx=2*xmax/60;
dy=dx;

for ix=1:1:61
x=-xmax + (ix-1)*dx;
    for iy=1:1:61
        y=-xmax + (iy-1)*dy;
        x1=x-abstand/2;
        x2=x+abstand/2;

 [phi1,r1]=cart2pol(x1,y);
 [phi2,r2]=cart2pol(x2,y);
 [phi3,r3]=cart2pol(x,y-abstand/2);
 [phi4,r4]=cart2pol(x,y+abstand/2);
 p = j*(exp(-j *2*pi*r1)./sqrt(r1) - exp(-j *2*pi*r2)./sqrt(r2));
 p = p + exp(-j *2*pi*r3)./sqrt(r3) - exp(-j *2*pi*r4)./sqrt(r4);

 [phi1,r1]=cart2pol(x1,y+0.01);
 [phi2,r2]=cart2pol(x2,y+0.01);
 [phi3,r3]=cart2pol(x,y-abstand/2+0.01);
 [phi4,r4]=cart2pol(x,y+abstand/2+0.01);
 py = j*(exp(-j *2*pi*r1)./sqrt(r1) - exp(-j *2*pi*r2)./sqrt(r2))
 py = py + exp(-j *2*pi*r3)./sqrt(r3) - exp(-j *2*pi*r4)./sqrt(r4

 [phi1,r1]=cart2pol(x1+0.01,y);
 [phi2,r2]=cart2pol(x2+0.01,y);
 [phi3,r3]=cart2pol(x+0.01,y-abstand/2);
 [phi4,r4]=cart2pol(x+0.01,y+abstand/2);
 px = j*(exp(-j *2*pi*r1)./sqrt(r1) - exp(-j *2*pi*r2)./sqrt(r2))
 px = px + exp(-j *2*pi*r3)./sqrt(r3) - exp(-j *2*pi*r4)./sqrt(r4

 vx(iy,ix) = (p-px)*10;
 vy(iy,ix) = (p-py)*10;

 if r1<0.1
     vx(iy,ix)=0;
     vy(iy,ix)=0;
 end
 if r2<0.1
     vx(iy,ix)=0;
     vy(iy,ix)=0;
 end
```

```
if r3<0.1
    vx(iy,ix)=0;
    vy(iy,ix)=0;
end
if r4<0.1
    vx(iy,ix)=0;
    vy(iy,ix)=0;
end
end
end

r=0.1;
dphi=2*pi/99.
for i=1:1:100
    phi=(i-1)*dphi;
    x=r*cos(phi);
    y=r*sin(phi);
xrefl(i)=x-abstand/2.;
yrefl(i)=y;

xrefl2(i)=x+abstand/2;
yrefl2(i)=y;

xrefl3(i)=x;
yrefl3(i)=y+abstand/2;

xrefl4(i)=x;
yrefl4(i)=y-abstand/2;
end

npoints=61;
M=particlequadru(vx,vy,npoints,xmax,
xrefl,yrefl,xrefl2,yrefl2,xrefl3,yrefl3,xrefl4,yrefl4);


% now the movie is shown:

function[M]=particlequadru(vx,vy,npoints,xmax,
xrefl,yrefl,xrefl2,yrefl2,xrefl3,yrefl3,xrefl4,yrefl4);

xmin=-xmax; ymin=xmin; ymax=xmax;

frames=50; scale=1;

point_style = 'k.'; [x,y]=meshgrid(1:npoints,1:npoints);
```

```
command ='axis off';

v=[vx,vy]; [vmaxval,vmaxpos]=mmax(abs(v));
[vmax,temppos]=max(real(v(vmaxpos)));
phase=angle(v(vmaxpos(temppos)));

dx=real(vx*exp(j*phase)); dy=real(vy*exp(j*phase));

figure('Position',[50 20 500 500],'color',[1 1 1]);

%Scale movie
answer='yes'; while answer=='yes'
   cla;
   plot(x+dx*scale,y+dy*scale,point_style,'Markersize',5)
   axis([-1 npoints+2 -1 npoints+2])
   hold on
   axis equal;
   axis manual;
   eval(command);
   answer=questdlg('Scale Particle Movement', ...
      'Continue Scaling?', ...
      'yes','no','yes');
   if strcmp(answer,'no'),break,end
   prompt={'Multiplication Factor:'};
   title='Scale Particle Movement';
   lineNo=1;
   def={num2str(scale)};
   scale=inputdlg(prompt,title,lineNo,def);

   if isempty(scale),break,end;
   scale=str2num(char(scale));
end scale
%Plot single frames of movie and combine them
M=moviein(frames); for k=0:frames-1;
   cla;
   axis equal;
   axis manual;
   eval(command);
   dx=real(vx*exp(j*2*pi/frames*k));
   dy=real(vy*exp(j*2*pi/frames*k));
   plot(x+dx*scale,y+dy*scale,point_style,'Markersize',5)

   % reflectors
```

```
   ax=(npoints-1)/(xmax-xmin);
   bx=1-ax*xmin;
   ay=(npoints-1)/(ymax-ymin);
   by=1-ay*ymin;
   xm=ax*xrefl + bx;
   ym=ay*yrefl + by;
   hp=plot(xm,ym);
   set(hp,'LineWidth',3.,'Color','k')

   xm=ax*xrefl2 + bx;
   ym=ay*yrefl2 + by;
   hp=plot(xm,ym);
   set(hp,'LineWidth',3.,'Color','k')

   xm=ax*xrefl3 + bx;
   ym=ay*yrefl3 + by;
   hp=plot(xm,ym);
   set(hp,'LineWidth',3.,'Color','k')

   xm=ax*xrefl4 + bx;
   ym=ay*yrefl4 + by;
   hp=plot(xm,ym);
   set(hp,'LineWidth',3.,'Color','k')

   M(:,k+1) = getframe;
end

%Play movie

answer='yes'; while answer=='yes'
   answer=questdlg('', ...
      'Play it again ?', ...
      'yes','no','yes');
   if strcmp(answer,'no'),break,end
    movie(M,8,30);  % 8 times, 30 pics/sec
end
```

```
function [m,i]=mmax(a)
%MMAX Matrix Maximum Value.
% MMAX(A) returns the maximum value in the matrix A.
% [M,I] = MMAX(A) in addition returns the indices of
% the maximum value in I = [row col].
```

```
% D.C. Hanselman, University of Maine, Orono ME 04469
% 1/4/95
% Copyright (c) 1996 by Prentice Hall, Inc.

if nargout==2,  %return indices
   [m,i]=max(a);
   [m,ic]=max(m);
   i=[i(ic) ic];
else,
   m=max(max(a));
end
```

Problem 8

By reforming the equation $\sin^2 x = 0.5 - 0.5 \cos 2x$, one can find the volume flow of the source within the time interval $0 < t < T_D$:

$$Q(t) = \frac{\pi}{2} \frac{S\xi_0}{T_D} \sin\left(\pi \frac{t}{T_D}\right)$$

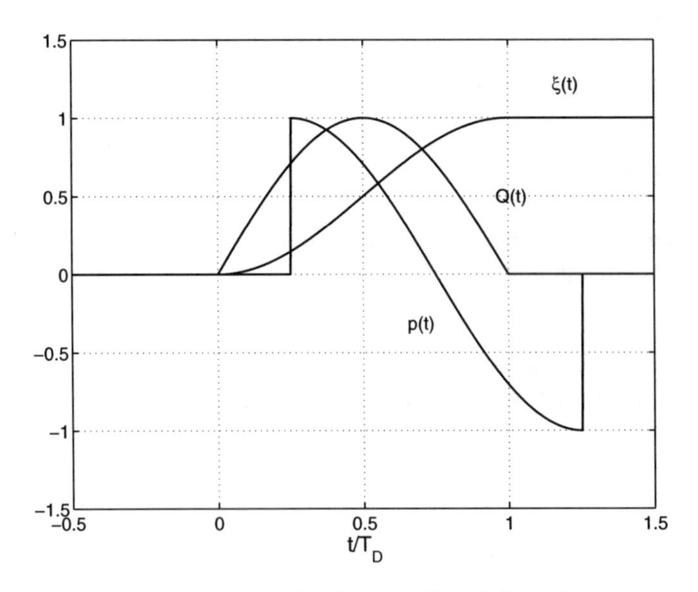

Fig. C.6. Depiction of the solution to Problem 8. Signal dependencies ξ, Q, and p for $r/c = T_D/4$, each divided by its corresponding maximum.

According to the law of outdoor volume sources, we subsequently obtain

$$p(r,t) = \frac{\pi}{8}\frac{\varrho}{r}\frac{S\xi_0}{T_D^2}\cos\left(\pi\frac{t-r/c}{T_D}\right)$$

for the time interval $0 < t < T_D$. For $t < 0$ and for $t > T_D$ is $p = 0$. All other time dependencies are presented in the image above.

Problem 9

In the far-field, the sound pressure resulting from the circular piston exists in

$$p_{\text{fern}}(R,\vartheta,\varphi) = \frac{j\omega\varrho v_0}{2\pi R}e^{-jkR}\int_{-b}^{b}\int_{-\sqrt{b^2-x_q^2}}^{\sqrt{b^2-x_q^2}} e^{jk(x_Q\sin\vartheta\cos\varphi + y_Q\sin\vartheta\sin\varphi)}\mathrm{d}y_Q\mathrm{d}x_Q\,,$$

representing an integration over the circular surface. Due to rotation symmetry (the sound field must be independent of the circumferential angle φ), it is only necessary to consider half of the plane. We therefore choose $\varphi = 0$ for our half-plane object of observation. This simplifies the far-field pressure to

$$p_{\text{fern}}(R,\vartheta,\varphi) = \frac{j\omega\varrho v_0}{2\pi R}e^{-jkR}\int_{-b}^{b}\int_{-\sqrt{b^2-x_q^2}}^{\sqrt{b^2-x_q^2}} e^{jkx_Q\sin\vartheta}\mathrm{d}y_Q\mathrm{d}x_Q\,.$$

Because the integrand of y_q is independent, we obtain

$$p_{\text{fern}}(R,\vartheta,\varphi) = \frac{j\omega\varrho v_0}{\pi R}\,e^{-jkR}\int_{-b}^{b} e^{jkx_Q\sin\vartheta}\,\sqrt{b^2-x_q^2}\,\mathrm{d}x_Q =$$

$$= \frac{j\omega\varrho v_0}{\pi R}e^{-jkR}\int_{-b}^{b}\left[\cos\left(kx_Q\sin\vartheta\right) + j\sin\left(kx_Q\sin\vartheta\right)\right]\sqrt{b^2-x_q^2}\,\mathrm{d}x_Q\,,$$

or, due to symmetry, we obtain

$$p_{\text{fern}}(R,\vartheta,\varphi) = \frac{2j\omega\varrho v_0}{\pi R}e^{-jkR}\int_{0}^{b}\cos\left(kx_Q\sin\vartheta\right)\sqrt{b^2-x_q^2}\,\mathrm{d}x_Q\,.$$

The substitution $u = x_q/b$ exhibits

$$p_{\text{fern}}(R,\vartheta,\varphi) = \frac{2j\omega\varrho v_0 b^2}{\pi R}e^{-jkR}\int_{0}^{1}\cos\left(kbu\sin\vartheta\right)\sqrt{1-u^2}\,\mathrm{d}u\,.$$

The integral contained therein is listed in Gradshteyn's table of integrals (refer to Gradshteyn, I.S.; Ryzhik,I.: Table of Integrals, Series, and Products. Academic Press, New York and London 1965; page 953, Nr. 8.411.8 with $\nu = 1$. Note: $\Gamma(3/2)\Gamma(1/2) = \pi/2$ applies). Using this, we obtain

$$p_{\text{fern}}(R, \vartheta, \varphi) = \frac{j\omega\varrho v_0 b^2}{R} e^{-jkR} \frac{J_1(kb\sin\vartheta)}{kb\sin\vartheta} ,$$

whereby J_1 refers to the first-order Bessel function. Check your calculations for accuracy by defining a point on the z-axis ($\vartheta = 0$). Use $J_1(x)/x = 1/2$ for small values of x and you should get the correct result given in eq.(3.69).

Notably, the directivity patterns of transmitters and receivers can possess remarkably similar characteristics (see Chapter 11.2 for a reference of the directivity patterns pertaining to this exercise).

Problem 10

It is only necessary to find the volume flow Q of the plate vibration to solve this problem. In the first approximation, the short-waved arrays can be considered to be omnidirectional volume sources. For the volume flow, one obtains

$$Q = v_0 \int_0^{l_y}\int_0^{l_x} \sin(n\pi x/l_x) \sin(m\pi y/l_y)\, dx\, dy =$$

$$= v_0 \frac{l_x l_y}{nm\pi^2} (\cos(n\pi) - 1)(\cos(m\pi) - 1) .$$

The volume flow is only a non-zero value if the orders n and m are both uneven numbers, resulting in

$$Q = v_0 \frac{4 l_x l_y}{nm\pi^2} .$$

The sound pressure in the far-field consequently comes out to be

$$p_{fern} = \frac{j\omega\varrho_0 Q}{2\pi R} e^{-jkR} .$$

Problem 11

- For $b/\lambda = 3.5$, we obtain the 3 pressure nodes at the locations $z/\lambda = 5.625$; 2.0625 and 0.5417 .
- For $b/\lambda = 4.5$, we obtain the 4 pressure nodes at the locations $z/\lambda = 9.625$; 4.0625; 1.875 and 0.5313 .
- For $b/\lambda = 5.5$, we obtain the 5 pressure nodes at the locations $z/\lambda =$ 14.625; 6.5625; 3.5417; 1.7813 and 0.525.

Problem 12

In the far-field (the only scenario where it makes sense to consider directivity patterns), we obtain

$$p_{far} = p_1[1 + \frac{Q_2}{Q_1} e^{jkhsin\vartheta_N}]$$

(p_1 is the field of the source Q_1 by itself), thus resulting in

$$p_{far} = p_1[1 - (1 + jkh)e^{jkhsin\vartheta_N}] \,.$$

With $e^z \cong 1 + z$ for $|z| << 1$, one obtains

$$p_{far} = p_1[1 - (1 + jkh)(1 + jkhsin\vartheta_N)] \cong -jkh(1 + sin\vartheta_N)p_1 \,,$$

whereby the summand of the second order (with $(kh)^2$) can be disregarded. The directivity pattern is kidney-shaped, characterized by a single nick at $\vartheta_N = -90°$. A half-plane depiction is shown in Fig. C.7. Of course, the directivity pattern is symmetrical, implying

$$p_{far}(180° - \vartheta_N) = p_{far}(\vartheta_N) \,.$$

It should be noted that this does not violate the general principle that states that for low frequencies, the sound field can be approximated using the sum of all sources at a single location. The source sum and sound field are namely both equal to zero in the first approximation.

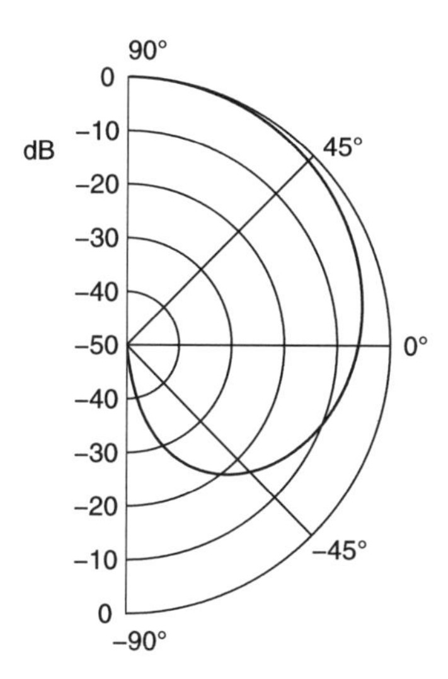

Fig. C.7. Directivity pattern of the source pair

C.4 Practice Exercises Chapter 4

Problem 1

The coincidence frequencies are

- for cement (gypsum) plates 8 cm thick ($c_L = 2000\,m/s$), 397 Hz,
- window panes 4 mm thick, 3241 Hz and for
- a door panel made of oak wood ($c_L = 3000\,m/s$) 25 mm thick, 847 Hz.

Problem 2

The somewhat cryptic expression k $\sqrt{B/m'}$ can be more easily expressed in terms of longitudinal wave velocity and beam thickness. With $B = Eh^3b/12$ and $m' = \varrho\, hb$, the expression becomes

$$\frac{B}{m'} = \frac{Eh^2}{12\varrho} = h^2 c_L^2/12\,.$$

This allows us to find the resonance frequencies for aluminum using $c_L = 5200\,m/s$

- for the mounted beam $f = 0.45\,m^2 h\,c_L/l^2$ ($m = 1, 2, 3, .$) and
- for the bilaterally supported beam $f = 0.45\,(m + 0,5)^2 h\,c_L/l^2$ ($m = 1, 2, 3, .$).

Next, we can find the resonance frequencies for the beam length of 1 m

- for the mounted beam of f/Hz = 11.7; 46.8; 105.3; 187.2; 292.5 and
- for the bilaterally supported beam of f/Hz = 26.3; 73.1; 143.3; 236.9; 353.9.

The resonance frequencies are four times as great for the beam lengths of 50 cm.

Problem 3

Here, too, the bending resistance and plate mass can be expressed in terms of longitudinal wave velocity and plate thickness:

$$f = 0.45[(\frac{n_x}{l_x})^2 + (\frac{n_y}{l_y})^2]h\,c_L$$

By varying $n_x = 1;2$ and $n_y = 1;2$, we obtain the following resonance frequencies:

- Window pane 4 mm thick with dimensions 50 cm and 100 cm: 44.1 Hz; 70.6 Hz; 149.9 Hz and 176.4 Hz.
- 10 cm thick plaster wall ($c_L = 2000 m/s$) with dimensions 3 m by 3 m: 20 Hz; 50 Hz; 50 Hz (double resonance) and 80 Hz.
- 2 mm enforced steel plate with dimensions 20 cm by 25 cm: 184.5 Hz; 400.5 Hz; 522 Hz and 738 Hz.

Problem 4

The beam is resting in the interval of $0 \leq x \leq l$ and is bilaterally supported at $x = 0$ ($v = 0$ and $dv/dx = 0$ at $x = 0$) and suspended at $x = l$ ($d^2v/dx^2 = 0$ and $d^3v/dx^3 = 0$ at $x = l$). Since no symmetry exists, the speed in this case must have four solution functions:

$$v = A \sin k_B x + B\, sh\, k_B x + C \cos k_B x + D\, ch\, k_B x$$

Due to $v(0) = 0$, $D = -C$ applies, and based on $dv/dx = 0$ at $x = 0$, it follows that $B = -A$. Thus resulting in the remaining solutions for the speed:

$$v = A[\sin k_B x - sh\, k_B x] + C[\cos k_B x - ch\, k_B x]$$

Based on the constraint $d^2v/dx^2 = 0$ at $x = l$, it follows that

$$v = A[\sin k_B l + sh\, k_B l] + C[\cos k_B l + ch\, k_B l]\,,$$

due to $d^3v/dx^3 = 0$, and subsequently,

$$v = A[\cos k_B l + ch\, k_B l] - C[\sin k_B l - sh\, k_B l]\,.$$

Resonance, in other words, vibration in the absence of excitation, occurs when the determinants of the last two equations disappear:

$$[\sin k_B l + sh\, k_B l][\sin k_B l - sh\, k_B l] + C[\cos k_B l + ch\, k_B l]^2 = 0$$

This results in the resonance frequencies expressed as

$$\cos k_B l = -\frac{1}{ch\, k_B l}.$$

This equation can be easily solved visually, as shown in Fig. C.8.

The lowest resonance frequency can obviously be accurately derived using $k_B l/2\pi = 0.3$, which is analogous to $k_B l/ = 0.6\pi$ (the exact value is $k_B l/ = 0.597\pi$. This difference, of course, being negligible). All higher resonances can be described as $\cos k_B l = 0$ and therefore, $k_B l = 3\pi/4 + n\pi$.

The modal forms exist in

$$v = [\sin k_B x - sh\, k_B x] + \frac{\cos k_B l + ch\, k_B l}{\sin k_B l - sh\, k_B l}[\cos k_B x - ch\, k_B x]\,,$$

which can be substituted for the eigen values mentioned above $k_B l$. The first four modes are depicted in the following Fig. C.9.

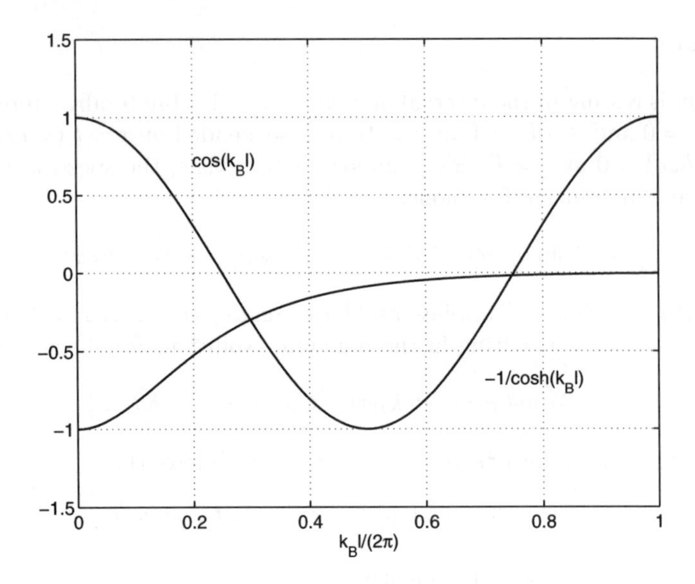

Fig. C.8. Graphical solution to the eigen value equation

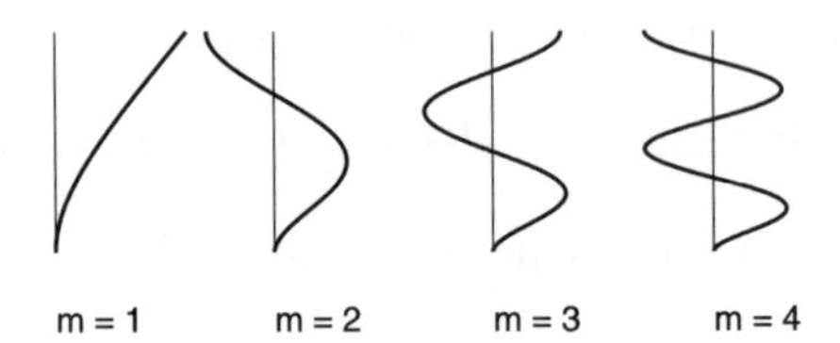

Fig. C.9. Vibration modes of a beam supported below and suspended at the top

Problem 5

The onset of the beam speed to the right of the point force for $x > 0$ contains a wave travelling to the right as well as a near-field diminishing in the x-direction

$$v = v_0(e^{-jk_B x} + Ae^{-k_B x}) .$$

The field must be symmetrical, meaning

$$v(-x) = v(x) .$$

As it vibrates, the beam is unable to crease under the point force, leading to the conclusion that at $x = 0$,

$$\beta = \frac{\partial v}{\partial x} = 0$$

is true. It follows then that $A = -j$ and therefore

$$v = v_0(e^{-jk_Bx} - je^{-k_Bx})$$

applies for the complex pointer of the beam speed. The time and spatial dependencies become, based on the time convention,

$$v(x,t) = v_0 Re\{(e^{-jk_Bx} - je^{-k_Bx})e^{j\omega t}\}\,.$$

The beam speed curves sought after in this problem are given in both figures below, separated by half-periods for more visual clarity. The displacement curves are equal due to $\xi = v/j\omega$, just phase-shifted in respect to the speed by 90°.

The speed at $x = 0$ is derived from

$$v(x,t) = v_0 Re\{(1-j)e^{j\omega t}\}\,.$$

With

$$1 - j = \sqrt{2}\,e^{-j\frac{\pi}{4}},$$

it becomes

$$v(x,t) = v_0\sqrt{2}\cos\left(\omega t - \frac{\pi}{4}\right).$$

The speed at $x = 0$ reaches its first maximum for all $t/T = 1/8$.

By integrating over time, the displacement $\xi(x = 0)$ subsequently becomes

$$\xi(x,t) = \frac{\sqrt{2}v_0}{\omega}\sin\left(\omega t - \frac{\pi}{4}\right).$$

The displacement at $x = 0$ therefore reaches its first maximum at $t/T = 3/8$.

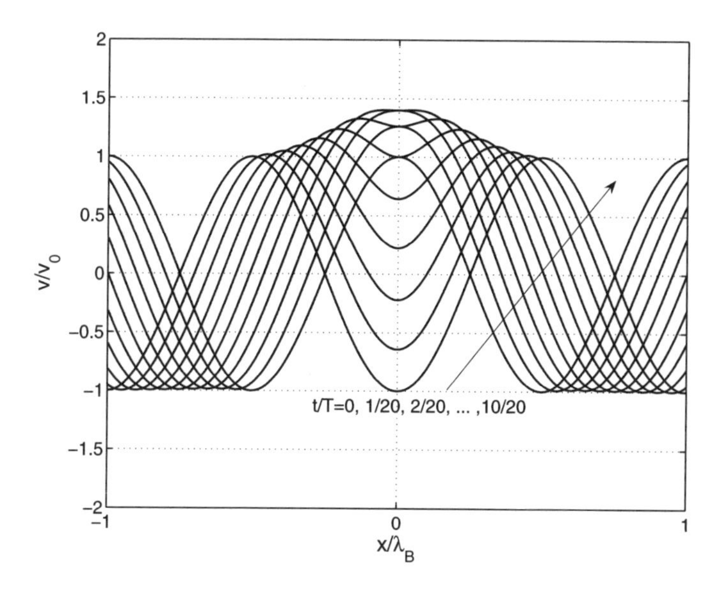

Fig. C.10. Beam speed for Problem 4

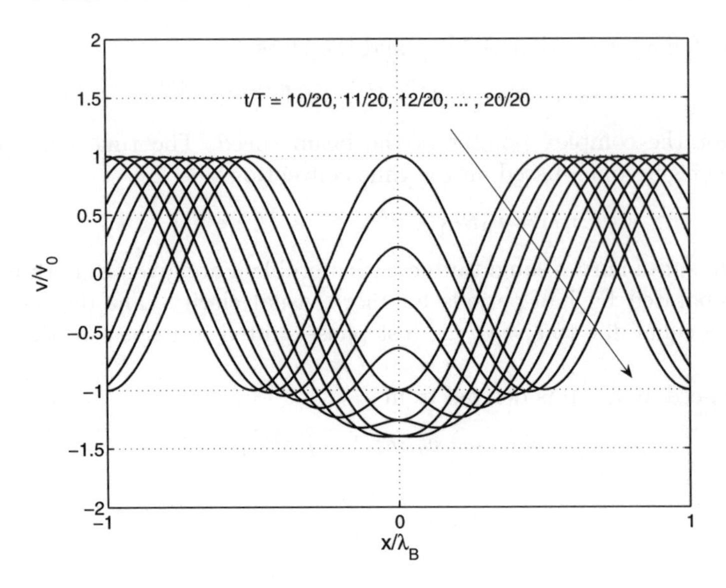

Fig. C.11. Beam speed for Problem 4

Problem 6

The first step to solving this problem is the same as the previous problem; it entails applying the homogenous bending wave equation

$$\frac{\partial^4 v}{\partial x^4} - k_B^4 v = 0\,,$$

which only takes into consideration a wave (travelling away from the source) and a near-field (diminishing away from the source)

$$v(x) = v_W e^{-jk_B x} + v_N e^{-k_B x}\,.$$

This equation only applies to $x > 0$. Of course, in this scenario, a symmetrical reverberation field $v(-x) = v(x)$ develops. Since the beam likewise does not bend in the middle, initially, the

$$\partial v(x)/\partial x = 0$$

must also applyfor $x = 0$. It then follows that $v_N = -jv_W$, simplifying the statement, once again, to

$$v(x) = v_W\left[e^{-jk_B x} - je^{-k_B x}\right].$$

The bending force to the right of the point force in the beam $F(x \rightarrowtail 0)$ then becomes

$$F(x \rightarrowtail 0) = \frac{B}{j\omega}\frac{\partial^3 v}{\partial x^3} = v_W \frac{2Bk_B^3}{\omega} = v_W \frac{2\omega m'}{k_B} \, ,$$

using $k_B^4 = m'\omega^2/B$ in the last step. The bending force to the left of the initial point force is the same. Because the sum of all three forces equal zero, it follows that

$$v_W = \frac{F_0 k_B}{4\omega m'} \, .$$

The speed is obtain by substitution

$$v(x) = \frac{F_0 k_B}{4\omega m'}[e^{-jk_B x} - je^{-k_B x}] = \frac{F_0}{4k_B\sqrt{m'B}}[e^{-jk_B x} - je^{-k_B x})] \, .$$

The following

$$\frac{F_0 k_B}{4\omega m'} = \frac{F_0 k_B^2}{4k_B \omega m'} = \frac{F_0 \omega}{4k_B \omega m'}\sqrt{\frac{m'}{B}} = \frac{F_0}{4k_B\sqrt{m'B}}$$

is applied in the last step in order to express the answer to this problem in terms of its frequency dependency.

C.5 Practice Exercises Chapter 5

Problem 1

The nuclear magnetic resonance tomograph weighs $10^4\, N$ (accounting for the acceleration of the earth at $g = 10\, m/s^2$), the flat bearing's surface is $0.36\, m^2$, the pressure exerted upon this surface is therefore $p_{stat} = 2.8\ 10^4\ N/m^2 = 0.028\, N/mm^2$. The e-module must be 20 times as great as, thus requiring $E = 0.56\, N/mm^2 = 56\ 10^4\ N/m^2$.

The layer thickness needed can be best construed from the static depression

$$x_{stat} = Mg/s$$

(M=mass) where the ratio s/M can be expressed as the the resonance frequency

$$x_{stat} = \frac{g}{\omega_{res}^2} .$$

Based on $x_{stat} = d/20$, it follows that

$$d = \frac{20g}{\omega_{res}^2} \approx \frac{g}{2f_{res}^2} .$$

A layer thickness of $2.6\, cm$ is therefore needed to achieve a resonance frequency of $14\, Hz$.

Problem 2

The decrease in level reduction is determined by

$$R_E = 20 \lg \frac{s_F}{s}.$$

For a $6\, dB$-level decrease, the spring stiffness of the foundation must be double the stiffness of the mounting itself $s = ES/d = 7.8\ 10^6\ N/m$. To achieve a $10\, dB$ reduction, the stiffness of the foundation must be 3.16 times greater, and for $20\, dB$, the foundation stiffness must be 10 times greater than the stiffness of the bearing.

Problem 3

When operating the tomograph (at Index T), the level difference was equal to 'receiver level - transmitter level':

f/Hz	$L_E(T)$ = transmitter level - receiver level dB
500	-33.3
1000	-33.0
2000	-31.1

Running the loudspeakers produced the following level difference:

f/Hz	$L_E(L)$ = receiver level - transmitter level dB
500	-39.9
1000	-41.2
2000	-41.4

As can be seen here, no significant sound level in the receiver room can be attributed to the sound transport in air; rather, it originates primarily from structure-borne transmission. Implementing an elastic bearing is therefore useful for reducing structure-borne transmission pathways. The effects of such measures are mitigated, however, when the structure-borne sound transmission has been reduced so much that airborne sound transmission takes precedence. The maximum possible level reduction due to elastic decoupling can therefore be summarized using $\Delta L = L_E(T) - L_E(L)$ producing the following results:

f/Hz	$L_E(T) - L_E(L)$ dB
500	6.6
1000	8.2
2000	10.3

The following would thus describe the level scenario in the receiver room when the tomograph is in operation:

f/Hz	Receiver level after elastic decoupling
500	25.4
1000	23.2
2000	20.1

The non-weighted net level would therefore come out to 28.2 dB.

Problem 4

The resonance frequency is multiplied by a factor of 1.22 (1.12: four times the mass and 1.06: eight times the mass) compared to a fixed foundation.

Problem 5

The attenuated resonance frequency $\omega_{0\eta}$ can be found by setting the first derivative of

$$|\frac{x}{F}|^2 = \frac{1}{(1-\omega^2/\omega_0^2)^2 + (\eta\omega/\omega_0)^2}$$

to zero in terms of ω. Insodoing, we obtain

$$\omega_{0\eta} = \omega_0\sqrt{1-\eta^2/2}\,,$$

where ω_0 represents the non-attenuated resonance, that is, the resonance for $\eta = 0$. It is of course always the attenuated resonance frequency which factors into real-life measurements. The critical attenuation is, of course, $\eta = \sqrt{2}$.

C.6 Practice Exercises Chapter 6

Problem 1

The following three images depict the solution to this problem. Represented here is the last duct segment, of which its length is equal to one wavelength. The reflector is located at $x = 0$. Central to our investigation is a sinusoid spatial dependency which travel from left to right as their amplitudes decrease and increase (in other words, a gradual fade-in and fade-out). The enveloping waves surrounding the curves and thus the spatial dependencies of the rms-values can easily be discerned in the graph. With increasing reflection factor, one can see how the larger the minima and maxima are, the farther apart they are from one another.

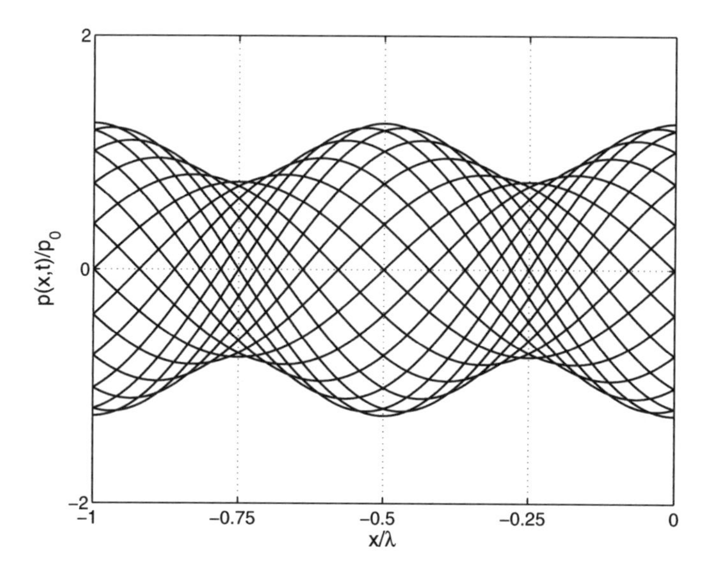

Fig. C.12. Spatial dependency of the sound pressure for specific times $t = nT/20(n = 0, 1, 2, 3, ..., 19)$ and at a reflection factor of $r = 0.25$ (p_0=amplitude of the impinging wave alone).

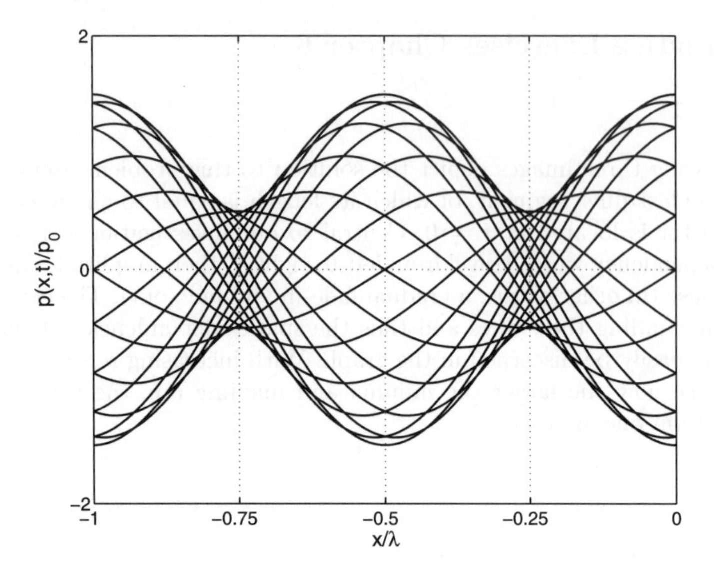

Fig. C.13. Spatial dependency of the sound pressure for specific times $t = nT/20(n = 0, 1, 2, 3, ..., 19)$ and for a reflection factor of $r = 0.5$ (p_0=amplitude of the impinging wave alone).

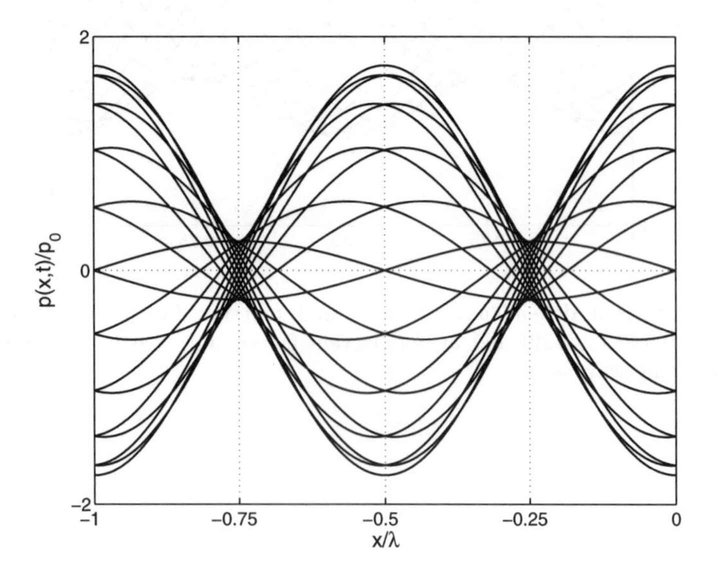

Fig. C.14. Spatial dependency of the sound pressure for specific times $t = nT/20(n = 0, 1, 2, 3, ..., 19)$ and for a reflection factor of $r = 0.75$ (p_0=amplitude of the impinging wave alone).

Problem 2

The frequency response function of the absorption coefficient and wall impedance of the structure consisting of an 8 cm–thick sheet made up of wood fiber and cement mixture, constructed in front of a non-reflective termination:

Frequency/Hz	α	Re$\{z/\varrho c\}$	Im$\{z/\varrho c\}$
200	0.52	1.52	-2.31
300	0.83	1.23	-0.98
400	0.96	1.40	-0.27
500	0.79	2.03	1.05
600	0.64	3.55	1.22
700	0.52	5.44	0.59
800	0.47	4.76	-2.71
900	0.53	4.05	-2.24
1000	0.57	2.32	-2.29
1100	0.64	2.09	-1.87
1200	0.79	1.54	- 1.16
1300	0.85	1.94	-0.69
1400	0.84	2.24	-0.41
1500	0.87	2.05	-0.34
1600	0.87	2.40	-0.86
1700	0.76	2.31	-1.10
1800	0.69	1.92	-1.61

Problem 3

The sought-after values are:

$z/\varrho c$	α	φ	$\lvert x_{min}\rvert/\lambda$
1+j	0.8	63.4^0	0.162
2+j	0.8	26.6^0	0.213
1+2j	0.5	45.0^0	0.188
3+j	0.706	12.5^0	0.233
1+3j	0.308	33.7^0	0.203

Problem 4

The following values can be calculated using the values $c = 340\,m/s$ and $\varrho = 1.21\,kg/m^3$:

for $\Xi = 10^4\ Ns/m^4$ and $\kappa = 2$

f/Hz	$z/\varrho c$	α
200	0.84 - j 2.58	0.33
400	0.93 - j 0.97	0.78
800	1.49 + j 0.10	0.96
1600	1.05 - j 0.57	0.93

for $\Xi = 10^4\ Ns/m^4$ and $\kappa = 1$

f/Hz	$z/\varrho c$	α
200	0.82 - j 2.71	0.31
400	0.86 - j 1.24	0.69
800	1.01 - j 0.42	0.96
1600	1.37 - j 0.49	0.94

for $\Xi = 2\ 10^4\ Ns/m^4$ and $\kappa = 2$

f/Hz	$z/\varrho c$	α
200	1.65 - j 2.73	0.46
400	1.73 - j 1.30	0.76
800	1.92 - j 0.78	0.84
1600	1.42 - j 0.56	0.92

for $\Xi = 2\ 10^4\ Ns/m^4$ and $\kappa = 1$

f/Hz	$z/\varrho c$	α
200	1.62 - j 2.85	0.43
400	1.61 - j 1.52	0.71
800	1.56 - j 0.97	0.83
1600	1.25 - j 0.74	0.89

Problem 5

The length-specific drag resistance is only significant in how it relates to the thickness of fluids; for water, a Ξ 826 times greater than for air must be set. The layer thickness must must be assumed to grow proportionally to the wavelengths and therefore increase along with the sound velocities. The layer thicknesses for water must therefore be 3.53 as great as for air.

Problem 6

Based on $\alpha = 1$ in the resonance frequency, an absorber materials with $\Xi d = \varrho c$ must be selected. The required surface mass can be ascertained by half-value width required as expressed in

$$m'' = \frac{\Xi d + \varrho c}{2\pi \Delta f} = \frac{\varrho c}{\pi \Delta f} .$$

Subsequently, the mass linings required can be calculated as $1.02\,kg/m^2$ (for $f_{res} = 250\,Hz$ and therefore $\Delta f = 125\,Hz$), $0.73\,kg/m^2$ (for $f_{res} = 350\,Hz$ and therefore $\Delta f = 175\,Hz$) and $0.51\,kg/m^2$ (for $f_{res} = 500\,Hz$ and therefore $\Delta f = 250\,Hz$) (with $\varrho c = 400\,kg/m^2 s$).

The cavity depth can be calculated by

$$a = \frac{\varrho c^2}{\omega_{res}^2 m''} = \frac{\varrho c^2}{4\pi^2 f_{res}^2 m''} ,$$

or, by substituting the above equation for the mass lining

$$a = \frac{c \Delta f}{4\pi f_{res}^2} = \frac{c}{8\pi f_{res}} ,$$

with $\Delta f / f_{res} = 0.5$ in the last step. This gives us $a = 5.4\,cm\,(f_{res} = 250\,Hz)$, $a = 3.9\,cm\,(f_{res} = 350\,Hz)$ and $a = 2.7\,cm\,(f_{res} = 500 Hz)$.

Problem 7

Based on

$$b = \frac{3}{5} \sigma_L \frac{m''}{\varrho} ,$$

we obtain $b = 1.26\,cm$ for $\sigma_L = 0.05$ and $b = 2.53\,cm$ for $\sigma_L = 0.1$.

Problem 8

Let the mid-point distance between two holes in a quadratic perforated grid pattern be l_a. It follows that the surface covering can be described as

$$\sigma_L = \frac{\pi b^2}{l_a^2} ,$$

and thereby

$$l_a = b \sqrt{\frac{\pi}{\sigma_L}} .$$

The factor $l_a/b = \sqrt{\frac{\pi}{\sigma_L}}$ is 7.93 for $\sigma_L = 0.05$ (5.6 for $\sigma_L = 0.1$).

Problem 9

The lowest cut-on frequency can be calculated from the larger cross-section measurements, resulting in 2429 Hz (1889 Hz).

Problem 10

Both diagrams illustrate the solution to this problem.

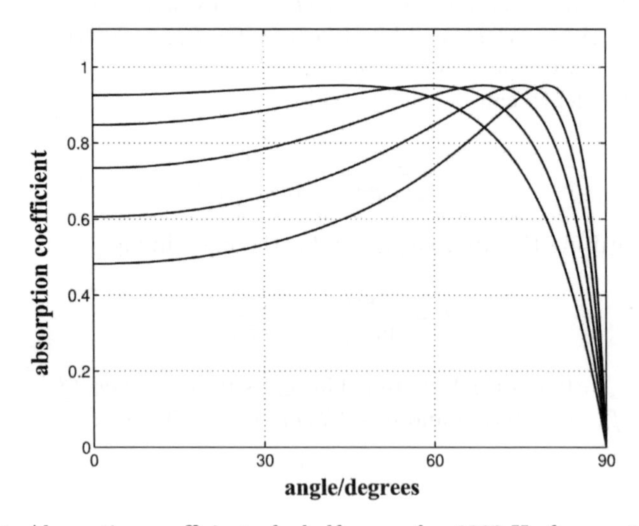

Fig. C.15. Absorption coefficient of a half-space $f = 1000\,Hz$ for $\kappa = 1$, 2, 4, 8 and 16 (from top to bottom).

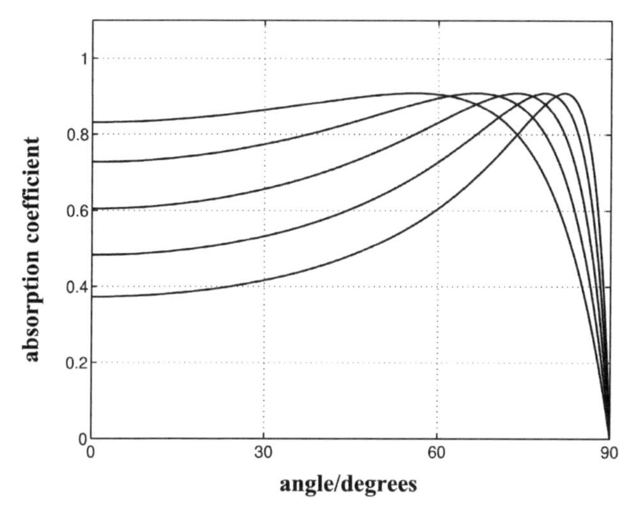

Fig. C.16. Absorption coefficient of a half-space $f = 500\,Hz$ for $\kappa = 1$, 2, 4, 8 and 16 (from top to bottom).

Conclusion: large structural factors exhibit directionally selective absorption characteristics.

Problem 11

The maximum hole covering is $\sigma_L = \pi/4$ (refer to Problem 6, this time using $b = l_a/2$).

Problem 12

See Chapter 9.2.1. for the answer to this problem.

Problem 13

The lowest cut-on frequencies of ducts with a circular cross-section are $4012\,Hz$, $2006\,Hz$ and $1337\,Hz$ for a diameter of $5\,cm$, $10\,cm$ and $15\,cm$ when using $c = 340\,m/s$ and the approximation equation $f_1 = 0.59\,c/d$ (d = diameter).

Problem 14

Based on eq.(6.49):

$$D(d) = \frac{4,35\sigma}{\sqrt{\kappa}} \frac{\Xi d}{\varrho c}$$

one obtains

$$\Xi = \frac{\sqrt{\kappa}}{4.35\sigma} \frac{D(d)}{d} \varrho c \,.$$

With $\varrho c = 400\, kg/m^2 s$, $\sigma = 0.95$, $\kappa = 2$ and $D(d)/d = 1/1\, cm$, it follows that $\Xi = 13.7\, 10^3\, Ns/m^4 = 13.7\, Rayl/cm$.

C.7 Practice Exercises Chapter 7

Problem 1

The A-weighted sound pressure level is $89\,dB(A)$.

The simplest way to proceed is to find the absorption areas and then the power levels using the Sabine formula. Insodoing, we obtain the following results:

f/Hz	A/m^2	$L_{P,thirdoctave}/dB$
400	6.5	80.5
500	6.8	82.9
630	8.0	82.2
800	9.1	83.6
1000	9.1	88
1250	9.3	87.9

The A-weighted sound power level is $92.5\,dB(A)$.

With the absorption area of the living room $A = 20.4\,m^2$, the sound pressure level is $L = 85.4\,dB(A)$ for the diffuse field in the living room.

Problem 2

Equivalent absorption areas and hall radii before renovation:

f/Hz	A/m^2	r_H/m
500	14.2	0.54
1000	16.8	0.59
2000	19.2	0.63

Equivalent absorption areas, reverberation times and level reduction ΔL after renovation when omitting the factor of covering any previous sound insulation with the new fixtures (in other words, we assume an initially reflective ceiling):

f/Hz	A/m^2	T/s	$\Delta L/dB$
500	80.2	0.7	7.5
1000	104.8	0.5	8
2000	129.2	0.4	8.3

Problem 3

f/Hz	$\Delta A/m^2$	α
500	4.6	0.46
630	5.4	0.54
800	7.4	0.74
1000	9.2	0.92

Problem 4

The first ten resonance frequencies are $28.3\,Hz$, $34\,Hz$, $42.5\,Hz$, $44.3\,Hz$, $51.1\,Hz$, $54.4\,Hz$, $56.7\,Hz$, $61.4\,Hz$, $66.1\,Hz$ and $68\,Hz$, if $c = 340\,m/s$ is used.

As in Chapter 1, the third-octave bandwidth is $\Delta f = 0.23\,f_m$. Consequently, the number of resonances in the third-octave is equal to

$$\Delta M = 0.92\pi V \left(\frac{f_m}{c}\right)^3 ,$$

leading to the following:

- $\Delta M = 71$ for the third-octave center frequency of $200\,Hz$,
- $\Delta M = 565$ for the third-octave center frequency of $400\,Hz$ and
- $\Delta M = 4518$ for the third-octave center frequency of $800\,Hz$.

Problem 5

Of course, it is N times the surface compared to simply operating one source. The equilibrium is maintained as long as N times the outflow is guaranteed for N times the inflow.

Problem 6

The sound pressure rms-value is $2\,N/m^2$, the energy density has a value of $2.94\ 10^{-5}\,Ws/m^3$ using $\varrho c = 400\,kg/m^2s$ and $c = 340\,m/s$. The net stored energy is $14.7\ 10^{-3}\,Ws$. The bulb would last for $0.0147\,s$, meaning it would only briefly flicker on.

Problem 7

Start by using the power balance analysis for both rooms to solve this problem. For room 1, the power flowing into the sum of the source power P_Q and the power flowing back to the room through the door opening in room 2 $p_2^2 S_T/4\varrho c$

(p_2: rms-value of the sound pressure in room 2). The power loss for room 1 can be ascertained by calculating the absorption area A_1 and the area of the door opening S_T, resulting in $p_1^2(A_1 + S_T)/4\varrho c$ (p_1: rms-value of the sound pressure in room 1). Therefore, in a state of equilibrium, we obtain

$$P_Q + \frac{p_2^2 S_T}{4\varrho c} = \frac{p_1^2(A_1 + S_T)}{4\varrho c} .$$

The power supplied to room 2 by room 1 is $p_1^2 S_T/4\varrho c$. The sum of room 2's absorption surface and the door opening constitutes the power loss, which is consequently $p_2^2(A_2 + S_T)/4\varrho c$. The balance is therefore summarized in

$$\frac{p_1^2 S_T}{4\varrho c} = \frac{p_2^2(A_2 + S_T)}{4\varrho c} .$$

The level difference between the two rooms can be ascertained using the aforementioned equation

$$\Delta L = L_1 - L_2 = 10 \lg (1 + A_2/S_T) .$$

By solving the power balance for room 2 for p_2^2 and the result substituted into the power balance for room 1, we get

$$P_Q = \frac{p_1^2 A_1}{4\varrho c}[1 + \frac{S_T}{A_1}\{1 - \frac{S_T}{S_T + A_2}\}] .$$

Thus it follows that the definition equation for the sound pressure level in room 1 is

$$L_1 = L_P - 10 \lg \frac{A_1}{m^2} - 10 \lg [1 + \frac{S_T}{A_1}\{1 - \frac{S_T}{S_T + A_2}\}] + 6 .$$

It is clear that room volumes are irrelevant in finding the answer to this problem. Using the values given in the problem, we obtain $\Delta L = 9.5\, dB$ and $L_1 = 87.6\, dB$.

Problem 8

First, find the equivalent absorption areas and then the A-weighted sound pressure levels in third octaves $L_{A,thirdoctave}$ based on the Sabine formula. The results are listed in the following table. The A-weighted net level of $L = 77,1\, dB(A)$ is obtained from the third octave sound pressure levels after first A-weighting, then using the level addition law.

f/Hz	A/m^2	$L_{A,thirdoctave}/dB$
400	7.24	70.6
500	8.15	69.7
630	9.31	68.4
800	10.87	69.8
1000	13.04	68.8
1250	13.04	68.4

C.8 Practice Exercises Chapter 8

Problem 1

f/Hz	L_S - L_E/dB	A/m^2	R/dB
400	30.2	11.4	29.6
500	32.8	12.7	31.8
630	37.0	13.4	35.7
800	39.0	14.3	37.4
1000	44.4	14.3	42.8
1250	42.6	15.2	40.8

Problem 2

With $\varrho = 1.21\,kg/m^3$ and $c = 340\,m/s$ one obtains the following resonance frequencies:

- for the surface mass $12.5\,kg/m^2$ at $5\,cm$ distance $f_{res} = 75.3\,Hz$,
- for the surface mass of $25\,kg/m^2$ at $5\,cm$ distance $f_{res} = 53.2\,Hz$,
- for the surface mass of $12,5\,kg/m^2$ at $10\,cm$ distance $f_{res} = 53,2\,Hz$ and
- for the surface mass of $25\,kg/m^2$ at $10\,cm$ distance $f_{res} = 37.6\,Hz$.

Problem 3

The very thin steel sheet metal results in extremely high coincidence frequencies of $25.4\,kHz$ (with longitudinal wave speed of steel being $c_L = 5000\,m/s$). At such given frequencies, steel can therefore be considered bendable. Given the density of steel $\varrho_{steel} = 7800\,kg/m^3$, the surface mass can be calculated at $m'' = 3.9\,kg/m^2$. If the critical impedance of air is defined by $\varrho\,c = 400\,kg/m^2s$, the transmission loss at $100\,Hz$ is only $R = 6.7\,dB$ (and increases by $6\,dB$ with each doubling frequency, therefore becoming $12.7\,dB$ at $200\,Hz$, $18.7\,dB$ at $400\,Hz$, etc).

Problem 4

The critical coincidence frequency of the wall is quite low at $53.4\,Hz$. The wall is therefore considered rigid. The transmission loss of the wall is therefore $R = 57.9\,dB$ with a mass of $m'' = 805\,kg/m^2$ at $200\,Hz$ and increases by $7.5\,dB$ per octave. It is therefore $R = 65.4\,dB$ at $400\,Hz$, and $R = 72,9\,dB$, at $800\,Hz$ and so forth.

Problem 5

The transmission coefficient of the entire construction is derived from adding the powers which are transmitted through the partial areas of the surface

$$S_{total}\tau_{total} = S_{window}\tau_{window} + S_{wall}\tau_{wall}\,.$$

Thus,

$$\tau_{total} = \frac{S_{window}}{S_{total}}\tau_{window} + \frac{S_{wall}}{S_{total}}\tau_{wall},$$

and finally, based on $R = -10\,lg(\tau)$,

$$R_{total} = -10\,lg(\frac{S_{window}}{S_{total}}10^{-R_{window}/10} + \frac{S_{wall}}{S_{total}}10^{-R_{wall}/10}).$$

For a window with area $3\,m^2$ in an entire wall with an area of $18\,m^2$, we obtain the total transmission loss of $R_{total} = 37.8\,dB$.

For the window which takes up half the entire area, it is $R_{total} = 33\,dB$.

Problem 6

The weighted transmission loss is $R_w = 45\,dB$.

C.9 Practice Exercises Chapter 9

Problem 1

For chamber silencers, the 3-dB width of the insertion loss peak and the center frequency of the peak are the same. Therefore, we must find the center frequency of 400 Hz. As long as the maximum insertion loss in the center frequency amounts to 10 dB, the conditions of this problem are fulfilled.

If the center frequency meets the condition $l/\lambda = 1/4$, the resulting chamber length l based on $\lambda = 0.85\,m$ is therefore $l = 0.213\,m$. The maximum of the insertion loss factor then fulfills

$$10^{R_{max}/10} = 1 + \frac{1}{4}(\frac{S_1}{S_2} - \frac{S_2}{S_1})^2 .$$

Since $R_{max} = 10\,dB$ are required, it follows that

$$\frac{S_1}{S_2} - \frac{S_2}{S_1} = 6 .$$

The solution to this quadratic equation in S_2/S_1 results in $S_2/S_1 = 6.16$ (or, of course $S_2/S_1 = 1/6.16$). Since the surfaces S_2 and S_1 behave like the squares of the diameters d_2 and d_1, $d_2 = 2.48\,d_1 = 12.4\,cm$ must also apply.

Problem 2

First, we can approximate the imaginary component of the wave number at low frequencies with a real impedance z as in

$$k_i = \frac{1}{2}\frac{\varrho c}{zh} ,$$

whereby, as outlined in the section 'Arbitrary change in cross-sectional area,' h must be substituted by the ratio of circumference to cross-sectional area S/U:

$$k_i = \frac{1}{2}\frac{\varrho c}{zS/U}$$

For the rectangular cross-section (the length of one side a) $S/U = a/4$, for the cylindrical cross-section with radius b, $S/U = b/2$ applies. It thus follows that

$$D_a = 8.7 k_i a = 17.4 \frac{\varrho c}{z}$$

and

$$D_b = 8.7 k_i b = 8.7 \frac{\varrho c}{z} .$$

Using these deductions, we obtain $D_a = 17.4\,dB$ and $D_b = 8.7\,dB$ for $z = \varrho c$. For $z = 2\varrho c$ we arrive at $D_a = 8.7\,dB$ and $D_b = 4.3\,dB$.

Problem 3

The maximum obtainable damping, based on the condition $z = 0$ in the resonance, is $D_h(max) = 13.5\,dB$.

In order to measure the required mass lining, it is first necessary to determine that the cavity depth d of the resonator in the relevant frequency range of approximately $50\,Hz$ remains much smaller than a quarter wavelength of about $1.7\,m$. For this reason, one can approximate the resonance frequency at

$$\omega_0^2 = \frac{\varrho c^2}{m'' d}.$$

The resulting mass lining m'' is $2.83\,kg/m^2$ (at cavity depth of $50\,cm$; at $100\,cm$: $1.42\,kg/m^2$).

Problem 4

The impedance of the non-damped resonator can be described as follows:

$$\frac{z}{\varrho\,c} = j\frac{\omega_0 m''}{\varrho\,c}\left[\frac{\omega}{\omega_0} - \frac{\omega_0}{\omega}\right]$$

($\omega_0 =$ resonance frequency). The factor $\frac{\omega_0 m''}{\varrho\,c}$ is 2.22 (at cavity depth of $50\,cm$; at $100\,cm$: 1.11).

If the frequency is reduced by 5 Hz to $45\,Hz$, the impedance becomes a stiffness impedance, causing the silencer to be completely ineffective.

If the frequency is increased by $5\,Hz$ to $55\,Hz$, we obtain a mass impedance. The bracketed term in the above equation then takes on a value of approximately 0.2. The channel wave number for mass impedances is

$$k_x = k\sqrt{1 - \frac{1}{\frac{|z|}{\varrho\,c}kh}} = -jk\sqrt{\frac{1}{\frac{|z|}{\varrho\,c}kh} - 1}\,,$$

resulting in a damping D_h of

$$D_h = 8,7k_ih = 8,7kh\sqrt{\frac{1}{\frac{|z|}{\varrho\,c}kh} - 1}\,.$$

The impedance is then equal to $z/\varrho\,c = j\,0.44$ for $d = 0.5\,m$ ($z/\varrho\,c = j\,0.22$ for $d = 0.5\,m$). Applying $kh = 0.25$ for $55\,Hz$ and $h = 0.25\,m$ one obtains $|z|kh/\varrho\,c = 0.11$ for $d = 0.5\,m$ (and $|z|kh/\varrho\,c = 0.055$ for $d = 1\,m$).

Based on this, we obtain the results $D_h = 6.2\,dB$ for $d = 0.5\,m$, and $D_h = 9\,dB$ for $d = 1\,m$. Therefore, greater construction depth results in a greater effective bandwidth.

C.10 Practice Exercises Chapter 10

Problem 1

Based on the aforementioned approximation equation

$$R_E = 20\lg\left(\frac{\sqrt{2\pi N}}{th(\sqrt{2\pi N})}\right) + 5\,dB,$$

one obtains the following level reductions in dB by changing the height of a sound reduction barrier:

Source distance:	$6.7\,m$	$10.5\,m$	$15\,m$
$4\,m$ to $5.5\,m$	2.5	2.6	2.7
$4\,m$ to $7.5\,m$	4.8	5.1	5.3
$5.5\,m$ to $7.5\,m$	2.3	2.5	2.6
$7.5\,m$ to $10\,m$	2.0	2.2	2.3

Problem 2

By applying the equation from the previous problem, we obtain the following insertion loss values for the sound reduction barrier in dB:

Source distance:	$6.7\,m$	$10.5\,m$	$15\,m$
$4\,m$ height	21.1	19.3	17.9
$5,5\,m$ height	23.6	22.0	20.6
$7,5\,m$ height	25.9	24.5	23.2
$10\,m$ height	28.0	26.7	25.5

Problem 3

Minimum and maximum level reduction in dB by changing the height of a sound reduction barrier:

	Minimum	Maximum
$4\,m$ to $5.5\,m$	1.4	2.8
$4\,m$ to $7.5\,m$	2.7	5.5
$5.5\,m$ to $7.5\,m$	1.3	2.7
$7.5\,m$ to $10\,m$	1.2	2.5

Problem 4

Based on the previous exercises, it follows that $R_{ges} = R_B - 7\,dB$.

Problem 5

The total pathway U can be logically obtained by the difference between the sum of the 3 perimeters K_1, K_2 and K_3 and the direct path D (see the following graph)

$$U = K_1 + K_2 + K_3 - D\,.$$

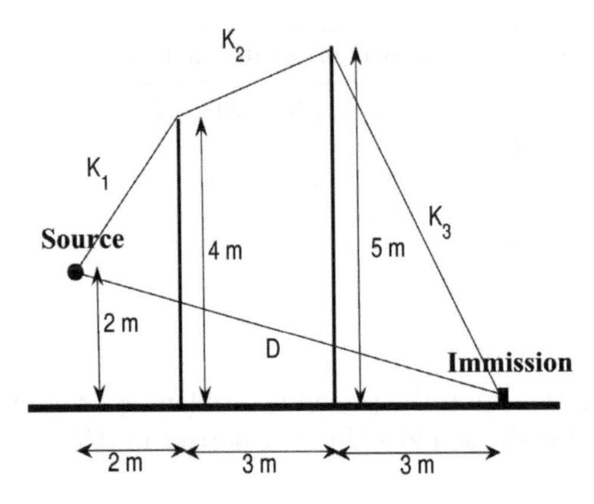

Fig. C.17. Perimeters K_1, K_2 and K_3 and direct pathway D

The partial pathways can each be calculated based on right triangles, resulting in

$$K_1 = \sqrt{2^2 + 2^2}\,m = 2.83\,m\,,$$

$$K_2 = \sqrt{1^2 + 3^2}\,m = 3.16\,m\,,$$

$$K_3 = \sqrt{5^2 + 3^2}\,m = 5.83\,m$$

and

$$D = \sqrt{2^2 + 8^2}\,m = 8.25\,m\,.$$

The total perimeter therefore amounts to $U = 3.57\,m$, the Fresnel number is $N = 2U/\lambda = 10.5$ for $500\,Hz$. Using the equation mentioned in the solution to Problem 1, we apply the above and arrive at $R_E = 23.2\,dB$.

Problem 6

Without the smaller of both barriers, the perimeter pathway becomes

$$K = 2\sqrt{5^2 + 3^2}\,m = 11,66\,m\,,$$

the direct pathway remaining $D = 8.25\,m$ as in Problem 5. The detour pathway is then $U = 3.41\,m$ and the Fresnel number $N = 2U/\lambda = 10.03$ for $500\,Hz$. The reduction in insertion loss ΔD can be calculated from the ration of the Fresnel numbers

$$\Delta D = 10\lg\frac{N(with)}{N(without)},$$

resulting in $\Delta D = 0.2\,dB$.

C.11 Practice Exercises Chapter 11

Problem 1

The frequencies are

- $kb = 2,5$: $f = 10,8\,kHz$,
- $kb = 5$: $f = 21,6\,kHz$ and
- $kb = 10$: $f = 43,2\,kHz$.

Problem 2

As mentioned in the previous chapters, the spring force is defined by

$$F_s = s(x - x_m)\ .$$

Here, the displacement x of the mass must be expressed through the spring force. According to the law of inertia

$$m\frac{d^2x}{dt^2} = -F_s\ ,$$

or, expressed in terms of complex amplitudes,

$$x = \frac{F_s}{m\omega^2}\ .$$

Thus follows that

$$F_s = F_s\frac{s}{m\omega^2} - sx_m\ ,$$

or

$$F_s(1 - \frac{s}{m\omega^2}) = -sx_m\ .$$

Finally, we obtain the following for the spring force

$$F_s = -\frac{sx_m}{1 - \frac{s}{m\omega^2}}\ ,$$

or, since the relation to the low end acceleration $a_m = -\omega^2 x_m$ is relevant here

$$F_s = -\frac{s\omega^2 x_m}{\omega^2 - \frac{s}{m}} = \frac{sa_m}{\omega^2 - \frac{s}{m}}\ .$$

Naturally, we are solely concerned the relationship of the frequency ω to the resonance frequency ω_0 (with $\omega_0^2 = s/m$) here:

$$F_s = \frac{-ma_m}{1 - (\omega/\omega_0)^2}$$

For frequencies far below the resonance frequency, the spring force $F_s = -ma_m$ is frequency independent. The sensitivity of the transducer increases with the mass m. At resonance, the frequency response is determined by attenuation. The frequencies far above the resonance frequency $\sqrt{s/m}$, $F_s = sa_m/\omega^2$ is true, in which case the frequency response of the transducer's sensitivity decreases by 12 dB/octave. Overall, the frequency response follows the gradient as shown in Fig. 11.1 (upper section).The frequency responses of a condenser microphone and acceleration capturer are therefore quite similar to one another.

Problem 3

The equivalent sound pressure is $2\ 10^{-3} N/m^2$, the equivalent sound pressure level is therefore $40\,dB$.

Problem 4

First, we need to know the first roots of the function $J_1(u)$. By referring to a table or simply by trying out different values using a computer programm such as Matlab, one obtains $J_1(u) = 0$ for u=3.83; 7.02; 10.2; 13.3 and 16.5. The angle of breaks can be obtained using

$$\sin\vartheta = \frac{u}{2\pi b/\lambda},$$

where u crosses the aforementioned values – the roots of the function $J_1(u)$. The following angles for breaks are:

- for b/λ=1: $\vartheta = 37.6°$;
- for b/λ=2: $\vartheta = 17.7°$; 34° and 54.3°;
- for b/λ=3: $\vartheta = 11.7°$; 21.9°; 32.8°, 44.9° and 61.1°.

  The frequencies are

- for $2b = 2,5\,cm$: 20.4 kHz ($b/\lambda = 1$); 40.8 kHz ($b/\lambda = 2$) and 61.2 kHz ($b/\lambda = 3$);
- for $2b = 1.25\,cm$: 40.8 kHz ($b/\lambda = 1$); 81.6 kHz ($b/\lambda = 2$) and 102.4 kHz ($b/\lambda = 3$);

The frequencies occupy various parts of the ultrasound spectrum.

C.12 Practice Exercises Chapter 12

Problem 1

It has already been established that a sound field consists of both pressure

$$p = p_+ e^{-jkx} + p_- e^{jkx}$$

and velocity

$$v = \frac{1}{\varrho\, c}(p_+ e^{-jkx} - p_- e^{jkx})\,.$$

The following describes the effective intensity:

$$I = \frac{1}{2}Re(pv^*) = \frac{1}{2\varrho\, c}Re((p_+ e^{-jkx} + p_- e^{jkx})(p_+^* e^{jkx} - p_-^* e^{-jkx}))$$

(Re = Real component of *: conjugate complex dimension). It thus follows that

$$I = \frac{1}{2\varrho\, c}(|p_+|^2 - |p_-|^2 + Re(p_+^* p_- e^{j2kx} - p_+ p_-^* e^{-j2kx}))$$

Based on $Re(z - z^*) = 0$, we arrive at

$$I = \frac{1}{2\varrho\, c}(|p_+|^2 - |p_-|^2)\,,$$

which indicates what is to be shown.

Problem 2

This problem was actually already solved in Section 12.2. The absorption coefficient must be defined in this case as $\alpha = -P_L/P_0$ (P_L is the additional power resulting from the loudspeaker, P_0 is the power resulting solely from the primary impinging wave). The maximum possible absorption coefficient is 0.5. At $x = 0$, where the secondary source is located, the primary source must therefore produce -0.5 times that of the primary wave. The sound pressure is half of that and to the right of the case where no secondary source is present. The power flowing to the right is therefore reduced to only a quarter of the primary power P_0 alone, $P_2 = P_0/4$. In the partial space $x < 0$, the power flows, as seen in the x direction

$$P_1 = P_0(1 - |r|^2)\,,$$

in the optimal case $|r| = 0.5$ producing a value of $P_1 = 3P_0/4$. The resulting difference between P_1 and P_2 – $P_0/2$ – is absorbed by the loudspeaker.

Problem 3

Basically, sound power is most easily measured in the far-field. For this reason, we first specify the far-field approximation for the net pressure. Apart from quadratically minuscule dimensions, the distances between the field points and the secondary sources in the far-field, based on the cosine law, are

$$r_1 = R - h\cos\vartheta$$

and

$$r_2 = R - h\cos(180^\circ - \vartheta) = R + h\cos\vartheta\,.$$

For the net pressure in the far-field, it thus follows that

$$p = \frac{j\omega\varrho Q_0}{4\pi}\frac{e^{-jkR}}{R}(1 - \beta(e^{jkh\cos\vartheta} + e^{-jkh\cos\vartheta}))$$

or

$$p = \frac{j\omega\varrho Q_0}{4\pi}\frac{e^{-jkR}}{R}(1 - 2\beta\cos(kh\cos\vartheta))\,.$$

The resulting intensity in the far-field can thus be defined as

$$I = \frac{1}{2}\frac{|p|^2}{\varrho c} = \frac{1}{2\varrho c}(\frac{\omega\varrho Q_0}{4\pi R})^2(1 - 2\beta\cos(kh\cos\vartheta))^2\,.$$

The sound power P is derived by integrating over the spherical surface in the far field (radius) R, resulting in

$$P = \int_0^{2\pi}\int_0^{\pi} IR^2 sin\vartheta\, d\vartheta\, d\varphi = 2\pi\int_0^{\pi} IR^2 sin\vartheta\, d\vartheta\,.$$

After substituting I, we obtain

$$P = P_0\int_0^{\pi/2}(1 - 2\beta\cos(kh\cos\vartheta))^2 sin\vartheta\, d\vartheta\,.$$

The last step in the calculations utilizes the fact that the power flowing through the upper and lower half-spheres are symmetrical and thus equal. The equation can be further simplified to

$$P_0 = \frac{1}{\varrho c}\frac{(\omega\varrho Q_0)^2}{8\pi}.$$

P_0 describes the power of the primary source alone, that is, the case $\beta = 0$).

The integral can be solved easily by using the variable substitution

$$u = cos\vartheta$$

$$du = -sin\vartheta \, d\vartheta$$

Proceeding as such, we arrive at

$$P = P_0 \int_0^1 (1 - 2\beta \cos{(khu)})^2 \, du \, .$$

Now it is prudent to use $cos^2 x = (1 + cos2x)/2$ in order to find

$$P/P_0 = 1 + 2\beta^2 - 4\beta \frac{sin(kh)}{kh} + 2\beta^2 \frac{sin(2kh)}{2kh} \, .$$

This power, for the case $\beta = 1/2$ without net volume flow for all three sources combined ($Q_0(1 - 2\beta) = 0$), is shown in the following graph. The other two curves shown are intended as a comparison to the case where only a single contrary source is present (see Chapter 3). In the first contrasting case, the secondary source is equal in negative value to the primary source, and the second contrasting case indicates the optimum scenario of reduced power.

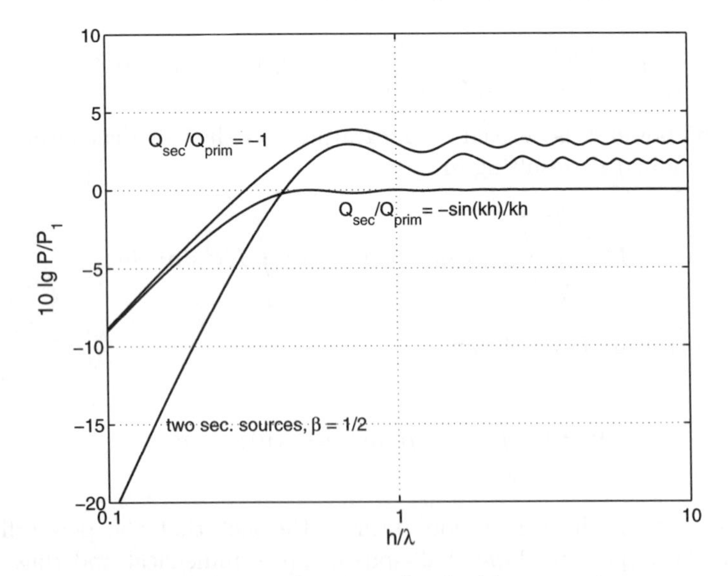

Fig. C.18. Reduction of the sound pressure level with one contrary source present (the two curves at the top represent gradients at low frequencies) and for two contrary sources with $\beta = 1/2$

This result shows that improved effectiveness can be achieved within the frequency band of the noise reduction by adding a secondary sound source, but does not change the bandwidth, however.

By setting the derivative of the power for β equal to zero, we arrive at the case of the least power emitted. Such a case must fulfill

$$\beta = \frac{\frac{sin(kh)}{kh}}{1 + \frac{sin(2kh)}{2kh}}.$$

C.13 Practice Exercises Chapter 13

Problem 1

The result is

$$A_n = \frac{\sin\left(n\pi T_D/T\right)}{2n\pi} .$$

As a rule, the amplitudes behave 'in the manner of $1/n$'. Fortunately, this is not always true, as the numerator dictates that sign changes occur with increasing n, which influences the convergence. Were A_n actually exact and not just approximately proportional to $1/n$, the function series would never converge. It is already a known fact that the series

$$\sum_{n=1}^{\infty} \frac{1}{n^a}$$

$(a > 0)$ is divergent for $a \leq 1$ and converges only in the case of $a > 1$.

Problem 2

The series expansion of the signal under investigation is expressed as

$$x(t) = \sum_{n=-\infty}^{\infty} A_n e^{j2\pi n \frac{t}{T}}$$

For the first derivative,

$$\frac{dx}{dt} = \frac{j2\pi}{T} \sum_{n=-\infty}^{\infty} n A_n e^{j2\pi n \frac{t}{T}} .$$

is true. Supposing x were continuous but the first derivative discontinuous, however, the solution in the previous problem $nA_n \sim 1/n$ would have to apply. In such a case, A_n would behave in the manner of $1/n^2$. This accounts for the main difference between the discontinuous case, shown in Figures 13.7 to 13.9 and the deviating curve, as illustrated by Figures 13.4 and 13.5. In the discontinuous case, the break A_n converges as in $1/n$, and in the deviating case, the deviation in the curve occurs in the manner of $1/n^2$. The implications for the number of corresponding summands in a good resolution of x_M are illustrated in a simple example. Suppose that the first one hundred summands are taken into account ($N = 100$). At the discontinuity, the last summand $n = N$ exists in the order of 0.01 times the first summand. In the deviating case, on the other hand, it will already have reached a multiple of 0.0001 times the first summand. This means, of course, that the expansion can be terminated much sooner in the deviating case than in the discontinuous case.

For the second derivative

$$\frac{dx}{dt} = (\frac{j2\pi}{T})^2 \sum_{n=-\infty}^{\infty} n^2 A_n e^{j2\pi n \frac{t}{T}}$$

is true. If the signal is continuous along with the first derivative, the second derivative however discontinuous, A_n behaves like $1/n^3$. If the signal is discontinuous along with the first m derivatives, but the $(m+1)$-th derivative is discontinuous, A_n will then behave like $1/n^{m+2}$.

Problem 3

For the physical dimension of the rectangular function $r_{\Delta T}(t)$

$$Dim[r_{\Delta T}(t)] = \frac{1}{Dim[t]} \, .$$

applies. This is of course also true for the delta function, which of course represents the marginal case of the rectangular function

$$Dim[\delta(t)] = \frac{1}{Dim[t]} \, .$$

Note that the delta function does not represent a dimensionless, but actually, a heavily dimensioned function.

Problem 4

The law of the convolution of the product of two signals states that the transform of the product of the time dependencies is equal to the convolution integral of both spectra

$$\int_{-\infty}^{\infty} x(t)g(t)e^{-j\omega t}dt = \frac{1}{2\pi}\int_{-\infty}^{\infty} X(\nu)G(\omega-\nu)d\nu \, .$$

Based on this, we obtain for the integral over the product of both signals

$$\int_{-\infty}^{\infty} x(t)g(t)dt = \frac{1}{2\pi}\int_{-\infty}^{\infty} X(\nu)G(-\nu)d\nu \, .$$

In order to prove the assertion of the energy law , the spectrum of $x^*(t)$ must be defined (*: complex conjugate). In order to find a general rule, we have made complex time functions permissible. For real-valued time functions, the conjugate sign * is omitted. Based on

$$x(t) = \frac{1}{2\pi}\int_{-\infty}^{\infty} X(\omega)e^{j\omega t}d\omega,$$

we arrive at

$$x^*(t) = \frac{1}{2\pi} \int_{-\infty}^{\infty} X^*(\omega) e^{-j\omega t} d\omega \ ,$$

or

$$x^*(t) = \frac{1}{2\pi} \int_{-\infty}^{\infty} X^*(-\omega) e^{j\omega t} d\omega$$

(which can also be expressed using the substitution $u = -\omega$). Thus follows that

$$\int_{-\infty}^{\infty} x(t) x^*(t) dt = \frac{1}{2\pi} \int_{-\infty}^{\infty} X(\nu) X^*(\nu) d\nu \ ,$$

verifying the assertion.

Problem 5

We approach this problem from the vantage point of the convolution law for the product of two spectra

$$\frac{1}{2\pi} \int_{-\infty}^{\infty} X(\omega) H(\omega) e^{j\omega t} d\omega = \int_{-\infty}^{\infty} x(\tau) h(t-\tau) d\tau \ .$$

The next step is to find the Fourier inverse transform belonging to $X^*(\omega)$. Based on

$$X(\omega) = \int_{-\infty}^{\infty} x(t) e^{-j\omega t} dt, \tag{C.1}$$

it follows that

$$X^*(\omega) = \int_{-\infty}^{\infty} x^*(t) e^{j\omega t} dt \ , \tag{C.2}$$

or

$$X^*(\omega) = \int_{-\infty}^{\infty} x^*(-t) e^{-j\omega t} dt \tag{C.3}$$

(which can also be shown informally using the substitution $u = -t$). The inverse transform of $X^*(\omega)$ is therefore $x^*(-t)$. We can thus conclude that

$$\frac{1}{2\pi} \int_{-\infty}^{\infty} X(\omega) X^*(\omega) e^{j\omega t} d\omega = \int_{-\infty}^{\infty} x(\tau) x^*(\tau - t) d\tau \ ,$$

which can also be written as

$$\frac{1}{2\pi}\int_{-\infty}^{\infty} X(\omega)X^*(\omega)e^{j\omega t}d\omega = \int_{-\infty}^{\infty} x(\tau+t)x^*(\tau)d\tau \ .$$

Problem 6

The task in this problem is to show that Fourier transforms and complex amplitudes can likewise be used to compute measurements for linear and time-invariant transmitters $y(t) = L\{x(t)\}$. The formal proof of this fundamental aspect can be easily undertaken:

Complex Amplitudes

Using

$$x(t) = \mathrm{Re}\{\underline{x}e^{j\omega t}\}$$

and

$$y(t) = \mathrm{Re}\{\underline{y}e^{j\omega t}\},$$

we arrive at by substitution

$$\mathrm{Re}\{\underline{y}e^{j\omega t}\} = L\{\mathrm{Re}\{\underline{x}e^{j\omega t}\}\} = \mathrm{Re}\{\underline{x}L\{e^{j\omega t}\}\}\,,$$

or, of course,

$$\underline{y}e^{j\omega t} = \underline{x}L\{e^{j\omega t}\}\,.$$

Fourier Transformation

Using

$$x(t) = \frac{1}{2\pi}\int_{-\infty}^{\infty} X(\omega)e^{j\omega t}d\omega$$

and

$$y(t) = \frac{1}{2\pi}\int_{-\infty}^{\infty} Y(\omega)e^{j\omega t}d\omega,$$

we obtain

$$\int_{-\infty}^{\infty} Y(\omega)e^{j\omega t}d\omega = L\{\int_{-\infty}^{\infty} X(\omega)e^{j\omega t}d\omega\} = \int_{-\infty}^{\infty} X(\omega)L\{e^{j\omega t}\}d\omega,$$

or, of course,

$$Y(\omega)e^{j\omega t} = X(\omega)L\{e^{j\omega t}\}\,.$$

Problem 7

Due to causality, $h(t < 0) = 0$ applies; for $t > 0$,

$$h(t) = \frac{\omega_0^2}{s\omega_d} e^{-\eta\omega_0 t/2} \sin \omega_d t \,,$$

whereby the attenuated resonance frequency ω_d is defined by

$$\omega_d = \omega_0 \sqrt{1 - \eta^2/4}$$

Problem 8

We begin with the general postulate

$$v(t) = \frac{1}{2\pi} \int\limits_{-\infty}^{\infty} V(\omega) e^{j\omega t} d\omega = \frac{1}{2\pi} \int\limits_{-\infty}^{\infty} [Re\{V(\omega)\} + jIm\{V(\omega)\}][\cos \omega t + j \sin \omega t] d\omega .$$

Based on the symmetries in the hint given in the problem

$$Re\{V(-\omega)\} = Re\{V(\omega)\}$$

and

$$Im\{V(-\omega)\} = -Im\{V(\omega)\},$$

the above equation becomes

$$v(t) = \frac{1}{\pi} \int\limits_{0}^{\infty} Re\{V(\omega)\} \cos \omega t - Im\{V(\omega)\} \sin \omega t d\omega \,.$$

This shows, once again, that the symmetries mentioned universally result in a real-valued inverse transform.

In this particular case,

$$v(t) = F \int\limits_{0}^{\infty} \frac{\cos(\alpha x \sqrt{\omega}) \cos(\omega t) + \sin(\alpha x \sqrt{\omega}) \sin(\omega t)}{\sqrt{\omega}} d\omega$$

$$+ \, F \int\limits_{0}^{\infty} \frac{e^{-\alpha x \sqrt{\omega}} \sin(\omega t)}{\sqrt{\omega}} d\omega$$

applies, whereby the abbreviations

$$\alpha = \sqrt[4]{\frac{m'}{B}}$$

and

$$F = \frac{F_0}{4\pi\alpha\sqrt{m'B}}$$

can be used to simplify the expressions. Using $\cos\alpha\cos\beta + \sin\alpha\sin\beta = \cos(\alpha - \beta)$,

$$v(t) = F\int_0^\infty \frac{\cos(\alpha x\sqrt{\omega} - t\omega)}{\sqrt{\omega}} d\omega$$

becomes

$$+ F\int_0^\infty \frac{e^{-\alpha x\sqrt{\omega}}\sin(t\omega)}{\sqrt{\omega}} d\omega .$$

The variable substitution $\sqrt{\omega} = u$ with $d\omega = 2udu$ and consequently $d\omega/\sqrt{\omega} = 2du$ yields

$$v(t) = F(I_1 + I_2),$$

with

$$I_1 = 2\int_0^\infty \cos(tu^2 - \alpha xu)du$$

and

$$I_2 = 2\int_0^\infty e^{-\alpha xu}\sin(tu^2)du .$$

For the second integral, we employ $-1 = j^2 = jj$:

$$I_2 = 2\int_0^\infty e^{jj\,\alpha xu}\sin(tu^2)du = 2\int_0^\infty [\cos(j\alpha xu) + j\sin(j\alpha xu)]\sin(tu^2)du$$

$$= \int_0^\infty \sin(tu^2 - j\alpha xu) + \sin(tu^2 + j\alpha xu)du$$

$$+ j\int_0^\infty \cos(tu^2 - j\alpha xu) - \cos(tu^2 + j\alpha xu)du ,$$

which can be easily shown utilizing the corresponding addition theorems. The integrals included in the I_1 and I_2 equations above are listed in a table (refer to Gradshteyn, I.S.; Ryzhik, M. : Table of Integrals, Series and Products, Academic Press, New York and London 1965, p. 397, numbers 3.693.1 and 3.693.2). The sought-after impulse response $v(t)$ thus is

$$v(t) = \frac{F_0}{2\sqrt{\pi t\sqrt{m'^3 B}}}\cos\left(\sqrt{\frac{m'}{B}}\frac{x^2}{4t} - \pi/4\right) .$$

A visual representation of the space and time dependency of these results can be found in Fig. 4.7.

Problem 9

We apply

$$F(\omega) = f_0 \int_{-\infty}^{\infty} e^{-\gamma t^2} e^{-j\omega t} dt =$$

$$= 2f_0 \int_{0}^{\infty} e^{-\gamma t^2} \cos \omega t dt$$

to this problem for reasons of symmetry. The integral is listed in a table (refer to Gradshteyn, I.S.; Ryzhik, M. : Table of Integrals, Series and Products, Academic Press, New York and London 1965, p. 480, number 3.987.1). Using the integral, we obtain

$$F(\omega) = f_0 \sqrt{\frac{\pi}{\gamma}} e^{-\omega^2/4\gamma}.$$

Most notably,

- the transform of the Gauss function itself is a Gauss function. The signal form remains, in principle, unchanged by the Fourier transformation and
- long-ranging, flat time dependencies (small γ) are narrow-banded while rapidly changing time signals (large γ) produce wide-banded transforms.

$f(t)$ is shown in the following graph. $F(\omega)$ possesses, as mentioned, the same signal shape. Only the bandwidth changes. It increases with decreasing T_0 (=$1/\gamma$).

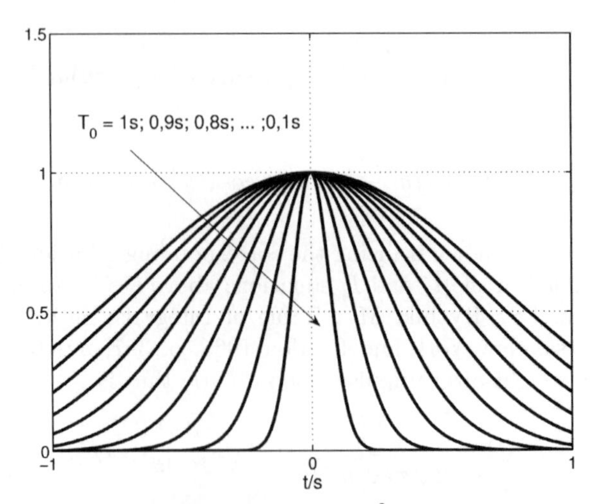

Fig. C.19. Gauss function $e^{-\gamma t^2}$ with $\gamma = 1/T_0^2$

Problem 10

First, we calculate the Fourier transform of the beam velocity:

$$V_y(k) = v_0 \int_{-\infty}^{\infty} e^{-|x|/x_0} e^{-jkx} dx$$

Since the sin function produces a point-symmetrical function, we are left only with

$$V_y(k) = 2v_0 \int_{0}^{\infty} e^{-x/x_0} coskx\, dx.$$

If the reader opts not to refer to the integral table, this can be transported to

$$V_y(k) = 2v_0 \mathrm{Re}\{\int_{0}^{\infty} e^{-x/x_0} e^{jkx} dx\},$$

resulting in

$$V_y(k) = 2v_0 x_0 \mathrm{Re}\{\frac{1}{1-jkx_0}\} = 2v_0 x_0 \mathrm{Re}\{\frac{1+jkx_0}{(1-jkx_0)(1+jkx_0)}\} = \frac{2v_0 x_0}{1+(kx_0)^2}.$$

This leaves us with the sound pressure in the far-field, according to eq.(13.81):

$$p_{\text{fern}} = \frac{j\omega\varrho\, b}{4\pi R} e^{-jk_0 R}\, V_y(k = -k_0 \sin\vartheta) \;= \frac{j\omega\varrho\, b}{4\pi R} e^{-jk_0 R} \frac{2v_0 x_0}{1+(k_0 x_0 sin\vartheta)^2};$$

The following graph illustrates some beam characteristics for small and large beam arrays.

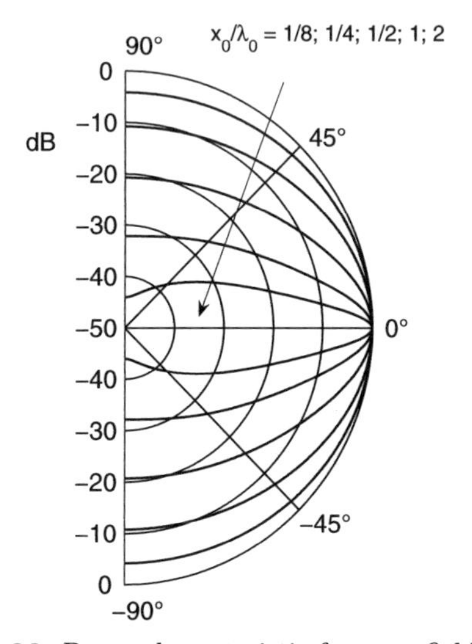

Fig. C.20. Beam characteristic for near-field oscillating sources

Problem 11

For $\varepsilon = 1$,

$$v = v_0\, cos \frac{2n\pi x}{l}$$

applies. This gradient has maxima ('antinodes') in $x = 0$ and $x = l$. For $\varepsilon = -1$,

$$v = jv_0\, sin \frac{2n\pi x}{l}$$

applies. This gradient has roots ('nodes') in $x = 0$ and $x = l$.

The Fourier transform is therefore

$$V(k) = \frac{jv_0 l}{2}(e^{-jkl} - 1)\frac{kl(1+\varepsilon) + 2n\pi(1-\varepsilon)}{(kl)^2 - (2n\pi)^2}.$$

This wave number spectrum becomes very small for small values of k ($|kl| << 2n\pi$) in the case of $\varepsilon = -1$ versus $\varepsilon = -1$, based on the absolute value. For small values of k,

$$|V(k)_{\varepsilon=-1}| << |V(k)_{\varepsilon=1}|$$

applies. This is why the power emission at low frequencies for $\varepsilon = -1$ is much less.

Problem 12

$$F_1(\omega) = \int_{-\infty}^{\infty} f_1(t)e^{-j\omega t}dt = \int_{-\infty}^{\infty} g(t)e^{-j(\omega-\omega_0)t}dt = G(\omega - \omega_0)\,,$$

whereby $G(\omega)$ is simply the transform of the enveloping function $g(t)$. The transform of the product is therefore equal to that of the transform of the enveloping function, simply shifted by ω_0.

For $F_2(\omega)$, based on $cosx = (e^{jx} + e^{-jx})/2$,

$$F_2(\omega) = \frac{1}{2}[G(\omega - \omega_0) + G(\omega + \omega_0)]$$

is true.

Problem 13

As verified in the previous problem, the Fourier transform of the signal located at $x = 0$ is

$$V(0,\omega) = \frac{1}{2}[G(\omega - \omega_0) + G(\omega + \omega_0)]\,,$$

whereby $G(\omega)$ constitutes the Fourier transform of the enveloping function $g(t)$.

The components with the frequency ω propagate along the wave guide with the wave number $k(\omega)$, validating the following for the Fourier transform at any arbitrary spatial location:

$$V(x,\omega) = V(0,\omega)e^{-jk(\omega)x}$$

The signal $v(x,t)$ of the oscillation results from inverse transformation to

$$v(x,t) = \frac{1}{2\pi}\int_{-\infty}^{\infty} V(x,\omega)e^{j\omega t}d\omega = \frac{1}{4\pi}\int_{-\infty}^{\infty}[G(\omega-\omega_0)+G(\omega+\omega_0)]e^{-jk(\omega)x}e^{j\omega t}d\omega.$$

Given that $G(\omega)$ invariably constitutes a small-banded function, only the frequency components $\omega \approx \omega_0$ and $\omega \approx -\omega_0$ influence the integral.

$\omega \approx \omega_0$:

In the small, significant frequency band around ω_0, $k(\omega)$ can be replaced by the first two terms of the Taylor series

$$k(\omega) \approx k(\omega_0) + (\omega - \omega_0)\frac{dk}{d\omega}|_{\omega=\omega_0} .$$

By simplifying

$$k_0 = k(\omega_0)$$

and

$$k_0' = \frac{dk}{d\omega}|_{\omega=\omega_0},$$

we obtain

$$k(\omega) \approx k_0 + \omega k_0' - \omega_0 k_0' .$$

The influence of this integration range on the integral $v_+(x,t)$ is therefore

$$v_+(x,t) = \frac{1}{4\pi}\int_{-\infty}^{\infty} G(\omega-\omega_0)e^{-jk_0x}e^{-j\omega k_0'x}e^{j\omega_0 k_0' x}e^{j\omega t}d\omega$$

$$= \frac{1}{2}e^{-jk_0x}e^{j\omega_0k_0'x}\frac{1}{2\pi}\int_{-\infty}^{\infty} G(\omega-\omega_0)e^{j\omega(t-k_0'x)}d\omega ,$$

which becomes, based on Problem 12,

$$\frac{1}{2\pi}\int_{-\infty}^{\infty} G(\omega-\omega_0)e^{j\omega t}d\omega = g(t)e^{j\omega_0 t} ,$$

therefore resulting in

$$\frac{1}{2\pi}\int_{-\infty}^{\infty} G(\omega-\omega_0)e^{j\omega(t-k_0'x)}d\omega = g(t-k_0'x)e^{j\omega_0(t-k_0'x)} ,$$

and accordingly,

$$v_+(x,t) \approx \frac{1}{2}e^{-jk_0x}e^{j\omega_0 k_0' x}g(t-k_0'x)e^{j\omega_0(t-k_0'x)} = \frac{1}{2}e^{j(\omega_0 t-k_0x)}g(t-k_0'x) .$$

$\omega \approx -\omega_0$:

Similarly, we can show

$$v_-(x,t) \approx \frac{1}{2}e^{-j(\omega_0 t-k_0x)}g(t-k_0'x) .$$

Net field:

For $v(x,t) = v_+ + v_-$, it follows, based on the above, that

$$v(x,t) \approx cos(\omega_0 t-k_0x)\, g(t-k_0'x) = cos(\omega_0(t-\frac{k_0}{\omega_0}x))\, g(t-k_0'x) .$$

This result can be interpreted as follows:

- The carrier signal $cos(\omega_0 t)$ disperses with the velocity of $c_0 = \omega_0/k_0$. This is the propagation velocity of an arbitrarily small-banded pure tone. The velocity c_0 is referred to a the phase velocity.
- The enveloping wave $g(t)$, on the other hand, propagates with a wave velocity of $c_g = 1/k_0'$, which is accordingly governed by

$$c_g = \frac{1}{\frac{dk}{d\omega}}|_{\omega=\omega_0} .$$

The velocity c_g is known as the group velocity. Because the carrier signal and the enveloping signal are travelling at different speeds, the net signal is distorted during the wave transport. The carrier signal and the enveloping signal shift in opposite directions to one another along the propagation pathway.

For bending waves,

$$k = \beta\sqrt{\omega}$$

(β is a constant). The phase velocity is

$$c_0 = \frac{\omega}{k} = \frac{\sqrt{\omega}}{\beta}$$

and the group velocity is

$$c_g = \frac{1}{\frac{dk}{d\omega}} = \frac{2\sqrt{\omega}}{\beta} .$$

For bending waves, the group velocity is double the phase velocity.

To further illustrate this point, an example of wave propagation along a dispersive wave guide is shown in Fig. C.21. Represented here are both time dependencies of a relevant field dimension occurring at two different locations $x = x_0$ and $x = x_0 + \Delta x$, such as the speed along a non-reflective bending wave guide. For the sake of clarity, the enveloping waves are shown as well. The carrier frequency does not propagate at the same speed as the enveloping wave.

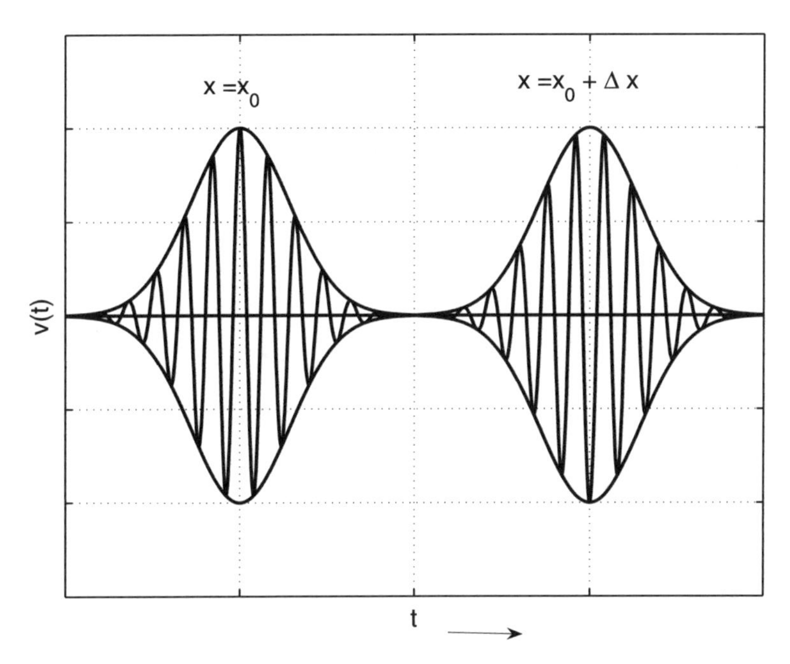

Fig. C.21. Two time signals along a dispersive wave guide

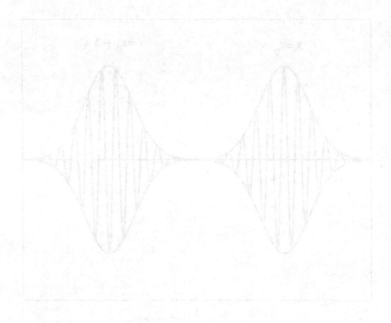

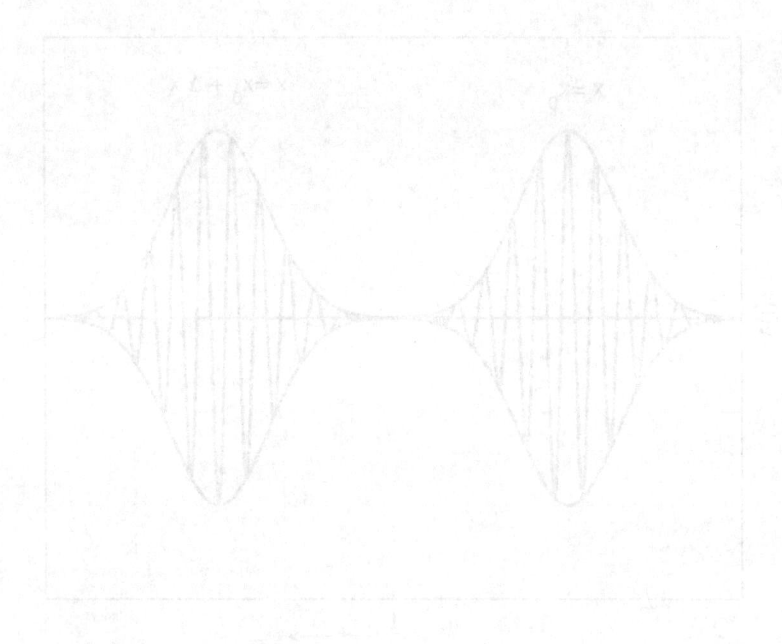

Index

$\lambda/4$-resonance, 294

A-filter, 10
  characteristics, 10
A-weighted sound pressure level, 9, 10, 69
Abrupt change in cross-section, 268
Absorbent noise barrier, 335
Absorber, 171
  attenuation, 191
  locally reacting, 186
  low-frequency, 201
  porous, 187
Absorbers, 401
Absorbing
  silencer, 267, 268
Absorption, 392
Absorption coefficient, 182, 228, 231, 232
Accuracy of replication, 402
Acoustic Antennae, 365
acoustic holography, 441
Acoustic screen, 311
Active field, 38
active level reduction, 389
Active noise compensation, 386
Active noise reduction, 385, 389
Active stabilization, 397
Additional lining, 252
Adiabatic, 20
  equation of state, 21
  linearised equation of state, 23
Aerodynamic force, 399
Air
  column, 24
  humidity, 229
  stiffness, 196
Airborne
  transmission loss, 244
Airborne sound transmission, 489
Aircraft engineering, 387
Aircraft noise, 385
Airfoil, 397
Alignment, in-phase, 37
Amplification factor, 458
Amplitude density, 437
Amplitude density function, 429, 436
Amplitude errors, 390
Angular velocity, 124
Anti-noise, 385
Arbitrary change in cross-sectional area, 283
Atmospheric pressure, 1, 6
Attenuation
  of the resonance, 353
  in ducts, 287
  optimum in a duct, 307
Attenuator
  splitter, 284
Automotive lightweight construction, 387
axis-symmetric, 432

Bandwidth, 7, 220
  3-dB, 279
  effective, 279

half-bandwidth, 164, 204
Barrier, 311
Basic equations, 28
basic modes, 271
Bassoon, 397
Beam displacement, 120
Beam forming, 95
Beam pattern, 81
of a dipole, 83
of a loudspeaker array, 90
of main and side lobes, 94
steered, 101
Beam resonances, 127
Beam steering
electronic, 96
Beam vibration modes, 130, 132, 134
Bending
angle, 120
free waves, 134
moment, 121, 124
science, 121
static, 120
stiffness, 118, 121, 135
of the beam, 134
of the plate, 134
wave equation in beams, 124
wave equation in plates, 134
wave equation in walls, 243
wave propagation, 119, 124
wave propagation speed, 125
wavelength, 125, 135
waves, 118, 123, 124, 136
in beams, 120, 135
in plates, 134
Bending waves, 439
Berger's mass law, 245
Bernoulli law, 397
Bessel function, 177
Bottle tone, 389
Boundary
soft, 288
Boyle-Mariotte equation, 18
Bridge, 397
Broadband signals, 7

Carbon dioxide, 465
causal, 433
Causality principle, 433
Cavity
damping of, 256
Cello, 397
Center frequency, 7
Chain of elements model, 24
Chamber
combinations, 279
silencer (expansion ch.), 276
Change
in cross-section, 268
in perception, 3
Circular piston, 107
clarinet, 397
Coherent, 453
Coincidence
dip, 250
frequency, 137, 244
coincidence critical frequency, 440
Coincidence effect, 246
Comb filter, 296
Complex amplitude, 34
Complex number, 455
Complex pointer, 455
Complex pointers, 34
Condenser microphone, 347
Conservation principle, 273
Conservative sound field, 40
Constraint, 127
Continuous, 421
Continuous sound level, 12
Contrabass, 397
Convergence property, 447
Converters, 345
Convolution, 415
Convolution integral, 415, 430
Convolution law, 429, 431, 444
Critical cycle, 399
Critical frequency, 137, 246
Cross array, 376
Cross coupling, 208
Cross-sectional area
arbitrary change, 283
Curves of equal loudness, 9
Cut-off
frequency, 287
Cut-off frequency, 174
cut-on effect, 178
Cylindrically symmetric sound field, 73

Damping

of cavity, 256
dB(A), 10
Dead-end termination, 272
Decadic logarithm, 451
Decomposition, 433
Decomposition into wavelengths, 444
Delay line, 395
Delta comb, 415
Delta function, 414
Delta operator, 39
Density, 18, 22
Detour law, 328
Difference threshold, 6
Differential operators, 39
Diffraction, 311
Diffraction angle, 313
Diffuse field, 220, 221
level, 229
microphone, 358
Dilatation, 27
Dipole, 72
Dirac function, 414
Directivity, 67
of microphones, 355
Discontinuity, 424
Dispersion, 126
Distance form source point to field point, 444
Distortion factor, 411
Divergence (div), 39
Doppler effect, 54
Doppler shift, 56
double Fourier transform, 443
Double window, 387
Double-glazed window, 256
Double-leaf partition, 252
Double-sphere characteristics, 83
Drag resistance, 495
Duct, 397
attenuation, 287, 304, 305
contraction, 270
expansion, 270
lined, 284
duct branch, 271
Duct branches, 271
Duct intersection, 274
Dynamic mass, 166

Echo
fluttering, 219
Effective mass per unit area, 207
Efficiency, 113
Eigenfrequency, 220
density of eigenfrequencies, 220
Eigenfunction, 173
Eigenvalue, 173
equation, 303
Eigenvibrations, 364
Elastic bearing, 149
Elastic decoupling, 147
Elastic deformation, 118
Elastic modulus, 27, 121, 153, 154
Electro-acoustic transducer, 345
Electro-acoustic transducers, 385
electrodynamic loudspeaker, 345, 362
electrodynamic microphone, 345, 358
Electronic beam steering, 96
Elongation, 27
Emission, 437
Emission of bending waves, 439
Emitting planes, 437
End correction, 279
Energy
density, 219
density of the sound field, 41
kinetic, 40
potential, 40
Energy law, 517
Energy reservoir, 398
Energy sink, 394
Energy-equivalent continuous sound level, 12
Engine noise, 386
Equal loudness, 9
Equivalent absorption area, 228, 240
Equivalent sound pressure, 384, 511
Equivalent sound pressure level, 384, 511
Excitation process, 399
Expansion chamber, 276

Far field, 80, 113
approximation, 80, 106
condition, 102, 106
far field, 445
far field approximation, 445
Far-field, 471
Far-field condition, 471

Feedback, 402
Figure-8 pattern, 83
Filter, 7
Flanking path, 237
Flanking transmission, 237
Flapping vibrations, 389
Flexible additional lining, 252
Flickering, 1
  frequency, 2
Floating cement floor, 261
Flow paths, 397
Flow resistance, 186
Flute, 397
Fluttering echo, 219
Folding frequency, 190, 351
Force
  of area, 134
  shear, 117, 124
Foundation, 147, 155
  compliance, 147, 155
  impedance, 155
Fourier, 417
Fourier acoustics, 435
Fourier decomposition, 417
Fourier pair, 431
Fourier series, 418, 426
Fourier sums, 419
Fourier transform, 427, 428, 431
Fourier transform, inverse, 428
Fourier transformation, 429
Free bending waves, 134
Free field
  microphone, 358
Frequency, 1
  centre, 7
  coincidence, 137, 244
  components, 2
  critical, 137, 246
  cut-off, 174, 287
  folding, 190, 351
  limiting, 137, 246
  resonance, 347
  typical, 350
Frequency domain measurements, 45
Frequency range
  of building acoustics, 240
Frequency response function, 3
  of a microphone, 349
Frequency variable, 428
Fresnel integral, 319, 342
Friction constant, 400
Function series, 418, 421

Gas
  temperature, 21
Geometric series, 461
Gibb, 426
Gibb phenomenon, 426
Gradient (grad), 39
Grazing incidence, 245
Group velocity, 526

Half-Bandwidth, 164, 204
Headphones, 386, 387
Hearing level, 1, 9
Hearing threshold, 6
Helmholtz equation, 436
Helmholtz resonator, 397
Hilbert Transformation, 433
Hilbert transformation, 435
Hollow building block, 251
Holography, 441
Holography, acoustic, 441
Hook's law, 27, 121, 148
Hybrid, 403
Hydrogen, 465

Impact sound
  level, 259
  reduction, 259, 261
Impact sound level, 259
Impedance
  mass, 293
  of a spring, 196
  optimum, 305, 307
  real, 293
  specific, 31
  stiffness, 292
  tube, 172
  wall, 184
Impedance tube, 171
Impulse response, 412, 415, 431, 433, 434
In-phase alignment, 37
Incidence angle, 219, 244, 320, 357
  critical, 245
Incident flow, 398
Incoherent signals, 453

Inertia law, 28
Infrasound, 1
Input impedance, 271
Input signal, 409
Insertion loss, 149, 158, 269, 277
of semi-infinite screen, 321, 328, 331, 334
Instable, 401
Insulating stripe, 262
Integral transformation, 429
Intensity, 42
Intensity level, 465
Intensity measurements, 44
Interconnection time delay, 395
Internal noise, 351
Internal automobile engine noise, 385
Interval, 3
Invariance principle, 416
Inverse Fourier transform, 428
Inverse transform, 426, 428
Irrotational, 40
Isobar, 21
Isochor, 21
Isotherms, 19

Junction, 270, 273

Kinetic energy, 40
Kundt's tube, 172

Law of compression, 29
Law of relative variation, 4
Leakage, 251, 258
Left-side limit, 424
Level
arithmetics, 451
inversion, 452
summation, 453
Light zone, 322, 325
Limiting frequency, 137, 246
Line array, 367
Line source, 68, 114, 471
Linear, 410
Linear level, 8
Linearity, 410
Lined duct, 284
Lined silencer, 267
Lining
soft, 288
Lobe
main, 92
side, 92
Localized impulse response, 445
Locally reacting absorber, 186
locally reactive, 209
Logarithm
decadic, 451
long-waved source elements, 438
Longitudinal wave propagation speed in beams, 135
Loss factor, 149, 164, 181, 246, 353
Loudness, 1, 4, 9
perception, 4
Loudspeaker, 362
array, 73
Loudspeaker pair, 395
Low-frequency absorber, 201
Lowest mode, 287, 291

Mach number, 56, 466
Main lobe, 92
Mapping, 426
Mass
characteristics, 289
impedance, 293
per unit area, 207
short circuit of, 72, 105, 362
Mass law, 245
Berger's, 245
mass-spring-mass resonance, 387
Matching, 192
Matching law, 184
Measured transmission loss, 241
Measurement
of sound power, 43
Microphone
condenser, 347
diffuse field, 358
electrodynamic, 358
free field, 358
frequency response function, 349
voice coil, 358
Microphone array, 367
Mirror source, 105, 217
Modal cut-off, 174
Modal damping, 287
Mode, 128, 173, 286
lowest, 287, 291

principal, in ducts, 304
Moment, 121
monochromatic, 439
Monopole source, 70
mounted, 127
Moving medium, 53
Musical instrument, 2

Near-field, 289
near-field, 437
Negative damper coefficient, 401
Net volume flow, 440
Newton's law, 27, 123, 147
Nitrogen, 466
Noise
internal, 351
Noise compensation, 386
Noise exposure, 1
Noise reduction, 385
Non-linearities, 410
Non-linearity, 59
Normal tension, 117, 120, 122
Normalized impact sound level, 259
Number
complex, 455

oblique sound incident, 208
Octave band
filter, 7
level, 8
Octave level, 461, 462
Omnidirectional radiation, 67
One-dimensional piston, 88
Optimum
attenuation, 307
impedance, 305, 307
Oscillation modes, 405
Output signal, 409
Overload protection, 5

Pair of sound incidents, 4
Particle displacement, 465
Particle motion, 58, 115
Partition
double-leaf, 252
single-leaf, 241
wall, 237
Pass band, 8
Perception
characteristics, 4
of loudness, 4, 5
of pitch, 3
of temperature, 5
of weight, 3, 5
Phase
function, 81
shift, 458
Phase errors, 390
Phase velocity, 526
Phase-shifter, 403
Phasor curve, 185
Phenomenon, Gibb, 426
Photo analogy, 427
pick-up characteristic
of microphones, 357
Piening formula, 302
Pipe elements, 279
Piston
circular, 107
one-dimensional, 88
Plane wave, 287
in plates, 135
Plate displacement, 119
Plate modes, 138
Plate resonance, 137
Plate resonance grids, 139
point-symmetric, 432
Pointer
complex, 455
Pointers, 34
Poisson's ratio, 134
Porosity, 188
Porous
absorber, 187
curtain, 198
sheet, 292
of finite thickness, 194
of infinite thickness, 191
Potential energy, 40
Power, 42
density, 42
Power level, 113, 465, 471
Pressure, 18, 22
reflection coefficient, 180
Primary, 385
Primary source, 385
Principal mode, 304
Progressive waves, 29, 50

Propagation
  of bending waves, 119, 124
  of sound, 67
Propagation speed, 25, 30, 97
  in absorbers, 190
  of bending waves, 125
  of longitudinal waves in beams, 135
Propeller noise, 386
Pulsating sphere, 70

Quarter-wavelength resonance, 294

Radiation, 67
  from plane surfaces, 104
  of line sources, 68
  of point sources, 67
  of sound, 67
  omnidirectional, 67
Radiation function, 90
Raised waves, 59
Rayleigh integral, 105, 107, 444
Rayleigh model, 188
Reactive field, 38
Real impedance, 293
Receiving room, 239
Rectangular duct, 176
Rectangular function, 412
Reference curve, 260
Reflecting silencer, 268
Reflection, 218, 355, 392
  coefficient, 180, 269
  zone, 325
Relative tonal impression, 3
Replication error, 389
Replication errors, 389
Residential noise, 237
Resistance force, 399
Resonance, 34, 37, 127
  $\lambda/4$, 294
  absorber, 201
  frequency, 347
  peak, 7
  quarter-wavelength, 294
Resonance density in plates, 141
Resonance frequencies, 37
Resonance grid of bending vibrations, 139
Resonance grid of plates, 139
Resonance phenomena, 34
Resonator, 289, 292
  lining, 297
Resonator in contact with air, 389
Reverberation, 224
  radius, 230
  room, 225, 231
  time, 224, 225, 240
Reversible converter, 394
Right-side limit, 424
Rigid screen, 312
Ring arrays, 379
Ripple parameter, 182
Room acoustics, 217
Root mean square, rms-value, 6, 43, 180, 223, 453
Rotation (rot), 40

Sabine equation, 240
  for reverberation time, 228
Saxophone, 397
Screen
  semi-infinite, 333
Second moment of area, 121
Secondary, 385
Secondary replication, 389
Secondary source, 385
Self-excitation, 389
Self-excited vibrations, 389
Self-induced, 397
Self-induced excitation vibration, 399
Self-induced vibration, 397
Semi-infinite screen, 333
Sensitivity
  angle-dependent, 355
  of a condenser microphone, 351
  of the human ear, 9
Shadow border, 326
Shadow region, 322, 323, 327
Shear force, 117, 124
Shear tension, 118, 120, 122
Short circuit
  of mass, 72, 105, 362
Short-circuit, 440
short-waved source element, 438
Side lobe, 92
  suppression, 93
Silencer, 267, 279
  absorbing, 268
  lined, 267

reflecting, 268
Simple source, 70
Sinc function, 89
Single value, 240
Single-leaf partitions, 241
Smokestack, 397
Soft boundary, 288
Soft lining, 288
Solid structure, 117
Sound
of a musical instrument, 2
Sound absorber, 171, 394
Sound absorption, 395
Sound barrier
wedge-shaped, 333
Sound beam, 290
Sound density, 22
Sound field
cylindrically symmetric, 73
diffuse, 220, 221
Sound impact, 259
protection, 260
Sound intensity, 42, 465
Sound occurrence, 401
Sound power, 42, 465
measurement, 43
transmission coefficient, 239, 244, 269, 277
Sound power level, 470
Sound pressure, 1, 5, 6, 22
level, 5, 6
transmission coefficient, 244, 269
Sound pressure level, 465
Sound propagation, 67
in the impedance tube, 171
Sound propagation speed, 29
Sound radiation, 67
from plane surfaces, 104
of line sources, 68
of point sources, 67
Sound reduction index, 240
Sound speed, 23
Sound temperature, 22
Sound velocity, 25, 29
Source
line, 68
monopole, 70
of order zero, 70
simple, 70
volume velocity, 70, 104
Source decomposition, 445
Source element, 104
Source length, 68
Source room, 239
Source wavelength, 98
Source wavenumber, 98
Spatial period, 98
Specific heat, 465
Specific impedance, 31
Specific resistance, 31
Specific stiffness, 121
Spectral components, 7
Speed, 465
Speed of sound, 23
Spherical waves, 68
Spiral wave, 471
Splitter attenuator, 284
Spring
impedance, 196
stiffness, 27
Squaring device, 410
Stability, 401
Stability boundary, 404
Stability chart, 403
Standing wave, 29, 50
Standing waves, 34
Static bending, 120
Static bending science, 120, 123
Steady state, 222
Steady-state
conditions, 226
Step function, 413
Stiffness
characteristics, 289
impedance, 292
of a spring, 27
of enclosed air, 196
specific, 121
Stimulus, 3
Strain, 119
waves, 118, 119
Stripe
insulating, 262
Structure coefficient, 188
Structure-borne
sound, 117
sound bridge, 258, 262
wave, 119

Structure-borne sound transmission, 489
Structured tube, 283
Substitute sources, 218
supported, 127
Suppression of side lobes, 93
suspended, 127
Swingset, 398
Symmetries, 431
System theory, 409

Tape recorder, 411
Tapping machine, 259
Temperature, 18, 22
tension
  normal, 117, 120, 122
  shear, 118, 120, 122
Test signal, 240
Third-octave band
  filter, 7
  level, 8
Third-octave level, 461, 462
Threshold stimulus, 4
Timbre, 1
Time convention, 34
Time domain, 45
Time signal, 3
Time-invariant, 411
time-invariant, 54
time-variant, 55
Torsional waves, 118, 119
Total level, 461
Total transmission, 246
Trace wavelength, 245
Traffic noise, 237
Transcendental equation, 297, 298, 304
Transducers, 3, 458
Transfer path, 161
Transformation, 426
Transmission
  between rooms, 237
  total, 246
Transmission coefficient, 182
  sound power, 239, 243, 244, 269, 277
  sound pressure, 244, 269
Transmission function, 429, 431
Transmission loss, 3, 137, 239, 247, 303
  airborne, 244
  of single-leaf partitions, 241
Transmitter, 409
transversal, 141
Typical frequency, 350

Ultrasound, 1
Universal gas constant, 19

Variation law, 4
Vector differential operators, 39
Vectorial intensity components, 44
Velocity, 25
Vibration equation, 400
Violin, 397
Voice coil microphone, 358
Volume flow, 112, 113
Volume velocity, 105
  source, 70, 104

Wall impedance, 184
Water, 467
Water waves, 401
Wave, 30
Wave decomposition, 445
Wave equation, 29
  of bending waves in beams, 124
  of bending waves in plates, 134
  of bending waves in walls, 243
  of the porous medium, 189
Wave form, 119, 127
Wave number spectrum, 435
Wave number variable, 436
Wave resistance, 31
Wave speed, 25
wave sum, 443
Wavelength
  bending, 125, 135
  source, 98
  trace, 245
Wavelength decomposition, 435
Wavenumber
  source, 98
Waves
  bending, 118, 123, 124, 136
    in beams, 120, 135
    in plates, 134
  free bending, 134
  plane, 287
  plate plane, 135
  spherical, 68

strain, 118, 119
torsional, 118, 119
Weber, 3
Weber-Fechner-Law, 5
Wedge-shaped sound barrier, 333
White noise, 7
Winding point, 307
Window
double-glazed, 256

Young's modulus, 121

Zero order source, 70

Printing: Krips bv, Meppel, The Netherlands
Binding: Stürtz, Würzburg, Germany